T0252868

MINI
Automotive Repair Manual

by Martynn Randall, Peter T Gill and John H Haynes

Member of the Guild of Motoring Writers

Models covered:

Mk I models:

    R50 - Cooper (2002 through 2006)
    R52 - Cooper Convertible/Cooper S Convertible (2005 through 2008)
    R53 - Cooper S (2002 through 2006)

Mk II models:

    R55 - Cooper Clubman/Clubman S (2008 through 2013)
    R56 - Cooper/Cooper S (2007 through 2013)
    R57 - Cooper Convertible/Cooper S Convertible (2009 through 2013)

Includes John Cooper Works (JCW) models (except twin-scroll turbo models).
Does not include Countryman models or convertible top information

(67020-9Y4)

J H Haynes & Co. Ltd.
Haynes North America, Inc.
www.haynes.com

Acknowledgements

2007 and later wiring diagrams provided exclusively for Haynes North America, Inc by Valley Forge Technical Information Services.

A book in the Haynes Automotive Repair Manual Series

ISBN-13: 978-1-62092-316-0

ISBN-10: 1-62092-316-5

Library of Congress Control Number: 2018940971

Contents

Haynes mechanic and photographer with a 2009 MINI Cooper S

About this manual

Its purpose

The purpose of this manual is to help you get the best value from your vehicle. It can do so in several ways. It can help you decide what work must be done, even if you choose to have it done by a dealer service department or a repair shop; it provides information and procedures for routine maintenance and servicing; and it offers diagnostic and repair procedures to follow when trouble occurs.

We hope you use the manual to tackle the work yourself. For many simpler jobs, doing it yourself may be quicker than arranging an appointment to get the vehicle into a shop and making the trips to leave it and pick it up. More importantly, a lot of money can be saved by avoiding the expense the shop must pass on to you to cover its labor and overhead costs. An added benefit is the sense of satisfaction and accomplishment that you feel after doing the job yourself.

Using the manual

The manual is divided into Chapters. Each Chapter is divided into numbered Sections, which are headed in bold type between horizontal lines. Each Section consists of consecutively numbered paragraphs.

At the beginning of each numbered Section you will be referred to any illustrations which apply to the procedures in that Section. The reference numbers used in illustration captions pinpoint the pertinent Section and the Step within that Section. That is, illustration 3.2 means the illustration refers to Section 3 and Step (or paragraph) 2 within that Section.

Procedures, once described in the text, are not normally repeated. When it's necessary to refer to another Chapter, the reference will be given as Chapter and Section number. Cross references given without use of the word "Chapter" apply to Sections and/or paragraphs in the same Chapter. For example, "see Section 8" means in the same Chapter.

References to the left or right side of the vehicle assume you are sitting in the driver's seat, facing forward.

Even though we have prepared this manual with extreme care, neither the publisher nor the author can accept responsibility for any errors in, or omissions from, the information given.

NOTE

A **Note** provides information necessary to properly complete a procedure or information which will make the procedure easier to understand.

CAUTION

A **Caution** provides a special procedure or special steps which must be taken while completing the procedure where the Caution is found. Not heeding a Caution can result in damage to the assembly being worked on.

WARNING

A **Warning** provides a special procedure or special steps which must be taken while completing the procedure where the Warning is found. Not heeding a Warning can result in personal injury.

Introduction

The MINI models covered by this manual include the MINI Cooper, Cooper Convertible, the extended wheelbase Cooper Clubman, and their "S" model equivalents (supercharged/turbocharged versions). Two generations of production are included and are identified throughout the manual as Mk I and Mk II models. Mk I models include 2006 and earlier Cooper/Cooper S models, and 2008 and earlier Convertible models. Mk II models include 2007 and later Cooper/Cooper S/Clubman/Clubman S, and 2009 and later Convertible models, 2011 Copper S hard top hatchback.

All models are equipped with a transversely mounted 1.6L inline four-cylinder engine. S models are equipped with a supercharger (Mk I models) or a turbocharger (Mk II models).

Transaxles include a 5- or 6-speed manual transaxle, a 6-speed "Agitronic" automatic transaxle with manual-shift mode, or a 5-speed CVT (continuously variable) transaxle.

The suspension is independent at all four corners. The front suspension is of MacPherson strut design, using strut/coil spring assemblies, lower control arms and a stabilizer bar. The rear suspension is a multi-link design, using trailing arms, upper and lower control arms, coil spring/shock absorber units and a stabilizer bar.

The steering system consists of a rack-and-pinion steering gear and two adjustable tie-rods. Mk I models are equipped with an electro-hydraulic power assist system. Mk II models are equipped with an electro-mechanical power assist system.

The brakes are disc at the front and rear, with power assist and an Anti-lock Brake System (ABS) as standard equipment.

Vehicle identification numbers

The Vehicle Identification Number (VIN) is located on a plate behind the lower left corner of the windshield

Modifications are a continuing and unpublicized process in vehicle manufacturing. Since spare parts manuals and lists are compiled on a numerical basis, the individual vehicle numbers are essential to correctly identify the component required.

Vehicle Identification Number (VIN)

This very important identification number is stamped on a plate behind the lower left corner of the windshield **(see illustration)**. It can also be found on the certification label located on the driver's side door post. The VIN also appears on the Vehicle Certificate of Title and Registration. It contains information such as where and when the vehicle was manufactured, the model year and the body style.

VIN engine and model year codes

Two particularly important pieces of information found in the VIN are the body/engine code and the model year code. Counting from the left, the engine code letter designation is the 6th digit and the model year code letter designation is the 10th digit.

On the models covered by this manual, the model year codes are:

2	2002
3	2003
4	2004
5	2005
6	2006
7	2007
8	2008
9	2009
A	2010
B	2011
C	2012
D	2013

The models covered by this manual are equipped with the following engines:

Mk I models:

2006 and earlier Cooper/2008 and earlier Convertible	W10B16A, SOHC, 16-valve
2006 and earlier Cooper S/2008 and earlier Convertible S	W11B16A, SOHC, 16-valve, supercharged

Mk II models:

2007 through 2010 Cooper/2009 and 2010 Convertible	N12, DOHC, 16-valve, dual VANOS, Valvetronic
2007 through 2010 Cooper S/2009 and 2010 Convertible S	N14, DOHC, 16-valve, single VANOS, direct fuel injection, turbocharged
2011 and 2012 Cooper JCW/JCW Clubman/JCW Convertible	N14, DOHC, 16-valve, single VANOS, direct fuel injection, turbocharged
2011 through 2013 Cooper/Convertible/Clubman	N16, DOHC, 16-valve, dual VANOS, Valvetronic
2011 through 2013 Cooper S/Clubman S/ Convertible S	N18, DOHC, 16-valve, dual VANOS, Valvetronic, direct fuel injection, turbocharged
2013 Cooper JCW/Clubman JCW/Convertible JCW	N18, DOHC, 16-valve, dual VANOS, Valvetronic, direct fuel injection, turbocharged

Certification label

The certification label is attached to the driver's door post (see illustration). The plate contains the name of the manufacturer, the month and year of production, the Gross Vehicle Weight Rating (GVWR), the Gross Axle Weight Rating (GAWR) and the certification statement.

Engine identification numbers

The engine serial number can be found on the front side of the engine **(see illustration)**.

The vehicle certification label is located on the driver's door post

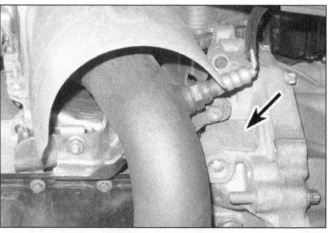

The engine serial number is located on the front side of the block, near the transaxle

Recall information

Vehicle recalls are carried out by the manufacturer in the rare event of a possible safety-related defect. The vehicle's registered owner is contacted at the address on file at the Department of Motor Vehicles and given the details of the recall. Remedial work is carried out free of charge at a dealer service department.

If you are the new owner of a used vehicle which was subject to a recall and you want to be sure that the work has been carried out, it's best to contact a dealer service department and ask about your individual vehicle - you'll need to furnish them your Vehicle Identification Number (VIN).

The table below is based on information provided by the National Highway Traffic Safety Administration (NHTSA), the body which oversees vehicle recalls in the United States. The recall database is updated constantly. For the latest information on vehicle recalls, check the NHTSA website at www.nhtsa.gov, www.safercar.gov, or call the NHTSA hotline at 1-888-327-4236.

Recall date	Recall campaign number	Model(s) affected	Concern
JUL 25, 2002	02V201000	2002 Cooper	On certain models equipped with manual transaxles, it is possible for the shift cable to detach from the transaxle shift linkage while the driver is attempting to change gears. If shift cable detachment occurs and the transaxle remains in any forward gear, the vehicle can still be driven in that gear only. However, if the transaxle moves into neutral and remains there, the ability to accelerate or maintain speed will be lost, increasing the risk of a crash.
FEB 28, 2003	03V086000	2003 Cooper, Cooper S	On certain models, the lower screw connection of the rear shock absorbers to the chassis may fail. The head (top) of the screw securing this section may break. The remaining portion of the screw will still hold the strut in place. If left unattended over a period of time, and depending on driving conditions, the remaining portion of the screw may break. If it broke, the chassis would lean directly on the tire, diminishing the driver's ability to control the vehicle, increasing the risk of a crash.
JUL 06, 2004	04V348000	2003, 2004 Cooper, Cooper S	On certain models, the flat tire monitoring system has not been correctly programmed. In the event of a flat tire, an audible signal will not sound to alert the driver to a flat tire. The driver may not be aware of a flat tire, which could increase the risk of a crash.
SEP 26, 2005	05V470000	2002 Cooper	Certain models fail to conform to the requirements of Federal Motor Vehicle Safety Standard No. 110, "tire selection and rims." The standard requires that the relationship between the tire size and the maximum tire pressure be indicated on the vehicle's tire information label. This relationship is not present. Too low or too high tire pressure can cause premature wear and/or tire damage, and could lead to unfavorable driving conditions.

Recall date	Recall campaign number	Model(s) affected	Concern
NOV 06, 2007	07V533000	2007 Cooper Convertible, Cooper S Convertible	Certain Convertible vehicles fail to conform to the requirements of Federal Motor Vehicle Safety Standard No. 110 "tire selection and rims." The standard requires that the tire size and corresponding tire pressure be indicated on the label and the label does not correspond to the actual tires mounted on the vehicle. Incorrect tire pressure can cause premature wear and tire damage, which could lead to unfavorable driving conditions, increasing the risk of a crash.
AUG 04, 2008	08E050000	2002, 2003, 2004, 2005 Cooper	Some models equipped with K2 Motor headlamps are missing the amber side reflex reflector, which fails to conform with the requirements of the Federal Motor Vehicle Safety Standard No. 108, "lamps, reflective devices, and associated equipment." Without the amber side reflex reflectors, the lighting visibility may be affected, possibly resulting in a vehicle crash.
SEP 15, 2008	08V507000	2009 JCW (John Cooper Works), JCW Clubman	Some JCW and JCW Clubman vehicles are equipped with 16-inch diameter front brake discs rather than 17-inch discs. Under conditions of intense brake usage, braking performance would be reduced. Depending on traffic and road conditions, as well as a driver's reactions, reduced brake performance could increase the risk of a crash.
DEC 12, 2008	08V657000	2007, 2008 Cooper S	On some vehicles, the centrally located tail pipe extension protrudes slightly beyond the rear bumper. It is possible for inadvertent contact to occur with a person's leg. If the tail pipe is hot during inadvertent contact, a burn could occur.
APR 05, 2009	09E025000	2002, 2003, 2004, 2005 Cooper	Some models equipped with Dope, Inc headlamps fail to conform to the requirements of Federal Motor Vehicle Safety Standard No. 108, "lamps, reflective devices, and associated equipment." The lamps do not contain the required amber side reflectors. Decreased visibility may result in a crash.
APR 17, 2009	09V143000	2009 Cooper Convertible	Some models fail to comply with the requirements of Federal Motor Vehicle Safety Standard No. 110, "tire selection and rims." The affected vehicles were delivered with 15-inch wheels instead of 16-inch wheels as intended, and the label indicates 16-inch wheels. The wheels on the vehicle do not conform to the label as required.

Recall date	Recall campaign number	Model(s) affected	Concern
DEC 04, 2009	09V474000	2010 Cooper, Cooper S	Some vehicles fail to comply with the requirements of Federal Motor Vehicle Safety Standard No. 110 "tire selection and rims." The affected vehicles were equipped with 17-inch wheels, but their label states that they were equipped with 16-inch wheels. Also the tire pressure label on the Cooper S is incorrect. Erroneous tire information could lead to improper tire fitment and inflation which could affect the durability of the tire and the stability of the vehicle, increasing the risk of a crash.
MAY 25, 2011	11V299000	2011, Mini Cooper Clubman, Cooper S Clubman	Some models manufactured from 2/8/11 through 5/11/11 are equipped with a European-specification tachometer that uses brake-system telltales/symbols that do not meet U.S. requirements. In the event of an ABS problem, only the outer circle of the symbol illuminates, not the ABS letters. As such, the driver may be unaware that there is a problem with the ABS, increasing the risk of a crash.
JAN 10, 2012	12V008000	2007, 2008, 2009 Cooper S, Cooper S Clubman, Cooper S convertible, Cooper JCW/convertible	On some models, the electric auxiliary water pump that cools the turbocharger has an electronic circuit board that can malfunction and overheat. The circuit board may smolder, which could result in a vehicle fire.
APR 06, 2015	15V205000	2005, 2006, Mini Cooper, Cooper S, 2005, 2006, 2007, 2008 Cooper convertible, Cooper S convertible	On some models, the front passenger seat occupant detection mat may not function properly and, as a result, the front passenger airbag may not deploy in a crash. An improperly functioning mat may cause the passenger frontal airbag to be inactive when the seat is occupied, and in the event of a crash, the airbag will not deploy, increasing the passenger's risk of injury.
OCT 15, 2015	15V660000	2002, 2003, 2004, 2005, Mini Cooper, Cooper S convertible	The affected vehicles may experience temporary or permanent loss of the electro-hydraulic steering assistance. If the vehicle experiences a loss of power steering assist, extra steering effort will be required at lower speeds, potentially increasing the risk of a crash.

Buying parts

Replacement parts are available from many sources, which generally fall into one of two categories - authorized dealer parts departments and independent retail auto parts stores. Our advice concerning these parts is as follows:

Retail auto parts stores: Good auto parts stores will stock frequently needed components which wear out relatively fast, such as clutch components, exhaust systems, brake parts, tune-up parts, etc. These stores often supply new or reconditioned parts on an exchange basis, which can save a considerable amount of money. Discount auto parts stores are often very good places to buy materials and parts needed for general vehicle maintenance such as oil, grease, filters, spark plugs, belts, touch-up paint, bulbs, etc. They also usually sell tools and general accessories, have convenient hours, charge lower prices and can often be found not far from home.

Authorized dealer parts department: This is the best source for parts which are unique to the vehicle and not generally available elsewhere (such as major engine parts, transmission parts, trim pieces, etc.).

Warranty information: If the vehicle is still covered under warranty, be sure that any replacement parts purchased - regardless of the source - do not invalidate the warranty!

To be sure of obtaining the correct parts, have engine and chassis numbers available and, if possible, take the old parts along for positive identification.

Maintenance techniques, tools and working facilities

Maintenance techniques

There are a number of techniques involved in maintenance and repair that will be referred to throughout this manual. Application of these techniques will enable the home mechanic to be more efficient, better organized and capable of performing the various tasks properly, which will ensure that the repair job is thorough and complete.

Fasteners

Fasteners are nuts, bolts, studs and screws used to hold two or more parts together. There are a few things to keep in mind when working with fasteners. Almost all of them use a locking device of some type, either a lockwasher, locknut, locking tab or thread adhesive. All threaded fasteners should be clean and straight, with undamaged threads and undamaged corners on the hex head where the wrench fits. Develop the habit of replacing all damaged nuts and bolts with new ones. Special locknuts with nylon or fiber inserts can only be used once. If they are removed, they lose their locking ability and must be replaced with new ones.

Rusted nuts and bolts should be treated with a penetrating fluid to ease removal and prevent breakage. Some mechanics use turpentine in a spout-type oil can, which works quite well. After applying the rust penetrant, let it work for a few minutes before trying to loosen the nut or bolt. Badly rusted fasteners may have to be chiseled or sawed off or removed with a special nut breaker, available at tool stores.

If a bolt or stud breaks off in an assembly, it can be drilled and removed with a special tool commonly available for this purpose. Most automotive machine shops can perform this task, as well as other repair procedures, such as the repair of threaded holes that have been stripped out.

Flat washers and lockwashers, when removed from an assembly, should always be replaced exactly as removed. Replace any damaged washers with new ones. Never use a lockwasher on any soft metal surface (such as aluminum), thin sheet metal or plastic.

Fastener sizes

For a number of reasons, automobile manufacturers are making wider and wider use of metric fasteners. Therefore, it is important to be able to tell the difference between standard (sometimes called U.S. or SAE) and metric hardware, since they cannot be interchanged.

All bolts, whether standard or metric, are sized according to diameter, thread pitch and length. For example, a standard 1/2 - 13 x 1 bolt is 1/2 inch in diameter, has 13 threads per inch and is 1 inch long. An M12 - 1.75 x 25 metric bolt is 12 mm in diameter, has a thread pitch of 1.75 mm (the distance between threads) and is 25 mm long. The two bolts are nearly identical, and easily confused, but they are not interchangeable.

In addition to the differences in diameter, thread pitch and length, metric and standard bolts can also be distinguished by examining the bolt heads. To begin with, the distance across the flats on a standard bolt head is measured in inches, while the same dimension on a metric bolt is sized in millimeters (the same is true for nuts). As a result, a standard wrench should not be used on a metric bolt and a metric wrench should not be used on a standard bolt. Also, most standard bolts have slashes radiating out from the center of the head to denote the grade or strength of the bolt, which is an indication of the amount of torque that can be applied to it. The greater the number of slashes, the greater the strength of the bolt. Grades 0 through 5 are commonly used on automobiles. Metric bolts have a property class (grade) number, rather than a slash, molded into their heads to indicate bolt strength. In this case, the higher the number, the stronger the bolt. Property class numbers 8.8, 9.8 and 10.9 are commonly used on automobiles.

Strength markings can also be used to distinguish standard hex nuts from metric hex nuts. Many standard nuts have dots stamped into one side, while metric nuts are marked with a number. The greater the number of dots, or the higher the number, the greater the strength of the nut.

Metric studs are also marked on their ends according to property class (grade). Larger studs are numbered (the same as metric bolts), while smaller studs carry a geometric code to denote grade.

It should be noted that many fasteners, especially Grades 0 through 2, have no distinguishing marks on them. When such is the case, the only way to determine whether it is standard or metric is to measure the thread pitch or compare it to a known fastener of the same size.

Standard fasteners are often referred to as SAE, as opposed to metric. However, it should be noted that SAE technically refers to a non-metric fine thread fastener only. Coarse thread non-metric fasteners are referred to as USS sizes.

Since fasteners of the same size (both standard and metric) may have different strength ratings, be sure to reinstall any bolts, studs or nuts removed from your vehicle in their original locations. Also, when replacing a fastener with a new one, make sure that the new one has a strength rating equal to or greater than the original.

Tightening sequences and procedures

Most threaded fasteners should be tightened to a specific torque value (torque is the twisting force applied to a threaded component such as a nut or bolt). Overtightening the fastener can weaken it and cause it to break, while undertightening can cause it to eventually come loose. Bolts, screws and studs, depending on the material they are made of and their thread diameters, have specific torque values, many of which are noted in the Specifications at the beginning of each Chapter. Be sure to follow the torque recommendations closely. For fasteners not assigned a

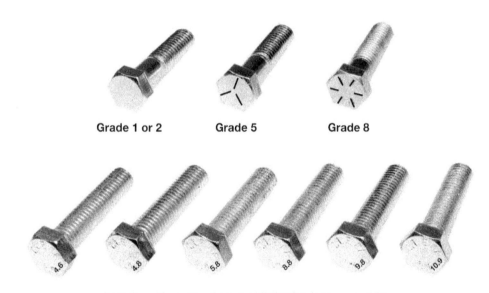

Grade 1 or 2   Grade 5   Grade 8

Bolt strength marking (standard/SAE/USS; bottom - metric)

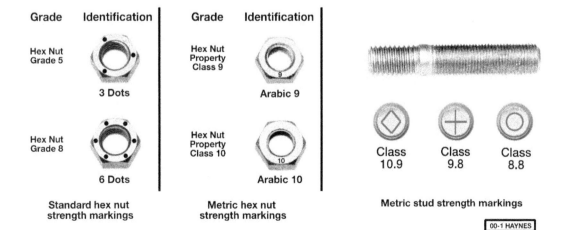

Standard hex nut strength markings

Metric hex nut strength markings

Metric stud strength markings

00-1 HAYNES

specific torque, a general torque value chart is presented here as a guide. These torque values are for dry (unlubricated) fasteners threaded into steel or cast iron (not aluminum). As was previously mentioned, the size and grade of a fastener determine the amount of torque that can safely be applied to it. The figures listed here are approximate for Grade 2 and Grade 3 fasteners. Higher grades can tolerate higher torque values.

Fasteners laid out in a pattern, such as cylinder head bolts, oil pan bolts, differential cover bolts, etc., must be loosened or tightened in sequence to avoid warping the component. This sequence will normally be shown in the appropriate Chapter. If a specific pattern is not given, the following procedures can be used to prevent warping.

Initially, the bolts or nuts should be assembled finger-tight only. Next, they should be tightened one full turn each, in a criss-cross or diagonal pattern. After each one has been tightened one full turn, return to the first one and tighten them all one-half turn, following the same pattern. Finally, tighten each of them one-quarter turn at a time until each fastener has been tightened to the proper torque. To loosen and remove the fasteners, the procedure would be reversed.

Component disassembly

Component disassembly should be done with care and purpose to help ensure that

Metric thread sizes	Ft-lbs	Nm
M-6	6 to 9	9 to 12
M-8	14 to 21	19 to 28
M-10	28 to 40	38 to 54
M-12	50 to 71	68 to 96
M-14	80 to 140	109 to 154
Pipe thread sizes		
1/8	5 to 8	7 to 10
1/4	12 to 18	17 to 24
3/8	22 to 33	30 to 44
1/2	25 to 35	34 to 47
U.S. thread sizes		
1/4 - 20	6 to 9	9 to 12
5/16 - 18	12 to 18	17 to 24
5/16 - 24	14 to 20	19 to 27
3/8 - 16	22 to 32	30 to 43
3/8 - 24	27 to 38	37 to 51
7/16 - 14	40 to 55	55 to 74
7/16 - 20	40 to 60	55 to 81
1/2 - 13	55 to 80	75 to 108

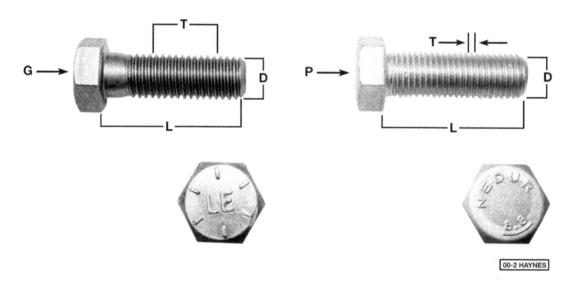

Standard (SAE and USS) bolt dimensions/grade marks

G   Grade marks (bolt strength)
L   Length (in inches)
T   Thread pitch (number of threads per inch)
D   Nominal diameter (in inches)

Metric bolt dimensions/grade marks

P   Property class (bolt strength)
L   Length (in millimeters)
T   Thread pitch (distance between threads in millimeters)
D   Diameter

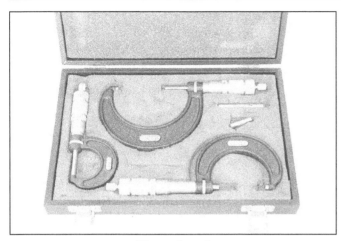

Micrometer set

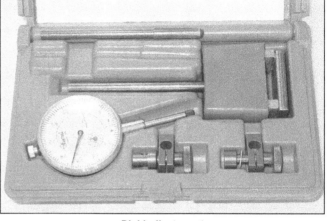

Dial indicator set

the parts go back together properly. Always keep track of the sequence in which parts are removed. Make note of special characteristics or marks on parts that can be installed more than one way, such as a grooved thrust washer on a shaft. It is a good idea to lay the disassembled parts out on a clean surface in the order that they were removed. It may also be helpful to make sketches or take instant photos of components before removal.

When removing fasteners from a component, keep track of their locations. Sometimes threading a bolt back in a part, or putting the washers and nut back on a stud, can prevent mix-ups later. If nuts and bolts cannot be returned to their original locations, they should be kept in a compartmented box or a series of small boxes. A cupcake or muffin tin is ideal for this purpose, since each cavity can hold the bolts and nuts from a particular area (i.e. oil pan bolts, valve cover bolts, engine mount bolts, etc.). A pan of this type is especially helpful when working on assemblies with very small parts, such as the carburetor, alternator, valve train or interior dash and trim pieces. The cavities can be marked with paint or tape to identify the contents.

Whenever wiring looms, harnesses or connectors are separated, it is a good idea to identify the two halves with numbered pieces of masking tape so they can be easily reconnected.

Gasket sealing surfaces

Throughout any vehicle, gaskets are used to seal the mating surfaces between two parts and keep lubricants, fluids, vacuum or pressure contained in an assembly.

Many times these gaskets are coated with a liquid or paste-type gasket sealing compound before assembly. Age, heat and pressure can sometimes cause the two parts to stick together so tightly that they are very difficult to separate. Often, the assembly can be loosened by striking it with a soft-face hammer near the mating surfaces. A regular hammer can be used if a block of wood is placed between the hammer and the part. Do not hammer on cast parts or parts that could be easily damaged. With any particularly stubborn part, always recheck to make sure that every fastener has been removed.

Avoid using a screwdriver or bar to pry apart an assembly, as they can easily mar the gasket sealing surfaces of the parts, which must remain smooth. If prying is absolutely necessary, use an old broom handle, but keep in mind that extra clean up will be necessary if the wood splinters.

After the parts are separated, the old gasket must be carefully scraped off and the gasket surfaces cleaned. Stubborn gasket material can be soaked with rust penetrant or treated with a special chemical to soften it so it can be easily scraped off. **Caution:** *Never use gasket removal solutions or caustic chemicals on plastic or other composite components.* A scraper can be fashioned from a piece of copper tubing by flattening and sharpening one end. Copper is recommended because it is usually softer than the surfaces to be scraped, which reduces the chance of gouging the part. Some gaskets can be removed with a wire brush, but regardless of the method used, the mating surfaces must be left clean and smooth. If for some reason the gasket surface is gouged, then a gasket sealer thick enough to fill scratches will have to be used during reassembly of the components. For most applications, a non-drying (or semi-drying) gasket sealer should be used.

Hose removal tips

Warning: *If the vehicle is equipped with air conditioning, do not disconnect any of the A/C hoses without first having the system depressurized by a dealer service department or a service station.*

Hose removal precautions closely parallel gasket removal precautions. Avoid scratching or gouging the surface that the hose mates against or the connection may leak. This is especially true for radiator hoses. Because of various chemical reactions, the rubber in hoses can bond itself to the metal spigot that the hose fits over. To remove a hose, first loosen the hose clamps that secure it to the spigot. Then, with slip-joint pliers, grab the hose at the clamp and rotate it around the spigot. Work it back and forth until it is completely free, then pull it off. Silicone or other lubricants will ease removal if they can be applied between the hose and the outside of the spigot. Apply the same lubricant to the inside of the hose and the outside of the spigot to simplify installation.

As a last resort (and if the hose is to be replaced with a new one anyway), the rubber can be slit with a knife and the hose peeled from the spigot. If this must be done, be careful that the metal connection is not damaged.

If a hose clamp is broken or damaged, do not reuse it. Wire-type clamps usually weaken with age, so it is a good idea to replace them with screw-type clamps whenever a hose is removed.

Tools

A selection of good tools is a basic requirement for anyone who plans to maintain and repair his or her own vehicle. For the owner who has few tools, the initial investment might seem high, but when compared to the spiraling costs of professional auto maintenance and repair, it is a wise one.

To help the owner decide which tools are needed to perform the tasks detailed in this manual, the following tool lists are offered: *Maintenance and minor repair, Repair/overhaul* and *Special.*

The newcomer to practical mechanics should start off with the *maintenance and minor repair* tool kit, which is adequate for the simpler jobs performed on a vehicle. Then, as confidence and experience grow, the owner can tackle more difficult tasks, buying additional tools as they are needed. Eventually the basic kit will be expanded into the *repair and overhaul* tool set. Over a period of time, the experienced do-it-yourselfer will assemble a tool set complete enough for most repair and overhaul procedures and will add tools from the special category when it is felt that the expense is justified by the frequency of use.

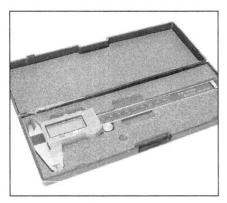

Dial caliper

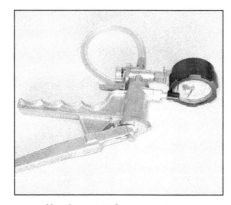

Hand-operated vacuum pump

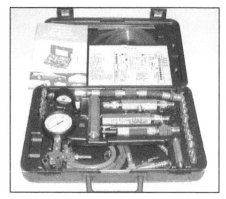

Fuel pressure gauge set

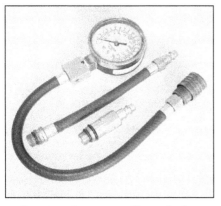

Compression gauge with spark plug
hole adapter

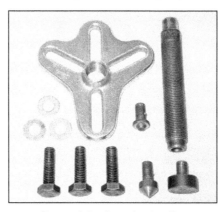

Damper/steering wheel puller

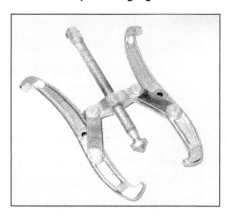

General purpose puller

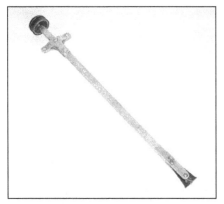

Hydraulic lifter removal tool

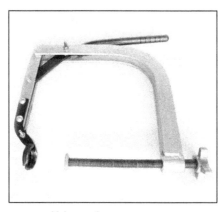

Valve spring compressor

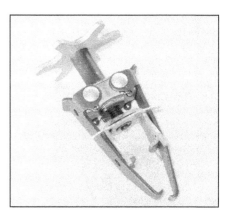

Valve spring compressor

Ridge reamer

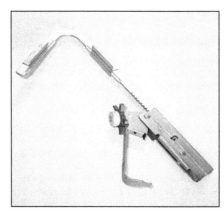

Piston ring groove cleaning tool

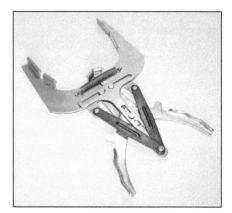

Ring removal/installation tool

Ring compressor

Cylinder hone

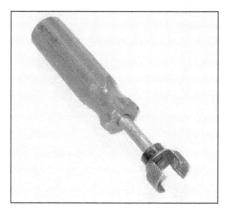

Brake hold-down spring tool

Torque angle gauge

Clutch plate alignment tool

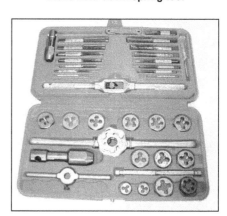

Tap and die set

Maintenance and minor repair tool kit

The tools in this list should be considered the minimum required for performance of routine maintenance, servicing and minor repair work. We recommend the purchase of combination wrenches (box-end and open-end combined in one wrench). While more expensive than open end wrenches, they offer the advantages of both types of wrench.

> *Combination wrench set (1/4-inch to
>   1 inch or 6 mm to 19 mm)*
> *Adjustable wrench, 8 inch*
> *Spark plug wrench with rubber insert*
> *Spark plug gap adjusting tool*
> *Feeler gauge set*
> *Brake bleeder wrench*
> *Standard screwdriver (5/16-inch x
>   6 inch)*
> *Phillips screwdriver (No. 2 x 6 inch)*
> *Combination pliers - 6 inch*
> *Hacksaw and assortment of blades*
> *Tire pressure gauge*
> *Grease gun*
> *Oil can*
> *Fine emery cloth*
> *Wire brush*
> *Battery post and cable cleaning tool*
> *Oil filter wrench*
> *Funnel (medium size)*
> *Safety goggles*
> *Jackstands (2)*
> *Drain pan*

Note: *If basic tune-ups are going to be part of routine maintenance, it will be necessary to purchase a good quality stroboscopic timing light and combination tachometer/dwell meter. Although they are included in the list of special tools, it is mentioned here because they are absolutely necessary for tuning most vehicles properly.*

Repair and overhaul tool set

These tools are essential for anyone who plans to perform major repairs and are in addition to those in the maintenance and minor repair tool kit. Included is a comprehensive set of sockets which, though expensive, are invaluable because of their versatility, especially when various extensions and drives are available. We recommend the 1/2-inch drive over the 3/8-inch drive. Although the larger drive is bulky and more expensive, it has the capacity of accepting a very wide range of large sockets. Ideally, however, the mechanic should have a 3/8-inch drive set and a 1/2-inch drive set.

> *Socket set(s)*
> *Reversible ratchet*
> *Extension - 10 inch*
> *Universal joint*
> *Torque wrench (same size drive as
>   sockets)*
> *Ball peen hammer - 8 ounce*
> *Soft-face hammer (plastic/rubber)*
> *Standard screwdriver (1/4-inch x 6 inch)*

> *Standard screwdriver (stubby -
>   5/16-inch)*
> *Phillips screwdriver (No. 3 x 8 inch)*
> *Phillips screwdriver (stubby - No. 2)*
> *Pliers - vise grip*
> *Pliers - lineman's*
> *Pliers - needle nose*
> *Pliers - snap-ring (internal and external)*
> *Cold chisel - 1/2-inch*
> *Scribe*
> *Scraper (made from flattened copper
>   tubing)*
> *Centerpunch*
> *Pin punches (1/16, 1/8, 3/16-inch)*
> *Steel rule/straightedge - 12 inch*
> *Allen wrench set (1/8 to 3/8-inch or
>   4 mm to 10 mm)*
> *A selection of files*
> *Wire brush (large)*
> *Jackstands (second set)*
> *Jack (scissor or hydraulic type)*

Note: *Another tool which is often useful is an electric drill with a chuck capacity of 3/8-inch and a set of good quality drill bits.*

Special tools

The tools in this list include those which are not used regularly, are expensive to buy, or which need to be used in accordance with their manufacturer's instructions. Unless these tools will be used frequently, it is not very economical to purchase many of them. A consideration would be to split the cost and use between yourself and a friend or friends. In addition, most of these tools can be obtained from a tool rental shop on a temporary basis.

This list primarily contains only those tools and instruments widely available to the public, and not those special tools produced by the vehicle manufacturer for distribution to dealer service departments. Occasionally, references to the manufacturer's special tools are included in the text of this manual. Generally, an alternative method of doing the job without the special tool is offered. However, sometimes there is no alternative to their use. Where this is the case, and the tool cannot be purchased or borrowed, the work should be turned over to the dealer service department or an automotive repair shop.

> *Valve spring compressor*
> *Piston ring groove cleaning tool*
> *Piston ring compressor*
> *Piston ring installation tool*
> *Cylinder compression gauge*
> *Cylinder ridge reamer*
> *Cylinder surfacing hone*
> *Cylinder bore gauge*
> *Micrometers and/or dial calipers*
> *Hydraulic lifter removal tool*
> *Balljoint separator*
> *Universal-type puller*
> *Impact screwdriver*
> *Dial indicator set*
> *Stroboscopic timing light (inductive*
>   *pick-up)*
> *Hand operated vacuum/pressure pump*
> *Tachometer/dwell meter*
> *Universal electrical multimeter*
> *Cable hoist*
> *Brake spring removal and installation*
>   *tools*
> *Floor jack*

Buying tools

For the do-it-yourselfer who is just starting to get involved in vehicle maintenance and repair, there are a number of options available when purchasing tools. If maintenance and minor repair is the extent of the work to be done, the purchase of individual tools is satisfactory. If, on the other hand, extensive work is planned, it would be a good idea to purchase a modest tool set from one of the large retail chain stores. A set can usually be bought at a substantial savings over the individual tool prices, and they often come with a tool box. As additional tools are needed, add-on sets, individual tools and a larger tool box can be purchased to expand the tool selection. Building a tool set gradually allows the cost of the tools to be spread over a longer period of time and gives the mechanic the freedom to choose only those tools that will actually be used.

Tool stores will often be the only source of some of the special tools that are needed,

but regardless of where tools are bought, try to avoid cheap ones, especially when buying screwdrivers and sockets, because they won't last very long. The expense involved in replacing cheap tools will eventually be greater than the initial cost of quality tools.

Care and maintenance of tools

Good tools are expensive, so it makes sense to treat them with respect. Keep them clean and in usable condition and store them properly when not in use. Always wipe off any dirt, grease or metal chips before putting them away. Never leave tools lying around in the work area. Upon completion of a job, always check closely under the hood for tools that may have been left there so they won't get lost during a test drive.

Some tools, such as screwdrivers, pliers, wrenches and sockets, can be hung on a panel mounted on the garage or workshop wall, while others should be kept in a tool box or tray. Measuring instruments, gauges, meters, etc. must be carefully stored where they cannot be damaged by weather or impact from other tools.

When tools are used with care and stored properly, they will last a very long time. Even with the best of care, though, tools will wear out if used frequently. When a tool is damaged or worn out, replace it. Subsequent jobs will be safer and more enjoyable if you do.

How to repair damaged threads

Sometimes, the internal threads of a nut or bolt hole can become stripped, usually from overtightening. Stripping threads is an all-too-common occurrence, especially when working with aluminum parts, because aluminum is so soft that it easily strips out.

Usually, external or internal threads are only partially stripped. After they've been cleaned up with a tap or die, they'll still work. Sometimes, however, threads are badly damaged. When this happens, you've got three choices:

1) *Drill and tap the hole to the next suitable oversize and install a larger diameter bolt, screw or stud.*
2) *Drill and tap the hole to accept a threaded plug, then drill and tap the plug to the original screw size. You can also buy a plug already threaded to the original size. Then you simply drill a hole to the specified size, then run the threaded plug into the hole with a bolt and jam nut. Once the plug is fully seated, remove the jam nut and bolt.*
3) *The third method uses a patented thread repair kit like Heli-Coil or Slimsert. These*

easy-to-use kits are designed to repair damaged threads in straight-through holes and blind holes. Both are available as kits which can handle a variety of sizes and thread patterns. Drill the hole, then tap it with the special included tap. Install the Heli-Coil and the hole is back to its original diameter and thread pitch.

Regardless of which method you use, be sure to proceed calmly and carefully. A little impatience or carelessness during one of these relatively simple procedures can ruin your whole day's work and cost you a bundle if you wreck an expensive part.

Working facilities

Not to be overlooked when discussing tools is the workshop. If anything more than routine maintenance is to be carried out, some sort of suitable work area is essential.

It is understood, and appreciated, that many home mechanics do not have a good workshop or garage available, and end up removing an engine or doing major repairs outside. It is recommended, however, that the overhaul or repair be completed under the cover of a roof.

A clean, flat workbench or table of comfortable working height is an absolute necessity. The workbench should be equipped with a vise that has a jaw opening of at least four inches.

As mentioned previously, some clean, dry storage space is also required for tools, as well as the lubricants, fluids, cleaning solvents, etc. which soon become necessary.

Sometimes waste oil and fluids, drained from the engine or cooling system during normal maintenance or repairs, present a disposal problem. To avoid pouring them on the ground or into a sewage system, pour the used fluids into large containers, seal them with caps and take them to an authorized disposal site or recycling center. Plastic jugs, such as old antifreeze containers, are ideal for this purpose.

Always keep a supply of old newspapers and clean rags available. Old towels are excellent for mopping up spills. Many mechanics use rolls of paper towels for most work because they are readily available and disposable. To help keep the area under the vehicle clean, a large cardboard box can be cut open and flattened to protect the garage or shop floor.

Whenever working over a painted surface, such as when leaning over a fender to service something under the hood, always cover it with an old blanket or bedspread to protect the finish. Vinyl covered pads, made especially for this purpose, are available at auto parts stores.

Jacking/dealing with a flat tire

Vehicles with the Compact spare wheel

Preparation

- When a puncture occurs, stop as soon as it is safe to do so.
- Park on firm level ground, if possible, and well out of the way of other traffic.
- Use hazard warning lights if necessary.
- If you have one, use a warning triangle to alert other drivers of your presence.
- Apply the handbrake and engage first or reverse gear.
- Chock the wheel diagonally opposite the one being removed – a chock is located beneath the jack under the luggage compartment lid.
- If the ground is soft, use a flat piece of wood to spread the load under the foot of the jack.

Changing the wheel

1   The spare wheel is stored under the floor of the luggage compartment. The tools are stored in the luggage compartment floor.

2   Lift the luggage compartment floor and remove the jack, wheel chock and wrench.
3   Place the chock behind or in front (as applicable) of the wheel diagonally opposite to the one to be removed.

4   Lift out the toolkit and, using the special wrench, remove the cover over the wheel carrier catch.
5   Screw the lifting handle from the toolkit onto the carrier catch thread, lift the handle slightly, squeeze together the retaining clips, and lower the wheel carrier.
6   Unscrew the valve extension from the spare wheel, remove the dust cap from the extension and fit it to the valve on the wheel.
7   Pulling by hand or using a screwdriver, remove the wheel trim/hub cap (as applicable) then, using the wrench from the toolkit, loosen each wheel bolt by a half turn.
8   If equipped with an anti-theft wheel bolt, pull the plastic cover from the bolt . . .

9   . . . then loosen it using the adapter supplied in the tool kit.

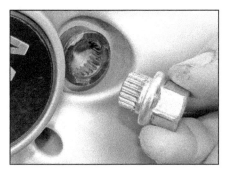

10   Locate the jack head under the jacking point nearest to the wheel that is to be removed.

As the jack is raised, the head must enter the rectangular recess in the jacking point.

11   Make sure the jack is located on firm ground then turn the jack handle clockwise until the wheel is raised clear of the ground. Unscrew the wheel bolts and remove the wheel. Install the spare wheel and screw in the wheel bolts. Lightly tighten the bolts with the wrench then lower the vehicle to the ground.
12   Securely tighten the wheel bolts in a diagonal sequence. Stow the punctured wheel and tools back in the luggage compartment and secure them in position. Note that the wheel bolts should be loosened and retightened to the specified torque at the earliest possible opportunity.

Finally . . .

- Remove the wheel chock.
- Stow the jack, chock and tools in the correct locations in the car.
- Check the tire pressure on the wheel just installed. If it is low, or if you don't have a pressure gauge with you, drive slowly to the next gas station and inflate the tire to the correct pressure.
- Observe any speed restrictions marked on the sidewall of the compact spare tire, and be aware that cornering and braking performance may be affected.
- Have the damaged tire or wheel repaired as soon as possible, or another puncture will leave you stranded.

Vehicles with MINI Mobility System

1   The MINI Mobility System consists of a bottle of puncture sealant and a compressor. Rather than change the punctured wheel, the system allows the tire to be sealed, enabling the journey to be resumed, albeit at a reduced speed.

2   The system is stored beneath the floor of the luggage compartment.

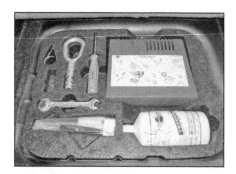

3   Remove the sealant bottle from the luggage compartment, and shake the contents well. Screw the filler hose onto the bottle.

4   Unscrew the valve cap from the punctured wheel, and using the valve removal tool (stored with the filler hose) unscrew the core from the valve.

5   Pull the stopper from the filler hose, attach the hose to the valve and, holding the bottle upside down, squeeze the entire contents into the tire.

6   Remove the filler hose, and screw the core back into the valve.

7   Retrieve the compressor from the toolkit, and insert the power plug into the vehicle's cigarette lighter/power outlet socket.

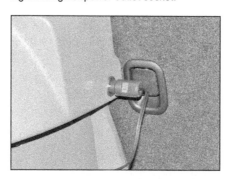

8   Connect the compressor hose to the tire valve and, with the ignition switch turned to position I, turn the compressor on and inflate

the tire to a pressure of between 1.8 and 26 to 36 psi (250 kPa/2.5 bar). If this pressure is not achieved within 6 minutes, turn off the compressor, disconnect it from the valve, and drive the vehicle forwards about 30 feet (10 m), then reverse back to place to redistribute the sealant, and repeat the inflation process. With the correct pressure achieved, disconnect the compressor and stow it in the tool kit.

9   Immediately drive the vehicle for approximately 10 minutes at a speed of between 12 and 35 mph to redistribute the sealant.

10   Stop the vehicle, connect the compressor, and check the tire pressure. If the pressure is less than 19 psi (130 kPa/1.3 bar), it's not safe to continue your journey, and the vehicle must be towed. If the pressure is above this, turn on the compressor and inflate the tire to the normal pressure for the vehicle, as specified on the sticker in the driver's door aperture.

11   With the tire inflated to the correct pressure, do not exceed the maximum speed of 50 mph. Have the tire repaired or replaced at the earliest opportunity.

Vehicles with Run Flat tires

1   These tires can be identified by the letters RSC moulded into the tire sidewall.

2   When used in conjunction with the special wheel rims, these tires are able to support the vehicle even when they are completely deflated. In the event of a puncture, the vehicle's road behavior will change (brak-

ing distance increased, directional stability reduced) but the journey can continue, albeit at a reduced speed of 50 mph.

3   If the vehicle is lightly loaded (1 to 2 persons without luggage), maximum range with the tire deflated is 150 miles.

4   If the vehicle has a medium load (2 per-

sons with full luggage, or 4 persons without luggage), maximum range is 90 miles.

5   If the vehicle is fully loaded (4 persons plus luggage), maximum range is 30 miles.

6   It's only possible to repair Run Flat tires which have had very little use. Consult a tire dealer at the earliest opportunity after a puncture.

Booster battery (jump) starting

Observe these precautions when using a booster battery to start a vehicle:

a) *Before connecting the booster battery, make sure the ignition switch is in the Off position.*

b) *Turn off the lights, heater and other electrical loads.*

c) *Your eyes should be shielded. Safety goggles are a good idea.*

d) *Make sure the booster battery is the same voltage as the dead one in the vehicle.*

e) *The two vehicles MUST NOT TOUCH each other!*

f) *Make sure the transaxle is in Neutral (manual) or Park (automatic).*

g) *If the booster battery is not a maintenance-free type, remove the vent caps and lay a cloth over the vent holes.*

The battery on these vehicles is located in various locations, depending on model and year (see Chapter 5 for access):

2006 and earlier MINI Cooper - left-rear corner of the engine compartment, beneath a cover.

2006 and earlier Cooper S - in the luggage compartment, under the floor mat (BUT - there is a remote positive terminal in the engine compartment for jump starting).

2007 and later models - right-rear corner of the engine compartment, beneath a cover.

Connect the red-colored jumper cable to the positive (+) terminal of the booster battery (or remote positive terminal) and the other end to the positive (+) terminal of the dead battery (or remote positive terminal). Then connect one end of the black jumper cable to the negative (-) terminal of the booster battery, and the other end of that cable to a good ground point on the engine of the disabled vehicle, preferably not too near the battery.

Start the engine using the booster battery and let the booster vehicle run at 2000 rpm for a few minutes to put some charge into the weak battery, then, with the engine running at idle speed, disconnect the jumper cables in the reverse order of connection.

On 2006 and earlier S models, there is a remote jump-start terminal in the engine compartment, adjacent to the air filter housing

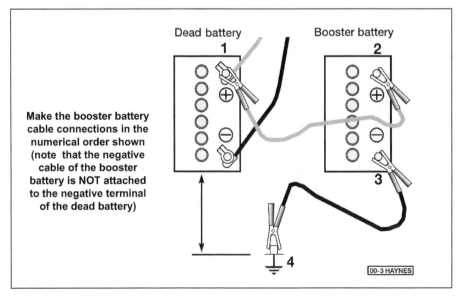

Make the booster battery cable connections in the numerical order shown (note that the negative cable of the booster battery is NOT attached to the negative terminal of the dead battery)

Dead battery
1

Booster battery
2

3

4

00-3 HAYNES

Automotive chemicals and lubricants

A number of automotive chemicals and lubricants are available for use during vehicle maintenance and repair. They include a wide variety of products ranging from cleaning solvents and degreasers to lubricants and protective sprays for rubber, plastic and vinyl.

Cleaners

Carburetor cleaner and choke cleaner is a strong solvent for gum, varnish and carbon. Most carburetor cleaners leave a dry-type lubricant film which will not harden or gum up. Because of this film it is not recommended for use on electrical components.

Brake system cleaner is used to remove brake dust, grease and brake fluid from the brake system, where clean surfaces are absolutely necessary. It leaves no residue and often eliminates brake squeal caused by contaminants.

Electrical cleaner removes oxidation, corrosion and carbon deposits from electrical contacts, restoring full current flow. It can also be used to clean spark plugs, carburetor jets, voltage regulators and other parts where an oil-free surface is desired.

Demoisturants remove water and moisture from electrical components such as alternators, voltage regulators, electrical connectors and fuse blocks. They are non-conductive and non-corrosive.

Degreasers are heavy-duty solvents used to remove grease from the outside of the engine and from chassis components. They can be sprayed or brushed on and, depending on the type, are rinsed off either with water or solvent.

Lubricants

Motor oil is the lubricant formulated for use in engines. It normally contains a wide variety of additives to prevent corrosion and reduce foaming and wear. Motor oil comes in various weights (viscosity ratings) from 0 to 50. The recommended weight of the oil depends on the season, temperature and the demands on the engine. Light oil is used in cold climates and under light load conditions. Heavy oil is used in hot climates and where high loads are encountered. Multi-viscosity oils are designed to have characteristics of both light and heavy oils and are available in a number of weights from 0W-20 to 20W-50.

Gear oil is designed to be used in differentials, manual transmissions and other areas where high-temperature lubrication is required.

Chassis and wheel bearing grease is a heavy grease used where increased loads and friction are encountered, such as for wheel bearings, balljoints, tie-rod ends and universal joints.

High-temperature wheel bearing grease is designed to withstand the extreme temperatures encountered by wheel bearings in disc brake equipped vehicles. It usually contains molybdenum disulfide (moly), which is a dry-type lubricant.

White grease is a heavy grease for metal-to-metal applications where water is a problem. White grease stays soft under both low and high temperatures (usually from -100 to +190-degrees F), and will not wash off or dilute in the presence of water.

Assembly lube is a special extreme pressure lubricant, usually containing moly, used to lubricate high-load parts (such as main and rod bearings and cam lobes) for initial start-up of a new engine. The assembly lube lubricates the parts without being squeezed out or washed away until the engine oiling system begins to function.

Silicone lubricants are used to protect rubber, plastic, vinyl and nylon parts.

Graphite lubricants are used where oils cannot be used due to contamination problems, such as in locks. The dry graphite will lubricate metal parts while remaining uncontaminated by dirt, water, oil or acids. It is electrically conductive and will not foul electrical contacts in locks such as the ignition switch.

Moly penetrants loosen and lubricate frozen, rusted and corroded fasteners and prevent future rusting or freezing.

Heat-sink grease is a special electrically non-conductive grease that is used for mounting electronic ignition modules where it is essential that heat is transferred away from the module.

Sealants

RTV sealant is one of the most widely used gasket compounds. Made from silicone, RTV is air curing, it seals, bonds, waterproofs, fills surface irregularities, remains flexible, doesn't shrink, is relatively easy to remove, and is used as a supplementary sealer with almost all low and medium temperature gaskets.

Anaerobic sealant is much like RTV in that it can be used either to seal gaskets or to form gaskets by itself. It remains flexible, is solvent resistant and fills surface imperfections. The difference between an anaerobic sealant and an RTV-type sealant is in the curing. RTV cures when exposed to air, while an anaerobic sealant cures only in the absence of air. This means that an anaerobic sealant cures only after the assembly of parts, sealing them together.

Thread and pipe sealant is used for sealing hydraulic and pneumatic fittings and vacuum lines. It is usually made from a Teflon compound, and comes in a spray, a paint-on liquid and as a wrap-around tape.

Chemicals

Anti-seize compound prevents seizing, galling, cold welding, rust and corrosion in fasteners. High-temperature ant-seize, usually made with copper and graphite lubricants, is used for exhaust system and exhaust manifold bolts.

Anaerobic locking compounds are used to keep fasteners from vibrating or working loose and cure only after installation, in the absence of air. Medium strength locking compound is used for small nuts, bolts and screws that may be removed later. High-strength locking compound is for large nuts, bolts and studs which aren't removed on a regular basis.

Oil additives range from viscosity index improvers to chemical treatments that claim to reduce internal engine friction. It should be noted that most oil manufacturers caution against using additives with their oils.

Gas additives perform several functions, depending on their chemical makeup. They usually contain solvents that help dissolve gum and varnish that build up on carburetor, fuel injection and intake parts. They also serve to break down carbon deposits that form on the inside surfaces of the combustion chambers. Some additives contain upper cylinder lubricants for valves and piston rings, and others contain chemicals to remove condensation from the gas tank.

Miscellaneous

Brake fluid is specially formulated hydraulic fluid that can withstand the heat and pressure encountered in brake systems. Care must be taken so this fluid does not come in contact with painted surfaces or plastics. An opened container should always be resealed to prevent contamination by water or dirt.

Weatherstrip adhesive is used to bond weatherstripping around doors, windows and trunk lids. It is sometimes used to attach trim pieces.

Undercoating is a petroleum-based, tar-like substance that is designed to protect metal surfaces on the underside of the vehicle from corrosion. It also acts as a sound-deadening agent by insulating the bottom of the vehicle.

Waxes and polishes are used to help protect painted and plated surfaces from the weather. Different types of paint may require the use of different types of wax and polish. Some polishes utilize a chemical or abrasive cleaner to help remove the top layer of oxidized (dull) paint on older vehicles. In recent years many non-wax polishes that contain a wide variety of chemicals such as polymers and silicones have been introduced. These non-wax polishes are usually easier to apply and last longer than conventional waxes and polishes.

Conversion factors

Length (distance)

Inches (in)	X	25.4	= Millimeters (mm)	X 0.0394	= Inches (in)
Feet (ft)	X	0.305	= Meters (m)	X 3.281	= Feet (ft)
Miles	X	1.609	= Kilometers (km)	X 0.621	= Miles

Volume (capacity)

Cubic inches (cu in; in³)	X	16.387	= Cubic centimeters (cc; cm³)	X 0.061	= Cubic inches (cu in; in³)
Imperial pints (Imp pt)	X	0.568	= Liters (l)	X 1.76	= Imperial pints (Imp pt)
Imperial quarts (Imp qt)	X	1.137	= Liters (l)	X 0.88	= Imperial quarts (Imp qt)
Imperial quarts (Imp qt)	X	1.201	= US quarts (US qt)	X 0.833	= Imperial quarts (Imp qt)
US quarts (US qt)	X	0.946	= Liters (l)	X 1.057	= US quarts (US qt)
Imperial gallons (Imp gal)	X	4.546	= Liters (l)	X 0.22	= Imperial gallons (Imp gal)
Imperial gallons (Imp gal)	X	1.201	= US gallons (US gal)	X 0.833	= Imperial gallons (Imp gal)
US gallons (US gal)	X	3.785	= Liters (l)	X 0.264	= US gallons (US gal)

Mass (weight)

Ounces (oz)	X	28.35	= Grams (g)	X 0.035	= Ounces (oz)
Pounds (lb)	X	0.454	= Kilograms (kg)	X 2.205	= Pounds (lb)

Force

Ounces-force (ozf; oz)	X	0.278	= Newtons (N)	X 3.6	= Ounces-force (ozf; oz)
Pounds-force (lbf; lb)	X	4.448	= Newtons (N)	X 0.225	= Pounds-force (lbf; lb)
Newtons (N)	X	0.1	= Kilograms-force (kgf; kg)	X 9.81	= Newtons (N)

Pressure

Pounds-force per square inch (psi; lbf/in²; lb/in²)	X	0.070	= Kilograms-force per square centimeter (kgf/cm²; kg/cm²)	X 14.223	= Pounds-force per square inch (psi; lbf/in²; lb/in²)
Pounds-force per square inch (psi; lbf/in²; lb/in²)	X	0.068	= Atmospheres (atm)	X 14.696	= Pounds-force per square inch (psi; lbf/in²; lb/in²)
Pounds-force per square inch (psi; lbf/in²; lb/in²)	X	0.069	= Bars	X 14.5	= Pounds-force per square inch (psi; lbf/in²; lb/in²)
Pounds-force per square inch (psi; lbf/in²; lb/in²)	X	6.895	= Kilopascals (kPa)	X 0.145	= Pounds-force per square inch (psi; lbf/in²; lb/in²)
Kilopascals (kPa)	X	0.01	= Kilograms-force per square centimeter (kgf/cm²; kg/cm²)	X 98.1	= Kilopascals (kPa)

Torque (moment of force)

Pounds-force inches (lbf in; lb in)	X	1.152	= Kilograms-force centimeter (kgf cm; kg cm)	X 0.868	= Pounds-force inches (lbf in; lb in)
Pounds-force inches (lbf in; lb in)	X	0.113	= Newton meters (Nm)	X 8.85	= Pounds-force inches (lbf in; lb in)
Pounds-force inches (lbf in; lb in)	X	0.083	= Pounds-force feet (lbf ft; lb ft)	X 12	= Pounds-force inches (lbf in; lb in)
Pounds-force feet (lbf ft; lb ft)	X	0.138	= Kilograms-force meters (kgf m; kg m)	X 7.233	= Pounds-force feet (lbf ft; lb ft)
Pounds-force feet (lbf ft; lb ft)	X	1.356	= Newton meters (Nm)	X 0.738	= Pounds-force feet (lbf ft; lb ft)
Newton meters (Nm)	X	0.102	= Kilograms-force meters (kgf m; kg m)	X 9.804	= Newton meters (Nm)

Vacuum

Inches mercury (in. Hg)	X	3.377	= Kilopascals (kPa)	X 0.2961	= Inches mercury
Inches mercury (in. Hg)	X	25.4	= Millimeters mercury (mm Hg)	X 0.0394	= Inches mercury

Power

Horsepower (hp)	X	745.7	= Watts (W)	X 0.0013	= Horsepower (hp)

Velocity (speed)

Miles per hour (miles/hr; mph)	X	1.609	= Kilometers per hour (km/hr; kph)	X 0.621	= Miles per hour (miles/hr; mph)

Fuel consumption*

Miles per gallon, Imperial (mpg)	X	0.354	= Kilometers per liter (km/l)	X 2.825	= Miles per gallon, Imperial (mpg)
Miles per gallon, US (mpg)	X	0.425	= Kilometers per liter (km/l)	X 2.352	= Miles per gallon, US (mpg)

Temperature

Degrees Fahrenheit = (°C x 1.8) + 32

Degrees Celsius (Degrees Centigrade; °C) = (°F - 32) x 0.56

*It is common practice to convert from miles per gallon (mpg) to liters/100 kilometers (l/100km),
where mpg (Imperial) x l/100 km = 282 and mpg (US) x l/100 km = 235

DECIMALS to MILLIMETERS

Decimal	mm	Decimal	mm
0.001	0.0254	0.500	12.7000
0.002	0.0508	0.510	12.9540
0.003	0.0762	0.520	13.2080
0.004	0.1016	0.530	13.4620
0.005	0.1270	0.540	13.7160
0.006	0.1524	0.550	13.9700
0.007	0.1778	0.560	14.2240
0.008	0.2032	0.570	14.4780
0.009	0.2286	0.580	14.7320
		0.590	14.9860
0.010	0.2540		
0.020	0.5080		
0.030	0.7620		
0.040	1.0160	0.600	15.2400
0.050	1.2700	0.610	15.4940
0.060	1.5240	0.620	15.7480
0.070	1.7780	0.630	16.0020
0.080	2.0320	0.640	16.2560
0.090	2.2860	0.650	16.5100
		0.660	16.7640
0.100	2.5400	0.670	17.0180
0.110	2.7940	0.680	17.2720
0.120	3.0480	0.690	17.5260
0.130	3.3020		
0.140	3.5560		
0.150	3.8100		
0.160	4.0640	0.700	17.7800
0.170	4.3180	0.710	18.0340
0.180	4.5720	0.720	18.2880
0.190	4.8260	0.730	18.5420
		0.740	18.7960
0.200	5.0800	0.750	19.0500
0.210	5.3340	0.760	19.3040
0.220	5.5880	0.770	19.5580
0.230	5.8420	0.780	19.8120
0.240	6.0960	0.790	20.0660
0.250	6.3500		
0.260	6.6040		
0.270	6.8580	0.800	20.3200
0.280	7.1120	0.810	20.5740
0.290	7.3660	0.820	21.8280
		0.830	21.0820
0.300	7.6200	0.840	21.3360
0.310	7.8740	0.850	21.5900
0.320	8.1280	0.860	21.8440
0.330	8.3820	0.870	22.0980
0.340	8.6360	0.880	22.3520
0.350	8.8900	0.890	22.6060
0.360	9.1440		
0.370	9.3980		
0.380	9.6520		
0.390	9.9060		
		0.900	22.8600
0.400	10.1600	0.910	23.1140
0.410	10.4140	0.920	23.3680
0.420	10.6680	0.930	23.6220
0.430	10.9220	0.940	23.8760
0.440	11.1760	0.950	24.1300
0.450	11.4300	0.960	24.3840
0.460	11.6840	0.970	24.6380
0.470	11.9380	0.980	24.8920
0.480	12.1920	0.990	25.1460
0.490	12.4460	1.000	25.4000

FRACTIONS to DECIMALS to MILLIMETERS

Fraction	Decimal	mm	Fraction	Decimal	mm
1/64	0.0156	0.3969	33/64	0.5156	13.0969
1/32	0.0312	0.7938	17/32	0.5312	13.4938
3/64	0.0469	1.1906	35/64	0.5469	13.8906
1/16	0.0625	1.5875	9/16	0.5625	14.2875
5/64	0.0781	1.9844	37/64	0.5781	14.6844
3/32	0.0938	2.3812	19/32	0.5938	15.0812
7/64	0.1094	2.7781	39/64	0.6094	15.4781
1/8	0.1250	3.1750	5/8	0.6250	15.8750
9/64	0.1406	3.5719	41/64	0.6406	16.2719
5/32	0.1562	3.9688	21/32	0.6562	16.6688
11/64	0.1719	4.3656	43/64	0.6719	17.0656
3/16	0.1875	4.7625	11/16	0.6875	17.4625
13/64	0.2031	5.1594	45/64	0.7031	17.8594
7/32	0.2188	5.5562	23/32	0.7188	18.2562
15/64	0.2344	5.9531	47/64	0.7344	18.6531
1/4	0.2500	6.3500	3/4	0.7500	19.0500
17/64	0.2656	6.7469	49/64	0.7656	19.4469
9/32	0.2812	7.1438	25/32	0.7812	19.8438
19/64	0.2969	7.5406	51/64	0.7969	20.2406
5/16	0.3125	7.9375	13/16	0.8125	20.6375
21/64	0.3281	8.3344	53/64	0.8281	21.0344
11/32	0.3438	8.7312	27/32	0.8438	21.4312
23/64	0.3594	9.1281	55/64	0.8594	21.8281
3/8	0.3750	9.5250	7/8	0.8750	22.2250
25/64	0.3906	9.9219	57/64	0.8906	22.6219
13/32	0.4062	10.3188	29/32	0.9062	23.0188
27/64	0.4219	10.7156	59/64	0.9219	23.4156
7/16	0.4375	11.1125	15/16	0.9375	23.8125
29/64	0.4531	11.5094	61/64	0.9531	24.2094
15/32	0.4688	11.9062	31/32	0.9688	24.6062
31/64	0.4844	12.3031	63/64	0.9844	25.0031
1/2	0.5000	12.7000	1	1.0000	25.4000

Safety first!

Regardless of how enthusiastic you may be about getting on with the job at hand, take the time to ensure that your safety is not jeopardized. A moment's lack of attention can result in an accident, as can failure to observe certain simple safety precautions. The possibility of an accident will always exist, and the following points should not be considered a comprehensive list of all dangers. Rather, they are intended to make you aware of the risks and to encourage a safety conscious approach to all work you carry out on your vehicle.

Essential DOs and DON'Ts

DON'T rely on a jack when working under the vehicle. Always use approved jackstands to support the weight of the vehicle and place them under the recommended lift or support points.

DON'T attempt to loosen extremely tight fasteners (i.e. wheel lug nuts) while the vehicle is on a jack - it may fall.

DON'T start the engine without first making sure that the transmission is in Neutral (or Park where applicable) and the parking brake is set.

DON'T remove the radiator cap from a hot cooling system - let it cool or cover it with a cloth and release the pressure gradually.

DON'T attempt to drain the engine oil until you are sure it has cooled to the point that it will not burn you.

DON'T touch any part of the engine or exhaust system until it has cooled sufficiently to avoid burns.

DON'T siphon toxic liquids such as gasoline, antifreeze and brake fluid by mouth, or allow them to remain on your skin.

DON'T inhale brake lining dust - it is potentially hazardous (see *Asbestos* below).

DON'T allow spilled oil or grease to remain on the floor - wipe it up before someone slips on it.

DON'T use loose fitting wrenches or other tools which may slip and cause injury.

DON'T push on wrenches when loosening or tightening nuts or bolts. Always try to pull the wrench toward you. If the situation calls for pushing the wrench away, push with an open hand to avoid scraped knuckles if the wrench should slip.

DON'T attempt to lift a heavy component alone - get someone to help you.

DON'T *rush or take unsafe shortcuts to finish a job.*

DON'T allow children or animals in or around the vehicle while you are working on it.

DO wear eye protection when using power tools such as a drill, sander, bench grinder, etc. and when working under a vehicle.

DO keep loose clothing and long hair well out of the way of moving parts.

DO make sure that any hoist used has a safe working load rating adequate for the job.

DO get someone to check on you periodically when working alone on a vehicle.

DO carry out work in a logical sequence and make sure that everything is correctly assembled and tightened.

DO keep chemicals and fluids tightly capped and out of the reach of children and pets.

DO remember that your vehicle's safety affects that of yourself and others. If in doubt on any point, get professional advice.

Steering, suspension and brakes

These systems are essential to driving safety, so make sure you have a qualified shop or individual check your work. Also, compressed suspension springs can cause injury if released suddenly - be sure to use a spring compressor.

Airbags

Airbags are explosive devices that can **CAUSE** injury if they deploy while you're working on the vehicle. Follow the manufacturer's instructions to disable the airbag whenever you're working in the vicinity of airbag components.

Asbestos

Certain friction, insulating, sealing, and other products - such as brake linings, brake bands, clutch linings, torque converters, gaskets, etc. - may contain asbestos or other hazardous friction material. Extreme care must be taken to avoid inhalation of dust from such products, since it is hazardous to health. If in doubt, assume that they do contain asbestos.

Fire

Remember at all times that gasoline is highly flammable. Never smoke or have any kind of open flame around when working on a vehicle. But the risk does not end there. A spark caused by an electrical short circuit, by two metal surfaces contacting each other, or even by static electricity built up in your body under certain conditions, can ignite gasoline vapors, which in a confined space are highly explosive. Do not, under any circumstances, use gasoline for cleaning parts. Use an approved safety solvent.

Always disconnect the battery ground (-) cable at the battery before working on any part of the fuel system or electrical system. Never risk spilling fuel on a hot engine or exhaust component. It is strongly recommended that a fire extinguisher suitable for use on fuel and electrical fires be kept handy in the garage or workshop at all times. Never try to extinguish a fuel or electrical fire with water.

Fumes

Certain fumes are highly toxic and can quickly cause unconsciousness and even death if inhaled to any extent. Gasoline vapor falls into this category, as do the vapors from some cleaning solvents. Any draining or pouring of such volatile fluids should be done in a well ventilated area.

When using cleaning fluids and solvents, read the instructions on the container carefully. Never use materials from unmarked containers.

Never run the engine in an enclosed space, such as a garage. Exhaust fumes contain carbon monoxide, which is extremely poisonous. If you need to run the engine, always do so in the open air, or at least have the rear of the vehicle outside the work area.

The battery

Never create a spark or allow a bare light bulb near a battery. They normally give off a certain amount of hydrogen gas, which is highly explosive.

Always disconnect the battery ground (-) cable at the battery before working on the fuel or electrical systems.

If possible, loosen the filler caps or cover when charging the battery from an external source (this does not apply to sealed or maintenance-free batteries). Do not charge at an excessive rate or the battery may burst.

Take care when adding water to a non maintenance-free battery and when carrying a battery. The electrolyte, even when diluted, is very corrosive and should not be allowed to contact clothing or skin.

Always wear eye protection when cleaning the battery to prevent the caustic deposits from entering your eyes.

Household current

When using an electric power tool, inspection light, etc., which operates on household current, always make sure that the tool is correctly connected to its plug and that, where necessary, it is properly grounded. Do not use such items in damp conditions and, again, do not create a spark or apply excessive heat in the vicinity of fuel or fuel vapor.

Secondary ignition system voltage

A severe electric shock can result from touching certain parts of the ignition system (such as the spark plug wires) when the engine is running or being cranked, particularly if components are damp or the insulation is defective. In the case of an electronic ignition system, the secondary system voltage is much higher and could prove fatal.

Hydrofluoric acid

This extremely corrosive acid is formed when certain types of synthetic rubber, found in some O-rings, oil seals, fuel hoses, etc. are exposed to temperatures above 750-degrees F (400-degrees C). The rubber changes into a charred or sticky substance containing the acid. *Once formed, the acid remains dangerous for years. If it gets onto the skin, it may be necessary to amputate the limb concerned.*

When dealing with a vehicle which has suffered a fire, or with components salvaged from such a vehicle, wear protective gloves and discard them after use.

Troubleshooting

Contents

This section provides an easy reference guide to the more common problems which may occur during the operation of your vehicle. These problems and their possible causes are grouped under headings denoting various components or systems, such as Engine, Cooling system, etc. They also refer you to the chapter and/or section which deals with the problem.

Remember that successful troubleshooting is not a mysterious black art practiced only by professional mechanics. It is simply the result of the right knowledge combined with an intelligent, systematic approach to the problem. Always work by a process of elimination, starting with the simplest solution and working through to the most complex - and never overlook the obvious. Anyone can run the gas tank dry or leave the lights on overnight, so don't assume that you are exempt from such oversights.

Finally, always establish a clear idea of why a problem has occurred and take steps to ensure that it doesn't happen again. If the electrical system fails because of a poor connection, check the other connections in the system to make sure that they don't fail as well. If a particular fuse continues to blow, find out why - don't just replace one fuse after another. Remember, failure of a small component can often be indicative of potential failure or incorrect functioning of a more important component or system.

Engine

1 Engine will not rotate when attempting to start

1 Battery terminal connections loose or corroded (Chapter 1).
2 Battery discharged or faulty (Chapter 1).
3 Automatic transmission not completely engaged in Park (Chapter 7) or clutch not completely depressed (Chapter 8).
4 Broken, loose or disconnected wiring in the starting circuit (Chapters 5 and 12).
5 Starter solenoid faulty (Chapter 5).
6 Starter motor faulty (Chapter 5).
7 Ignition switch faulty (Chapter 12).
8 Starter pinion or flywheel teeth worn or broken (Chapter 5).

2 Engine rotates but will not start

1 Fuel tank empty.
2 Battery discharged (engine rotates slowly) (Chapter 5).
3 Battery terminal connections loose or corroded (Chapter 1).
4 Faulty fuel pump, pressure regulator, etc. (Chapter 4).
5 Faulty Crankshaft Position (CKP) sensor or Camshaft Position (CMP) sensor (Chapter 6).
6 Fouled spark plugs (Chapter 1).

7 Broken or stripped timing belt/broken timing chain (Chapter 2).
8 Defective fuel pump relay and/or harness at relay (Chapter 4)

3 Engine hard to start when cold

1 Battery discharged or low (Chapter 1).
2 Malfunctioning fuel system (Chapter 4).

4 Engine hard to start when hot

1 Air filter clogged (Chapter 1).
2 Malfunctioning fuel system (Chapter 4).
3 Injector(s) leaking (Chapter 4).
4 Corroded battery connections.
5 Malfunctioning EVAP system (Chapter 6)

5 Starter motor noisy or excessively rough in engagement

1 Pinion or flywheel gear teeth worn or broken (Chapter 5).
2 Starter motor mounting bolts loose or missing (Chapter 5).

6 Engine starts but stops immediately

1 Insufficient fuel reaching the fuel injector(s) (Chapter 4).
2 Vacuum leak at the gasket between the intake manifold and throttle body (Chapters 1 and 4).

7 Oil puddle under engine

1 Oil pan gasket and/or oil pan drain bolt washer leaking (Chapter 2).
2 Oil pressure sending unit leaking (Chapter 2).
3 Valve cover leaking (Chapter 2).
4 Engine oil seals leaking (Chapter 2).

8 Engine lopes while idling or idles erratically

1 Vacuum leakage (Chapters 2 and 4).
2 Defective EGR valve (Chapter 6).
3 Air filter clogged (Chapter 1).
4 Fuel pump not delivering sufficient fuel to the fuel injection system (Chapter 4).
5 Leaking head gasket (Chapter 2).
6 Timing belt/chain and/or sprockets worn (Chapter 2).
7 Camshaft lobes worn (Chapter 2).
8 Problem in Engine Mount Control System (Chapter 2).

9 Engine misses at idle speed

1 Spark plugs worn or not gapped properly (Chapter 1).
2 Vacuum leaks (Chapter 1).
3 Uneven or low compression (Chapter 2).

10 Engine misses throughout driving speed range

1 Fuel filter clogged and/or impurities in the fuel system (Chapter 4).
2 Low fuel pressure (Chapter 4).
3 Faulty or incorrectly gapped spark plugs (Chapter 1).
4 Faulty emission system components (Chapter 6).
5 Low or uneven cylinder compression pressures (Chapter 2).
6 Weak or faulty ignition system (Chapter 5).
7 Vacuum leak.

11 Engine stumbles on acceleration

1 Spark plugs fouled (Chapter 1).
2 Fuel injection system faulty (Chapter 4).
3 Fuel filter clogged (Chapters 1 and 4).
4 Incorrect ignition timing (Chapter 5).
5 Intake air leak (Chapters 2 and 4).

12 Engine surges while holding accelerator steady

1 Intake air leak (Chapter 4).
2 Fuel pump faulty (Chapter 4).
3 Loose fuel injector wire harness connectors (Chapter 4).
4 Defective PCM or information sensor (Chapter 6).

13 Engine stalls

1 Fuel filter clogged and/or water and impurities in the fuel system (Chapters 1 and 4).
2 Faulty emissions system components (Chapter 6).
3 Faulty or incorrectly gapped spark plugs (Chapter 1).
4 Vacuum leak in the intake manifold or vacuum hoses (Chapters 2 and 4).
5 Valve clearances incorrectly set (Chapters 1 and 2).

14 Engine lacks power

1 Faulty or incorrectly gapped spark plugs (Chapter 1).
2 Fuel injection system malfunction (Chapter 4).
3 Faulty coil(s) (Chapter 5).
4 Brakes binding (Chapter 9).

5 Clutch slipping (Chapter 8).
6 Fuel filter clogged and/or impurities in the fuel system (Chapter 4).
7 Emission control system not functioning properly (Chapter 6).
8 Low or uneven cylinder compression pressures (Chapter 2).
9 Obstructed exhaust system (Chapter 4).

15 Engine backfires

1 Emission control system not functioning properly (Chapter 6).
2 Fuel injection system malfunctioning (Chapter 4).
3 Vacuum leak at fuel injector(s), intake manifold, air control valve or vacuum hoses (Chapters 2 and 4).
4 Valve clearances incorrectly set and/or valves sticking (Chapters 1 and 2).

16 Pinging or knocking engine sounds during acceleration or uphill

1 Incorrect grade of fuel.
2 Fuel injection system faulty (Chapter 4).
3 Improper or damaged spark plugs (Chapter 1).
4 Vacuum leak (Chapters 2 and 4).

17 Engine runs with oil pressure light on

1 Low oil level (Chapter 1).
2 Short in wiring circuit (Chapter 12).
3 Faulty oil pressure sender (Chapter 2).
4 Worn engine bearings and/or oil pump (Chapter 2).

18 Engine diesels (continues to run) after switching off

1 Defective ignition switch or Stop/Start switch (Chapter 12).
2 Faulty Powertrain Control Module (Chapter 6).
3 Faulty Body Control Module.
4 Leaking fuel injector(s) (Chapter 4).

Engine electrical system

19 Battery will not hold a charge

1 Drivebelt defective (Chapter 1).
2 Battery electrolyte level low (Chapter 1).
3 Battery terminals loose or corroded (Chapter 1).
4 Alternator not charging properly (Chapter 5).

5 Loose, broken or faulty wiring in the charging circuit (Chapter 5).
6 Internally defective battery (Chapters 1 and 5).

20 Alternator light fails to go out

1 Faulty alternator or charging circuit (Chapter 5).
2 Drivebelt defective (Chapter 1).
3 Alternator voltage regulator inoperative (Chapter 5).

21 Alternator light fails to come on when key is turned on

1 Instrument cluster module defective (Chapter 5).
2 Fault in the Powertrain Control Module.

Fuel system

22 Excessive fuel consumption

Dirty or clogged air filter element (Chapter 1).

23 Fuel leakage and/or fuel odor

1 Leaking fuel line (Chapters 1 and 4).
2 Tank overfilled.
3 Evaporative canister defective (Chapter 6).

Cooling system

24 Overheating

1 Insufficient coolant in system (Chapter 1).
2 Radiator core blocked or grille restricted (Chapter 3).
3 Thermostat faulty (Chapter 3).
4 Electric coolant fan problem (Chapter 3).
5 Cooling system pressure cap not maintaining proper pressure (Chapter 3).

25 Overcooling

1 Faulty thermostat (Chapter 3).
2 Electric coolant fan problem (Chapter 3).

26 External coolant leakage

1 Deteriorated/damaged hoses; loose clamps (Chapters 1 and 3).
2 Water pump defective (Chapter 3).
3 Leakage from radiator core or coolant

expansion tank (Chapter 3).
4 Engine drain or water jacket core plugs leaking (Chapter 2).

27 Internal coolant leakage

1 Leaking cylinder head gasket (Chapter 2).
2 Cracked cylinder bore or cylinder head (Chapter 2).

28 Coolant loss

1 Too much coolant in system (Chapter 1).
2 Coolant boiling away because of overheating (Chapter 3).
3 Internal or external leakage (Chapter 3).
4 Faulty radiator cap (Chapter 3).

29 Poor coolant circulation

1 Inoperative water pump (Chapter 3).
2 Restriction in cooling system (Chapters 1 and 3).
3 Thermostat sticking (Chapter 3).

Clutch

30 Pedal travels to floor - no pressure or very little resistance

1 No fluid in reservoir (Chapter 1).
2 Faulty clutch master cylinder, release cylinder or hydraulic line (Chapter 8).
3 Broken release bearing or fork (Chapter 8).

31 Unable to select gears

1 Faulty transaxle (Chapter 7).
2 Faulty clutch disc (Chapter 8).
3 Release lever and bearing not assembled properly (Chapter 8).
4 Faulty pressure plate (Chapter 8).
5 Pressure plate-to-flywheel bolts loose (Chapter 8).

32 Clutch slips (engine speed increases with no increase in vehicle speed)

1 Clutch plate worn (Chapter 8).
2 Clutch plate is oil soaked by leaking rear main seal (Chapter 8).
3 Warped pressure plate or flywheel (Chapter 8).
4 Weak diaphragm spring (Chapter 8).
5 Clutch plate overheated. Allow to cool.

33 Grabbing (chattering) as clutch is engaged

1   Oil on clutch plate lining, burned or glazed facings (Chapter 8).
2   Worn or loose engine or transaxle mounts (Chapters 2 and 7).
3   Worn splines on clutch plate hub (Chapter 8).
4   Warped pressure plate or flywheel (Chapter 8).
5   Burned or smeared resin on flywheel or pressure plate (Chapter 8).

34 Transaxle rattling (clicking)

1   Release lever loose (Chapter 8).
2   Clutch plate damper spring failure (Chapter 8).
3   Low engine idle speed (Chapter 1).

35 Noise in clutch area

1   Release fork improperly installed (Chapter 8).
2   Faulty bearing (Chapter 8).

36 Clutch pedal stays on floor

1   Faulty clutch master or release cylinder (Chapter 8).
2   Broken release bearing or fork (Chapter 8).

37 High pedal effort

1   Piston binding in bore of clutch master or release cylinder (Chapter 8).
2   Pressure plate faulty (Chapter 8).

Manual transaxle

38 Knocking noise at low speeds

1   Worn driveaxle constant velocity (CV) joints (Chapter 8).
2   Worn driveaxle bore in differential case (Chapter 7A).*

39 Noise most pronounced when turning

Differential gear noise (Chapter 7A).*

40 Clunk on acceleration or deceleration

1   Loose engine or transaxle mounts (Chapters 2 and 7A).
2   Worn differential pinion shaft in case.*
3   Worn driveaxle bore in differential case (Chapter 7A).*
4   Worn or damaged driveaxle inboard CV joints (Chapter 8).

41 Clicking noise in turns

Worn or damaged outboard CV joint (Chapter 8).

42 Vibration

1   Rough wheel bearing (Chapters 1 and 10).
2   Damaged driveaxle (Chapter 8).
3   Out of round tires (Chapter 1).
4   Tire out of balance (Chapters 1 and 10).
5   Worn CV joint (Chapter 8).

43 Noisy in neutral with engine running

1   Damaged input gear bearing (Chapter 7A).*
2   Damaged clutch release bearing (Chapter 8).

44 Noisy in one particular gear

1   Damaged or worn constant mesh gears (Chapter 7A).*
2   Damaged or worn synchronizers (Chapter 7A).*
3   Bent reverse fork (Chapter 7A).*
4   Damaged fourth speed gear or output gear (Chapter 7A).*
5   Worn or damaged reverse idler gear or idler bushing (Chapter 7A).*

45 Noisy in all gears

1   Insufficient lubricant (Chapter 7A).
2   Damaged or worn bearings (Chapter 7A).*
3   Worn or damaged input gear shaft and/or output gear shaft (Chapter 7A).*

46 Slips out of gear

1   Worn or improperly adjusted linkage (Chapter 7A).
2   Transaxle loose on engine (Chapter 7A).
3   Shift linkage does not work freely, binds

(Chapter 7A).
4   Input gear bearing retainer broken or loose (Chapter 7A).*
5   Worn shift fork (Chapter 7A).*

47 Leaks lubricant

1   Driveaxle oil seals worn (Chapter 7).
2   Excessive amount of lubricant in transaxle (Chapters 1 and 7A).
3   Loose or broken input gear shaft bearing retainer (Chapter 7A).*
4   Input gear bearing retainer O-ring and/or lip seal damaged (Chapter 7A).*

48 Locked in gear

Lock pin or interlock pin missing (Chapter 7A).*
Although the corrective action necessary to remedy the symptoms described is beyond the scope of the home mechanic, the above information should be helpful in isolating the cause of the condition so that the owner can communicate clearly with a professional mechanic.

Automatic transaxle
Note: *Due to the complexity of the automatic transaxle, it is difficult for the home mechanic to properly diagnose and service this component. For problems other than the following, the vehicle should be taken to a dealer or transmission shop.*

49 Fluid leakage

1   Automatic transmission fluid is a deep red color. Fluid leaks should not be confused with engine oil, which can easily be blown onto the transaxle by air flow.
2   To pinpoint a leak, first remove all built-up dirt and grime from the transaxle housing with degreasing agents and/or steam cleaning. Then drive the vehicle at low speeds so air flow will not blow the leak far from its source. Raise the vehicle and determine where the leak is coming from.

50 Transaxle fluid brown or has a burned smell

Transaxle fluid burned (Chapter 1).

51 General shift mechanism problems

1   Chapter 7, Part B, deals with checking and adjusting the shift linkage on automatic transaxles. Common problems which may be attributed to poorly adjusted linkage are:
a) *Engine starting in gears other than Park or Neutral.*
b) *Indicator on shifter pointing to a gear other than the one actually being used.*
c) *Vehicle moves when in Park.*
2   Refer to Chapter 7B for the shift linkage adjustment procedure.

52 Transaxle will not downshift with accelerator pedal pressed to the floor

Since these transmissions are electronically controlled, check for any diagnostic trouble codes stored in the PCM. The actual repair will most likely have to be performed by a qualified repair shop with the proper equipment.

53 Engine will start in gears other than Park or Neutral

Transmission range switch malfunctioning (Chapter 6).

54 Transaxle slips, shifts roughly, is noisy or has no drive in forward or reverse gears

There are many probable causes for the above problems, but the home mechanic should be concerned with only one possibility - fluid level. Before taking the vehicle to a repair shop, check the level and condition of the fluid as described in Chapter 1. Correct the fluid level as necessary or change the fluid and filter if needed. If the problem persists, have a professional diagnose the cause.

Driveaxles

55 Clicking noise in turns

Worn or damaged outboard CV joint (Chapter 8).

56 Shudder or vibration during acceleration

1   Excessive toe-in (Chapter 10).
2   Incorrect spring heights (Chapter 10).

3   Worn or damaged inboard or outboard CV joints (Chapter 8).
4   Sticking inboard CV joint assembly (Chapter 8).

57 Vibration at highway speeds

1   Out of balance front wheels/tires (Chapters 1 and 10).
2   Out of round front tires (Chapters 1 and 10).
3   Worn CV joint(s) (Chapter 8).

Brakes

Note: *Before assuming that a brake problem exists, make sure that:*
a) *The tires are in good condition and properly inflated (Chapter 1).*
b) *The front end alignment is correct (Chapter 10).*
c) *The vehicle is not loaded with weight in an unequal manner.*

58 Vehicle pulls to one side during braking

1   Incorrect tire pressures (Chapter 1).
2   Front end out of line (have the front end aligned).
3   Front, or rear, tires not matched to one another.
4   Restricted brake lines or hoses (Chapter 9).
5   Malfunctioning caliper assembly (Chapter 9).
6   Loose suspension parts (Chapter 10).
7   Excessive wear of pad material or disc on one side.

59 Noise (high-pitched squeal when the brakes are applied)

Disc brake pads worn out. Replace pads with new ones immediately (Chapter 9).

60 Brake roughness or chatter (pedal pulsates)

1   Excessive lateral runout (Chapter 9).
2   Defective disc (Chapter 9).

61 Excessive brake pedal effort required to stop vehicle

1   Malfunctioning power brake booster (Chapter 9).
2   Partial system failure (Chapter 9).
3   Excessively worn pads (Chapter 9).
4   Piston in caliper stuck or sluggish (Chapter 9).
5   Brake pads contaminated with oil or

grease (Chapter 9).
6   New pads installed and not yet seated. It will take a while for the new material to seat against the disc.

62 Excessive brake pedal travel

1   Partial brake system failure (Chapter 9).
2   Insufficient fluid in master cylinder (Chapters 1 and 9).
3   Air trapped in system (Chapters 1 and 9).

63 Dragging brakes

1   Incorrect adjustment of brake light switch (Chapter 9).
2   Master cylinder pistons not returning correctly (Chapter 9).
3   Restricted brakes lines or hoses (Chapters 1 and 9).
4   Incorrect parking brake adjustment (Chapter 9).

64 Grabbing or uneven braking action

1   Contaminated pad lining material (Chapter 9).
2   Excessively worn brake pads (Chapter 9).

65 Brake pedal feels spongy when depressed

1   Air in hydraulic lines (Chapter 9).
2   Master cylinder mounting bolts loose (Chapter 9).
3   Master cylinder defective (Chapter 9).

66 Brake pedal travels to the floor with little resistance

1   Little or no fluid in the master cylinder reservoir caused by leaking caliper piston(s) (Chapter 9).
2   Loose, damaged or disconnected brake lines (Chapter 9).

67 Parking brake does not hold

Parking brake improperly adjusted (Chapters 1 and 9).

Suspension and steering systems

Note: *Before attempting to diagnose the suspension and steering systems, perform the following preliminary checks:*

a) *Tires for wrong pressure and uneven wear.*

b) *Steering universal joints from the column to the steering gear for loose connectors or wear.*

c) *Front and rear suspension and the steering gear assembly for loose or damaged parts.*

d) *Out-of-round or out-of-balance tires, bent rims and loose and/or rough wheel bearings.*

68 Vehicle pulls to one side

1   Mismatched or uneven tires (Chapter 10).
2   Broken or sagging springs (Chapter 10).
3   Wheel alignment (Chapter 10).
4   Front brake dragging (Chapter 9).

69 Abnormal or excessive tire wear

1   Wheel alignment (Chapter 10).
2   Sagging or broken springs (Chapter 10).
3   Tire out of balance (Chapter 10).
4   Worn strut/shock absorber (Chapter 10).
5   Overloaded vehicle.
6   Tires not rotated regularly.

70 Wheel makes a thumping noise

1   Blister or bump on tire (Chapter 10).
2   Worn strut/shock absorber (Chapter 10).

71 Shimmy, shake or vibration

1   Tire or wheel out-of-balance or out-of-round (Chapter 10).
2   Loose or worn front hub or wheel bearings (Chapters 1, 8 and 10).
3   Worn tie-rod ends (Chapter 10).
4   Worn lower balljoints (Chapters 1 and 10).
5   Excessive wheel runout (Chapter 10).
6   Blister or bump on tire (Chapter 10).

72 Hard steering

1   Worn balljoints and/or tie-rod ends (Chapters 1 and 10).
2   Front wheels out of alignment (Chapter 10).
3   Low tire pressure(s) (Chapters 1 and 10).

73 Poor returnability of steering to center

1   Worn balljoints and/or tie-rod ends (Chapters 1 and 10).

2   Worn steering gear assembly (Chapter 10).
3   Front wheels out of alignment (Chapter 10).

74 Abnormal noise at the front end

1   Worn balljoints and/or tie-rod ends (Chapters 1 and 10).
2   Damaged strut mount (Chapter 10).
3   Worn control arm bushings or tie-rod ends (Chapter 10).
4   Loose stabilizer bar (Chapter 10).
5   Loose wheel nuts (Chapters 1 and 10).
6   Loose suspension bolts (Chapter 10)

75 Wander or poor steering stability

1   Mismatched or uneven tires (Chapter 10).
2   Lack of lubrication at balljoints and tie-rod ends (Chapters 1 and 10).
3   Worn strut/shock absorber/coil spring assemblies (Chapter 10).
4   Broken or sagging springs (Chapter 10).
5   Wheels out of alignment (Chapter 10).

76 Erratic steering when braking

1   Front hub bearings worn (Chapter 10).
2   Broken or sagging springs (Chapter 10).
3   Leaking caliper (Chapter 10).
4   Warped discs (Chapter 10).

77 Excessive pitching and/or rolling around corners or during braking

1   Loose stabilizer bar (Chapter 10).
2   Worn strut/shock absorber/coil spring assemblies or mountings (Chapter 10).
3   Broken or sagging springs (Chapter 10).
4   Overloaded vehicle.

78 Suspension bottoms

1   Overloaded vehicle.
2   Worn strut/shock absorber/coil spring assemblies (Chapter 10).
3   Incorrect, broken or sagging springs (Chapter 10).

79 Cupped tires

1   Front wheel or rear wheel alignment (Chapter 10).
2   Worn strut/shock absorber/coil spring assemblies (Chapter 10).
3   Wheel bearings worn (Chapter 10).

4   Excessive tire or wheel runout (Chapter 10).
5   Worn balljoints (Chapter 10).

80 Excessive tire wear on outside edge

1   Inflation pressures incorrect (Chapter 1).
2   Excessive speed in turns.
3   Front end alignment incorrect (excessive toe-in). Have professionally aligned.
4   Suspension arm bent or twisted (Chapter 10).

81 Excessive tire wear on inside edge

1   Inflation pressures incorrect (Chapter 1).
2   Front end alignment incorrect (toe-out). Have professionally aligned.
3   Loose or damaged steering or suspension components (Chapter 10).

82 Tire tread worn in one place

1   Tires out of balance.
2   Damaged wheel. Inspect and replace if necessary.
3   Defective tire (Chapter 1).

83 Excessive play or looseness in steering system

1   Front hub bearing(s) worn (Chapter 10).
2   Tie-rod end loose (Chapter 10).
3   Steering gear loose or worn (Chapter 10).
4   Worn or loose steering intermediate shaft (Chapter 10).

84 Rattling or clicking noise in steering gear

1   Steering gear loose (Chapter 10).
2   Steering gear defective.

Notes

Chapter 1
Tune-up and routine maintenance

Contents

Specifications

Note: *Throughout this Chapter you will find references to "Mk I" and "Mk II" models; this is done to simplify which specifications and procedures apply to which models. Mk I models include 2006 and earlier Cooper/Cooper S models, and 2008 and earlier Convertible models. Mk II models include 2007 and later Cooper/Cooper S/Clubman/Clubman S and 2009 and later Convertible models.*

Recommended lubricants and fluids

Engine oil

Mk I models .... MINI Longlife 04, BMW Longlife 01 or BMW Longlife 98 or equivalent (ACEA A3 may be used for topping-up between oil changes) - SAE 0W-30 or SAE 5W-30, synthetic

Mk II models .... MINI Longlife (rating LL-01) or equivalent - SAE 5W-30, synthetic

Power steering fluid (Mk I models) .... Pentosin CHF 11 S, or equivalent

Brake/clutch fluid .... DOT 4

Manual transaxle lubricant

Mk I models .... MTF 94 or equivalent

Mk II models .... MTF-LT-4 or equivalent

Automatic transaxle fluid

CVT .... ESSO CVT EZL 799A or equivalent

Agitronic 6-speed .... ESSO JWS-3309 (MINI part no. 83 22 0 402 413) or equivalent

Engine coolant .... 50/50 mixture of BMW phosphate free/nitrate free (or equivalent) antifreeze and distilled water

Capacities*

Engine oil (including filter)

2004 and earlier models...	4.7 qts (4.5 liters)

2005 and later Mk I models

Cooper...	4.7 qts (4.5 liters)
Cooper S...	5.0 qts (4.8 liters)
Mk II models...	4.4 qts (4.2 liters)

Cooling system

Cooper	5.6 qts (5.3 liters)
Cooper S..	6.3 qts (6.0 liters)

Manual transaxle

Cooper	1.8 qts (1.7 liters)
Cooper S..	2.0 qts (1.9 liters)

Automatic transaxle

CVT	4.8 qts (4.5 liters)
Agitronic 6-speed	4.8 qts (4.5 liters)

* All capacities approximate. Add as necessary to bring to the appropriate level.

FRONT OF
VEHICLE ❶②③④

↓

Cylinder locations

Cooling system

Antifreeze mixture

50% antifreeze	Protection down to -35-degrees F (-37°C)
55% antifreeze	Protection down to -49-degees F (-45°C)

Note: *Refer to antifreeze manufacturer for latest recommendations.*

Ignition system

Spark plug type...	Consult your local auto parts store or MINI dealer parts department
Gap..	Electrode gap not adjustable on these plugs.
Firing order	1-3-4-2

Brakes

Friction material minimum thickness, front and rear brake pads.............	1/8-inch (3.0 mm)

Parking brake lever travel

Mk I models	8 clicks maximum
Mk II models	6 clicks maximum

Torque specifications

	Ft-lbs	Nm
Engine oil drain plug		
Mk I models	18	25
Mk II models	22	30
Oil filter housing cap..	18	25
Wheel bolts		
Mk I models	89	120
Mk II models	103	140
Spark plugs		
Mk I models	20	27
Mk II models	17	23

Underhood view of a MINI Cooper - Mk I models

1	Brake/clutch fluid reservoir	5	Spark plug wires	8	Power steering fluid reservoir
2	Battery (under cover)		(spark plugs underneath)	9	Engine oil filler cap
3	Air filter housing	6	Windshield washer fluid reservoir	10	Cooling system pressure cap
4	Engine oil dipstick	7	Coolant expansion tank		

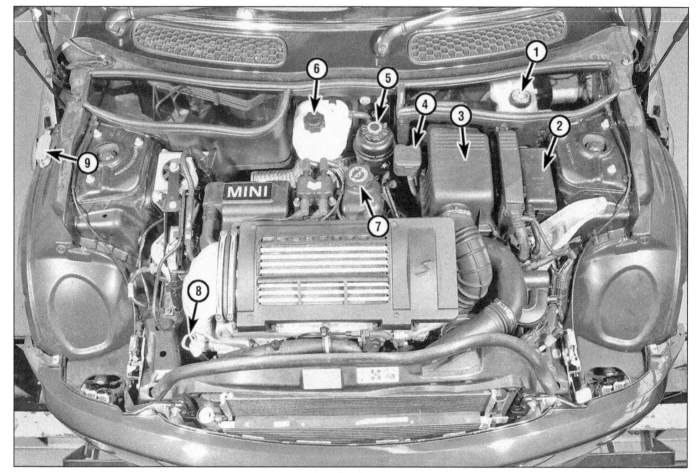

Underhood view of a MINI Cooper S - Mk I models

1	Brake/clutch fluid reservoir	4	Remote battery positive terminal	7	Engine oil filler cap
2	Underhood fuse/relay box	5	Power steering fluid reservoir	8	Engine oil dipstick
3	Air filter housing	6	Coolant expansion tank	9	Windshield washer fluid reservoir

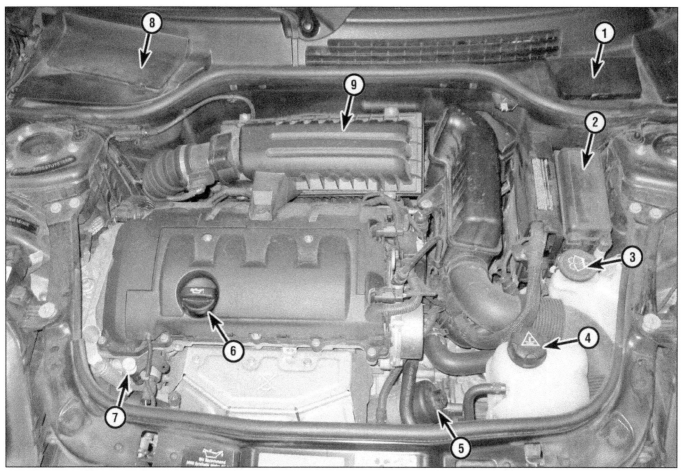

Underhood view of a MINI Cooper - Mk II models

1	Brake/clutch fluid reservoir (under cover)	4	Coolant expansion tank	7	Engine oil dipstick
2	Underhood fuse/relay box	5	Oil filter housing	8	Battery (under cover)
3	Washer fluid reservoir	6	Engine oil filler cap	9	Air filter housing

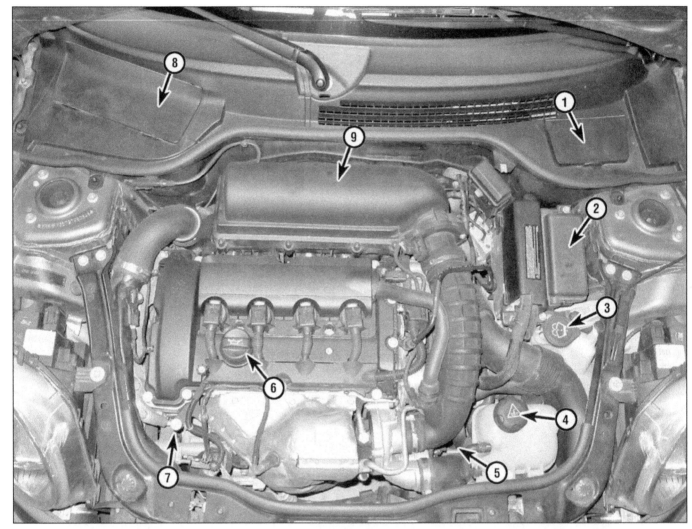

Underhood view of a MINI Cooper S - Mk II models

1	Brake/clutch fluid reservoir (under cover)	4	Coolant expansion tank	7	Engine oil dipstick
2	Underhood fuse/relay box	5	Oil filter housing (not visible)	8	Battery (under cover)
3	Washer fluid reservoir	6	Engine oil filler cap	9	Air filter housing

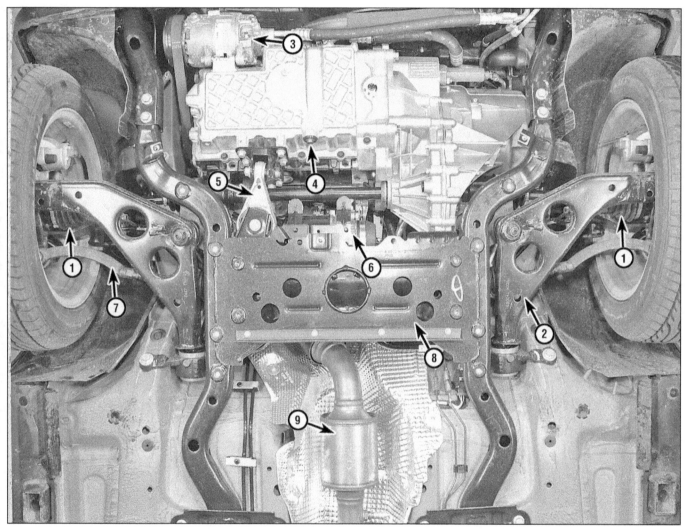

Front underbody view (Mk I Cooper model shown - other models similar)

1	Outer driveaxle boot	4	Engine oil drain plug	7	Tie-rod end
2	Lower suspension arm	5	Engine lower stabilizer	8	Subframe
3	Air conditioning compressor	6	Electric power steering pump	9	Catalytic converter

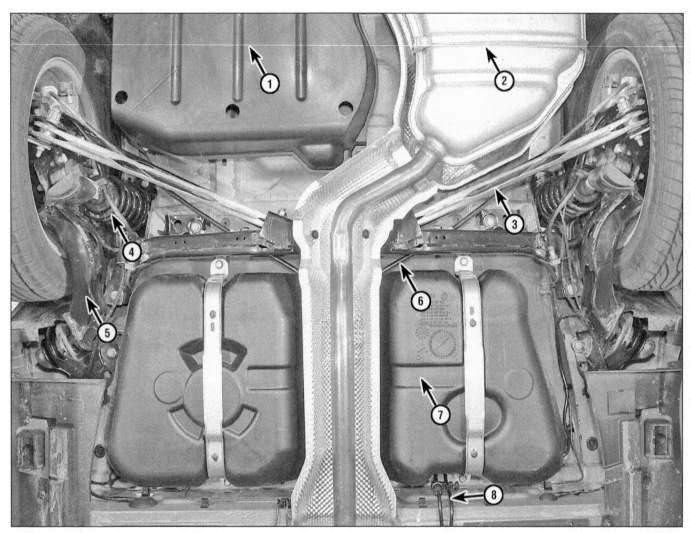

Rear underbody view (Mk I Cooper model shown - other models similar)

1   *Spare wheel carrier*
2   *Exhaust rear muffler*
3   *Lower control arm*

4   *Shock absorber*
5   *Trailing arm*
6   *Parking brake cable*

7   *Fuel tank*
8   *Fuel/EVAP lines*

Maintenance schedule

Note: *Throughout this Chapter you will find references to "Mk I" and "Mk II" models; this is done to simplify which specifications and procedures apply to which models. Mk I models include 2006 and earlier Cooper/Cooper S models, and 2008 and earlier Convertible models. Mk II models include 2007 and later Cooper/Cooper S/Clubman/Clubman S and 2009 and later Convertible models.*

All MINI models are equipped with a service display in the center of the instrument panel, which shows the type of service next due, and the distance remaining until the service is required. Once that distance is reduced to zero, the display then shows the distance since the service was due. On Mk I models, two types of service are specified: an Oil Service and an Inspection Service. On Mk II models, a similar system is used, but operations are broken down into different intervals, and the specific services are indicated on the display. The occurrence of the service on the display unit will depend on how the vehicle is being used (number of starts, length of journeys, vehicle speeds, brake pad wear, hood opening frequency, fuel consumption, oil level and oil temperature). For example, if a vehicle is being used under extreme driving conditions, the service may occur at 10,000 miles, whereas, if the vehicle is being used under moderate driving conditions, it may occur at 20,000 miles. It is important to realize that this system is completely variable according to how the vehicle is being used, and therefore the service should be carried out when indicated on the display. For more details, refer to the owner's manual supplied with the vehicle.

Mk I models

There are two different inspection services, Inspection I and Inspection II, these should be carried out alternately with some additional items to be included at the specified time intervals. If you are unclear as to which inspection schedule was carried out last time, start with Inspection II (including the additional items).

Every 250 miles (400 km) or weekly

Check the engine oil level (Section 2)
Check the engine coolant level (Section 2)
Check the windshield washer fluid level (Section 2)
Check the tires and tire pressures (Section 2)

Oil service

Change the engine oil and filter (Section 3)
Reset the service interval display (Section 4)
Check the front and rear brake pad thickness (Section 5)
Check the operation of the parking brake (Section 7)
Replace the cabin air filter (Section 8)

Inspection I

All items listed above, plus:
Check all underhood components and hoses for fluid leaks (Section 9)
Check the steering and suspension components for condition and security (Section 10)
Check the exhaust system and mountings (Section 11)

Check the condition and operation of the seat belts (Section 12)
Lubricate all hinges and locks (Section 13)
Check the headlight beam alignment (Section 14)
Check the operation of the windshield/headlight washer system(s) (as applicable) (Section 15)
Check the engine management system (Section 16)
Change the Continuously Variable Transmission (CVT) fluid (Chapter 7B)
Carry out a road test (Section 17)

Inspection II

All items listed above, plus:
Replace the drivebelt (Section 18)
Replace the spark plugs (Section 19)
Replace the air filter element (Section 20)
Check the condition of the driveaxle boots (Section 21)

Every 2 years

Note: *The manufacturer states that the following should be carried out regardless of mileage.*
Replace the brake fluid (Section 22)

Every 4 years

Note: *The manufacturer states that the following should be carried out regardless of mileage.*
Replace the coolant (Mk I models) (Section 23)*
Replace the Mobility System puncture sealant (if equipped)
* The manufacturer states that routine coolant changes are not necessary for Mk II models. The coolant replacement procedure can be used, however, for procedures that require draining and refilling the cooling system.

Mk II models

Engine oil service

Change the engine oil and filter (Section 3)
Check the tires and tire pressures (Section 2), reset the tire pressure monitor
Check the engine management system (Section 16)
Check the parking brake adjustment (Chapter 9)
Check the windshield washer fluid level (Section 2)

Condition Based Service (CBS) operations

Replace the front and rear brake pads (Chapter 9)
Replace the brake fluid (Section 22)

Every other oil change (or when prompted by the CBS display)

Check the engine coolant (Section 2 and Chapter 3)
Check the battery and charging system (Section 2 and Chapter 5)
Check the brake system hydraulic lines and hoses (Chapter 9), the brake pads (Section 5), the power brake booster (Section 6), and the parking brake (Section 7)
Check the operation of all interior and exterior lights, turn signal and hazard flasher operation, horn operation
Check the steering and suspension systems (Section 10)
Check the exhaust system (Section 11)
Check the operation of the windshield wipers and washers (Section 15)
Check the drivebelt (Section 18)
Check the seat belts (Section 12)
Check the operation of the instrument cluster and all its functions

Mileage-based operations

Every 30,000 miles (48,000 km):
Replace the spark plugs (JCW models only) (Section 19)

Every 45,000 miles (72,500 km):
Replace the air filter element (more often if driven in dusty conditions) (Section 20)

Every 60,000 miles (96,500 km):
Replace the spark plugs (all Cooper S and 2011 Cooper models) (Section 19)

Every 100,000 miles (160,000 km):
Replace the spark plugs (2007 through 2010 Cooper models) (Section 19)
Replace the automatic transaxle fluid (Chapter 7B)

Every 120,000 miles:
Replace the oxygen sensors (Chapter 6)

---

1   Tune-up and routine maintenance - general information

Note: *Throughout this Chapter you will find references to "Mk I" and "Mk II" models; this is done to simplify which specifications and procedures apply to which models. Mk I models include 2006 and earlier Cooper/Cooper S models, and 2008 and earlier Convertible models. Mk II models include 2007 and later Cooper/Cooper S/Clubman/Clubman S and 2009 and later Convertible models.*

Routine maintenance

This Chapter is designed to help the home mechanic maintain the MINI for peak performance, economy, safety and long life.

Included in this Chapter is a master maintenance schedule, followed by Sections dealing specifically with each item on the schedule. Visual checks, adjustments, component replacement and other helpful items are included. Refer to the **accompanying illustrations** of the engine compartment and the underside of the vehicle for the location of various components.

Servicing your vehicle in accordance with the mileage/time maintenance schedule and the following Sections will provide it with a planned maintenance program that should result in a long and reliable service life. This is a comprehensive plan, so maintaining some items but not others at the specified service intervals will not produce the same results.

As you service your vehicle, you will discover that many of the procedures can, and should, be grouped together because of the nature of the particular procedure you're per-forming or because of the close proximity of two otherwise unrelated components to one another.

For example, if the vehicle is raised for any reason, you should inspect the exhaust, suspension, steering and fuel systems while you're under the vehicle. When you're rotating the tires, it makes good sense to check the brakes and wheel bearings since the wheels are already removed.

Finally, let's suppose you have to borrow or rent a torque wrench. Even if you only need to replace the spark plugs, you might as well check the torque of as many critical fasteners as time allows.

The first step of this maintenance program is to prepare yourself before the actual work begins. Read through all Sections pertinent to the procedures you're planning to do, then make a list of and gather together all the parts and tools you will need to do the job. If it looks as if you might run into problems during a particular segment of some procedure, seek advice from your local auto parts stores or dealer service department.

Tune-up

The term tune-up is used in this manual to represent a combination of individual operations rather than one specific procedure.

If, from the time the vehicle is new, the routine maintenance schedule is followed closely and frequent checks are made of fluid levels and high wear items, as suggested throughout this manual, the engine will be kept in relatively good running condition and the need for additional work will be minimized.

More likely than not, however, there will be times when the engine is running poorly due to lack of regular maintenance. This is even more likely if a used vehicle, which has not received regular and frequent maintenance checks, is purchased. In such cases, an engine tune-up will be needed outside of the regular routine maintenance intervals.

The first steps to take in the event of a poor running engine would be to see if there are any stored trouble codes (see Chapter 6) and to perform a cylinder compression check (see Chapter 2B). A compression check will help determine the condition of internal engine components and should be used as a guide for tune-up and repair procedures. If, for instance, the compression check indicates serious internal engine wear, a major tune-up won't improve the performance of the engine and would be a waste of time and money. Because of its importance, the compression check should be done by someone with the right equipment and the knowledge to use it properly.

Minor tune-up

Check all engine-related fluids (see Section 2).
Clean, inspect and test the battery (see Section 2).
Check the condition of all hoses, and check for fluid leaks (Section 9).
Check the condition of the air filter, and replace if necessary (Section 20).
Check the ignition system (Chapter 5).

Major tune-up

All items listed under Minor tune-up, plus the following:
Check the fuel system (Chapter 4).
Check the charging system (Chapter 5).
Replace the spark plugs (Section 19).

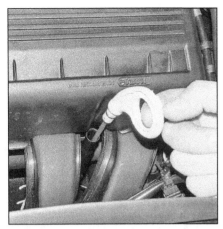

2.2a The dipstick handle is often brightly colored for easy identification - this is an Mk I model . . .

2.2b . . . and this is an Mk II model

2.4a Pull the dipstick out, wipe it off with a clean cloth or paper towel, then reinsert it and pull it out again

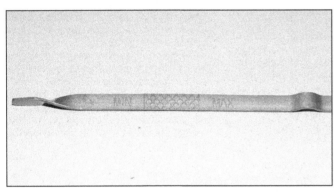

2.4b Note the oil level on the end of the dipstick. On Mk I models, it should be within the cross-hatched area on the dipstick. Approximately one quart (1.0L) of oil will raise the level from the lower mark to the upper mark

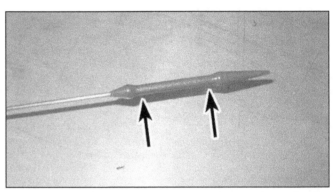

2.4c On Mk II models, the oil level should be in this range, if this type of dipstick is used

2 Weekly checks

Introduction

There are some very simple checks which need only take a few minutes to carry out, but which could save you a lot of inconvenience and expense. These *Weekly checks* require no great skill or special tools, and the small amount of time they take to perform could prove to be very well spent, for example:

• Keeping an eye on tire condition and pressures will not only help to stop them wearing out prematurely, but could also save your life.
• Many breakdowns are caused by electrical problems. Battery-related faults are particularly common, and a quick check on a regular basis will often prevent the majority of these.
• If your car develops a brake fluid leak, the first time you might know about it is when your brakes don't work properly. Checking the level regularly will give advance warning of this kind of problem.

• If the oil or coolant levels run low, the cost of repairing any engine damage will be far greater than fixing the leak, for example.

Engine oil level

Refer to illustrations 2.2a, 2.2b, 2.4a, 2.4b, 2.4c and 2.4d

1   Make sure the car is on level ground.
2   The oil level is checked with a dipstick **(see illustrations)**.
3   The oil level should be checked before the vehicle has been driven, or about 5 minutes after the engine has been shut off. If the oil is checked immediately after driving the vehicle, some of the oil will remain in the upper part of the engine, resulting in an inaccurate reading on the dipstick.
4   Pull the dipstick out of the tube and wipe all the oil from the end with a clean rag or paper towel **(see illustration)**. Insert the clean dipstick all the way back into the tube and pull it out again. Note the oil at the end of the dipstick. The level should be between the MIN and MAX marks on the dipstick **(see illustrations)**.
5   It takes about one quart of oil to raise

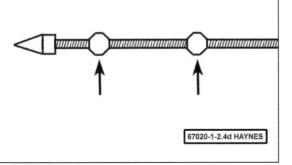

2.4d Some Mk II models have a dipstick that looks like this; the level should be within the indicated range

67020-1-2.4d HAYNES

2.9 The coolant level should be kept between the minimum and maximum marks on the side of the expansion tank

2.15 The MAX and MIN marks are indicated on the side of the brake fluid reservoir. The fluid level must be kept between the marks at all times

the level from the MIN mark to the MAX mark on the dipstick. Do not allow the level to drop below the MIN mark or oil starvation may cause engine damage. Conversely, overfilling the engine (adding oil above the MAX mark) may cause oil fouled spark plugs, oil leaks or oil seal failures. Maintaining the oil level above the MAX mark can cause excessive oil consumption.

6   To add oil, remove the filler cap. After adding oil, wait a few minutes to allow the level to stabilize, then pull out the dipstick and check the level again. Add more oil if required. Install the filler cap and tighten it by hand only.

7   Checking the oil level is an important preventive maintenance step. A consistently low oil level indicates oil leakage through damaged seals, defective gaskets or past worn rings or valve guides. If the oil looks milky in color or has water droplets in it, the cylinder head gasket may be blown or the head or block may be cracked. The engine should be checked immediately. The condition of the oil should also be checked. Whenever you check the oil level, slide your thumb and index finger up the dipstick before wiping off the oil. If you see small dirt or metal particles clinging to the dipstick, the oil should be changed (see Section 3).

Coolant level

Refer to illustration 2.9

Warning: *Do not allow antifreeze to come in contact with your skin or painted surfaces of the vehicle. Flush contaminated areas immediately with plenty of water. Don't store new coolant or leave old coolant lying around where it's accessible to children or pets - they're attracted by its sweet smell. Ingestion of even a small amount of coolant can be fatal! Wipe up garage floor and drip pan spills immediately. Keep antifreeze containers covered and repair cooling system leaks as soon as they're noticed.*

8   All vehicles covered by this manual are equipped with a pressurized coolant recovery system. A plastic expansion tank, located at

the rear of the engine compartment (Mk I models) or at the left front corner of the engine compartment (Mk II models), is connected by hoses to the cooling system. As the engine heats up during operation, the expanding coolant fills the tank.

9   The coolant level in the tank should be checked regularly. **Warning:** *Do not remove the expansion tank cap to check the coolant level when the engine is warm!* The level in the tank varies with the temperature of the engine. When the engine is cold, the coolant level should be at the MIN mark on the reservoir. If it isn't, remove the cap from the tank and add a 50/50 mixture of ethylene glycol-based antifreeze and water **(see illustration)**.

10   Drive the vehicle, let the engine cool completely then recheck the coolant level. Don't use rust inhibitors or additives. If only a small amount of coolant is required to bring the system up to the proper level, water can be used. However, repeated additions of water will dilute the antifreeze and water solution. In order to maintain the proper ratio of antifreeze and water, always top up the coolant level with the correct mixture. An empty plastic milk jug or bleach bottle makes an excellent container for mixing coolant.

11   If the coolant level drops consistently, there may be a leak in the system. Inspect the radiator, hoses, filler cap, drain plugs and water pump. If no leaks are noted, have the expansion tank cap pressure tested by a service station.

12   If you have to remove the expansion tank cap wait until the engine has cooled completely, then wrap a thick cloth around the cap and unscrew it slowly, stopping if you hear a hissing noise. If coolant or steam escapes, let the engine cool down longer, then remove the cap.

13   Check the condition of the coolant as well. If it's brown or rust colored, the system should be drained, flushed and refilled. Even if the coolant appears to be normal, the corrosion inhibitors wear out, so it must be replaced at the specified intervals.

Brake and clutch fluid

Refer to illustration 2.15

14   The brake master cylinder is mounted on the front of the power booster unit in the engine compartment. The hydraulic clutch master cylinder used on manual transaxle vehicles also uses fluid from this reservoir.

15   To check the fluid level of either system, simply look at the MAX and MIN marks on the brake fluid reservoir **(see illustration)**.

16   If the level is low, wipe the top of the reservoir cover with a clean rag to prevent contamination of the brake system before unscrewing the cover.

17   Add only the specified brake fluid to the reservoir (refer to *Recommended lubricants and fluids* in this Chapter's Specifications or to your owner's manual). Mixing different types of brake fluid can damage the system. Fill the brake master cylinder reservoir only to the MAX line. **Warning:** *Use caution when filling the reservoir - brake fluid can harm your eyes and damage painted surfaces. Do not use brake fluid that is more than one year old or has been left open. Brake fluid absorbs moisture from the air. Excess moisture can cause a dangerous loss of braking.*

18   While the reservoir cap is removed, inspect the master cylinder reservoir for contamination. If deposits, dirt particles or water droplets are present, the system should be drained and refilled.

19   After filling the reservoir to the proper level, make sure the cap is properly seated to prevent fluid leakage.

20   The fluid in the brake master cylinder will drop slightly as the brake pads at each wheel wear down during normal operation. If the master cylinder requires repeated replenishing to keep it at the proper level, this is an indication of leakage in the brake or clutch system, which should be corrected immediately. If the brake system shows an indication of leakage, check all brake lines and connections, along with the calipers and booster. If the hydraulic clutch system shows an indication of leakage check all clutch lines and connections, along with the clutch release cylinder (see Chapter 8 for more information).

21   If, upon checking the brake master cylinder fluid level, you discover the reservoir empty or nearly empty, the brake and clutch hydraulic systems should be bled and checked for leaks (see Chapters 8 and 9).

Power steering fluid level (Mk I models)

Refer to illustrations 2.26a and 2.26b

22   Check the power steering fluid level periodically to avoid steering system problems, such as damage to the pump. **Caution:** *DO NOT hold the steering wheel against either stop (extreme left or right turn) for more than five seconds. If you do, the power steering pump could be damaged.*

23   The power steering reservoir is located at the center rear of the engine compartment.

24   Park the vehicle on level ground and

2.26a The power steering fluid reservoir is located at the rear of the engine compartment. Wipe the area around the cap clean, then unscrew the cap from the reservoir and wipe off the dipstick

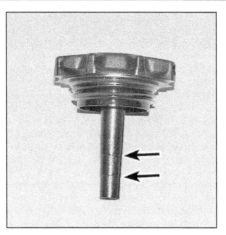

2.26b Insert the power steering fluid dipstick into the reservoir (without screwing on the cap), then remove it. The fluid level should be between the MIN and MAX marks

2.29a On Mk I models, the windshield washer fluid reservoir is located at the right rear of the engine compartment

apply the parking brake.

25   Run the engine until it has reached normal operating temperature. With the engine at idle, turn the steering wheel back and forth about 10 times to get any air out of the steering system. Shut the engine off with the wheels in the straight-ahead position.

26   Remove the cap, wipe off the dipstick, insert it back into the reservoir (without screwing on the cap), then pull it out and note the fluid level on the dipstick. It should be between the two marks **(see illustrations)**.

27   Add small amounts of fluid until the level is correct. **Caution:** *Do not overfill the reservoir. If too much fluid is added, remove the excess with a clean syringe or suction pump.*

28   Check the power steering hoses and connections for leaks and wear.

Windshield washer fluid level

Refer to illustrations 2.29a and 2.29b

29   Fluid for the windshield washer system is stored in a plastic reservoir located at the right rear of the engine compartment on Mk I models and in the left rear corner of the engine compartment on Mk II models **(see illustrations)**.

30   In milder climates, plain water can be used in the reservoir, but it should be kept no more than 2/3 full to allow for expansion if the water freezes. In colder climates, use windshield washer system antifreeze, available at any auto parts store, to lower the freezing point of the fluid. Mix the antifreeze with water in accordance with the manufacturer's directions on the container. **Caution:** *Do not use cooling system antifreeze - it will damage the vehicle's paint.*

Wiper blades

Refer to illustrations 2.33 and 2.34

31   The wiper blades should be inspected periodically for damage, loose components and cracked or worn blade elements. Road

film can build up on the wiper blades and affect their efficiency, so they should be washed regularly with a mild detergent solution.

32   If the wiper blade elements are cracked, worn or warped, or no longer clean adequately, they should be replaced with new ones.

33   To remove a front wiper blade, lift the arm assembly away from the glass, turn the blade 90-degrees to the arm and press on the release lever, then slide the wiper blade out of the hook in the end of the arm **(see illustration)**.

34   To remove a rear wiper blade, pull the arm away from the window, turn it 90-degees to the arm, then press the pivot pin from the recess in the arm **(see illustration)**.

35   Attach the new wiper to the arm. Connection can be confirmed by an audible click.

2.29b On Mk II models, the windshield washer fluid reservoir is located at the left side of the engine compartment

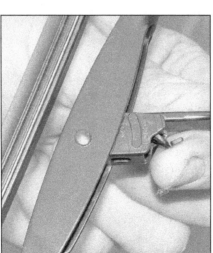

2.33 To remove a front wiper blade, push the release lever and pull the wiper blade away from the end of the arm

2.34 To remove a rear wiper blade, turn the blade 90-degrees to the arm and press the pin

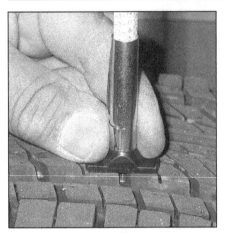

2.37 A tire tread depth indicator should be used to monitor tire wear - they are available at auto parts stores and service stations and cost very little

Tire and tire pressure checks

Refer to illustrations 2.37, 2.38, 2.39a, 2.39b and 2.43

36   Periodic inspection of the tires may spare you the inconvenience of being stranded with a flat tire. It can also provide you with vital information regarding possible problems in the steering and suspension systems before major damage occurs.

37   The original tires on this vehicle are equipped with 1/2-inch wide bands that will appear when tread depth reaches 1/16-inch, at which point they can be considered worn out. Tread wear can be monitored with a simple, inexpensive device known as a tread depth indicator **(see illustration)**.

38   Note any abnormal tread wear **(see illustration)**. Tread pattern irregularities such as cupping, flat spots and more wear on one side than the other are indications of front end alignment and/or balance problems. If any of these conditions are noted, take the vehicle to a tire shop or service station to correct the problem.

39   Look closely for cuts, punctures and embedded nails or tacks. Sometimes a tire will hold air pressure for a short time or leak down very slowly after a nail has embedded itself in the tread. If a slow leak persists, check the valve stem core to make sure it is tight **(see illustration)**. Examine the tread for an object that may have embedded itself in the tire or for a plug that may have begun to leak (radial tire punctures are repaired with a plug that is installed in a puncture). If a puncture is suspected, it can be easily verified by spraying a solution of soapy water onto the puncture area **(see illustration)**. The soapy solution will bubble if there is a leak. Unless the puncture is unusually large, a tire shop or service station can usually repair the tire.

40   Carefully inspect the inner sidewall of each tire for evidence of brake fluid leakage. If you see any, inspect the brakes immediately.

41   Correct air pressure adds miles to the life span of the tires, improves mileage and enhances overall ride quality. Tire pressure cannot be accurately estimated by looking at a tire, especially if it's a radial. A tire pressure gauge is essential. Keep an accurate gauge in the glove compartment. The pressure gauges attached to the nozzles of air hoses at gas stations are often inaccurate.

42   Always check tire pressure when the tires are cold. Cold, in this case, means the vehicle has not been driven over a mile in the three hours preceding a tire pressure check. A pressure rise of four to eight pounds is not uncommon once the tires are warm.

43   Unscrew the valve cap protruding from the wheel or hubcap and push the gauge firmly onto the valve stem **(see illustration)**. Note the reading on the gauge and compare the figure to the recommended tire pressure shown on the tire placard on the driver's side door. Be sure to reinstall the valve cap to keep dirt and moisture out of the valve stem mechanism. Check all four tires and, if necessary, add enough air to bring them up to the recommended pressure.

44   On models so equipped, don't forget to keep the spare tire inflated to the specified pressure (refer to the pressure molded into the tire sidewall).

UNDERINFLATION

CUPPING

Cupping may be caused by:

• Underinflation and/or mechanical irregularities such as out-of-balance condition of wheel and/or tire, and bent or damaged wheel.
• Loose or worn steering tie-rod or steering idler arm.
• Loose, damaged or worn front suspension parts.

OVERINFLATION

INCORRECT TOE-IN OR EXTREME CAMBER

FEATHERING DUE TO MISALIGNMENT

2.38 This chart will help you determine the condition of your tires, the probable cause(s) of abnormal wear and the corrective action necessary

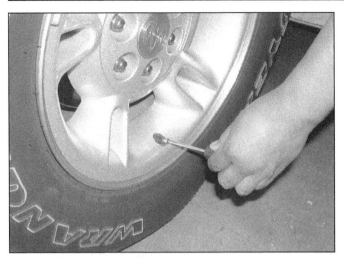

2.39a If a tire loses air on a steady basis, check the valve core first to make sure it's snug (special inexpensive wrenches are commonly available at auto parts stores)

2.39b If the valve core is tight, raise the corner of the vehicle with the low tire and spray a soapy water solution onto the tread as the tire is turned slowly - slow leaks will cause small bubbles to appear

Battery

Refer to illustrations 2.45, 2.49a, 2.49b, 2.49c, 2.50a, 2.50b, 2.51a and 2.51b

Warning: *Certain precautions must be followed when checking and servicing the battery. Hydrogen gas, which is highly flammable, is always present in the battery cells, so keep lighted tobacco and all other open flames and sparks away from the battery. The electrolyte inside the battery is actually diluted sulfuric acid, which will cause injury if splashed on your skin or in your eyes. It will also ruin clothes and painted surfaces. When removing the battery cables, always detach the negative cable first and hook it up last!*

45   A routine preventive maintenance program for the battery in your vehicle is the only way to ensure quick and reliable starts. But before performing any battery maintenance,

make sure that you have the proper equipment necessary to work safely around the battery **(see illustration)**.

46   There are also several precautions that should be taken whenever battery maintenance is performed. Before servicing the battery, always turn the engine and all accessories off and disconnect the cables from the negative terminal of the battery (see Chapter 5).

47   The battery produces hydrogen gas, which is both flammable and explosive. Never create a spark, smoke or light a match around the battery. Always charge the battery in a ventilated area.

48   Electrolyte contains poisonous and corrosive sulfuric acid. Do not allow it to get in your eyes, on your skin on your clothes. Never ingest it. Wear protective safety glasses when working near the battery. Keep children away from the battery.

2.43 To extend the life of your tires, check the air pressure at least once a week with an accurate gauge (don't forget the spare!)

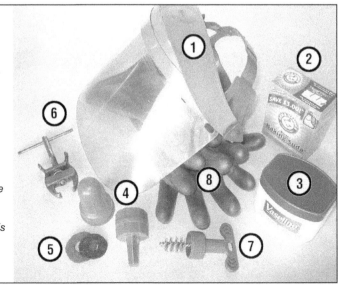

2.45 Tools and materials required for battery maintenance

1   **Face shield/safety goggles** - *When removing corrosion with a brush, the acidic particles can easily fly up into your eyes*

2   **Baking soda** - *A solution of baking soda and water can be used to neutralize corrosion*

3   **Petroleum jelly** - *A layer of this on the battery posts will help prevent corrosion*

4   **Battery post/cable cleaner** - *This wire brush cleaning tool will remove all traces of corrosion from the battery posts and cable clamps*

5   **Treated felt washers** - *Placing one of these on each post, directly under the cable clamps, will help prevent corrosion*

6   **Puller** - *Sometimes the cable clamps are very difficult to pull off the posts, even after the nut/bolt has been completely loosened. This tool pulls the clamp straight up and off the post without damage*

7   **Battery post/cable cleaner** - *Here is another cleaning tool which is a slightly different version of number 4, but it does the same thing*

8   **Rubber gloves** - *Another safety item to consider when servicing the battery; remember that's acid inside the battery*

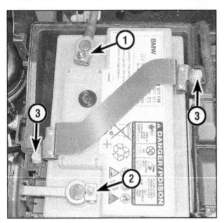

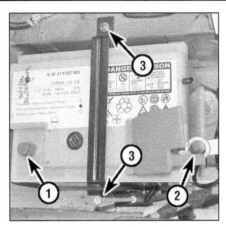

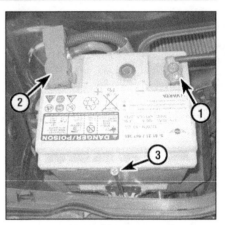

2.49a On Mk I Cooper models, the battery is located at the left rear corner of the engine compartment, under a cover

1   *Negative terminal*
2   *Positive terminal*
3   *Hold-down bolts and clamp*

2.49b On Mk I S models, the battery is located in the luggage compartment, under a cover

1   *Negative terminal*
2   *Positive terminal*
3   *Hold-down bolts and clamp*

2.49c On Mk II models, the battery is located at the right rear of the engine compartment, under a cover

1   *Negative terminal*
2   *Positive terminal*
3   *Hold-down bolt and clamp*

49   Note the external condition of the battery **(see illustrations)**. If the positive terminal and cable clamp on your vehicle's battery is equipped with a rubber or plastic protector, make sure that it's not torn or damaged. It should completely cover the terminal. Look for any corroded or loose connections, cracks in the case or cover or loose hold-down clamps. Also check the entire length of each cable for cracks and frayed conductors.
50   If corrosion, which looks like white, fluffy deposits **(see illustration)** is evident, particularly around the terminals, the battery should be removed for cleaning. Loosen the cable clamp bolts with a wrench, being careful to remove the negative cable first, and slide them off the terminals **(see illustration)**. Then disconnect the hold-down clamp bolt and nut,

remove the clamp and lift the battery from the engine compartment.
51   Clean the cable clamps thoroughly with a battery brush or a terminal cleaner and a solution of warm water and baking soda **(see illustration)**. Wash the terminals and the top of the battery case with the same solution but make sure that the solution doesn't get into the battery. When cleaning the cables, terminals and battery top, wear safety goggles and rubber gloves to prevent any solution from coming in contact with your eyes or hands. Wear old clothes too - even diluted, sulfuric acid splashed onto clothes will burn holes in them. If the terminals have been extensively corroded, clean them up with a terminal cleaner **(see illustration)**. Thoroughly wash all cleaned areas with plain water.
52   Make sure that the battery tray is in good

condition and the hold-down clamp fasteners are tight **(see illustration 2.49a, 2.49b or 2.49c)**. If the battery is removed from the tray, make sure no parts remain in the bottom of the tray when the battery is reinstalled. When reinstalling the hold-down clamp bolts, do not overtighten them.
53   Information on removing and installing the battery can be found in Chapter 5. If you disconnected the cable(s) from the negative and/or positive battery terminals, the Powertrain Control Module (PCM) must relearn its idle and fuel trim strategy for optimum driveability and performance (see Chapter 5 for this procedure). Information on jump starting can be found at the front of this manual. For more detailed battery checking procedures, refer to the *Haynes Automotive Electrical Manual*.

2.50a Battery terminal corrosion usually appears as light, fluffy powder

2.50b Removing a cable from the battery post with a wrench - sometimes a pair of special battery pliers are required for this procedure if corrosion has caused deterioration of the nut hex. Always remove the ground (-) cable first and hook it up last!

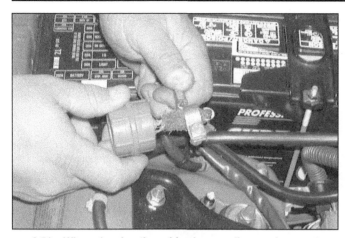

2.51a When cleaning the cable clamps, all corrosion must be removed

2.51b Regardless of the type of tool used to clean the battery posts, a clean, shiny surface should be the result

Cleaning

54 Corrosion on the hold-down components, battery case and surrounding areas can be removed with a solution of water and baking soda. Thoroughly rinse all cleaned areas with plain water.

55 Any metal parts of the vehicle damaged by corrosion should be covered with a zinc-based primer, then painted.

Charging

Warning: *When batteries are being charged, hydrogen gas, which is very explosive and flammable, is produced. Do not smoke or allow open flames near a charging or a recently charged battery. Wear eye protection when near the battery during charging. Also, make sure the charger is unplugged before connecting or disconnecting the battery from the charger.*

56 Slow-rate charging is the best way to restore a battery that's discharged to the point where it will not start the engine. It's also a good way to maintain the battery charge in a vehicle that's only driven a few miles between starts. Maintaining the battery charge is particularly important in the winter when the battery must work harder to start the engine and

electrical accessories that drain the battery are in greater use.

57 It's best to use a one or two-amp battery charger (sometimes called a "trickle" charger). They are the safest and put the least strain on the battery. They are also the least expensive. For a faster charge, you can use a higher amperage charger, but don't use one rated more than 1/10th the amp/hour rating of the battery. Rapid boost charges that claim to restore the power of the battery in one to two hours are hardest on the battery and can damage batteries not in good condition. This type of charging should only be used in emergency situations.

58 The average time necessary to charge a battery should be listed in the instructions that come with the charger. As a general rule, a trickle charger will charge a battery in 12 to 16 hours.

3 Engine oil and filter replacement

Refer to illustrations 3.3a, 3.3b, 3.5, 3.8, 3.9, 3.11a, 3.11b and 3.14

1 Frequent oil changes are the best preventive maintenance the home mechanic can

give the engine, because aging oil becomes diluted and contaminated, which leads to premature engine wear.

2 Make sure that you have all the necessary tools before you begin this procedure. You should also have plenty of rags or newspapers handy for mopping up any spills. Access to the underside of the vehicle is greatly improved if the vehicle can be lifted on a hoist, driven onto ramps or supported by jackstands.

3 Working in the engine compartment, locate the oil filter housing:

   Mk I models: on the rear of the engine on the right-hand side **(see illustration)**.

   Mk II models: on the front of the engine on the left-hand side **(see illustration)**.

4 Place a shop rag around the bottom of the housing to absorb any spilled oil.

5 Using a special oil filter removal tool or socket, unscrew and remove the cover, and lift the filter cartridge out. It is possible to unscrew the cover using a strap wrench or wrench **(see illustration)**. The oil will drain from the housing back into the oil pan as the cover is removed. Note that Cooper models manufactured up to 07/2004 have a guide sleeve and spring fitted between the element

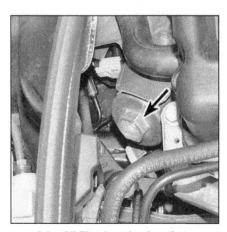

3.3a Oil filter housing location - Mk I models

3.3b Oil filter housing location - Mk II models

3.5 Unscrew the filter cover with a wrench

3.8 Replace the filter cover O-ring

3.9 Install the new element to the filter cover

and cover, Cooper S models and vehicles manufactured after this date do not have the sleeve or spring. Do not remove the guide sleeve and spring from the cover.

3.11a Loosen and remove the oil pan drain plug - Mk I model

6   Remove the O-ring from the cover.
7   Using a clean rag, wipe the mating faces of the housing and cover.
8   Install a new O-ring on the cover **(see illustration)**.
9   Install the new element into the cover **(see illustration)**.
10   Smear a little clean engine oil on the O-ring, install the cover and tighten it to 18 ft-lbs (25 Nm) if using a socket, or securely if using a strap wrench/regular wrench. Note that some force will be required to push the element far enough into the housing to engage the cover threads.
11   Working under the vehicle, loosen the oil pan drain plug about half a turn. Position the draining container under the drain plug, then remove the plug completely **(see illustrations)**. If possible, try to keep the plug pressed into the oil pan while unscrewing it by hand the last couple of turns.
12   Discard the sealing washer - a new one must be used. If the sealing ring is integral with the plug, install a new plug.
13   Allow some time for the old oil to drain, noting that it may be necessary to reposi-

tion the container as the oil flow slows to a trickle.
14   After all the oil has drained, clean the area around the drain plug opening, then install and tighten the plug **(see illustration)**.
15   Remove the old oil and all tools from under the vehicle, then lower the vehicle to the ground (if applicable).
16   Remove the dipstick then unscrew the oil filler cap. Fill the engine, using the correct grade and type of oil (see Section 2). Pour in half the specified quantity of oil first, then wait a few minutes for the oil to run to the oil pan. Continue adding oil a small quantity at a time until the level is up to the lower mark on the dipstick. Finally, bring the level up to the upper mark on the dipstick. Insert the dipstick, and install the filler cap.
17   Start the engine and run it for a few minutes; check for leaks around the oil filter seal and the oil pan drain plug. Note that there may be a delay of a few seconds before the oil pressure warning light goes out when the engine is first started, as the oil circulates through the engine oil galleries and the new oil filter, before the pressure builds-up.
18   Switch off the engine, and wait a few minutes for the oil to settle in the oil pan once more. With the new oil circulated and the filter completely full, recheck the level on the dipstick, and add more oil as necessary.
19   Reset the service interval display (see Section 4)
20   During the first few trips after an oil change, make it a point to check frequently for leaks and proper oil level.
21   The old oil drained from the engine cannot be reused in its present state and should be disposed of. Check with your local auto parts store, disposal facility or environmental agency to see if they will accept the oil for recycling. After the oil has cooled it can be drained into a container (capped plastic jugs, topped bottles, milk cartons, etc.) for transport to one of these disposal sites. Don't dispose of the oil by pouring it on the ground or down a drain!

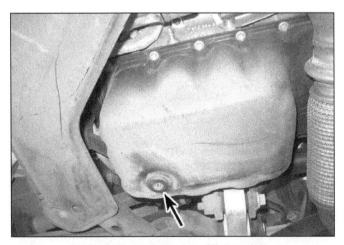

3.11b Oil pan drain plug - Mk II models

3.14 Note that some drain plugs have an integral seal (it is recommended that these be replaced with a new one whenever removed)

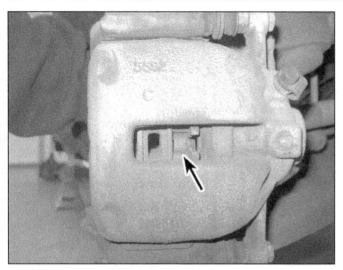

5.5a You will find an inspection hole like this in each caliper through which you can view the thickness of remaining friction material for the inner pad

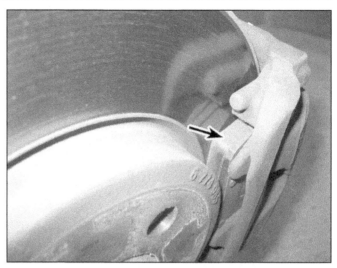

5.5b Be sure to check the thickness of the outer pad material, too

4   Resetting the service interval display

Mk I models

1   With the ignition turned off, press and hold the trip reset button.
2   Ensure that all electrical items are switched off, then turn on the ignition switch to position I. **Note:** *Do not start the engine.*
3   After 5 seconds the words "Oil Service" or "Inspection" are shown with the word "reset" or "re."
4   Release the button, immediately re-press and hold it. After 5 seconds OIL SERVICE or INSPECTION will be displayed.
5   Release the button, then press and hold the button for 5 seconds to change the service reset mode.
6   Press and release the button one more time. The service interval is now displayed for a few seconds. Turn off the ignition.

Mk II models

7   Sit in the driver's seat, then close any open doors.
8   Turn the ignition On (but don't crank the engine).
9   Locate the trip reset button on the tachometer; depress the button for about ten seconds.
10   Service operations will be displayed in the upper part of the tachometer, the time or mileage interval will be displayed at the bottom of the tach. To navigate through each service item on the display, push the button on the end of the turn signal lever.
11   To reset the desired item, depress the button on the end of the turn signal lever until the display on the lower part of the tach reads "RESET."

5   Brake check

Refer to illustrations 5.5a and 5.5b

Warning: *Dust created by the brake system is harmful to your health. Never blow it out with compressed air and don't inhale any of it. An approved filtering mask should be worn when working on brakes. Do not, under any circumstances, use petroleum-based solvents to clean brake parts. Use brake system cleaner only!*

1   The brakes should be inspected every time the wheels are removed or whenever a defect is suspected. Indications of a potential brake system problem include the vehicle pulling to one side when the brake pedal is depressed, noises coming from the brakes when they are applied, excessive brake pedal travel, a pulsating pedal and leakage of fluid, usually seen on the inside of the tire or wheel. **Note:** *It is normal for a vehicle equipped with an Anti-lock Brake System (ABS) to exhibit brake pedal pulsations during severe braking conditions.*
2   Disc brakes can be visually checked without removing any parts except the wheels. Remove the hub caps (if applicable) and loosen the wheel bolts a quarter turn each.
3   Raise the vehicle and place it securely on jackstands. **Warning:** *Never work under a vehicle that is supported only by a jack!*
4   Remove the wheels. Now visible is the disc brake caliper which contains the pads. There is an outer brake pad and an inner pad. Both must be checked for wear. **Note:** *Usually the inner pad wears faster than the outer pad.*
5   Measure the thickness of the outer pad at each end of the caliper and the inner pad through the inspection hole in the caliper body **(see illustrations)**. Compare the measurement with the limit given in this Chapter's Specifications; if any brake pad thickness is

less than specified, then all brake pads must be replaced (see Chapter 9).
6   If you're in doubt as to the exact pad thickness or quality, remove them for measurement and further inspection (see Chapter 9).
7   Check the disc for score marks, wear and burned spots. If any of these conditions exist, the disc should be removed for servicing or replacement (see Chapter 9).
8   Before installing the wheels, check all the brake lines and hoses for damage, wear, deformation, cracks, corrosion, leakage, bends and twists, particularly in the vicinity of the rubber hoses and calipers.
9   Install the wheels, lower the vehicle and tighten the wheel lug nuts to the torque given in this Chapter's Specifications.

6   Brake booster check

Check the operation of the power brake booster as described in Chapter 9.

7   Parking brake check

1   Slowly pull up on the parking brake and count the number of clicks you hear until the handle is up as far as it will go. The adjustment is correct if you hear the specified number of clicks (see this Chapter's Specifications). If you hear more clicks, it's time to adjust the parking brake (see Chapter 9).
2   An alternative method of checking the parking brake is to park the vehicle on a steep hill with the engine running (so you can apply the brakes if necessary), the parking brake set and the transaxle in Neutral. If the parking brake cannot prevent the vehicle from rolling, it needs adjustment (see Chapter 9).

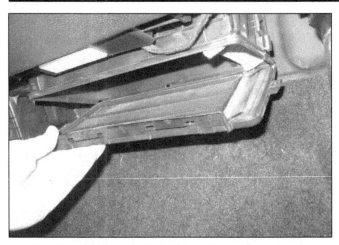

8.1a On some models the cabin air filter cover is clipped into place . . .

8.1b . . . while on others the cover is retained by screws

8   Cabin air filter replacement

Refer to illustrations 8.1a, 8.1b and 8.3

1   Working in the passenger's side footwell, unclip the plastic cover from the cabin air filter. On some models, the plastic cover may be retained by screws **(see illustrations)**.

2   Pull down the lower end of the filter element, and slide it from the housing.

3   Install the new filter element into the housing, noting how it curves around to fit in the guide slots **(see illustration)**.

4   Install the filter cover, and secure it in place with the retaining clips. If the clips break, it is possible to secure the cover using self-tapping screws through the holes provided in the cover.

9   Hose and fluid leak check

1   Visually inspect the engine joint faces, gaskets and seals for any signs of water or oil leaks. Pay particular attention to the areas around the valve cover, cylinder head, oil filter and oil pan joint faces. Bear in mind that, over a period of time, some very slight seepage from these areas is to be expected - what you are really looking for is any indication of a serious leak. Should a leak be found, replace

the gasket or oil seal by referring to the appropriate Chapters in this manual.

2   Also check the security and condition of all the engine-related pipes and hoses. Ensure that all cable-ties or securing clamps are in place and in good condition. Clamps which are broken or missing can lead to chafing of the hoses, pipes or wiring, which could cause more serious problems in the future.

3   Carefully check the radiator hoses and heater hoses along their entire length. Replace any hose which is cracked, swollen or deteriorated. Cracks will show up better if the hose is squeezed. Pay close attention to the hose clips that secure the hoses to the cooling system components. Hose clamps can pinch and puncture hoses, resulting in cooling system leaks.

4   Inspect all the cooling system components (hoses, joint faces, etc) for leaks. Where any problems of this nature are found on system components, replace the component or gasket with reference to Chapter 3.

5   With the vehicle raised, inspect the fuel tank and filler neck for punctures, cracks and other damage. The connection between the filler neck and tank is especially critical. Sometimes a rubber filler neck or connecting hose will leak due to loose retaining clamps or deteriorated rubber.

6   Carefully check all rubber hoses and metal fuel lines leading away from the gaso-

line tank. Check for loose connections, deteriorated hoses, crimped lines, and other damage. Pay particular attention to the vent pipes and hoses, which often loop up around the filler neck and can become blocked or crimped. Follow the lines to the front of the vehicle, carefully inspecting them all the way. Replace damaged sections as necessary.

7   Closely inspect the metal brake lines which run along the vehicle underbody. If they show signs of excessive corrosion or damage they must be replaced.

8   From within the engine compartment, check the security of all fuel hose attachments and pipe fittings, and inspect the fuel hoses and vacuum hoses for kinks, chafing and deterioration.

9   On models with hydraulic power steering, check the condition of the power steering fluid hoses and pipes.

10   Steering and suspension check

Front suspension and steering

Refer to illustration 10.4

1   Raise the front of the vehicle, and support it securely on jackstands.

2   Visually inspect the balljoint dust covers and the steering rack-and-pinion boots for splits, chafing or deterioration. Any wear of these components will cause loss of lubricant, then dirt and water entry, resulting in rapid deterioration of the balljoints or steering gear.

3   Check the power steering fluid hoses for chafing or deterioration, and the pipe and hose fittings for fluid leaks. Also check for signs of fluid leakage under pressure from the steering gear rubber boots, which would indicate failed fluid seals within the steering gear.

4   Grasp the wheel at the 12 o'clock and 6 o'clock positions, and try to rock it **(see illustration)**. Very slight freeplay may be felt, but if the movement is significant, further investigation is necessary to determine the source. Continue rocking the wheel while an assistant

8.3 Install the new air filter element, noting how it curves around to fit into the guide slots

10.4 Check for wear in the hub bearings by grasping the wheel and trying to rock it

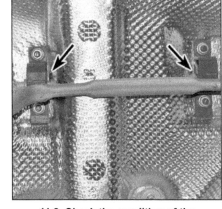

11.2 Check the condition of the exhaust mounts

15.1 Use a small screwdriver to adjust the aim of the washer jets

depresses the brake. If the movement is now eliminated or significantly reduced, it is likely that the hub bearings are at fault. If the free-play is still evident with the brake depressed, then there is wear in the suspension joints or mountings.

5   Grasp the wheel at the 9 o'clock and 3 o'clock positions, and try to rock it as before. Any movement felt may be caused by wear in the hub bearings or the steering tie-rod ends. If the inner or outer balljoint is worn, the visual movement will be obvious.

6   Using a large screwdriver or flat bar, check for wear in the suspension mounting bushings by prying between the relevant suspension component and its attachment point. Some movement is to be expected as the mountings are made of rubber, but excessive wear should be obvious. Also check the condition of any visible rubber bushings, looking for splits, cracks or contamination of the rubber.

7   With the vehicle on the ground, have an assistant turn the steering wheel back-and-forth about an eighth of a turn each way. There should be very little, if any, lost movement between the steering wheel and wheels. If this is not the case, closely observe the joints and mountings previously described, but in addition, check the steering column universal joints for wear, and the rack-and-pinion steering gear itself.

Strut/shock absorber

8   Check for any signs of fluid leakage around the suspension strut/shock absorber body, or from the rubber boot around the piston rod. Should any fluid be noticed, the suspension strut/shock absorber is defective internally, and should be replaced. **Note:** *Suspension struts/shock absorbers should always be replaced in pairs on the same axle.*

9   The efficiency of the suspension strut/shock absorber may be checked by bouncing the vehicle at each corner. Generally speaking, the body will return to its normal position and stop after being depressed. If

it rises and returns on a rebound, the suspension strut/shock absorber is probably suspect. Examine also the suspension strut/shock absorber upper and lower mountings for any signs of wear.

11   Exhaust system check

Refer to illustration 11.2

1   With the engine cold (at least an hour after the vehicle has been driven), check the complete exhaust system from the engine to the end of the tailpipe. The exhaust system is most easily checked with the vehicle raised on a hoist, or suitably supported on jackstands, so that the exhaust components are readily visible and accessible.

2   Check the exhaust pipes and connections for evidence of leaks, severe corrosion and damage. Make sure that all brackets and mounts are in good condition, and that all relevant nuts and bolts are tight **(see illustration)**. Leakage at any of the joints or in other parts of the system will usually show up as a black sooty stain in the vicinity of the leak.

3   Rattles and other noises can often be traced to the exhaust system, especially the brackets and mounts. Try to move the pipes and mufflers. If the components are able to come into contact with the body or suspension parts, secure the system with new mountings. Otherwise separate the joints (if possible) and twist the pipes as necessary to provide additional clearance.

12   Seat belt check

1   Carefully examine the seat belt webbing for cuts or any signs of serious fraying or deterioration. If the seat belt is of the retractable type, pull the belt all the way out, and examine the full extent of the webbing.

2   Fasten and unfasten the belt, ensuring that the locking mechanism holds securely and releases properly when intended. If the

belt is of the retractable type, check also that the retracting mechanism operates correctly when the belt is released.

3   Check the security of all seat belt mountings and attachments which are accessible, without removing any trim or other components, from inside the vehicle.

13   Hinge and lock lubrication

Lubricate the hinges of the hood, doors and tailgate with a light general-purpose oil. Similarly, lubricate all latches, locks and lock strikers. At the same time, check the security and operation of all the locks, adjusting them if necessary (see Chapter 11).

Lightly lubricate the hood release mechanism and cable with a suitable grease.

14   Headlight beam alignment check

Accurate adjustment of the headlight beam is only possible using optical beam-setting equipment, and this work should therefore be carried out by a MINI dealer or service station with the necessary facilities.

15   Windshield/headlight washer system(s) check

Refer to illustration 15.1

Check that each of the washer jet nozzles are clear and that each nozzle provides a strong jet of washer fluid. The washer jets should be aimed to spray at a point slightly above the center of the screen. Use a small screwdriver (or similar) to adjust the aim of the jets **(see illustration)**. The aim of the rear washer jet is not adjustable.

To adjust the headlight washer jets, pull the jet out and adjust the aim using a pin or length of fine, stiff wire. Take care not to damage the water channels in the jets.

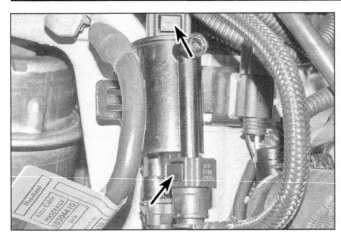

18.1 Press in the buttons to disconnect the hoses from the tank vent valve

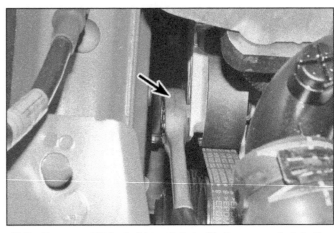

18.5a Engage a 3/8-inch drive ratchet with the square hole in the drivebelt tensioner arm

16 Engine management system check

1   This check is part of the manufacturer's maintenance schedule, and involves testing the engine management system using special dedicated test equipment. Such testing will allow the test equipment to read any Diagnostic Trouble Codes stored in the electronic control unit memory.

2   Unless a fault is suspected, this test is not essential, although it should be noted that it is recommended by the manufacturers.

3   If access to suitable test equipment is not possible, make a thorough check of all ignition, fuel and emission control system components, hoses and wiring for security and obvious signs of damage. Further details of the fuel system, emission control system, ignition system and engine management system can be found in Chapters 4, 5 and 6.

17 Road test

Instruments and electrical equipment

1   Check the operation of all instruments and electrical equipment.

2   Make sure that all instruments read correctly, and switch on all electrical equipment in turn, to check that it functions properly.

Steering and suspension

3   Check for any abnormalities in the steering, suspension, handling or road feel.

4   Drive the vehicle, and check that there are no unusual vibrations or noises.

5   Check that the steering feels positive, with no excessive sloppiness, or roughness, and check for any suspension noises when cornering and driving over bumps.

Drivetrain

6   Check the performance of the engine, clutch, transmission and driveaxles.

7   Listen for any unusual noises from the engine, clutch and transmission.

8   Make sure that the engine runs smoothly when idling, and that there is no hesitation when accelerating.

9   Check that the clutch action is smooth and progressive, that the drive is taken up smoothly, and that the pedal travel is not excessive. Also listen for any noises when the clutch pedal is depressed.

10   Check that all gears can be engaged smoothly without noise, and that the gear lever action is smooth and not abnormally vague or notchy.

Braking system

11   Make sure that the vehicle does not pull to one side when braking, and that the wheels do not lock when braking hard.

12   Check that there is no vibration through the steering when braking (if there is, suspect disc runout; see Chapter 9).

18 Drivebelt replacement

Mk I models

Refer to illustration 18.1

1   Open the hood, depress the retaining clips and disconnect both hoses and the electri-

cal connector from the tank vent valve at the right-hand end of the engine **(see illustration)**.

2   Remove the right-hand front engine mount retaining bolt, and move the tank vent valve and bracket to one side.

3   Loosen the right-hand wheel bolts, then raise the front of the vehicle and support securely on jackstands. Remove the wheel, then remove the screws, pry out the plastic rivets and remove the wheelwell liner.

4   Remove the screws and remove the engine undershield.

Cooper

Refer to illustrations 18.5a, 18.5b, 18.6 and 18.7

5   Insert a 3/8" drive ratchet into the square hole in the auxiliary belt tensioner arm **(see illustrations)**.

6   Using the ratchet, rotate the tensioner arm counterclockwise, until a 4 mm diameter rod/bolt approximately 30 mm long can be inserted into the hole in the casing to lock the tensioner arm in place **(see illustration)**. Access to the locking hole is restricted. Attaching a length of welding rod to the end to the rod/bolt may allow greater control when inserting the locking rod/bolt. Slide the belt from the pulleys. If the belt is to be re-used, mark its direction of rotation, and ensure it's

18.5b Drivebelt tensioner arm square hole (engine removed for clarity)

18.6 Lock the tensioner arm in place using a 4 mm bolt. We attached a length of welding rod to the bolt to give greater control in the restricted access

installed in the same direction of travel.

7   Install the drivebelt around the pulleys, hold the arm in place using the ratchet, then withdraw the locking rod and allow the tensioner to rotate and tension the belt **(see illustration)**.

MINI Cooper S

Refer to illustrations 18.9a, 18.9b, 18.9c, 18.9d, 18.10 and 18.12

8   Pull the engine oil level dipstick from the guide tube.

9   In order to relieve the tension on the belt, a MINI special tool (11 8 410) is available to compress the tensioner spring. In the absence of this tool, fabricate a home-made equivalent using two lengths flat steel bar. Drill a 20 mm hole and bolt the two lengths together as shown. Fit the hole in the tool over the tensioner arm pivot bolt and lever against the lower edge of the tensioner arm **(see illustrations)**. Take great care to ensure the lower end of the tool doesn't slip when compressing the spring.

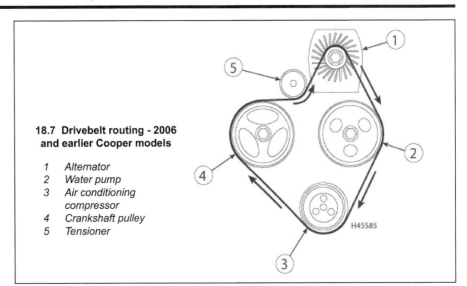

18.7 Drivebelt routing - 2006 and earlier Cooper models

1   Alternator
2   Water pump
3   Air conditioning compressor
4   Crankshaft pulley
5   Tensioner

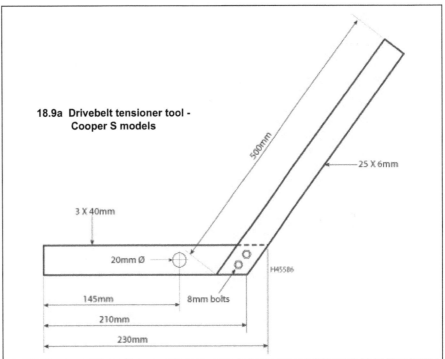

18.9a Drivebelt tensioner tool - Cooper S models

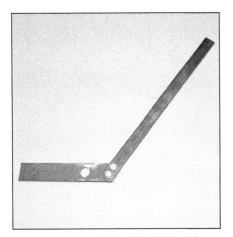

18.9b The tool should look like this

18.9c Locate the tool over the tensioner arm pivot bolt head . . .

18.9d . . . ensure the lower end of the tool doesn't slip from the bottom of the arm

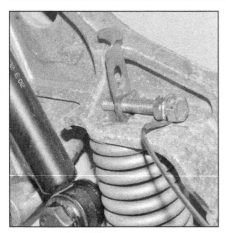

18.10 Use a 4.0 mm drill bit/bolt/rod to lock the tensioner in position (shown with the tensioner removed for clarity). We attached a length of welding rod to the bolt for greater control in the limited access

10   With the tensioner spring compressed, lock the tensioner in place using a 4.0 mm diameter drill bit/rod/bolt **(see illustration)**.

11   Slide the belt from the pulleys. If the belt is to be reinstalled, mark its direction of rotation, and ensure it's installed in the same direction of travel.

12   Install the drivebelt around the pulleys, hold the tensioner arm in place, then withdraw the locking pin/drill, and allow the belt to be tensioned **(see illustration)**.

All models

13   Install the engine undershield, wheelwell liner, and wheel, then lower the vehicle to the ground. Tighten the wheel bolts to the specified torque.

14   Install the tank vent valve, and reconnect the hoses and wiring plug. On Cooper S models, install the engine oil level dipstick.

Mk II models

Refer to illustrations 18.18 and 18.20

15   Loosen the right-hand wheel bolts, then raise the front of the vehicle and support it

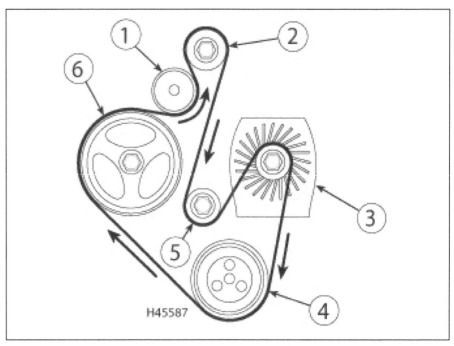

18.12 Drivebelt routing - Mk I Cooper S models

1	Tensioner	4	Air conditioning compressor
2	Supercharger	5	Idler pulley
3	Alternator	6	Crankshaft pulley

securely on jackstands. Remove the wheel, then remove the screws, pry out the plastic rivets and remove the wheelwell liner.

16   Remove the right-side headlight (see Chapter 12).

17   Unbolt and tilt forward the radiator support panel (see Chapter 11).

18   Place a wrench on the hex casting of the tensioner arm and rotate the tensioner away from the belt **(see illustration)**.

19   Slide the belt from the pulleys. If the belt is to be reinstalled, mark its direction of rotation, and ensure it's installed in the same direction of travel.

20   Install the drivebelt around the pulleys. Rotate the tensioner back slightly to allow the

pin to snap back into place and allow the belt to be tensioned **(see illustration)**.

21   The remainder of installation is the reverse of removal.

19   Spark plug replacement

Refer to illustrations 19.3, 19.6, 19.11a and 19.11b

Caution: *Wait until the engine is cool before removing the spark plugs.*

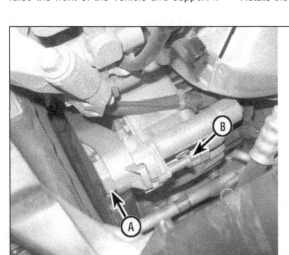

18.18 Place a wrench on the cast-in hex on the tensioner arm and rotate it away from the belt (A), then push in the pin (B) to hold it in position (Mk II models)

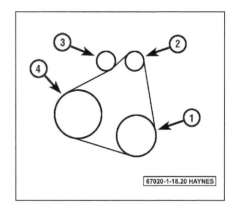

67020-1-18.20 HAYNES

18.20 Drivebelt routing - Mk II models

1   Air conditioning compressor
2   Alternator
3   Tensioner
4   Crankshaft pulley

19.3 Pull the spark plug wire from the plug by gripping the end fitting, not the wire

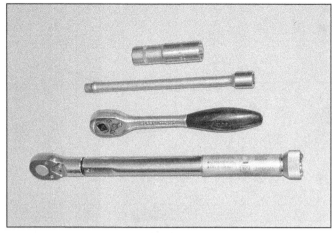

19.6 Use a ratchet, long extension and spark plug socket for spark plug removal. The torque wrench is for installation

1   The correct functioning of the spark plugs is vital for the correct running and efficiency of the engine. It is essential that the plugs installed are appropriate for the engine. The spark plugs should not need attention between scheduled replacement intervals.
2   The spark plugs are located in the top of the cylinder head.
3   **Models with spark plug wires:** If the marks on the original-equipment spark plug wires cannot be seen, mark the wires 1 to 4, corresponding to the cylinder the wire serves (No 1 cylinder is at the timing chain/right-hand end of the engine). Pull the wires from the plugs by gripping the end fitting, not the wire, otherwise the connection may be fractured **(see illustration)**.
4   **Models with coil-over-plug ignition:** Remove the ignition coils (see Chapter 5).
5   It is advisable to remove the dirt from the spark plug recesses, using a clean brush, vacuum cleaner or compressed air before removing the plugs, to prevent dirt dropping into the cylinders.

6   Unscrew the plugs using a spark plug wrench, suitable box wrench, or a deep socket and extension bar **(see illustration)**. Keep the socket aligned with the spark plug - if it is forcibly moved to one side, the ceramic insulator may be broken off. As each plug is removed, examine it as follows.
7   Examination of the spark plugs will give a good indication of the condition of the engine. If the insulator nose of the spark plug is clean and white, with no deposits, this is indicative of a weak mixture or too hot a plug (a hot plug transfers heat away from the electrode slowly, a cold plug transfers heat away quickly).
8   If the tip and insulator nose are covered with hard black-looking deposits, then this is indicative that the mixture is too rich. Should the plug be black and oily, then it is likely that the engine is fairly worn, as well as the mixture being too rich.
9   If the insulator nose is covered with light tan to grayish-brown deposits, then the mixture is correct, and it is likely that the

engine is in good condition. See the inside back cover of this manual for more spark plug conditions.
10   The recommended spark plugs are of the multi-electrode type, and the gap between the center electrode and the ground electrodes cannot be adjusted.
11   Before installing the spark plugs, check that the plug exterior surfaces and threads are clean. It is very often difficult to insert spark plugs into their holes without cross-threading them. To avoid this possibility, fit a short length of hose over the end of the spark plug **(see illustration)**. Also, it's a good idea to apply a thin film of anti-seize compound to the threads of the plug **(see illustration)**.
12   Remove the rubber hose (if used), and tighten the plug to the specified torque (see this Chapter's Specifications) using the spark plug socket and a torque wrench. Install the remaining plugs in the same way.
13   Connect the spark plug wires in the correct order, or install the ignition coils.

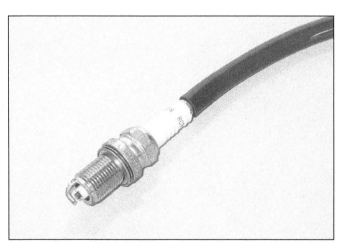

19.11a A length of snug-fitting rubber hose will save time and prevent damaged threads when installing the spark plugs

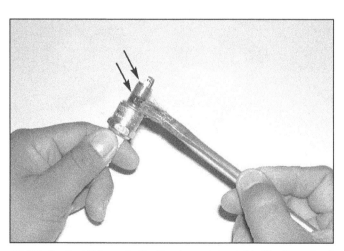

19.11b Apply a thin coat of anti-seize compound to the threads of the spark plug, being careful not to get any near the end of the plug

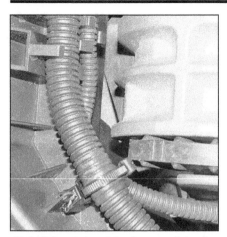

20.2 Release the wiring harness from the air filter cover

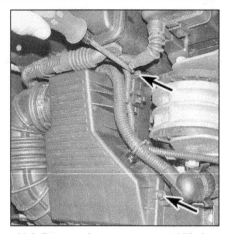

20.3 Remove the two screws and lift the filter cover up and to the right

20.4 Lift out the air filter element

20   Air filter element replacement

Mk I models

Cooper

Refer to illustrations 20.2, 20.3 and 20.4

1   The air cleaner assembly is located at the front left-hand corner of the engine compartment.

2   Unclip the wiring harness from the plastic clips on the air filter cover **(see illustration)**.

3   Loosen the two screws and slide the filter cover to the right **(see illustration)**.

4   Lift out the filter element **(see illustration)**.

5   Wipe out the air cleaner housing and the cover.

6   Lay the new filter element in position, then install the cover and secure with the screws.

Cooper S

Refer to illustrations 20.7, 20.8 and 20.12

7   Open the hood, and slide the battery jump starting terminal up from its bracket **(see illustration)**.

8   Release the clamp and disconnect the air intake duct from the filter cover **(see illustration)**.

9   Loosen the two screws, lift up the front edge of the filter cover, and remove it.

10   Lift the air filter element from the housing.

11   Clean the air filter housing, removing all debris.

12   Install the new filter element **(see illustration)**.

13   The remainder of installation is the reverse of removal.

Mk II models

Refer to illustrations 20.15a, 20.15b and 20.18

14   If you're working on an S model, loosen the clamp and remove the intake air duct from the air filter housing. Also unplug the electrical connector from the Mass Airflow (MAF) sensor.

15   Remove the filter housing cover screws. On Cooper models, there are four screws securing the cover to the housing, plus one holding the cover to the valve cover **(see illustration)**. On S models, there are four screws along the front edge of the housing **(see illustration)**.

20.7 Slide the remote jump starting terminal up from its bracket

16   Lif the cover up and remove the old filter.

17   Clean the air filter housing, removing all debris.

18   Install the new filter element, making sure it seats in the housing **(see illustration)**.

19   The remainder of installation is the reverse of removal.

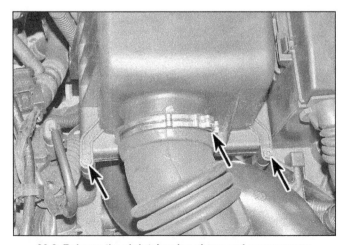

20.8 Release the air intake pipe clamp and cover screws

20.12 Note the rubber seal is at the top of the filter element

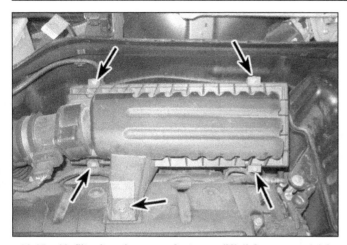

20.15a Air filter housing cover fasteners (Mk II Cooper models)

20.15b Air filter housing cover fasteners (Mk II Cooper S models)

21 Driveaxle boot check

Refer to illustration 21.1

1   With the vehicle raised and securely supported on stands, slowly rotate the front wheel. Inspect the condition of the outer constant velocity (CV) joint rubber boots, squeezing the boots to open out the folds **(see illustration)**. Check for signs of cracking, splits or deterioration of the rubber, which may allow the grease to escape, and lead to water and grit entry into the joint. Also check the security and condition of the retaining clips. Repeat these checks on the inner CV joints. If any damage is found, the boots should be replaced (see Chapter 8).

2   At the same time, check the general condition of the CV joints themselves by first holding the driveaxle and attempting to rotate the wheel. Repeat this check by holding the inner joint and attempting to rotate the driveaxle. Any noticeable movement indicates wear in the joints, wear in the driveaxle splines, or a loose driveaxle retaining nut.

20.18 Make sure the rubber lip of the element seats in the groove in the housing

22 Brake fluid replacement

Warning: *Brake fluid can harm your eyes and damage painted surfaces, so use extreme caution when handling and pouring it. Do not use fluid that has been standing open for some time, as it absorbs moisture from the air. Excess moisture can cause a dangerous loss of braking effectiveness.*

1   The procedure is similar to that for the bleeding of the hydraulic system as described in Chapter 9, except that the brake fluid reservoir should be emptied by siphoning, using a clean poultry baster or similar before starting, and allowance should be made for the old fluid to be expelled when bleeding a section of the circuit.

2   Working as described in Chapter 9, open the first bleed screw in the sequence, and pump the brake pedal gently until nearly all the old fluid has been emptied from the master cylinder reservoir.

3   Top-up to the MAX level with new fluid, and continue pumping until only the new fluid remains in the reservoir, and new fluid can be

21.1 Check the driveaxle boot for signs of cracking, splitting, and deterioration

seen emerging from the bleed screw. Tighten the screw, and top the reservoir level up to the MAX level line.

4   Work through all remaining bleed screws in the sequence until new fluid can be seen at all of them. Be careful to keep the master cylinder reservoir topped-up to above the MIN level at all times, or air may enter the system and increase the length of the task.

5   When the operation is complete, check that all bleed screws are securely tightened, and that their dust caps are refitted. Wash off all traces of spilled fluid, and recheck the master cylinder reservoir fluid level.

6   Check the operation of the brakes before taking the vehicle on the road.

23 Cooling system servicing (draining, flushing and refilling)

Warning: *Wait until the engine is cold before starting this procedure. Do not allow antifreeze to come in contact with your skin, or with the painted surfaces of the vehicle. Rinse off spills immediately with plenty of water. Never leave antifreeze lying around in an open container, or in a puddle in the driveway or on the garage floor. Children and pets are attracted by its sweet smell, but antifreeze can be fatal if ingested.*

Draining

Refer to illustration 23.5

1   With the engine completely cold, cover the filler neck cap (Mk I Cooper models) or expansion tank cap (all other models) with a wad of rag, and slowly turn the cap counter-clockwise to relieve the pressure in the cooling system (a hissing sound may be heard). Wait until any pressure in the system is released, then continue to turn the cap until it can be removed.

2   Raise the front of the vehicle and support it securely on jackstands.

3   Remove the under-vehicle splash shield.

23.5 The cylinder block drain plug is above the starter motor - shown with the exhaust manifold removed (Mk I models)

23.17a Open the bleed screw in the radiator top hose (Mk I models)

4   Position a suitable container beneath the bottom hose connection on the base of the radiator. Release the clamp and disconnect the hose from the radiator. Be prepared for coolant spillage, and allow the coolant to drain.

5   **Mk I models:** A cylinder block drain plug is located on the rear face of the cylinder block, adjacent to the starter motor **(see illustration)**.

6   If the coolant has been drained for a reason other than replacement, then provided it is clean and less than two years old, it can be re-used, though this is not recommended.

7   Once all the coolant has drained, reconnect the hose and secure it in place with the clamp.

Flushing

8   If coolant replacement has been neglected, or if the antifreeze mixture has become diluted, then in time, the cooling system may gradually lose efficiency, as the coolant passages become restricted due to rust, scale deposits, and other sediment. The cooling system efficiency can be restored by

flushing the system clean.

9   The radiator should be flushed independently of the engine, to avoid unnecessary contamination.

Radiator flushing

10   To flush the radiator, disconnect the top and bottom hoses and any other relevant hoses from the radiator (see Chapter 3).

11   Insert a garden hose into the radiator top inlet. Direct a flow of clean water through the radiator, and continue flushing until clean water emerges from the radiator bottom outlet.

12   If after a reasonable period, the water still does not run clear, the radiator can be flushed with a good proprietary cooling system cleaning agent. It is important that their manufacturer's instructions are followed carefully. If the contamination is particularly bad, insert the hose in the radiator bottom outlet, and reverse-flush the radiator.

Engine flushing

13   Reattach any disconnected hoses.

14   Refill the cooling system with plain water. On Mk I models, follow the bleeding

procedure described under *Refilling*. Start the engine and allow it to reach normal operating temperature, then allow it to cool completely.

15   Drain the cooling system, then repeat Step 14 until the water draining from the system is clear, with no sediment or discoloration.

Refilling

16   Before attempting to fill the cooling system, make sure that all hoses and clamps are in good condition, and that the clamps are tight and the radiator and cylinder block drain plugs are securely tightened. Note that an antifreeze mixture must be used all year round, to prevent corrosion of the engine components (see following sub-Section).

Mk I models

Refer to illustrations 23.17a, 23.17b and 23.17c

17   Loosen the bleed screw(s) **(see illustrations)**.

18   Turn on the ignition, and set the heater control to maximum temperature, with the fan speed set to low. This opens the heating valves.

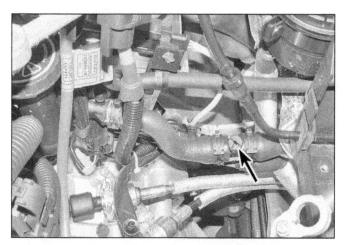

23.17b On Mk I Cooper models, an additional bleed screw is located in the heater hoses, beneath the battery tray

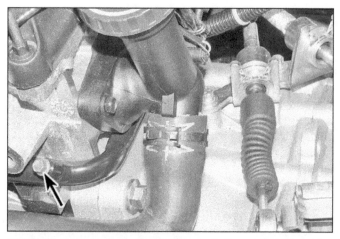

23.17c If necessary, open the bleed screw in the coolant pipe at the left-front corner of the cylinder head - all Mk I models

**23.19 Coolant filler neck pressure cap -
Mk I Cooper models**

**23.21 Expansion tank cap -
Mk I Cooper models**

Cooper models

Refer to illustrations 23.19 and 23.21

19 Remove the filler neck pressure cap **(see illustration)**. Fill the system by slowly pouring the coolant into the filler neck, closing the bleed screws as bubble-free coolant emerges. With the coolant to the top of the filler neck, install the pressure cap.

20 If the coolant is being replaced, begin by pouring in a couple of liters of water, followed by the correct quantity of antifreeze, then top-up with more water.

21 Now fill the expansion tank to the MAX mark **(see illustration)**.

22 Start the engine and run it at idle speed. Remove the filler neck pressure cap and continue running the engine until the coolant level in the filler neck ceases to drop. Turn off the engine and add coolant to the top of the filler neck.

23 Start the engine and allow it to reach normal operating temperature, then allow it to cool completely.

24 Check for leaks, particularly around disturbed components. Check the coolant level

in the expansion tank and filler neck, and top-up if necessary. Note that the system must be cold before an accurate level is indicated in the expansion tank. If the filler neck pressure cap is removed while the engine is still warm, cover the cap with a thick cloth, and unscrew the cap slowly to gradually relieve the system pressure (a hissing sound will normally be heard). Wait until any pressure remaining in the system is released, then continue to turn the cap until it can be removed.

Mk I Cooper S models and all Mk II models

Refer to illustrations 23.25a and 23.25b

25 Remove the expansion tank pressure cap **(see illustrations)**. Fill the system by slowly pouring the coolant into the expansion tank, closing the bleed screws as bubble-free coolant emerges.

26 With the coolant to the MAX mark on the expansion tank, start the engine, and allow it to run at idle speed. Add coolant as the level in the tank drops, and stop the engine once

the level ceases to drop.

27 Start the engine and allow it to reach normal operating temperature, then allow it to cool completely.

28 Check for leaks, particularly around components that have been disconnected. Check the coolant level in the expansion tank, and top-up if necessary. Note that the system must be cold before an accurate level is indicated in the expansion tank. If the expansion tank pressure cap is removed while the engine is still warm, cover the cap with a thick cloth, and unscrew the cap slowly to gradually relieve the system pressure (a hissing sound will normally be heard). Wait until any pressure remaining in the system is released, then continue to turn the cap until it can be removed.

Antifreeze mixture

29 The antifreeze should always be replaced at the specified intervals. This is necessary not only to maintain the antifreeze properties, but also to prevent corrosion which would otherwise occur as the corrosion inhibitors become progressively less effective.

30 Always use the type of antifreeze recommended in this Chapter's Specifications. The quantity of antifreeze and levels of protection are also indicated in the Specifications.

31 Before adding antifreeze, the cooling system should be completely drained, preferably flushed, and all hoses checked for condition and security.

32 After filling with antifreeze, a label should be attached to the expansion tank, stating the type and concentration of antifreeze used, and the date installed. Any subsequent topping-up should be made with the same type and concentration of antifreeze.

33 Do not use engine antifreeze in the windshield/tailgate washer system, as it will damage the vehicle paint. A windshield washer additive should be added to the washer system in the quantities stated on the bottle.

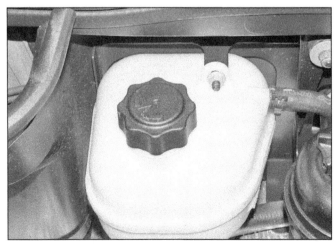

**23.25a Coolant expansion tank pressure cap
(Mk I Cooper S models)**

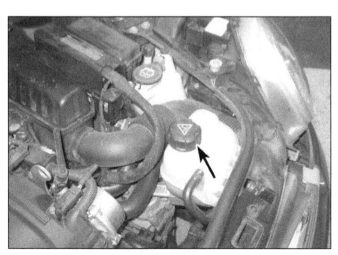

**23.25b Coolant expansion tank pressure cap
(Mk II models)**

Notes

Chapter 2 Part A
Mk I engine

Contents

Specifications

Note: *This Chapter applies to 2006 and earlier Cooper/Cooper S models, and 2008 and earlier Convertible models.*

General

Engine code	
Cooper	W10B16A
Cooper S..	W11B16A
Bore	3.03 inches (77.00 mm)
Stroke	3.37 inches (85.80 mm)
Displacement..	97.5 cubic inches (1.6 liters)
Direction of engine rotation..	Clockwise (viewed from right side of vehicle)
No 1 cylinder location	Timing chain end
Firing order	1-3-4-2
Minimum compression pressure	
Cooper	167 psi (11.5 bar)
Cooper S..	130 psi (9.0 bar)
Compression ratio	
Cooper	10.5 to 1
Cooper S..	8.33 to 1

Camshaft

Endplay	0.002 inch to 0.004 inch (0.05 to 0.095 mm)

Torque specifications

Ft-lbs (unless otherwise indicated) **Nm**

Note: *One foot-pound (ft-lb) of torque is equivalent to 12 inch-pounds (in-lbs) of torque. Torque values below approximately 15 foot-pounds are expressed in inch-pounds, because most foot-pound torque wrenches are not accurate at these smaller values.*

	Ft-lbs	Nm
Camshaft bearing cap/rocker arm shaft bolts	21	28
Camshaft sprocket bolt	75	102
Crankshaft pulley/sprocket bolt*	85	115
Cylinder head bolts*		
Stage 1	30	40
Stage 2	Tighten an additional 1/4 turn (90-degrees)	
M8	20	28
Driveplate bolts*		
Step 1	71 in-lbs	8
Step 2	22	30
Step 3	Tighten an additional 1/4 turn (90-degrees)	
Flywheel bolts*		
Cooper	59	80
Cooper S	66	90
Front suspension control arm bracket-to-body bolts*	See Chapter 10	
Intake manifold fasteners	19	26
Exhaust manifold fasteners	18	24
Subframe bolts	74	100
Subframe-to-crush tube bolts	74	100
Oil filter housing-to-cylinder block bolts	18	25
Oil pan-to-engine bolts	23	31
Oil pan-to-transaxle bolts	63	85
Timing chain cover		
M6 bolts	108 in-lbs	12
M8 bolts	156 in-lbs	18
Timing chain guide rail-to-cylinder head bolts	21	28
Timing chain idle pulley bolt (Cooper S)	33	45
Timing chain tensioner cap	46	63
Timing chain tensioning rail-to-cylinder head bolts	21	28
Valve cover fasteners	108 in-lbs	12
Valve cover caps	156 in-lbs	18

* *Do not re-use*

3.3 Loosen the screw, then pry out the plastic expansion rivet

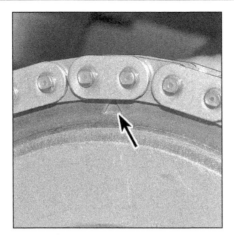

3.5 Set the triangular mark on the camshaft sprocket at the 12 o'clock position in relation to the engine block

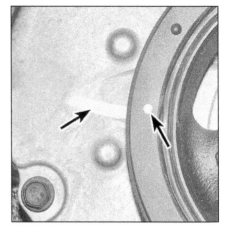

3.6 Paint alignment marks between the crankshaft pulley and the timing cover

1   General information

How to use this Chapter

This Part of Chapter 2 is devoted to repair procedures possible while the engine is still installed in the vehicle. Since these procedures are based on the assumption that the engine is installed in the vehicle, if the engine has been removed from the vehicle and mounted on a stand, some of the steps outlined will not apply.

Information concerning engine/transaxle removal and replacement and engine overhaul can be found in Part C of this Chapter.

Engine description

This four-cylinder engine is of a single overhead camshaft design, mounted transversely, with the transaxle bolted to the left end. A single-row timing chain drives the single camshaft, and the valves are operated via rocker arms with hydraulic lash adjuster units incorporated into the ends of the arms. There is no valve clearance adjustment facility. The camshaft is supported by bearings machined directly in the cylinder head.

The crankshaft is supported in five main bearings of the usual shell-type. Endplay is controlled by thrust bearing shells on No 3 main bearing. No balancer shaft is installed.

The pistons are selected to be of matching weight, and incorporate wrist pins pressed into the connecting rods.

The rotor-type oil pump is located at the front of the engine, and is driven directly by two machined flats on the crankshaft.

2   Repair operations possible with the engine in the vehicle

Many major repair operations can be accomplished without removing the engine from the vehicle.

Clean the engine compartment and

the exterior of the engine with some type of degreaser before any work is done. It will make the job easier and help keep dirt out of the internal areas of the engine.

Depending on the components involved, it may be helpful to remove the hood to improve access to the engine as repairs are performed (see Chapter 11 if necessary). Cover the fenders to prevent damage to the paint. Special pads are available, but an old bedspread or blanket will also work.

If vacuum, exhaust, oil or coolant leaks develop, indicating a need for gasket or seal replacement, the repairs can generally be made with the engine in the vehicle. The intake and exhaust manifold gaskets, oil pan gasket, crankshaft oil seals (except the rear main oil seal on models equipped with a CVT transaxle) and cylinder head gasket are all accessible with the engine in place.

Exterior engine components, such as the intake and exhaust manifolds, the oil pan, the oil pump, the water pump, the starter motor, the alternator and the fuel system components can be removed for repair with the engine in place.

Since the camshaft(s) and cylinder head can be removed without pulling the engine, valve component servicing can also be accomplished with the engine in the vehicle. Replacement of the timing chain and sprockets is also possible with the engine in the vehicle.

In extreme cases caused by a lack of necessary equipment, repair or replacement of piston rings, pistons, connecting rods and rod bearings is possible with the engine in the vehicle. However, this practice is not recommended because of the cleaning and preparation work that must be done to the components involved.

3   Reference position for the crankshaft and camshaft

Refer to illustrations 3.3, 3.5 and 3.6

1   Loosen the driver's side front wheel bolts, then raise the front of the vehicle and

support it securely on jackstands. Remove the driver's side front wheel.

2   This procedure is used to position the crankshaft and camshaft prior to removing the camshaft sprocket during the cylinder head or camshaft removal procedures, etc. This does not set any of the pistons at TDC (Top Dead Center), instead it sets the all pistons down in their cylinder bores, so there is no possibility of accidental valve-to-piston contact during repair procedures.

3   Remove the screws, pry out the plastic rivets **(see illustration)**, and remove the driver's side wheelwell liner (see Chapter 11). Note the screw securing the liner to the front bumper.

4   Remove the valve cover (see Section 4).

5   Using a wrench or socket on the crankshaft pulley bolt, turn the crankshaft clockwise until the triangular timing mark on the front of the camshaft sprocket is pointing vertically upwards (in relation to the engine block) **(see illustration)**.

6   Using white paint, make alignment marks between the triangular mark and the timing chain, and the crankshaft vibration damper/ pulley and the timing cover **(see illustration)**. The engine is now set in the reference position.

4   Valve cover - removal and installation

Note: *A new gasket and/or seals may be required on installation.*

Removal

Refer to illustrations 4.3, 4.5, 4.6a, 4.6b, 4.8a and 4.8b

1   On Cooper S models, remove the intercooler (see Chapter 4).

2   On Cooper S models, remove the retaining bolts and remove the intercooler mounting brackets.

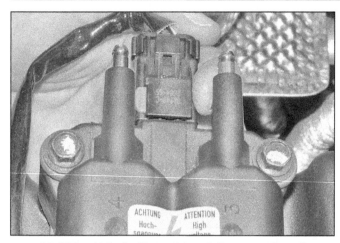

4.3 Slide out the locking catch and disconnect the coil electrical connector

4.5 Remove the four bolts and remove the ignition coil

4.6a Pull the breather hose from the right end of the valve cover . . .

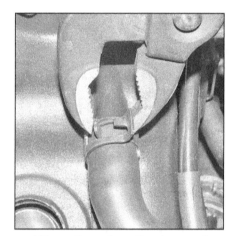

4.6b . . . then release the clamp and disconnect the left breather hose

8   On Cooper models, slide up the locking catch, squeeze the top of the connector and disconnect the electrical connectors from the top of the fuel injectors. Release the clips and slide the connector rail up from the fuel rail to allow access to the front valve cover screws **(see illustrations)**.
9   Detach the wiring harness clips from the valve cover screws.
10   Remove the 12 screws/studs and lift off the valve cover. Note that a deep 8 mm socket will be required to unscrew the studs. Remove the rubber gasket.

Installation

Refer to illustrations 4.12a, 4.12b and 4.14
11   Thoroughly clean the gasket faces on the valve cover and the engine.
12   The manufacturer states that the valve cover rubber gasket and spark plug tube seals must be replaced **(see illustrations)**. Lay the gasket in position in the valve cover. Pry the tube seals out from the underside of the cover, then push the new seals into place with the rounded side facing downwards.
13   Position the cover on the cylinder head,

3   Slide the locking catch out and disconnect the ignition coil electrical connector **(see illustration)**.
4   Pull the spark plug wire caps from the spark plugs. Pull only on the boot, not on the wires.

5   Remove the four bolts and remove the ignition coil **(see illustration)**.
6   Disconnect the engine breather hoses from the valve cover **(see illustrations)**.
7   On Cooper models, release the clips and lift off the plastic cover over the fuel injectors.

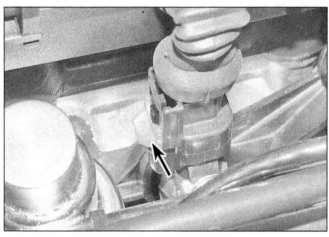

4.8a Slide up the fuel injector connector locking catch

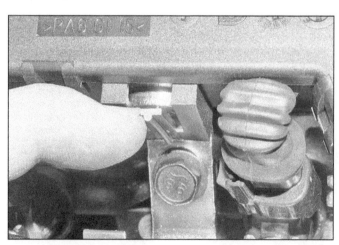

4.8b Release the retaining clips and remove the connector rail

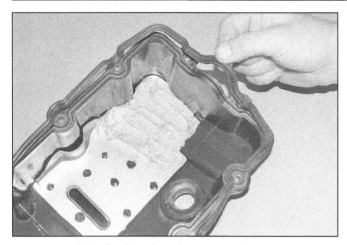

4.12a Replace the valve cover gasket . . .

4.12b . . . and the spark plug tubes seals

ensuring that the lug on the gasket engages with the corresponding cut-out in the rear of the cylinder head.

14   Install the cover securing bolts, and tighten them in two stages to the specified torque, in sequence **(see illustration)**.

15   Install the injector wiring rail and ignition coil. Reconnect the electrical connector and spark plug caps, then tighten the coil screws securely.

16   On Cooper S models, install the mounting brackets and intercooler (see Chapter 4).

5   Crankshaft vibration damper/ pulley - removal and installation

Removal

Refer to illustrations 5.3a and 5.3b

1   Remove the drivebelt (see Chapter 1).

2   Remove the crankshaft pulley bolt. To prevent the crankshaft from rotating, use a large pin spanner engaged with the pulley hub, or remove the starter (see Chapter 5)

and wedge a large screwdriver into the flywheel/driveplate ring gear teeth. Discard the pulley bolt; a new one must be installed.

3   Insert a 10 mm diameter rod approximately 110 mm long into the end of the crank-

shaft, and attach a puller to the pulley **(see illustrations)**, acting on the end of the rod. Pull the vibration damper/pulley from the crankshaft.

4   The pulley is not keyed to the crankshaft.

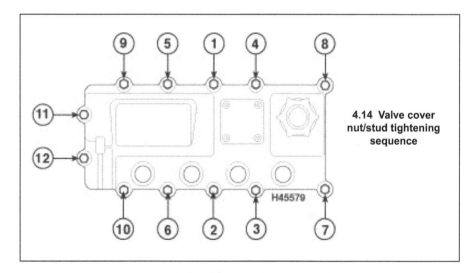

4.14 Valve cover nut/stud tightening sequence

5.3a Insert a 110 mm length of 10 mm diameter rod into the end of the crankshaft, then use a three-jaw puller to remove the pulley - Cooper models

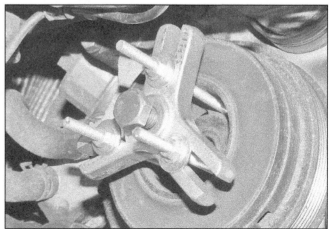

5.3b Insert a 110 mm length of 10 mm diameter rod into the end of the crankshaft, then attach a puller to the pulley using three lengths of 6 mm studding with washers and nuts - Cooper S models

5.5 Use a length of 12 mm studding with a nut and washer to draw the crankshaft pulley into place

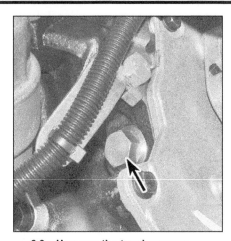

6.3a Unscrew the tensioner cap . . .

6.3b . . . then remove the tensioner piston

Installation

Refer to illustration 5.5

5   Install the pulley to the crankshaft. Note that the pulley may be a tight fit on the

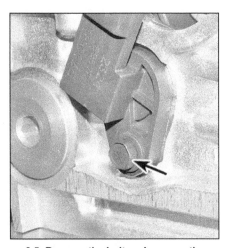

6.5 Remove the bolt and remove the CMP sensor

crankshaft. If necessary, use a length of 12 mm studding with a nut and washer to draw the pulley onto the crankshaft **(see illustration)**.

6   Insert the new pulley retaining bolt, and tighten it to the specified torque.

7   Install the drivebelt (see Chapter 1).

---

6   Timing chain, sprockets and tensioner - removal, inspection and installation

---

Removal

Refer to illustrations 6.3a, 6.3b, 6.5, 6.8a, 6.8b, 6.9a and 6.9b

1   Position the crankshaft in the reference position (see Section 3).

2   Remove the crankshaft vibration damper/pulley (see Section 5).

3   Working at the rear of the engine, unscrew the cap and remove the timing chain tensioner piston **(see illustrations)**. Be prepared for oil spillage.

4   Support the engine oil pan with a floor jack, and remove the right engine mount assembly (see Section 15).

5   Remove the bolt and remove the Camshaft Position (CMP) sensor **(see illustration)**.

6   Loosen and remove the camshaft sprocket bolt. Use a homemade tool (or similar) to prevent the camshaft from rotating. **Note:** *To make a sprocket holding tool, obtain two lengths of steel strip about 6 mm thick by 30 mm wide or similar: one 600 mm long, the other about 200 mm long (all dimensions are approximate). Bolt the two strips together to form a forked end, leaving the bolt loose so that the shorter end can pivot freely. At the end of each prong of the fork, secure a bolt with a nut and locknut, to act as the fulcrums; these will engage with the cut-outs in the sprocket, and should protrude about 30 mm.*

7   Pull the camshaft sprocket from the camshaft, then disengage the sprocket from the chain.

8   Unscrew the two cover caps, then remove the bolts **(see illustrations)**. Remove the timing chain guide and tensioner rails.

9   Remove the bolts and remove the drive-

6.8a Remove the two cover caps . . .

6.8b . . . then remove the guide rail bolts

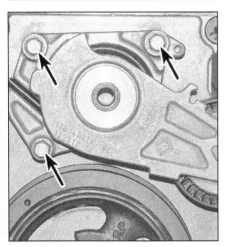

6.9a Drivebelt tensioner bolt location - Cooper models . . .

belt tensioner assembly **(see illustrations)**.
10   Place the Modular Front End (MFE) in the service position (see Chapter 11).

Cooper models
11   On 2005 and later models, unscrew the alternator mounting bolts, and pull the alternator forwards a little.
12   Remove the bolts securing the coolant pump to the cylinder block, and move the pump forwards a little. There's no need to disconnect the coolant pipes.

Cooper S models
13   Remove the retaining bolt and remove the drivebelt idler pulley from the timing cover.

All models
Refer to illustrations 6.14a and 6.14b

14   Remove the bolts and remove the timing chain cover. Note the locations of the two Torx bolts, and the central bolt with the O-ring seal. Recover the timing chain cover seal and the oil pump seals **(see illustrations)**.
15   Remove the timing chain.

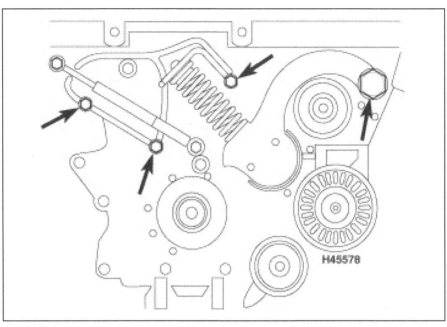

6.9b . . . and Cooper S models

Inspection
Refer to illustration 6.19

16   The chain should be replaced if the sprockets are worn or if the chain is worn (indicated by excessive lateral play between the links, and excessive noise in operation). It is wise to replace the chain in any case if the engine is disassembled for overhaul. Note that the rollers on a very badly worn chain may be slightly grooved. To avoid future problems, if there is any doubt at all about the condition of the chain, replace it.
17   Examine the teeth on the sprockets for wear. Each tooth forms an inverted V. If worn, the side of each tooth under tension will be slightly concave in shape when compared with the other side of the tooth (the teeth will have a hooked appearance). If the teeth appear worn, the sprockets must be replaced. Also check the chain guide and tensioner rail

contact surfaces for wear, and replace any worn components as necessary.
18   If the crankshaft sprocket is worn, it must be removed from the crankshaft by means of a puller. If necessary, there are special MINI tools (11 8 300 and 11 2 000) available especially for this task.
19   To install the crankshaft sprocket, heat the sprocket up to a temperature of 300 degrees F (150 degrees C), and push it over the end of the crankshaft with the timing marks on the outside face, ensuring it is positioned correctly over the pin in the crankshaft **(see illustration)**.

Installation
Refer to illustrations 6.20, 6.21, 6.22, 6.24, 6.26a, 6.26b, 6.28 and 6.29

20   Install the new timing chain to the crankshaft sprocket, aligning the two copper-colored

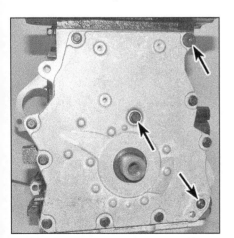

6.14a Timing cover Torx bolts and central bolt with integral O-ring seal

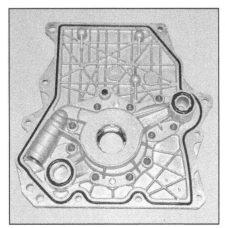

6.14b Replace the timing cover and oil pump seals

6.19 Ensure the crankshaft sprocket positions correctly over the pin

6.20 Align the two copper links on the chain with the two arrow marks on the crankshaft sprocket

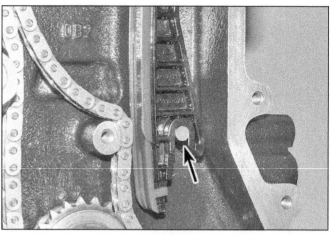

6.21 Install the guide rail, ensuring the lower end engages correctly with the mounting pin (timing cover removed for clarity)

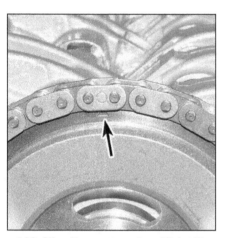

6.22 Align the copper-colored link with the triangular timing mark on the camshaft sprocket

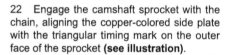

6.24 Using only hand pressure, fully compress the piston

22   Engage the camshaft sprocket with the chain, aligning the copper-colored side plate with the triangular timing mark on the outer face of the sprocket **(see illustration)**.
23   Install the sprocket to the camshaft, aligning the dowel on the shaft with the corresponding hole in the sprocket. Check that the two copper-colored chain side plates are still aligned with the timing marks on the crankshaft sprocket, then tighten the camshaft sprocket retaining bolt to the specified torque.
24   Hold the chain tensioner piston upright and, using hand pressure, fully compress the piston **(see illustration)**. Note that it may take two or three attempts before the piston remains in its fully compressed state.
25   Insert the piston into the hole in the cylinder block, then install the cap and tighten it to the specified torque.
26   Use a length of flat steel bar to lever the chain tensioner rail backwards, which compresses the tensioner piston and releases it from its compressed state and tensions the chain **(see illustrations)**. Ensure the bar levers against the tensioner guide rail, and not the timing chain.

link plates with the timing marks on the sprocket's outer face **(see illustration)**. Pull the chain upwards through the casting, and secure it in place using a length of wire (or similar).

21   Install the chain tensioner rail and guide rail, ensuring the lower ends of the guide rail engage correctly **(see illustration)**. Install the upper rail retaining bolts and valve cover caps, then tighten them to the specified torque.

6.26a Use a flat steel bar to lever the rail backwards, releasing the piston (timing cover and engine removed for clarity)

6.26b Ensure the steel bar levers against the edge of the tensioner rail and not the chain

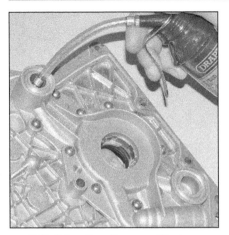

6.28 Prime the oil pump with clean engine oil

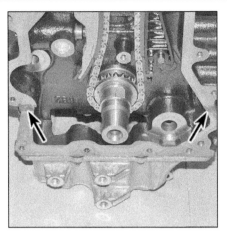

6.29 Apply a 1/8-inch (3 mm) wide bead of sealant to the joint between the lower crankcase and the cylinder block

7.3 Depress the collar, and pull the booster vacuum hose from the manifold

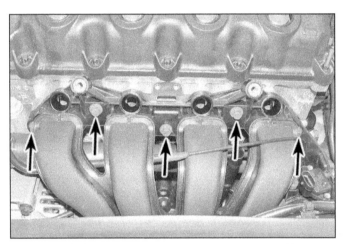

7.6 Remove the five bolts securing the intake manifold

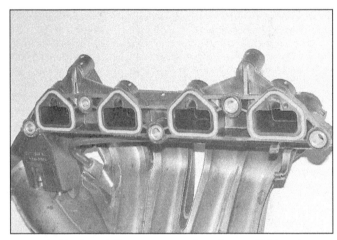

7.7 Check and, if necessary, replace the manifold seals

27   Check the condition of the timing cover seal, crankshaft seal (see Section 13), and the two oil pump seals. If in doubt, replace them.
28   Prime the oil pump with clean engine oil **(see illustration)**.
29   Apply a 1/8-inch (3 mm) bead of sealant (Loctite RTV 5999 or similar) to the joint between the lower crankcase and the cylinder block **(see illustration)**.
30   Install the timing cover, aligning the oil pump rotor flats with those on the crankshaft, and tighten the bolts to the specified torque. Note that the bolt with the seal is installed on the center of the cover **(see illustration 6.14a)**.
31   The remainder of installation is the reverse of removal.

---

7   Manifolds - removal and installation

Intake manifold
Cooper

Refer to illustrations 7.3, 7.6 and 7.7

1   Ensure the ignition is switched off.
2   Remove the throttle body (see Chapter 4).

3   Depress the retaining collar and disconnect the brake booster vacuum hose from the intake manifold **(see illustration)**.
4   Remove the fuel rail and injectors (see Chapter 4).
5   Pull the oil level dipstick from the guide tube.
6   Remove the five bolts securing the manifold to the cylinder head **(see illustration)**.
7   Unclip the coolant hose from the underside of the manifold, and maneuver the manifold from place. Remove the manifold seals **(see illustration)**.
8   Installation is a reversal of removal, bearing in mind the following points:
  a) *Check the condition of the seals, and replace if necessary.*
  b) *Ensure that all wires and hoses are correctly routed and reconnected as noted before removal.*

Cooper S

Refer to illustrations 7.13, 7.15 and 7.18

9   Ensure the ignition is switched off.
10   Remove the intercooler (see Chapter 4).

11   Remove the throttle body (see Chapter 4).
12   Remove the fuel rail and injectors (see Chapter 4).
13   Disconnect the engine breather hose from the valve cover **(see illustration)**.

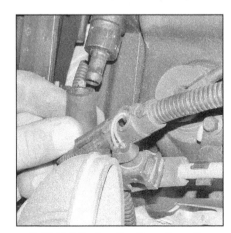

7.13 Pull the breather hose from the valve cover

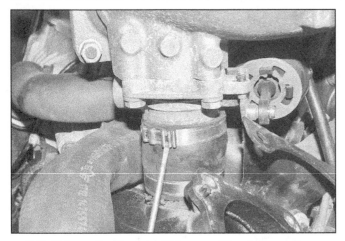

7.15 Release the clamp and disconnect the air bypass hose

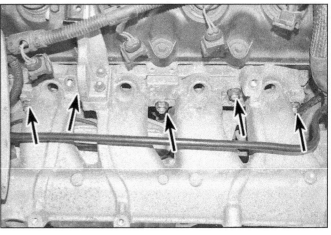

7.18 Intake manifold nuts - Cooper S

14   Disconnect the MAP sensor and knock sensor electrical connectors.

15   Release the clamps and disconnect the hose from the air bypass valve **(see illustration)**.

16   Release any wiring harnesses from their retaining clips on the intake manifold.

17   Detach the radiator upper hose support bracket from the manifold.

18   Remove the manifold retaining nuts and maneuver it from position **(see illustration)**. Discard the manifold gasket, a new one must be installed. Note that when installing the new gasket, the side marked TOP should face away from the cylinder head.

19   When installing the manifold, tighten the retaining nuts gradually, starting from outside-in, to the specified torque.

20   The remainder of installation is a reversal of removal, bearing in mind the following points:

a) Check the condition of all seals/gaskets, and replace if necessary.

b) Ensure that all wires and hoses are correctly routed and reconnected as noted before removal.

Exhaust manifold

21   Raise the front of the vehicle, and support it securely on jackstands.

22   Remove the nuts and separate the exhaust pipe from the manifold and catalytic converter.

23   Working underneath the vehicle, remove the nuts/bolts securing the heat shield above the exhaust pipe to access the oxygen sensor electrical connectors.

24   Trace back the wiring from the oxygen sensor(s), and disconnect the electrical connectors. Label the connectors to ensure correct installation. Unclip the cable harness from any retainers on the manifolds.

25   Remove the two bolts and remove the heat shield from above the manifold (if equipped).

26   Remove the manifold-to-cylinder head bolts.

27   Maneuver the manifold from under the vehicle.

28   Installation is a reversal of removal, noting the following points:

a) Apply some anti-seize compound to the manifold bolts.

b) Always replace the manifold gaskets, with the raised edges of the gasket facing towards the manifold.

8   Camshaft and rocker arms - removal and installation

Removal

Refer to illustration 8.8

1   Position the crankshaft/camshaft in the reference position (see Section 3).

2   Working at the rear of the engine, unscrew the cap and remove the timing chain tensioner piston **(see illustrations 6.3a and 6.3b)**. Be prepared for oil spillage.

3   Support the engine oil pan with a floor jack and block of wood, and remove the right engine mount assembly.

4   Remove the Camshaft Position (CMP) sensor **(see illustration 6.5)**.

5   Loosen and remove the camshaft sprocket bolt. Use a homemade tool (or similar) to prevent the camshaft from rotating (see Section 6).

6   Pull the camshaft sprocket from the camshaft, then disengage the sprocket from the chain.

7   Working from the outside-in, and in a spiral pattern, gradually and evenly loosen and remove the rocker shaft retaining bolts. Lift out the rocker arms and shafts. If the rocker arms are to be reinstalled, make a note of their locations - it's essential they are reinstalled into the same positions.

8   Check the camshaft bearing caps for identification marks. The bearing caps should be marked 1 to 5 from the timing chain end of the engine **(see illustration)**. Make suitable

8.8 The camshaft bearing caps are marked 1 to 5 starting at the timing chain end

marks if necessary.

9   Remove the bearing caps.

10   Lift the camshaft from the cylinder head.

Inspection

Refer to illustration 8.12

11   Clean all the components, including the bearing surfaces in the bearing castings and bearing caps. Examine the components carefully for wear and damage. In particular, check the bearing and cam lobe surfaces of the camshaft(s) for scoring and pitting. Examine the surfaces of the cam followers for signs of wear or damage. If the camshaft shows signs of wear or damage, the manufacturer insists that the rockers arms must also be replaced. Replace components as necessary.

12   The camshaft endplay can be checked by laying the camshaft in position on the cylinder head, and inserting feeler gauges between the end of the camshaft and the cylinder head casting **(see illustration)**. Compare the gap with that given in the Specifications, and replace the camshaft if the gap is too small.

8.12 To check the camshaft endplay, insert a feeler gauge in the gap between the end of the camshaft and the cylinder head

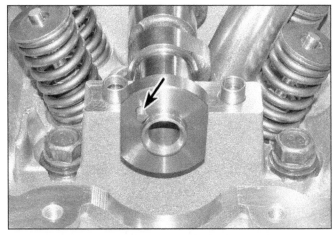

8.13 With the camshaft in the correct position, the flat sides of the flange must be vertical and the sprocket locating pin at the 11 o'clock position

Installation

Refer to illustrations 8.13 and 8.17

13 Lubricate the bearing surfaces in the cylinder head, then lay the camshaft in position so that the flat sides of the camshaft flange are vertical, and the sprocket locating pin in the end of the camshaft is in approximately the 11 o'clock position **(see illustration)**.

14 Lubricate the bearing surfaces in the bearing caps.

15 Install the bearing caps to their correct locations as noted before removal.

16 Install the rocker arms and shafts to the camshaft bearing caps, insert the bolts but only finger-tighten them at this stage. Ensure the rocker arms are equally spaced, and install either side of the camshaft bearing caps.

17 Working in sequence, gradually and evenly tighten the rocker shaft/camshaft bearing cap bolts to the specified torque **(see illustration)**.

18 Check that the flat sides of the camshaft flange are still positioned as described in Step 13. If not, use a wrench to carefully turn the camshaft to the correct position.

19 Lubricate the rocker arm rollers with clean engine oil.

20 Install the sprocket to the camshaft, aligning the dowel on the shaft with the corresponding hole in the sprocket. Ensure the triangular timing mark on the sprocket faces outwards. Tighten the camshaft sprocket retaining bolt to the specified torque, counterholding the sprocket using the same method employed during removal.

21 Hold the chain tensioner piston upright and, using hand pressure, fully compress the piston **(see illustration 6.24)**.

22 Insert the piston into the hole in the cylinder block, then install the cap and tighten it to the specified torque. Ensure the timing chain is correctly positioned within the channels of the guides.

23 Use a length of flat steel bar to lever the guide rail backwards, which compresses the tensioner piston and releases it from its compressed state **(see illustrations 6.26a and 6.26b)**. Only pry against the tensioner guide rail, and not the timing chain.

24 The remainder of installation is the reverse of removal.

9 Cylinder head - removal and installation

Warning: *Wait until the engine is completely cool before beginning this procedure.*
Note: *New cylinder head bolts and a new cylinder head gasket will be required on installation.*

Removal

Refer to illustrations 9.5, 9.9, 9.11, 9.12, 9.13, 9.15 and 9.25

1 Relieve the fuel system pressure (see Chapter 4).

2 Position the camshaft and crankshaft in the reference position (see Section 3).

3 On Cooper models, remove the battery and battery tray (see Chapter 5).

4 Drain the cooling system (see Chapter 1).

5 Depress the retaining clips and disconnect the hoses and the electrical connector from the tank vent valve at the right end of the engine **(see illustration)**.

6 Remove the intake manifold (see Section 7).

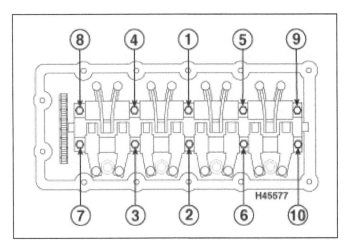

8.17 Camshaft bearing cap/rocker arm bolt tightening sequence

H45577

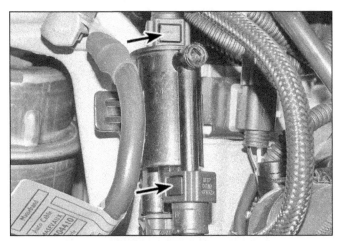

9.5 Depress the buttons then disconnect the hoses and the wiring plug from the tank vent valve

9.9 Disconnect the oxygen sensor wiring plug and detach the support bracket

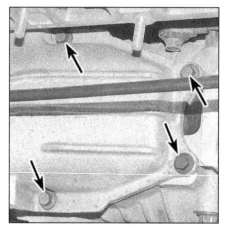

9.11 Remove the bolts and remove the supercharger outlet duct

9.12 Slide down the locking catch and disconnect the coolant temperature sensor electrical connector

7   Loosen the clamps and disconnect the coolant hoses from the thermostat housing at the left end of the cylinder head.

8   Remove the bolts and detach the exhaust manifold from the cylinder head (see Section 7). If care is taken, it isn't necessary to detach the exhaust pipe from the manifold.

9   Disconnect the oxygen sensor electrical connector and detach the plug support bracket from the left end of the cylinder head **(see illustration)**.

10   On Cooper models, disconnect the coolant hose from the filler neck so that the engine wiring harness can be routed around the thermostat housing. Be prepared for coolant spillage.

11   On Cooper S models, remove the retaining bolts, and remove the supercharger outlet duct. Discard the gasket; a new one must be installed **(see illustration)**.

12   Release the locking catch, and disconnect the coolant temperature sensor electrical connector **(see illustration)**.

13   Remove the screw securing the coolant distributor pipe to the cylinder head **(see illustration)**.

14   Position a floor jack under the engine oil pan, and place a block of wood between the jack head and the oil pan casing to prevent any damage. Take up the weight of the engine with the jack.

Models up to 12/2003

15   Unclip the fuel lines from the engine carrier bracket, then remove the two bolts and remove the bracket **(see illustration)**.

16   Remove the nuts and detach the ground straps from the engine mounting bracket, then remove the four bolts and one nut, and remove the bracket.

17   Detach the hydraulic mount from the right inner fender **(see illustration 15.9)**.

Models from 12/2003

18   Remove the ground strap nut, then remove the four bolts and one nut securing the engine mounting bracket at the right end of the engine **(see illustration 15.14)**. Remove the bracket.

19   Remove the Torx screw on the underside of the body member, and the bolt at the rear, and remove the right engine mount from the vehicle body **(see illustrations 15.15 and 15.16)**.

All models

20   Remove the bolt and remove the Camshaft Position (CMP) sensor **(see illustration 6.5)**.

21   Remove the two cover caps from the right end of the cylinder head **(see illustration 6.8a)**.

22   Loosen, but do not remove, the bolt securing the chain sprocket to the camshaft. Use a homemade tool to prevent the camshaft from rotating as the bolt is loosened (see Section 6).

23   Working at the right rear corner of the engine, detach the wiring harness bracket from the cylinder block, then slowly remove the timing chain tensioner cap to relieve the tension on the chain **(see illustrations 6.3a and 6.3b)**. With the cap removed, extract the tensioner piston assembly.

9.13 Remove the coolant distributor pipe

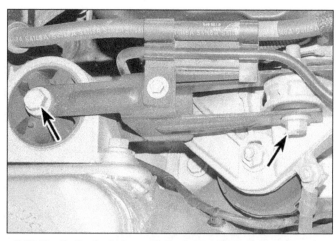

9.15 Unclip the fuel pipe, remove the two bolts and remove the engine carrier bracket

9.25 The tensioner blade and guide rails are secured by two bolts accessible through the cover cap holes in the end of the cylinder head

9.38 Ensure the locating dowels are in place

24   Remove the camshaft sprocket bolt, and disengage the sprocket from the chain. Do not allow the crankshaft to rotate. Use a length of wire or similar to support the chain and prevent it from falling down into the timing cover and becoming disengaged from the crankshaft sprocket.

25   Remove the two bolts **(see illustration)**, and remove the timing chain tensioner blade and guide rail **(see illustration 6.8b)**.

26   Working in the **reverse** of the sequence in **illustration 9.42**, gradually and evenly loosen then remove the cylinder head bolts.

27   Make a final check to ensure that all relevant hoses and wires have been disconnected to allow cylinder head removal.

28   Release the cylinder head from the cylinder block and locating dowels by rocking it. Do not pry between the mating faces of the cylinder head and block, as this may damage the gasket faces.

29   Ideally, two assistants will now be required to help remove the cylinder head. Have one assistant hold the timing chain up, clear of the cylinder head, making sure that tension is kept on the chain. With the aid of another assistant, lift the cylinder head from the block - take care, as the cylinder head is heavy. Support the timing chain from the cylinder block using wire.

30   Remove the cylinder head gasket. Do not rest the cylinder head on its sealing face as this could damage the valves.

Inspection

31   The mating faces of the cylinder head and block must be perfectly clean before installing the head. Use a scraper to remove all traces of gasket and carbon, and also clean the tops of the pistons. Take particular care with the aluminium cylinder head, as the soft metal is easily damaged. The surface must not be scratched.

32   Also make sure that debris is not allowed to enter the oil and water passages. Using adhesive tape and paper, seal the water, oil

and bolt holes in the cylinder block. To prevent carbon entering the gap between the pistons and bores, smear a little grease in the gap. After cleaning each piston, rotate the crankshaft so that the piston moves down the bore, then wipe out the grease and carbon with a cloth rag. Take care not to disengage the timing chain from the crankshaft sprocket.

33   Check the block and head for nicks, deep scratches and other damage. If very slight, they may be removed from the cylinder block carefully with a file. More serious damage may be repaired by machining, but this is a specialist job.

34   If warpage of the cylinder head is suspected, use a straight-edge and feeler gauge to check it for distortion.

35   Clean out the bolt holes in the block using a pipe cleaner or thin rag and a screwdriver. Make sure that all oil and water is removed, otherwise there is a possibility of the block being cracked by hydraulic pressure when the bolts are tightened.

36   Examine the bolt threads and the threads in the cylinder block for damage. If necessary, use the correct size tap to chase out the threads in the block.

Installation

Refer to illustrations 9.38 and 9.42

37   Ensure that the mating faces of the cylinder block and head are spotlessly clean, that the new cylinder head bolt threads are clean and dry, and that they screw in and out of their locations.

38   Check that the cylinder head locating dowels are correctly positioned in the cylinder block **(see illustration)**.

39   Install a new cylinder head gasket to the block, locating it over the dowels. Make sure that it is the correct way up. Note that 0.3 mm thicker-than-standard gaskets are available for use if the cylinder head has been machined.

40   Lower the cylinder head into position. As the cylinder head is lowered, feed the timing chain up through the head, keeping it in tension to prevent the chain from becoming disengaged from the crankshaft sprocket. Ensure that the cylinder head engages with the locating dowels.

41   Install the new cylinder head bolts, and tighten the bolts as far as possible by hand.

42   Tighten the bolts in order **(see illustration)**, and in the stages given in the Specifi-

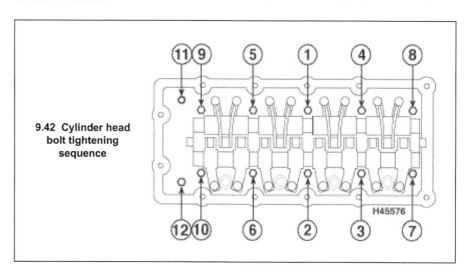

9.42 Cylinder head bolt tightening sequence

H45576

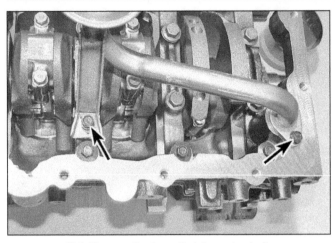

10.9 Remove the two oil pick-up pipe bolts

10.10 Install a new O-ring seal to the oil pick-up pipe

cations. Note that the M8 bolts in positions 11 and 12 only have a single stage torque setting.

43   Install the timing chain tensioner rail and guide rails down into position in the timing cover, then install and tighten the bolts securing the timing chain tensioner rail and the chain guide to the cylinder head to their specified torque. Ensure the lower end of the guide rail locates correctly **(see illustration 6.21)**.

44   Ensure the crankshaft is still positioned correctly, aligning the previously-made marks.

45   Push back the top end of the chain tensioner rail, as during removal, to allow the camshaft sprocket to be installed. Engage the camshaft sprocket with the chain, in it's original position, and manipulate the sprocket so that the timing arrow aligns with the previously-made painted mark on the chain side plate **(see illustration 3.5)**.

46   Install the sprocket to the camshaft, aligning the locating pin, and finger-tighten the retaining bolt.

47   Counterhold the sprocket using the same method employed on removal, and tighten the sprocket bolt to the specified torque.

48   Remove the cap from the timing chain tensioner piston assembly, then place the piston upright on a hard, clean, level surface and, using the palm of your hand, exert continuous pressure until the piston is completely compressed **(see illustration 6.24)**.

49   Install the timing chain tensioner piston to the cylinder block, and tighten the cap to the specified torque.

50   Inspect the timing chain routing to ensure the chain lies between the tensioner rail and the guide rail then, using a flat steel bar, push the tensioner rail rearwards to release the piston from its compressed state and tension the chain **(see illustrations 6.26a and 6.26b)**. Do not lever directly on the chain as damage may result.

51   The remainder of installation is the reverse of removal, noting the following points:

a)   Ensure that all hoses and wires are correctly reconnected and routed as noted before removal.

b)   Install the intake manifold (see Section 7).

c)   Install the air filter assembly (see Chapter 4).

d)   Reconnect the exhaust manifold (see Section 7).

e)   Install the spark plugs (if removed) (see Chapter 1).

f)   Refill the cooling system (see Chapter 1).

---

10   Oil pan - removal and installation

Removal

1   Drain the engine oil (see Chapter 1).
2   Remove the drivebelt (see Chapter 1).

Models with air conditioning

3   Remove the front bumper and bumper carrier (see Chapter 11).

4   Place the Modular Front End (MFE) in the service position (see Chapter 11).

5   Remove the retaining bolts, and detach the air conditioning compressor from the engine. There is no need to disconnect the refrigerant pipes. Position the compressor clear of the engine and secure it with a cable tie to the MFE.

All models

Refer to illustration 10.9

6   Remove the bolts and remove the rear lower stabilizer arm and mounting bracket from the oil pan casing.

7   Progressively unscrew and remove all the oil pan securing bolts, including the one at the rear of the oil pan adjacent to the transaxle bellhousing, and the three bolts securing the bellhousing to the oil pan.

8   Lower the oil pan, and maneuver it out from under the vehicle.

9   Remove the oil pan gasket, and discard it. If required, remove the two bolts and pull out the oil pick-up pipe **(see illustration)**.

Installation

Refer to illustrations 10.10, 10.11, 10.13 and 10.14

10   Thoroughly clean the mating surfaces of the oil pan and cylinder block. If removed, install the oil pick-up pipe (with a new O-ring) to the oil pan **(see illustration)**. Tighten the bolts securely.

11   Place a new gasket in position on the lower crankcase. Note the gasket lugs which locate on the lower crankcase and hold it in place **(see illustration)**.

12   Raise the oil pan up to the cylinder block, ensuring that the gasket stays in place.

13   Install the oil pan securing bolts, tightening them finger-tight only at this stage. If installing the oil pan with the transaxle removed, ensure the edge of the oil pan is flush with the end of the cylinder block **(see illustration)**.

14   Tighten the oil pan-to-cylinder block bolts, in sequence, to the specified torque **(see illustration)**.

15   Tighten the oil pan-to-transaxle bolts to the specified torque.

16   The remainder of installation is the

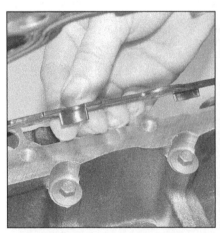

10.11 Note how the oil pan gasket lugs locate with the lower crankcase

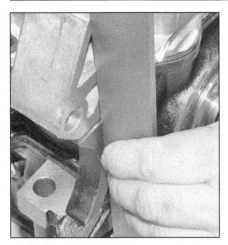

10.13 Use a straight-edge or ruler to check that the end of the oil pan is flush with the end of the cylinder block

reverse of removal. Refill the engine with oil (see Chapter 1).

---

11 Front subframe - removal and installation

Note: *After installing the front subframe, it is essential that the alignment of the wheels is checked and, if necessary, adjusted.*

Removal

Refer to illustrations 11.4, 11.9 and 11.12

1 Loosen the front wheel bolts, then raise the front of the vehicle and support it securely on jackstands. Remove the front wheels.
2 Remove the screws and remove the engine undershield.
3 Remove the front bumper and bumper carrier (see Chapter 11).
4 Remove the clamp bolt securing the power steering reservoir to the engine firewall

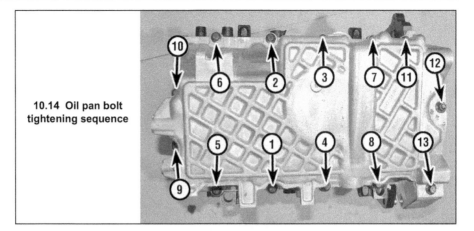

10.14 Oil pan bolt tightening sequence

and detach the reservoir from the clamp **(see illustration)**.
5 Working underneath the vehicle, remove the bolts/nuts and remove the engine stabilizer link from the right rear of the engine **(see illustration 15.24)**.
6 Disconnect the tie-rod ends and control arms from the steering knuckles (see Chapter 10).
7 Disconnect the stabilizer bar links from the stabilizer bar (see Chapter 10).
8 Disconnect the wiring plugs from the power steering pump.
9 Remove the pinch-bolt and nut, slide the steering column joint rearwards, and detach it from the steering gear pinion shaft **(see illustration)**.
10 Place a floor jack under the subframe to support it.
11 Remove the bolts securing the control arm brackets to the vehicle body.
12 Remove the bolts securing the subframe to the vehicle body, and lower the subframe with the floor jack, feeding the power steering reservoir through the engine compartment as the subframe is lowered **(see illustration)**.
13 If the subframe is to be replaced, remove the stabilizer bar, steering gear, steering pump, and control arms (see Chapter 10).

11.4 Remove the bolt securing the power steering reservoir to the firewall bracket

Installation

Refer to illustration 11.14

14 Installation is the reverse of removal, noting the following points:

a) *The subframe must engage correctly with the lugs on the underside of the chassis members* **(see illustration)**.

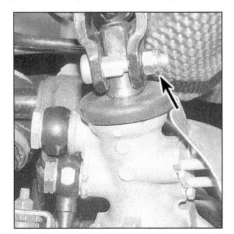

11.9 Remove the pinch-bolt, slide the joint rearwards and detach it from the steering gear pinion shaft

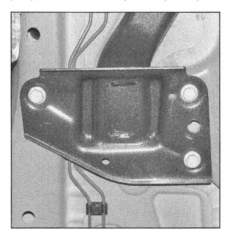

11.12 Subframe rear mounting bolts

11.14 The subframe must engage correctly with the lugs on the underside of the chassis members

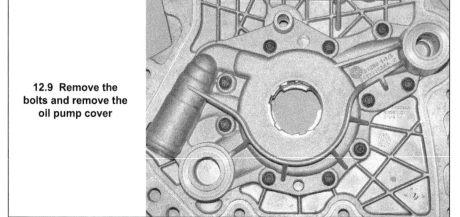

12.9 Remove the bolts and remove the oil pump cover

12.10a Check the oil pump rotors for identification marks

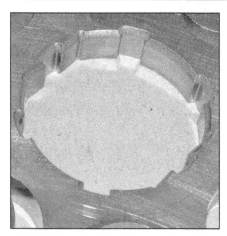

12.10b The chamfered inner edge of the inner rotor must face the timing chain

12.16a Press down the plug and extract the snap-ring

b) *Tighten the subframe mounting bolts before tightening the control arm bracket bolts.*
c) *Tighten all fasteners to the specified torque, where given.*

12 Oil pump - removal, inspection and installation

Removal

Refer to illustration 12.9

1   The oil pump is integral with the timing chain cover.
2   Remove the crankshaft vibration damper/

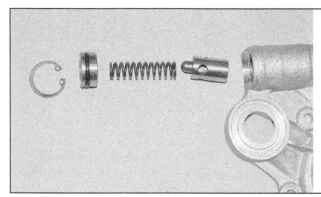

12.16b Oil pressure relief valve snap-ring, plug, spring and piston

pulley (see Section 5).
3   Place the Modular Front End (MFE) in the service position (see Chapter 11).
4   On 2005 and later Cooper models, remove the alternator mounting bolts and pull the alternator forwards a little.
5   On all Cooper models, remove the bolts securing the coolant pump to the cylinder block, and move the pump forwards slightly. There's no need to disconnect the coolant pipes.
6   On Cooper S models, remove the retaining bolt and detach the drivebelt idler pulley from the timing cover.
7   Remove the bolts and remove the drivebelt tensioner from the timing cover **(see illustrations 6.9a and 6.9b)**.

8   Remove the bolts and remove the timing chain cover. Note the locations of the two Torx bolts, and the central bolt with the O-ring seal. Discard the timing chain cover seal and the oil pump seals; new ones must be installed **(see illustration 6.14a and 6.14b)**.
9   Remove the Torx bolts and remove the oil pump cover from the inner face of the timing chain cover **(see illustration)**.

Inspection

Refer to illustrations 12.10a, 12.10b, 12.16a and 12.16b

10   Check the rotors for identification marks **(see illustration)**, and if necessary mark the rotors to ensure they are installed in their original positions (mark the top faces of both rotors to ensure they are installed the correct way up). The chamfered edge on the inner rotor must face the timing chain **(see illustration)**. Remove the rotors from the housing.
11   Clean the housing and the rotors thoroughly, then install the rotors to the housing, ensuring that they are returned to their original positions.
12   Examine the rotors, checking for any signs of wear or damage. No specific pump wear tolerances are given by MINI.
13   If there is any sign of damage or wear, consult a MINI dealer regarding the availability of spare parts. The rotors should always be replaced as a matched pair. It may be necessary to replace the complete rotor/housing assembly as a unit.
14   If the rotors appear serviceable, remove the rotors, then pour a small amount of engine oil into the housing. Install the rotors and turn them to lubricate all the contact surfaces.
15   Install the oil pump cover, and tighten the retaining bolts securely.
16   To check the oil pressure relief valve, push the plug down slightly, extract the snapring and remove the plug, spring and piston **(see illustrations)**. Note that the spring will force the plug from the housing as the snapring is removed. As no specifications are available concerning the spring free length, compare the spring with a new one.
17   Reassemble the pressure relief valve by reversing the disassembly procedure.

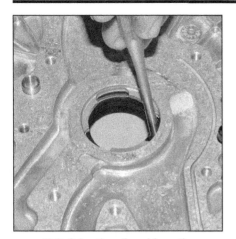

13.3 Drive the oil seal from the timing cover

13.5 Using a tubular spacer (socket) which bears only on the hard, outer edge, drive the seal into place

13.8a Drill a small hole in the oil seal . . .

Installation

18 Replace the timing cover gasket and oil seals. Prime the oil pump with clean engine oil prior to installing the cover, aligning the pump rotor flats with those on the crankshaft. Note the center cover bolt has an integral seal.

19 Installation is the reverse of removal.

---

13 Oil seals - replacement

Crankshaft front oil seal

Refer to illustrations 13.3 and 13.5

1 Remove the crankshaft vibration damper/pulley (see Section 5).

2 Note the installed depth of the oil seal in the cover.

3 Pull the oil seal from the cover using a hooked instrument. Alternatively, drill small holes in the oil seal, and screw self-tapping screws into the holes, then use pliers to pull the seal out **(see illustrations 13.8a and 13.8b)**. Take great care not to damage the crankshaft or the sealing surface of the timing cover. If you are unable to pull the seal out using these methods, the timing cover must

be removed (see Section 6), and the seal driven out **(see illustration)**.

4 Clean the oil seal housing and the crankshaft sealing surface.

5 Apply a small amount of clean engine oil to the seal lips, then carefully drive the seal into place until the outer edge is almost flush with the outer surface of the timing cover. Use a seal driver or a tubular spacer that bears evenly on the hard, outer edge of the seal **(see illustration)**.

6 Install the crankshaft vibration damper/pulley (see Section 5).

Crankshaft rear main oil seal

Refer to illustrations 13.8a, 13.8b, 13.10a and 13.10b

7 Remove the flywheel/driveplate (see Section 14).

8 Drill several small holes in the oil seal, and screw in self-tapping screws. Pull the screws, complete with seal, from the housing **(see illustrations)**.

9 Clean the oil seal housing and the crankshaft sealing surface.

10 In order to install the new seal over

13.8b . . . then pull the seal from the cylinder block

the end of the crankshaft, a guide sleeve is required. There is a MINI special tool (11 8 220), or improvise a guide sleeve using a section from a plastic bottle of suitable size. Slide the sleeve, with the seal fitted, over the crankshaft **(see illustrations)**. Do not touch the seal inner lip with your fingers.

13.10a Improvise a guide sleeve using a section from a plastic bottle

13.10b Slide the sleeve and seal over the end of the crankshaft

14.2 Lock the flywheel using a homemade locking tool

14.3 Remove the flywheel/driveplate bolts

11   Remove the guide sleeve, then carefully drive the seal into the housing until it is flush with the cylinder block surface, using a seal driver or tubular spacer that bears evenly on the hard, outer edge of the seal.
12   Install the flywheel/driveplate (see Section 14).

14   Flywheel/driveplate - removal and installation

Refer to illustrations 14.2, 14.3 and 14.4

1   On manual transaxle models, remove the clutch assembly (see Chapter 8).
2   Use a tool to lock the flywheel/driveplate in place **(see illustration)**. In the absence of a suitable tool, use a flat-bladed screwdriver to lock the flywheel teeth.
3   Remove the bolts and remove the flywheel/driveplate **(see illustration)**. Discard the bolts; new ones must be installed.
4   When installing, align the dot (Cooper models) or the cut-out (Cooper S) on the flywheel/driveplate with the dot on the rear of the crankshaft **(see illustration)**.

5   Install the new bolts, and tighten them evenly to the specified torque.
6   On manual transaxle models, install the clutch (see Chapter 8).

15   Powertrain mounts - inspection and replacement

Inspection

1   Three engine mounts are used. One at the left end of the transaxle, one at the right upper end of the engine, and one under the engine at the right rear corner.
2   Check the mount rubber to see if it is cracked, hardened, or separated from the metal at any point. Replace the mount if any such damage is evident.
3   Check that all the mount fasteners are securely tightened.
4   Using a large screwdriver or a crowbar, check for wear in the mount by carefully levering against it to check for free play. Where this is not possible, enlist the aid of an assistant to move the engine/transaxle back-and-forth, or side-to-

side, while you observe the mount. While some free play is to be expected, even from new components, excessive wear should be obvious. If excessive free play is found, check first that the fasteners are correctly secured, then replace any worn components as required.

Replacement

Refer to illustrations 15.9, 15.10, 15.13, 15.14, 15.15, 15.16, 15.22 and 15.24

5   Raise the front of the vehicle and support it securely on jackstands. Remove the screws and remove the engine undershield.
6   Position a floor jack under the engine oil pan, with a block of wood between the jack head and the oil pan casting to prevent any damage.

Right mount on models up to 12/2003

7   Unclip the fuel hoses from the engine carrier bracket, remove the two bolts and remove the bracket from the mounts **(see illustration 9.15)**.
8   Remove the bolts/nut and remove the engine mount bracket from the engine cylinder block. Note the installed position of the ground strap.

14.4 The dot (Cooper models) or cut-out (Cooper S) on the flywheel/driveplate must align with the dot on the crankshaft

15.9 Use a strap wrench to counterhold the mount, while unscrewing the Torx bolt on the underside of the inner fender

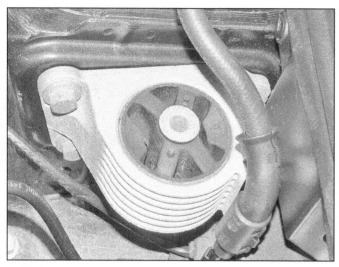

15.10 Remove the two bolts and remove the mount from the inner fender

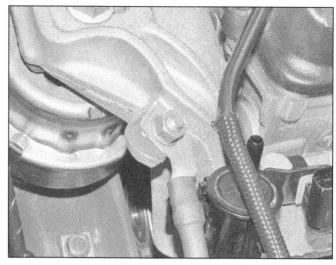

15.13 Remove the nut and detach the ground strap

9   Using a strap wrench (or similar), counterhold the mount while removing the Torx screw from the underside of the inner fender **(see illustration)**.
10   If required, remove the bolts and remove the engine carrier bracket mount from the inner fender **(see illustration)**.
11   Installation is the reverse of removal.

Right mount on models from 12/2003
12   Remove the right wheel. Remove the screws, pry out the plastic expansion rivets, and remove the right wheelwell liner.
13   Remove the nut(s) and detach the ground strap from the mount bracket **(see illustration)**.
14   Remove the four bolts and one nut, then lift the mount bracket from position **(see illustration)**.

15   Working underneath the inner fender, remove the mount retaining screw **(see illustration)**.
16   Remove the screw securing the mount to the inner fender **(see illustration)**.
17   Installation is the reverse of removal.

Left mount
18   On Cooper models, remove the battery and battery carrier (see Chapter 5).
19   On Cooper S models, remove the air filter housing (see Chapter 4).
20   On late Cooper and Cooper S models, remove the retaining screws and remove the clutch release cylinder from the transaxle bellhousing. There is no need to disconnect the hydraulic pipe.
21   Place a floor jack under the transaxle casing, with a block of wood between the

15.14 Remove the four bolts and one nut, then remove the engine mounting bracket

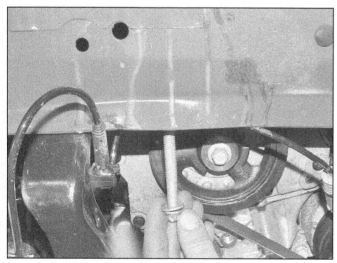

15.15 Reach under the inner fender and unscrew the engine mount Torx screw

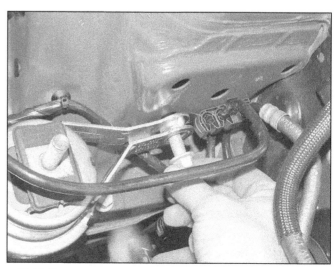

15.16 Remove the bolt securing the mount to the inner fender

15.22 Remove the mount through-bolt

15.24 Remove the two bolts securing the lower engine stabilizer

jack head and the casting to prevent any damage. Take up the weight of the transaxle with the jack.

22   Remove the bolts securing the mount to the transaxle casing and the through-bolt secur-ing the mount to the bracket **(see illustration)**.

23   Installation is the reverse of removal.

Lower engine stabilizer

24   Working underneath the vehicle, loosen and remove the two bolts securing the stabi-lizer link, and maneuver it from position **(see illustration)**.

25   Installation is the reverse of removal.

Notes

Notes

Chapter 2 Part B
Mk II engine

Contents

Specifications

Note: *This Chapter applies to 2007 and later Cooper/Cooper S/Clubman S, and 2009 and later Convertible models.*

General

Engine code
  Cooper
    2010 and earlier models N12
    2011 and later models N16
  Cooper S
    2010 and earlier models N14
    2011 and later models N18
Bore .... 3.03 inches (77.00 mm)
Stroke .... 3.37 inches (85.80 mm)
Displacement .... 97.5 cubic inches (1.6 liters)
Direction of engine rotation .... Clockwise (viewed from right side of vehicle)
No 1 cylinder location .... Timing chain end
Firing order 1-3-4-2
Minimum compression pressure
  Cooper .... 167 psi (11.5 bar)
  Cooper S .... 130 psi (9.0 bar)
Compression ratio
  Cooper .... 11.0 to 1
  Cooper S .... 10.5 to 1

Torque specifications

Note: *One foot-pound (ft-lb) of torque is equivalent to 12 inch-pounds (in-lbs) of torque. Torque values below approximately 15 foot-pounds are expressed in inch-pounds, because most foot-pound torque wrenches are not accurate at these smaller values.*

	Ft-lbs (unless otherwise indicated)	Nm
Camshaft bearing cap bolts	84 in-lbs	10
Exhaust camshaft sprocket bolt (N14 engine)*		
Step 1	15	20
Step 2	Tighten an additional 1/4 turn (90-degrees)	
Camshaft VANOS unit bolt*		
Step 1	15	20
Step 2	Tighten an additional 1/2 turn (180-degrees)	
Crankshaft pulley-to-hub bolts	21	28
Crankshaft sprocket hub bolt*		
Step 1	37	50
Step 2	Tighten an additional 180-degrees	
Cylinder head bolts* (in sequence)		
M10 x 145 mm bolts		
Step 1	22	30
Step 2	Tighten an additional 1/4 turn (90-degrees)	
Step 3	Tighten an additional 1/4 turn (90-degrees)	
M8 x 95 mm bolts		
Step 1	132 in-lbs	15
Step 2	Tighten an additional 1/4 turn (90-degrees)	
Step 3	Tighten an additional 1/4 turn (90-degrees)	
M8 x 35 mm bolts	22	30
Valve cover fasteners	84 in-lbs	10
Flywheel/driveplate bolts*		
Step 1	42 in-lbs	8
Step 2	22	30
Step 3	Tighten an additional 1/4 turn (90-degrees)	
Intake manifold-to-cylinder head		
Bolts	132 in-lbs	15
Nuts	15	20
Bracket fasteners	15	20
Exhaust manifold-to-cylinder head*	18.5	25
Turbocharger		
To exhaust manifold nuts	15	20
Coolant line banjo bolts	25	35
Oil feed line banjo bolt	22	30
Oil return line bolt	71 in-lbs	8
Support bracket fasteners		
To turbocharger	18.5	25
To engine block	168 in-lbs	19
Catalytic converter-to-turbocharger nuts	See Chapter 6	
Subframe bolts	74	100
Subframe-to-crush tube bolts	74	100
Front suspension control arm bracket-to-body bolts*	44	59
Oil pan-to-engine bolts	106 in-lbs	12
Oil pump sprocket-to-oil pump bolt*		
Step 1	44 in-lbs	5
Step 2	Tighten an additional 1/4 turn (90-degrees)	
Timing chain guide rail-to-cylinder head bolt	15	20
Timing chain tensioner-to-cylinder head bolt	48	65
Timing chain guide/cassette bolts	22	30

* Do not re-use

---

1   General information

How to use this Chapter

This Part of Chapter 2 is devoted to repair procedures possible while the engine is still installed in the vehicle. Since these procedures are based on the assumption that the engine is installed in the vehicle, if the engine has been removed from the vehicle and mounted on a stand, some of the steps outlined will not apply.

Information concerning engine/transaxle removal and installation and engine overhaul can be found in Part C of this Chapter.

Engine description

This four-cylinder engine is of dual overhead camshaft design, mounted transversely, with the transaxle bolted to the left end.

A single-row timing chain drives the two camshafts, and the valves are operated via rocker arms with hydraulic lash adjuster units incorporated into the ends of the arms. There is no provision for valve clearance adjustment. The camshaft is supported by bearings machined directly in the cylinder head. All models except the N14 engine incorporate a VANOS (variable camshaft timing) unit on the end of each camshaft. The N14 engine has a VANOS unit on the intake camshaft and a conventional sprocket on the exhaust camshaft. Additionally, all models

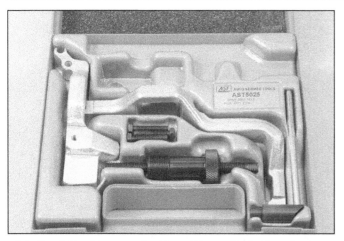

3.1 Special tools are required to set the engine in the reference position. Kits like this are available from specialty tool dealers and online tool retailers. Be sure to obtain the proper kit for your engine

3.3a Rotate the crankshaft until the IN mark on the intake camshaft . . .

except the N14 are equipped with Valvetronic, which is an electric motor-controlled variable valve lift apparatus that alters intake valve lift during varying operating conditions to maximize performance, and to minimize pumping losses to increase fuel efficiency.

Lubrication is by means of an oil pump, which is chain driven off the crankshaft right-hand end. It draws oil through a strainer located in the oil pan, then forces it through an externally mounted filter into galleries in the cylinder block/crankcase

The pistons are selected to be of matching weight, and incorporate wrist pins pressed into the connecting rods.

2   Repair operations possible with the engine in the vehicle

Many major repair operations can be accomplished without removing the engine from the vehicle.

Clean the engine compartment and the exterior of the engine with some type of degreaser before any work is done. It will make the job easier and help keep dirt out of the internal areas of the engine.

Depending on the components involved, it may be helpful to remove the hood to improve access to the engine as repairs are performed (see Chapter 11 if necessary). Cover the fenders to prevent damage to the paint. Special pads are available, but an old bedspread or blanket will also work.

If vacuum, exhaust, oil or coolant leaks develop, indicating a need for gasket or seal replacement, the repairs can generally be made with the engine in the vehicle. The intake and exhaust manifold gaskets, oil pan gasket, crankshaft oil seals and cylinder head gasket are all accessible with the engine in place.

Exterior engine components, such as the intake and exhaust manifolds, the oil pan, the oil pump, the water pump, the starter motor, the alternator and the fuel system components can

3.3b . . . and the EX mark on the exhaust camshaft are pointing upward

be removed for repair with the engine in place.

Removal and installation of the cylinder head and replacement of the timing chain and sprockets is also possible with the engine in the vehicle.

In extreme cases caused by a lack of necessary equipment, repair or replacement of piston rings, pistons, connecting rods and rod bearings is possible with the engine in the vehicle. However, this practice is not recommended because of the cleaning and preparation work that must be done to the components involved.

3   Reference position for the crankshaft and camshaft

Refer to illustrations 3.1, 3.3a, 3.3b, 3.4, 3.5a and 3.5b

1   This procedure positions the crankshaft in the "90-degree" position so that the pistons are all half-way down in their cylinder bores.

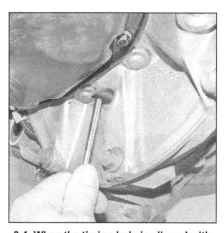

3.4 When the timing hole is aligned with the hole in the flywheel/driveplate, the pin can be inserted to lock the crankshaft in position

This ensures that there will be no valve-to-piston contact when removing the VANOS units, timing chain, or cylinder head. Special tools are required to perform this task (see illustration). The special tools include a pin that locks the crankshaft in position and fixtures that are bolted down to the top of the cylinder head, engaging flats on the camshafts, locking them in position.

2   Disconnect the cable from the negative terminal of the battery (see Chapter 5). Remove the valve cover (see Section 4).

3   Using the crankshaft balancer center bolt, rotate the engine clockwise until the IN and EX markings at the center of each camshaft become visible (see illustrations).

4   Locate the hole in the lower cylinder block near the left end of the oil pan (see illustration). Insert MINI special tool no. 11 9 950 through the hole and into its corresponding hole in the flywheel/driveplate. If it won't go in, try turning the crankshaft back and forth a little until the tool can be inserted (it shouldn't take much movement of the crankshaft).

3.5a Here the special camshaft holding tools are bolted in place on an N14 engine . . .

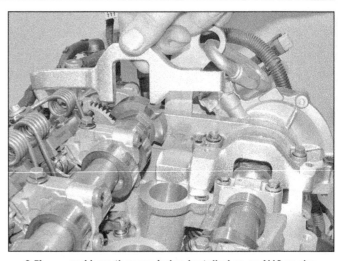

3.5b . . . and here they are being installed on an N12 engine with Valvetronic

5   At this point, the camshafts must be locked securely in position. While there is no danger of the valves contacting the pistons, there *is* a danger of intake valves and exhaust valves contacting each other if the VANOS unit(s)/camshaft sprocket are loosened. Also, any procedure requiring the engine to be set at this position will also require the camshafts to be timed to each other and to the crankshaft.

   a) *If you're working on an N12 or N16 engine, MINI special tool no. 11 9 540 will be required.*
   b) *If you're working on an N14 engine, MINI special tool nos. 11 9 551 and 11 9 552 will be required.*
   c) *If you're working on an N18 engine, MINI special tool no. 11 7 440 will be required.*

These tools fit over machined flats near the left ends of the camshaft and bolt to the cylinder head to retain the camshafts in their proper positions **(see illustrations)**. Bolt the camshaft holding tools to the cylinder head.
6   If the camshaft holding tool cannot be installed over one or both camshafts, the VANOS unit or camshaft sprocket bolt at the front of the camshaft will have to be loosened and the camshaft turned slightly so the tool fits. To do this, place a 27 mm open-end wrench on the hex at the left end of the camshaft to counterhold the camshaft while the bolt is being loosened, then turn the camshaft with the wrench until the holding tool engages with the flats on the camshaft **(see illustration 6.8)**. **Caution:** *Whenever a VANOS unit/ camshaft sprocket bolt is loosened, it must be replaced with a new one.*
7   Work can now be performed without worry of valve or piston damage, and the camshafts are in proper phasing in relation to the crankshaft. This is the ONLY way to properly set the valve timing for these engines, and if not done correctly, severe and costly engine damage will result. **Note:** *Do not attempt to rotate the engine while the crankshaft/camshafts are locked in position. If the engine is to be left in this state for a long period of time, it is a good idea to place suit-* *able warning notices inside the vehicle, and in the engine compartment. This will reduce the possibility of the engine being accidentally cranked on the starter motor, which is likely to cause damage with the locking rods/ tool in place.*

4   Valve cover - removal and installation

Removal

Refer to illustrations 4.3, 4.4a, 4.4b and 4.6

1   If you're working on a Cooper model, remove the air filter housing (see Chapter 4).
2   Remove the ignition coils (see Chapter 5).
3   Pry up the ignition coil wiring harness cover **(see illustration)**, then detach the wiring harness from the clips underneath. Position the harness aside.
4   Detach the crankcase breather hoses from the valve cover **(see illustrations)**.

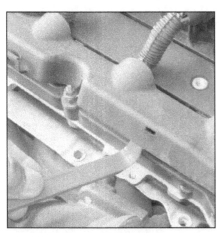

4.3 Pry up this cover, then disconnect the ignition coil wiring harness from the clips

4.4a Squeeze the retainer and disconnect the crankcase breather hose from the right rear corner of the valve cover

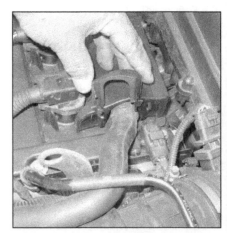

4.4b Pull off the retainer and detach the crankcase breather hose from the left end of the valve cover

4.6 Detach the wiring harness guide from its bracket at the left end of the cylinder head

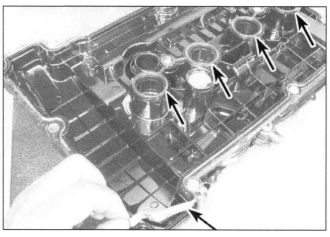

4.9 The valve cover gasket and spark plug hole seals can be reused if in good condition. Make sure they are seated in their grooves properly

5   Remove the Camshaft Position (CMP) sensor(s) from the valve cover (see Chapter 6).

6   Slide the wiring harness guide up and off its bracket at the left end of the cylinder head **(see illustration)**. Detach any other wiring harnesses or hoses from the valve cover and disconnect any electrical connectors that would interfere with removal of the cover.

7   Starting at the center of the valve cover and working outwards, remove the valve cover retaining fasteners. Remove the cover from the cylinder head.

Installation

Refer to illustrations 4.9 and 4.11

8   Thoroughly clean the gasket faces on the valve cover and the engine.

9   Check the valve cover gasket and spark plug tube seals **(see illustration)**. If they are in good condition, they can be reused.

10   Apply a thin coat of RTV sealant to the semi-circular cutout at the right end of the cylinder head.

11   Install the valve cover and bolts. Tighten the bolts a little at a time, in sequence **(see illustration)**, to the torque listed in this Chapter's Specifications.

12   The remainder of installation is the reverse of removal.

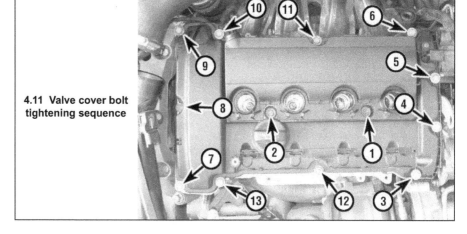

4.11 Valve cover bolt tightening sequence

turning while the pulley retaining bolts are being loosened on manual transaxle models, select 5th gear and have an assistant apply the brakes firmly. On automatic transaxle models, it will be necessary to remove the starter motor (see Chapter 5) and lock the driveplate with a suitable tool. If the engine has been removed from the vehicle, lock the flywheel ring gear as described in Section 12.

Do not attempt to lock the pulley by inserting a bolt/drill through the timing hole. If the locking pin is in position, temporarily remove it prior to loosening the pulley bolt, then install it once the bolt has been loosened.

3   Unscrew the three crankshaft pulley retaining bolts and remove the pulley from the center hub on the end of the crankshaft **(see illustrations)**.

---

5   Crankshaft vibration damper/ pulley - removal and installation

Caution: *This procedure is only for removing the vibration damper/pulley. DO NOT loosen the crankshaft pulley hub bolt unless the timing chain is to be removed, as crankshaft/ camshaft timing will be disturbed and will have to be reset as described in Section 3.*

Removal

Refer to illustrations 5.3a and 5.3b

1   Remove the drivebelt (see Chapter 1).

2   If required, to prevent the crankshaft

5.3a Unscrew the retaining bolts . . .

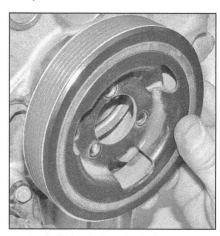

5.3b . . . and remove the crankshaft pulley

6.4a Insert the tool into the cylinder head where the chain tensioner fits

6.4b Screw the threaded center part of the tool in until it contacts the timing chain guide and hand-tighten

Installation

4   Locate the pulley on the hub on the end of the crankshaft, install the three retaining bolts and tighten them to the specified torque.
5   Install the drivebelt (see Chapter 1).

---

6   Timing chain, VANOS unit(s)/ sprocket and tensioner - general information, removal and installation

---

General information

Refer to illustrations 6.4a, 6.4b and 6.4c

1   The timing chain drives the camshafts from a toothed sprocket on the end of the crankshaft.
2   The chain should be replaced if a sprocket or chain is worn, indicated by excessive lateral play between the links and excessive noise in operation. It is wise to replace the chain in any case if the engine is to be disassembled for overhaul. Note that the rollers on a very badly worn chain may be slightly

grooved. To avoid future problems, if there is any doubt at all about the condition of the chain, replace it.
3   The timing chain and guides are removed as a complete assembly and are withdrawn out through the top of the cylinder head.
4   To check the wear of the timing chain, a special tool is required. Insert the tool into the cylinder head where the tensioner fits **(see illustration)**, and tighten the outer part. Run the threaded center part of the tool in until it contacts the timing chain guide and tighten it hand-tight **(see illustration)**. Lock the nut on the center thread and remove the special tool. Check the length of the tool to make sure that it does not exceed 68 mm **(see illustration)**.

Removal

Refer to illustrations 6.8, 6.11, 6.12, 6.13, 6.14a, 6.14b, 6.15a, 6.15b, 6.17a, 6.17b, 6.17c, 6.18a and 6.18b

5   Remove the valve cover (see Section 4).
6   Set the engine in the reference position for the crankshaft and camshaft (see Section 3).
7   Loosen the crankshaft pulley center hub retaining bolt. To prevent the crankshaft turn-

ing while the bolt is being loosened on manual transaxle models, select 5th gear and have an assistant apply the brakes firmly. On automatic transaxle models it will be necessary to remove the starter motor (see Chapter 5) and lock the driveplate with a suitable tool. If the engine has been removed from the vehicle, lock the flywheel ring gear as described in Section 12. *Do not* attempt to lock the pulley by inserting a bolt/drill through the timing hole. If the locking pin is in position (as shown in **illustration 3.4**), temporarily remove it prior to loosening the bolt, then install it once the bolt has been loosened. **Caution:** *If the crankshaft pulley hub bolt is extremely tight, it will be necessary to obtain a special holding tool with a long handle that bolts in place of the pulley.*
8   Holding the camshafts in position with an open-ended wrench on the flats on the camshaft, loosen the camshaft sprocket/ VANOS unit retaining bolts **(see illustration)**. **Caution:** *Obtain new VANOS unit/camshaft sprocket bolts; the old ones must NOT be reused.*
9   Remove the crankshaft pulley (see Section 5).

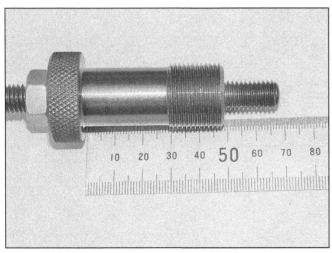

6.4c Make sure the length does not exceed 68 mm

6.8 Hold the camshafts in position and loosen the camshaft sprocket/VANOS unit retaining bolts

6.11 Remove the timing chain tensioner

6.12 Remove the guide rail from the cylinder head

6.13 Remove the engine mounting bracket

6.14a Remove the upper guide securing bolts . . .

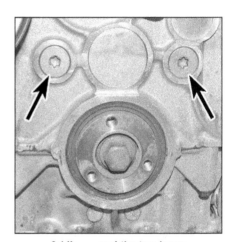

6.14b . . . and the two lower securing bolts

10   Remove the throttle body (see Chapter 4).
11   Remove the timing chain tensioner from the right-rear corner of the cylinder head **(see illustration)**. Before the tensioner is removed, make sure the camshafts and crankshaft are locked in position as described in Section 3.
12   Remove the chain guide rail from the top of the cylinder head **(see illustration)**.

13   Position a floor jack under the engine, placing a block of wood between the jack head and the oil pan. Take up the weight of the engine, remove the bolts/nuts and remove the right-hand engine mount and bracket **(see illustration)**.
14   Remove the timing chain guide retaining bolts **(see illustrations)**.

15   Remove the crankshaft pulley center bolt and hub **(see illustrations)**.
16   Withdraw the dipstick and remove it from the dipstick tube. The dipstick goes down through the chain guide; if this is not removed, it will prevent the timing chain assembly from being withdrawn out through the cylinder head.
17   Remove the camshaft sprocket/VANOS

6.15a Remove the crankshaft hub retaining bolt . . .

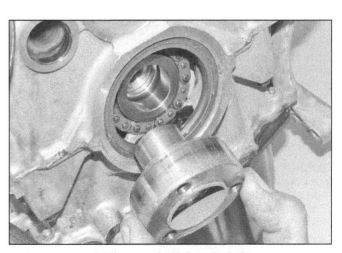

6.15b . . . and withdraw the hub

6.17a Remove the retaining bolts . . .

6.17b . . . then remove the intake . . .

6.17c . . . and exhaust camshaft VANOS units

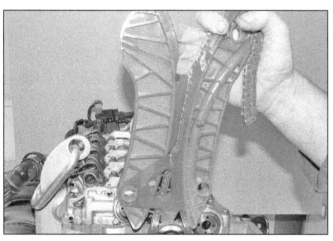

6.18a Withdraw the timing chain guide assembly . . .

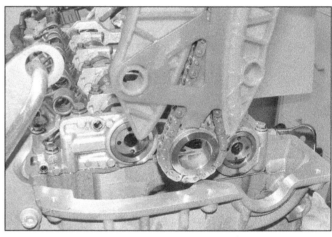

6.18b . . . complete with crankshaft sprocket

units from the ends of the camshafts, keeping the timing chain held in position **(see illustrations)**. **Caution:** *Do not mix-up the VANOS units.* **Note:** *N14 engines only have a VANOS unit on the intake camshaft. The exhaust camshaft uses a traditional timing sprocket.*

18 Lift the timing chain and guide assembly out through the top of the cylinder head, complete with crankshaft sprocket **(see illustrations)**.

19 Check the timing chain assembly carefully for any signs of wear. Replace it if there is the slightest doubt about its condition. If the engine is undergoing an overhaul, it is advisable to replace the chain as a matter of course, regardless of its apparent condition.

Installation

Refer to illustrations 6.21, 6.22, 6.26 and 6.27

20 Before installing, thoroughly clean the crankshaft and camshaft sprockets/VANOS units; they are a compression fit. The mating faces need to be clean and free from any oil when assembling. Remove all traces of oil with brake system cleaner.

21 Check that the VANOS units are in the rest position **(see illustration)**. If these are not aligned they will have to be replaced.

22 Lower the timing chain assembly down through the top of the cylinder head, complete with crankshaft sprocket **(see illustration)**.

23 Install the crankshaft pulley hub; making sure it engages correctly with the crankshaft timing chain and oil pump sprockets. Tighten the bolt to the torque listed in this Chapter's

Specifications, preventing the crankshaft from turning as carried out for removal. If the pin was removed from the lower crankcase/flywheel, reinstall it.

24 Install the timing chain guide retaining bolts and replace the seals/washer.

25 Check that the camshafts and crankshaft

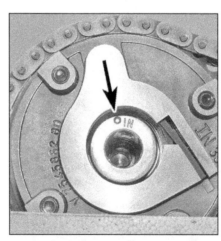

6.21 Check that the alignment marks are in position

6.22 Lower the crankshaft sprocket into position

6.26 Install the camshaft VANOS units/sprockets

6.27 Install the guide rail and bolts

are still locked in their correct positions as described in Section 3.

26 Install the camshaft sprocket/VANOS units to the ends of the camshafts, install the NEW bolts and tighten to the correct torque setting. Make sure the timing chain is located correctly around the sprockets **(see illustration)**. **Note:** *The camshaft sprockets are marked IN for intake and EX for exhaust.*

27 Install the chain guide rail to the top of the cylinder head **(see illustration)**.

28 Install the timing chain tensioner and tighten it to the torque listed in this Chapter's Specifications.

29 Remove the camshaft and crankshaft locking tools, and rotate the crankshaft four complete revolutions clockwise (viewed from the right-hand end of the engine). Realign the engine and install the locking tools to confirm that the timing is still aligned. If not, refer to Section 3. When correct, remove the timing locking tools from the engine. **Caution:** *Do not attempt to rotate the engine while the locking tools are in position.*

30 Install the right-hand engine mount (see Section 13).

31 Install the throttle body (see Chapter 4).

32 Install the crankshaft pulley (see Section 5).

33 Install the valve cover (see Section 4).

34 Install the engine oil dipstick and reconnect the battery negative cable.

---

7 Manifolds - removal and installation

Intake manifold

1 Disconnect the cable from the negative terminal of the battery (see Chapter 5).

2 Remove the air filter housing (see Chapter 4).

Cooper models

Refer to illustration 7.3

3 Disconnect the electrical connector from the VANOS solenoid at the right rear of the cylinder head **(see illustration)**.

4 Disconnect the electrical connector from the throttle body.

5 Raise the vehicle and support it securely on jackstands.

6 Remove the EVAP purge solenoid/fuel

tank vent valve (see Chapter 6).

7 Free the wires from the guide on the Valvetronic actuating motor at the left end of the manifold.

8 Remove the wiring harness bracket from the starter motor.

9 Remove the intake manifold nuts **(see illustration 7.3)** and detach the manifold from the cylinder head.

10 Clean the manifold and cylinder head mating surfaces. Install new seals around the intake manifold runner ports.

11 Installation is the reverse of removal. Tighten the manifold fasteners to the torque listed in this Chapter's Specifications.

Cooper S models

Refer to illustration 7.13

12 Disconnect the electrical connector from the intake manifold pressure sensor (see Chapter 6).

13 Remove the intake manifold nuts and detach the manifold from the head **(see illustration)**.

14 Mark and disconnect all hoses and electrical connectors from the underside of the manifold.

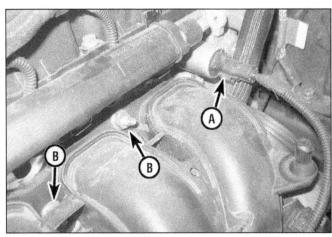

7.3 VANOS solenoid/electrical connector (A) and intake manifold mounting nuts (B, two of five shown) (Cooper models)

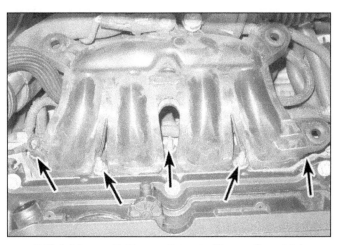

7.13 Intake manifold mounting nuts (Cooper S models)

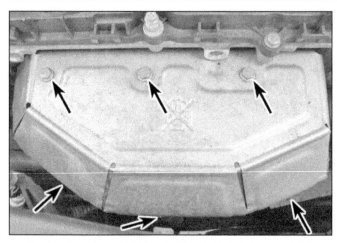

7.20 Exhaust manifold head shield bolts (Cooper models)

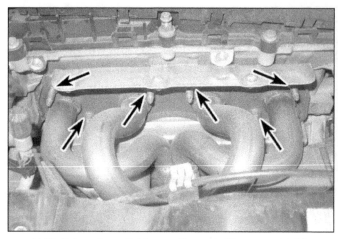

7.22 Exhaust manifold mounting nuts (Cooper models)

15 Clean the manifold and cylinder head mating surfaces. Install new seals around the intake manifold runner ports.

16 Installation is the reverse of removal. Tighten the manifold fasteners to the torque listed in this Chapter's Specifications.

Exhaust manifold

Warning: *Wait until the engine is completely cool before beginning this procedure.*

Note: *Apply penetrating oil to all exhaust system fasteners and allow it to soak in for awhile before attempting to remove them. When reassembling, apply anti-seize compound to the threads of the fasteners.*

17 Disconnect the cable from the negative terminal of the battery (see Chapter 5).

18 Set the radiator support panel in the service position (see Chapter 11).

Cooper models

Refer to illustrations 7.20 and 7.22

Note: *On these models, the primary catalytic converter is part of the exhaust manifold.*

19 Remove the upstream oxygen sensor (see Chapter 6).

20 Remove the exhaust manifold heat shield **(see illustration)**.

21 Raise the front of the vehicle and support it securely on jackstands, then remove the downstream oxygen sensor. Detach the exhaust pipe from the catalytic converter, then remove the fasteners and detach the converter from the engine block.

22 Remove the exhaust manifold nuts and detach the manifold from the cylinder head **(see illustration)**.

23 Remove the old gasket, then clean the manifold and cylinder head mating surfaces.

24 Installation is the reverse of the removal procedure. Be sure to use a new gasket. If the mounting nuts were difficult to remove, replace them with new ones (also replace the studs if necessary).

Cooper S models

Refer to illustration 7.29, 7.30 and 7.32

Note: *This procedure includes removal of the turbocharger.*

25 Drain the cooling system (see Chapter 1) and remove the coolant expansion tank (see Chapter 3).

26 Remove the (primary) catalytic converter (see Chapter 6).

27 Loosen the clamps and detach the intake duct and charge air duct from the turbocharger.

28 Detach the radiator hose from the thermostat housing (see Chapter 3).

29 Detach the coolant and oil lines from the turbocharger **(see illustration)**.

30 Disconnect the vacuum line and electrical connector from the left side of the turbocharger. Also remove the bracket bolt for the coolant pipe **(see illustration)**.

31 Remove the support bracket from the underside of the exhaust manifold.

32 Remove the exhaust manifold mounting nuts and detach the manifold and turbocharger assembly from the cylinder head **(see illustration)**.

33 If necessary, remove the nuts and detach the turbocharger from the manifold.

34 Installation is the reverse of the removal

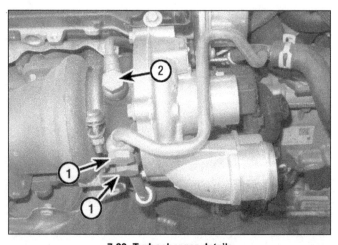

7.29 Turbocharger details

1   *Coolant line fittings*
2   *Oil feed line fitting (return line fitting on underside of turbo)*

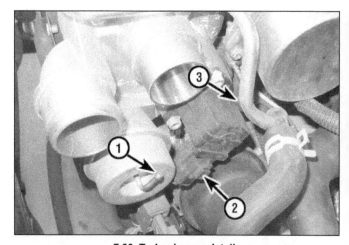

7.30 Turbocharger details

1   *Vacuum hose fitting*   3   *Coolant pipe bracket bolt*
2   *Electrical connector*

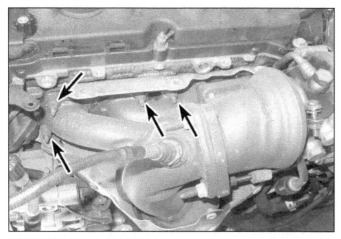

7.32 Exhaust manifold mounting nuts (four of ten shown here)

8.4 Disconnect the vacuum pipe

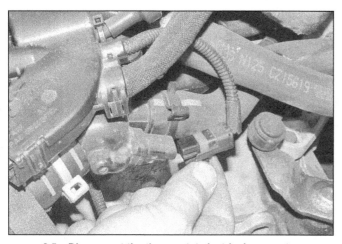

8.5a Disconnect the thermostat electrical connector

8.5b Disconnect the oil pressure sensor and temperature sensor wiring connectors

procedure. Be sure to use a new gasket(s). If the mounting nuts were difficult to remove, replace them with new ones (also replace the studs if necessary).

35 Refill the cooling system (see Chapter 1).

8 Cylinder head - removal and installation

Warning: *Wait until the engine is completely cool before beginning this procedure.*

Caution: *Do not attempt to remove the camshafts or Valvetronic mechanism from the cylinder head - special tools are required.*

Removal

Refer to illustrations 8.4, 8.5a, 8.5b, 8.6, 8.7, 8.8, 8.9, 8.10, 8.11a, 8.11b, 8.14, 8.15a, 8.15b, 8.16 and 8.17

1 Disconnect the cable from the negative terminal of the battery (see Chapter 5).

2 Apply the parking brake, then raise the front of the vehicle and support it on jackstands. Remove the engine undershield.

3 Drain the cooling system as described in Chapter 1.

4 Disconnect the pipe from the vacuum pump, release the securing clip and move the pipe to one side **(see illustration)**.

5 Disconnect the wiring connectors from the thermostat, temperature sensor and oil pressure sensor on the end of the cylinder head **(see illustrations)**.

6 Disconnect the wiring connectors from the camshaft position sensors at the transaxle end of the valve cover **(see illustration)**.

7 Slide the wiring harness bracket upwards from the end of the valve cover and move it to one side **(see illustration)**.

8 Disconnect the four coolant hoses from

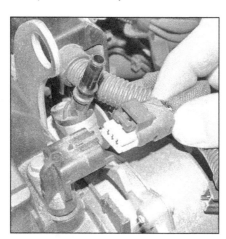

8.6 Disconnect the camshaft sensor electrical connector(s)

8.7 Unclip the wiring harness mounting bracket

8.8 Disconnect the coolant hoses

8.9 Release the wire retaining clip at the rear of the housing

8.10 Detach the thermostat housing from the cylinder head

8.11a Remove the retaining bolts . . .

8.11b . . . and, using an Allen key, withdraw the Valvetronic actuator from the cylinder head

the thermostat housing **(see illustration)**.

9   Release the retaining clip at the rear of the cylinder head, where the thermostat housing joins the pipe to the water pump **(see illustration)**.

10   Remove the thermostat housing from the cylinder head and detach it from the water pump feed pipe **(see illustration)**.

11   Disconnect the wiring connector from the Valvetronic actuator at the left-hand rear of the cylinder head. Remove the actuator retaining bolts and detach it from the cylinder head by using a 4 mm Allen key to turn the center shaft counterclockwise while withdrawing the actuator from the cylinder head **(see illustrations)**.

12   Remove intake and exhaust manifolds (see Section 7).

13   Support the engine with a floor jack and block of wood positioned under the oil pan, then remove the timing chain assembly (see Section 6).

14   If not already done, remove the mounting bracket from the right-hand end of the cylinder head. Release the wiring harness and fuel lines from the securing clips where required **(see illustration)**.

15   Disconnect the electrical connectors from the intake and exhaust VANOS solenoids at the front and rear of the cylinder head at the timing chain end **(see illustrations)**.

16   Remove the cylinder head bolt at the right-hand rear of the cylinder head **(see illustration)**.

17   Remove the two cylinder head bolts at the timing chain end of the cylinder head **(see illustration)**.

18   Progressively loosen and remove the ten main cylinder head bolts, working in an order opposite that of the tightening sequence **(see illustration 8.30)**.

19   With all the cylinder head bolts removed, the joint between the cylinder head and gas-

8.14 Remove the engine mounting bracket

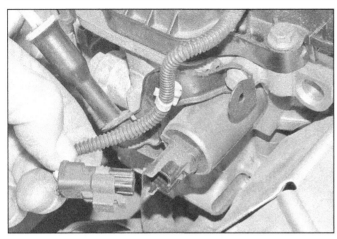

8.15a Disconnect the exhaust VANOS solenoid electrical connector . . .

8.15b . . . and the intake VANOS solenoid electrical connector

ket, and the cylinder block/crankcase must now be broken. Carefully rock the cylinder head free towards the front of the car. Do not try to swivel the head on the cylinder block/crankcase; it is located by dowels. When the joint is broken, lift the cylinder head away. Use a hoist or seek assistance if possible, as it is a heavy assembly. Remove the gasket from the top of the block, noting the two locating dowels. If the locating dowels are a loose fit, remove them and store them with the head for safe-keeping. Do not discard the gasket; it will be needed for identification purposes.

Preparation for installation

Refer to illustration 8.22

20   The mating faces of the cylinder head and cylinder block/crankcase must be perfectly clean before refitting the head. Use a hard plastic or wooden scraper to remove all traces of gasket and carbon and also clean the piston crowns. Make sure that the carbon is not allowed to enter the oil and water passages - this is particularly important for the lubrication system, as carbon could block the oil supply to the engine's components. Using

adhesive tape and paper, seal the water, oil and bolt holes in the cylinder block/crankcase. To prevent carbon entering the gap between the pistons and bores, smear a little grease in the gap. After cleaning each piston, use a small brush to remove all traces of grease and carbon from the gap, and then wipe away the remainder with a clean rag. Clean all the pistons in the same way.

21   Check the mating surfaces of the cylinder block/crankcase and the cylinder head for nicks, deep scratches and other damage. If slight, they may be removed carefully with a file, but if excessive, machining may be the only alternative to replacement. If warpage of the cylinder head gasket surface is suspected, use a straight-edge to check it for distortion.

22   Obtain a new cylinder head gasket before starting the installation procedure. There are different thicknesses available - check the old head gasket for markings on the rear right-hand side of the gasket **(see illustration)**.

23   Due to the stress that the cylinder head bolts are under, it is imperative that they be replaced, regardless of their apparent condition.

Installation

Refer to illustration 8.30

24   Wipe clean the mating surfaces of the cylinder head and cylinder block/crankcase.

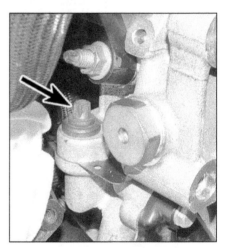

8.16 Right-rear cylinder head securing bolt

8.17 Right-end cylinder head securing bolts

8.22 Cylinder head gasket thickness markings

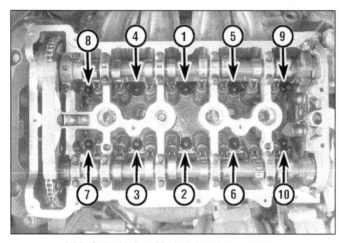

8.30 Cylinder head bolt tightening sequence

9.4 Remove these two bolts at the rear of the pan

Check that the two locating dowels are in position at each end of the cylinder block/crankcase surface.

25 Position a new gasket on the cylinder block/crankcase surface.

26 Check that the crankshaft pulley and camshafts are still at their locked positions (see Section 3). If the camshafts have been removed (for example, at a machine shop to have cylinder head work performed), position them with their IN and EX marks facing up and attach the camshaft holding fixtures.

27 With the aid of an assistant, carefully lower the cylinder head assembly onto the block, aligning it with the locating dowels.

28 Place the cylinder head bolt washers on the new bolts. Don't wipe the new bolts off or clean off the coating.

29 Carefully enter each of the new bolts and washers into their relevant hole (*do not drop them in*), then screw them in finger-tight.

30 Working in the indicated sequence **(see illustration),** tighten the ten main cylinder head bolts to their Step 1 torque setting, then tighten the two bolts at the timing chain end of the cylinder head and the one at the rear of the head to their Step 1 torque setting.

31 Once all the bolts have been tightened to

their Step 1 torque setting, proceed to tighten them through the remaining steps as given in the Specifications. It is recommended that an angle-measuring gauge be used for the angle-tightening stages. However, if a gauge is not available, use white paint to make alignment marks between the bolt head and cylinder head prior to tightening; the marks can then be used to check that the bolt has rotated sufficiently. Each step must be completed in one movement without stopping.

32 The remainder of the installation procedure is a reversal of removal, referring to the relevant Chapters or Sections as required. On completion, refill the cooling system and change the engine oil and filter (see Chapter 1).

---

9 Oil pan - removal and installation

Removal

Refer to illustrations 9.4 and 9.6

1 Apply the parking brake, then raise the front of the vehicle and support it securely on jackstands. Remove the under-vehicle splash shield.

2 Drain the engine oil, then clean and install the engine oil drain plug, tightening it securely. If the engine is nearing its service interval when the oil and filter are due for replacement, it is recommended that the filter is also removed, and a new one installed. After reassembly, the engine can then be refilled with fresh oil. Refer to Chapter 1 for further information.

3 Withdraw the engine oil dipstick from the guide tube.

4 Remove the mounting plate from the rear of the oil pan **(see illustration).**

5 Progressively loosen and remove all of the oil pan retaining bolts. Make a note of the installed positions of the bolts, as they may be different lengths. This will avoid the possibility of installing the bolts in the wrong locations.

6 Break the joint by striking the oil pan with the palm of your hand. Lower the oil pan, and withdraw it from underneath the vehicle **(see illustration).** While the oil pan is removed, take the opportunity to check the oil pump pick-up/strainer for signs of clogging or splitting. If necessary, remove the pump as described in Section 10, and clean or replace the strainer. If required, unbolt the oil baffle plate from the bottom of the lower crankcase, noting which way it is installed.

Installation

Refer to illustration 9.10

7 If removed, install the baffle plate to the lower crankcase and tighten the bolts securely.

8 If removed, install the oil pump and pick-up/strainer with reference to Section 10.

9 Clean all traces of sealant/gasket from the mating surfaces of the lower crankcase and oil pan, then use a clean rag to wipe out the oil pan and the engine's interior.

10 Ensure that the oil pan mating surfaces are clean and dry, then apply a thin coating of RTV sealant to the oil pan mating surface **(see illustration).**

11 Install the oil pan to the lower crankcase, insert the bolts and finger-tighten them at this

9.6 Remove the oil pan

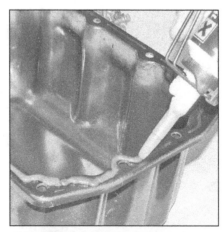

9.10 Apply a thin bead of RTV sealant to the oil pan mating surface

10.2 Unclip the cover . . .

10.3 . . . and remove the sprocket retaining bolt

stage, so that it is still possible to move the oil pan. Make sure the bolts are installed in their correct locations.

12   Install the mounting plate to the rear of the oil pan and tighten all the retaining bolts.

13   Check that the oil drain plug is tightened securely, then install the engine undershield (where applicable) and lower the vehicle.

14   Install the dipstick and refill the engine with oil (see Chapter 1).

10 Oil pump - removal, inspection and installation

Removal

Refer to illustrations 10.2, 10.3 and 10.4

1   Remove the oil pan as described in Section 9.

2   Unclip the cover from the oil pump drive sprocket **(see illustration)**.

3   Remove the drive sprocket retaining bolt **(see illustration)**, then withdraw the sprocket

from the drive chain. If you're working on a 2011 model (N16 or N18 engine), unplug the electrical connector from the wiring harness, then feed the harness and grommet into the engine block.

4   Remove the oil pump retaining bolts and withdraw the pump from the bottom of the main bearing ladder **(see illustration)**.

Inspection

Refer to illustration 10.7

5   At the time of writing, checking specifications for the oil pump were not available. Clean the pump and inspect it for damage and excessive wear.

6   Thoroughly clean the oil pump strainer with a suitable solvent, and check it for signs of clogging or splitting. If the strainer is damaged, the strainer and cover assembly must be replaced.

7   If the oil pump drive chain needs replacing, the timing chain assembly will need to be removed first as described in Section 6. The chain is around a sprocket on the end of

the crankshaft, at the rear of the crankshaft sprocket for the timing chain **(see illustration)**.

Installation

8   Clean the mating surfaces of the oil pump and cylinder block and install the retaining bolts. Tighten them to the torque listed in this Chapter's Specifications.

9   Locate the sprocket onto the pump, making sure it is located correctly in the drive chain. Tighten the sprocket retaining bolt securely.

10   Install the sprocket cover to the end of the oil pump.

11   Install the oil pan as described in Section 9.

12   Before starting the engine, prime the oil pump as follows. Disconnect the electrical connectors from the ignition coils, then turn the engine over on the starter motor until the oil pressure warning light goes out. Reconnect the electrical connectors on completion.

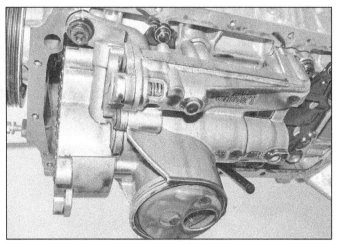

10.4 Remove the oil pump housing retaining bolts

10.7 Oil pump drive chain sprocket

11.2 Check the installed depth of the oil seal

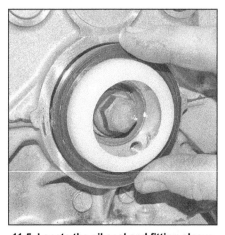

11.5 Locate the oil seal and fitting sleeve over the end of the crankshaft

11.10a Drill a hole . . .

11   Crankshaft oil seals - replacement

Crankshaft front oil seal

Refer to illustrations 11.2 and 11.5

1   Remove the crankshaft pulley, with reference to Section 5.
2   Check the depth of the seal in the engine casing before removing **(see illustration)**.
3   Punch or drill two small holes opposite each other in the seal. Screw a self-tapping screw into each, and pull on the screws with pliers to extract the seal **(see illustrations 11.10a and 11.10b)**. Alternatively, the seal can be levered out of position. Use a flat-bladed screwdriver, and take great care not to damage the crankshaft shoulder or seal housing.
4   Clean the seal housing, and polish off any burrs or raised edges which may have caused the seal to fail in the first place.
5   Lubricate the lips of the new seal with clean engine oil, and carefully locate the seal on the end of the crankshaft. The new seal will normally be supplied with a plastic fitting sleeve to protect the seal lips as the seal is installed.

If so, lubricate the sleeve and locate it over the end of the crankshaft **(see illustration)**.
6   Install the new seal using a seal driver or suitable tubular drift, which bears only on the hard outer edge of the seal. Tap the seal into position, to the same depth in the housing as the original was prior to removal.
7   Wash off any traces of oil, then install the crankshaft pulley as described in Section 5.

Rear main oil seal

Refer to illustrations 11.10a, 11.10b, 11.11, 11.12a, 11.12b and 11.13

8   Remove the flywheel/driveplate and crankshaft timing plate (see Section 12).
9   Make a note of the correct installed depth of the seal in its housing **(see illustration 11.2)**.
10   Punch or drill two small holes opposite each other in the seal. Screw a self-tapping screw into each, and pull on the screws with pliers to extract the seal **(see illustrations)**.
11   Clean the seal housing, and polish off any burrs or raised edges which may have caused the seal to fail in the first place **(see illustration)**.

11.10b . . . then use a self-tapping screw and pliers to extract the oil seal

12   Lubricate the lips of the new seal with clean engine oil, and carefully locate the seal on the end of the crankshaft. The new seal will normally be supplied with a plastic fitting sleeve to protect the seal lips as the seal is installed. If so, lubricate the sleeve and locate it over the end of the crankshaft **(see illustrations)**.
13   Before driving the seal fully into position,

11.11 Clean out the recess

11.12a The new oil seal comes with a protective sleeve . . .

11.12b . . . which fits over the end of the crankshaft

11.13 Apply a small amount of sealant at the casing joints

12.2 Use a tool to lock the flywheel ring gear and prevent rotation

put a small amount of sealant at each side of the seal, where the upper and lower crankcases meet **(see illustration)**. Drive the seal into position, to the same depth in the housing as the original was prior to removal.

14   Clean off any excess sealant or oil, then install the crankshaft timing plate and flywheel/driveplate (see Section 12).

12 Flywheel/driveplate - removal, inspection and installation

Removal
Flywheel

Refer to illustrations 12.2, 12.3a, 12.3b and 12.4

1   Remove the transaxle as described in Chapter 7A, then remove the clutch assembly as described in Chapter 8.

2   Prevent the crankshaft from turning by using a flywheel locking tool **(see illustration)**. Alternatively, lock the flywheel with a wide-bladed screwdriver between the ring gear teeth and a bolt installed into the engine

block. *Do not* attempt to lock the flywheel in position using the crankshaft locking pin described in Section 3.

3   Remove the flywheel retaining bolts, then remove the flywheel from the end of the crankshaft **(see illustrations)**. Be careful not to drop it; it is heavy. If the flywheel locating dowel is a loose fit in the crankshaft end, remove it and store it with the flywheel for safe-keeping. Discard the flywheel bolts; new ones must be used on installation.

4   If required, remove the crankshaft TDC timing plate from the end of the crankshaft **(see illustration)**.

Driveplate

5   Remove the transaxle as described in Chapter 7B. Lock the driveplate as described in Step 2. Mark the relationship between the torque converter plate and the driveplate, and loosen all the driveplate retaining bolts.

6   Remove the retaining bolts, along with the torque converter plate and (if equipped) the two shims (one on each side of the torque converter plate). Note that the shims are of different thicknesses, the thicker one being on the outside of the torque converter plate. Dis-

12.3a Remove the securing bolts . . .

card the driveplate retaining bolts; new ones must be used on installation.

7   Remove the driveplate from the end of the crankshaft. If the locating dowel is a loose fit in the crankshaft end, remove it and store it with the driveplate for safe-keeping.

12.3b . . . and remove the flywheel

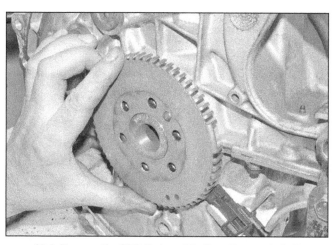

12.4 Remove the TDC timing plate from the crankshaft

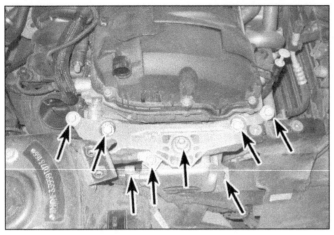

13.6 Right-side engine mount and bracket fasteners (horizontal bracket-to-cylinder head bolts not visible)

13.7 Remove the engine mounting bracket

8   If required, remove the crankshaft TDC timing plate from the end of the crankshaft.

Inspection

9   On models with a manual transaxle, examine the flywheel for scoring of the clutch face, and for wear or chipping of the ring gear teeth. If the clutch face is scored, the flywheel may be surface-ground, but replacement is preferable. If the ring gear is worn or damaged, the flywheel must be replaced, as it is not possible to replace the ring gear separately.

10   On models with an automatic transaxle, check the driveplate carefully for signs of distortion. Look for any hairline cracks around the bolt holes or radiating outwards from the center, and inspect the ring gear teeth for signs of wear or chipping. If any sign of wear or damage is found, the driveplate must be replaced.

11   Check the crankshaft TDC timing plate for any damage; make sure that none of the teeth around the circumference of the disc are bent.

Installation

Flywheel

12   Clean the mating surfaces of the flywheel and crankshaft. Remove any remaining locking compound from the threads of the crankshaft holes, using the correct-size tap, if available.

13   If the new flywheel retaining bolts are not supplied with their threads already pre-coated, apply a suitable thread-locking compound to the threads of each bolt.

14   If removed, install the crankshaft TDC timing plate to the end of the crankshaft.

15   Ensure the locating dowel is in position. Install the flywheel, locating it on the dowel, and install the new retaining bolts.

16   Lock the flywheel and tighten the retaining bolts to the specified torque and angle.

17   Install the clutch as described in Chapter 8. Remove the flywheel locking tool, and

install the transaxle as described in Chapter 7A.

Driveplate

18   Carry out the operations described in Steps 12 and 13.

19   If removed, install the crankshaft TDC timing plate to the end of the crankshaft.

20   Locate the driveplate on its locating dowel.

21   Install the torque converter plate, with the thinner shim positioned behind the plate and the thicker shim on the outside, and align the marks made prior to removal.

22   Install the new retaining bolts, then lock the driveplate using the method employed on disassembly. Tighten the retaining bolts to the specified torque wrench setting and angle.

23   Remove the driveplate locking tool, and install the transaxle as described in Chapter 7B.

---

13 Powertrain mounts - inspection and replacement

Inspection

1   If improved access is required, raise the front of the car and support it securely on jackstands.

2   Check the rubber portion of each mount to see if it is cracked, hardened or separated from the metal at any point; replace the mount if any such damage or deterioration is evident.

3   Check that all the mount fasteners are securely tightened.

4   Using a large screwdriver or a crowbar, check for wear in each mount by carefully prying against it to check for freeplay. While some freeplay is to be expected even from new components, excessive wear should be obvious. If excessive freeplay is found, check first that the fasteners are secure, and then replace any worn components.

Replacement

Right-hand mount

Refer to illustrations 13.6 and 13.7

5   Place a jack beneath the engine, with a block of wood on the jack head. Raise the jack until it is supporting the weight of the engine. On Cooper S models, remove the right-side charge air duct.

6   Remove the bolts securing the mount to the body, and the mounting bracket to the bracket bolted to the cylinder head **(see illustration)**.

7   If required, remove the bracket from the cylinder head **(see illustration)**. Release the wiring harness from the retaining clips as it is removed.

8   Check for signs of wear or damage on all components, and replace as necessary.

9   On reassembly, install the bracket to the cylinder head, tightening the bolts to the specified torque.

10   Install the mount and mounting bracket and tighten its retaining bolts securely.

11   Remove the jack from under the engine.

Left-hand (transaxle mount)

Refer to illustration 13.13

12   Place a jack beneath the transaxle, with a block of wood on the jack head. Raise the jack until it is supporting the weight of the transaxle. Disconnect the cable from the negative terminal of the battery (see Chapter 5), then unbolt and reposition the underhood fuse and relay box.

13   Remove the mounting bracket-to-body bolts, the refrigerant line bracket and the bracket-to-mount nut **(see illustration)**.

14   Remove the four mount-to-transaxle bolts and detach the mount from the transaxle.

15   Check carefully for signs of wear or damage on all components, and replace them where necessary.

16   Installation is the reverse of removal, tightening all fasteners securely.

17   Remove the jack from underneath the transaxle, then install the battery (see Chapter 5).

Rear lower engine stabilizer (lower link)

Refer to illustration 13.19

18   If not already done, firmly apply the parking brake, then raise the front of the vehicle and support it securely on jackstands.

19   Remove the bolts securing the lower link to the bracket on the transaxle and subframe **(see illustration)**, then remove the link.

20   Check carefully for signs of wear or damage on all components, and replace them where necessary.

21   Install the lower link, tightening both bolts securely.

22   Lower the vehicle.

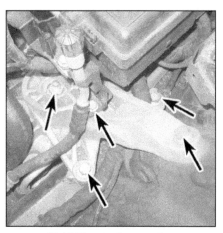

13.13 Transaxle mounting bracket fasteners

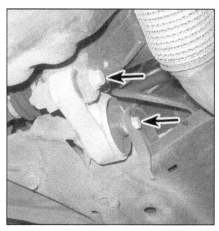

13.19 Rear lower engine stabilizer (lower link) mounting bolts

Notes

Chapter 2 Part C
General engine overhaul procedures

Contents

Specifications

General

Bore x Stroke	3.03 x 3.38 inches (77.00 x 85.80 mm)
Minimum compression pressure	
Cooper	166 psi (11.5 bar)
Cooper S	116 psi (8.0 bar)
Minimum oil pressure at normal operating temperature	
2006 and earlier models	
Idle speed	3.5 psi (0.25 bar)
At 3000 rpm	24.5 to 80 psi (1.7 to 5.5 bar)
2007 and later models	
Idle speed	10 psi (0.7 bar)
At 3000 rpm	
N12 engine	21.5 to 43.5 psi (1.5 to 3.0 bar)
N14 engine	24.5 to 43.5 psi (1.7 to 3.0 bar)
N16 engine	Not available
N18 engine	16.5 to 93.5 psi (1.15 to 6.45 bar)

Torque specifications

Ft-lbs (unless otherwise indicated) **Nm**

Note: *One foot-pound (ft-lb) of torque is equivalent to 12 inch-pounds (in-lbs) of torque. Torque values below approximately 15 foot-pounds are expressed in inch-pounds, because most foot-pound torque wrenches are not accurate at these smaller values.*

2006 and earlier models
   Connecting rod bearing cap bolts*

	Ft-lbs	Nm
Stage 1	15	20
Stage 2	Tighten an additional 90 degrees	
Main bearing bedplate bolts*		
M10...	44	60
M8...	26	35
2007 and later models		
Connecting rod bearing cap bolts*		
Step 1	43 in-lbs	5
Step 2	132 in-lbs	15
Step 3	Tighten an additional 130 degrees	
Main bearing bedplate bolts		
M9 bolts		
Step 1	22	30
Step2..	Tighten an additional 150 degrees	
M6 bolts	79 in-lbs	9
Oil spray nozzles (turbo models).......................................	15	20

* *Use new bolts*

An engine block being bored. An engine rebuilder will use
special machinery to recondition the cylinder bores

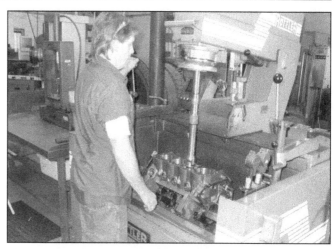

If the cylinders are bored, the machine shop will normally hone
the engine on a machine like this

1   What is an overhaul and when is it needed?

An engine overhaul is the reconditioning of internal components to like-new condition. If properly maintained, a modern engine can last 200,000 miles or more. If neglected, abused or overheated, an engine can fail very quickly. Here are some signs of a worn or damaged engine:

Oil consumption - If you're adding oil frequently between oil changes, and there are no leaks, it is a likely sign of excessive internal engine wear. You might also see bluish gray smoke from the tailpipe at times.

Low oil pressure - See Section 3
Low compression - See Section 4

Poor running and/or lack of power - These symptoms are generally caused by problems with the ignition, fuel, or engine-management systems. However, worn or damaged piston rings, camshaft lobes, valves or valvetrain components can also cause an engine to run poorly. A compres-

sion test will often indicate these internal engine problems.

Noises - Knocking or tapping noises from inside the engine indicate worn components. While valve components can sometimes be adjusted to eliminate tapping noises, knocking noises generally indicate the need for an overhaul.

Overhauling the internal components on today's engines is a difficult and time-consuming task which requires a significant number of specialty tools. It is best left to a professional engine rebuilder who will handle the inspection of your old parts and help you choose the best course.

2   Engine rebuilding alternatives

Depending on the time and money available, you have a number of options for getting a strong-running engine back in your car:

Individual parts - If inspection reveals most of your engine's internal parts to be in re-useable condition, purchasing individual

A crankshaft having a main bearing
journal ground

parts and having a rebuilder overhaul your engine may be the most economical alternative. However, overhauling an engine takes time - maybe weeks. If you're in a hurry, it

A machinist checks for a bent connecting
rod, using specialized equipment

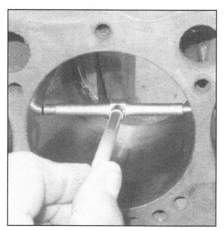

A bore gauge being used to check the
main bearing bore

Uneven piston wear like this indicates a
bent connecting rod

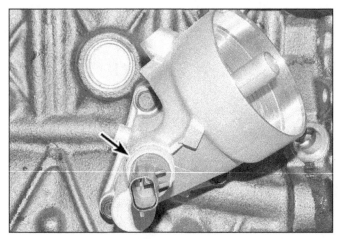

3.2a Oil pressure sending unit location - Mk I models

3.2b Oil pressure sending unit location - Mk II models

may make more sense to exchange your worn engine for one that's ready to go.

Short block - A short block consists of an engine block with a crankshaft and piston/ connecting rod assemblies already installed. All new bearings are used and all clearances will be correct. The existing camshafts, valve train components, cylinder head and external parts are bolted to the short block with little or no machine shop work necessary.

Long block - A long block consists of a short block plus an oil pump, oil pan, cylinder head, valve cover, camshaft and valve train components, and timing components. All components are installed with new bearings, seals and gaskets. The installation of manifolds and external parts is all that's necessary.

Low mileage used engines - Some companies offer low-mileage used engines, which is a very cost-effective way to get your vehicle running again. These engines often come from vehicles that have been wrecked or from other countries that have a higher vehicle turnover rate. Used engines frequently have warranties.

Give careful thought to which alternative is best for you and discuss the situation with local automotive machine shops, auto parts dealers and experienced rebuilders before ordering or purchasing replacement parts.

3   Oil pressure check

Refer to illustrations 3.2a and 3.2b

1   Low engine oil pressure can be a sign of an engine in need of rebuilding. A low oil pressure indicator (often called an "idiot light") is not a test of the oiling system. Such indicators only come on when the oil pressure is dangerously low. Even a factory oil pressure gauge in the instrument panel is only a relative indication, although much better for driver information than a warning light. A better test is with a mechanical (not electrical) oil pressure gauge.

2   Locate the oil pressure indicator sending unit:

a) *On Mk I models, the sending unit is located on the oil filter housing* **(see illustration)**.

b) *On Mk II models, the sending unit is located on the left end of the cylinder head* **(see illustration)**.

3   Unscrew and remove the oil pressure sending unit, then screw in the hose for your oil pressure gauge. If necessary, install an adapter fitting. Use Teflon tape or thread sealant on the threads of the adapter and/or the fitting on the end of your gauge's hose.

4   Check the oil pressure with the engine running (normal operating temperature) at the specified engine speed, and compare it to this Chapter's Specifications. If it's extremely low, the bearings and/or oil pump are probably worn out.

4   Cylinder compression check

Refer to illustration 4.5

1   A compression check will tell you what mechanical condition the upper end of your engine (pistons, rings, valves, head gaskets) is in. Specifically, it can tell you if the compression is down due to leakage caused by worn piston rings, defective valves and seats or a blown head gasket. **Note:** *The engine must be at normal operating temperature and the battery must be fully charged for this check.*

2   Begin by cleaning the area around the spark plugs before you remove them (compressed air should be used, if available). The idea is to prevent dirt from getting into the cylinders as the compression check is being done.

3   Remove all of the spark plugs from the engine (see Chapter 1).

4   Disable the fuel pump circuit by removing the fuel pump fuse (see Chapter 4, Section 2).

5   Install a compression gauge in the spark plug hole **(see illustration)**.

6   Crank the engine over at least seven compression strokes and watch the gauge.

The compression should build up quickly in a healthy engine. Low compression on the first stroke, followed by gradually increasing pressure on successive strokes, indicates worn piston rings. A low compression reading on the first stroke, which doesn't build up during successive strokes, indicates leaking valves or a blown head gasket (a cracked head could also be the cause). Deposits on the undersides of the valve heads can also cause low compression. Record the highest gauge reading obtained.

7   Repeat the procedure for the remaining cylinders and compare the results to this Chapter's Specifications.

8   Add some engine oil (about three squirts from a plunger-type oil can) to each cylinder, through the spark plug hole, and repeat the test.

9   If the compression increases after the oil is added, the piston rings are definitely worn. If the compression doesn't increase significantly, the leakage is occurring at the valves or head gasket. Leakage past the valves may be caused by burned valve seats and/or faces or warped, cracked or bent valves.

10   If two adjacent cylinders have equally low compression, there's a strong possibility that the head gasket between them is blown. The appearance of coolant in the combustion chambers or the crankcase would verify this condition.

11   If one cylinder is slightly lower than the others, and the engine has a slightly rough idle, a worn lobe on the camshaft could be the cause.

12   If the compression is unusually high, the combustion chambers are probably coated with carbon deposits. If that's the case, the cylinder head(s) should be removed and decarbonized.

13   If compression is way down or varies greatly between cylinders, it would be a good idea to have a leak-down test performed by an automotive repair shop. This test will pinpoint exactly where the leakage is occurring and how severe it is.

5   Vacuum gauge diagnostic checks

Refer to illustration 5.6

1   A vacuum gauge provides inexpensive but valuable information about what is going on in the engine. You can check for worn rings or cylinder walls, leaking head or intake manifold gaskets, incorrect carburetor adjustments, restricted exhaust, stuck or burned valves, weak valve springs, improper ignition or valve timing and ignition problems.

2   Unfortunately, vacuum gauge readings are easy to misinterpret, so they should be used in conjunction with other tests to confirm the diagnosis.

3   Both the absolute readings and the rate of needle movement are important for accurate interpretation. Most gauges measure vacuum in inches of mercury (in-Hg). The following references to vacuum assume the diagnosis is being performed at sea level. As elevation increases (or atmospheric pressure decreases), the reading will decrease. For every 1,000 foot increase in elevation above approximately 2,000 feet, the gauge readings will decrease about one inch of mercury.

4   Connect the vacuum gauge directly to the intake manifold vacuum, not to ported (throttle body) vacuum. Be sure no hoses are left disconnected during the test or false readings will result.

5   Before you begin the test, allow the engine to warm up completely. Block the wheels and set the parking brake. With the transmission in Park, start the engine and allow it to run at normal idle speed. **Warning:** *Keep your hands and the vacuum gauge clear of the fans.*

6   Read the vacuum gauge; an average, healthy engine should normally produce about 17 to 22 in-Hg with a fairly steady needle **(see illustration)**. Refer to the following vacuum gauge readings and what they indicate about the engine's condition:

7   A low steady reading usually indicates a leaking gasket between the intake manifold and cylinder head(s) or throttle body, a leaky vacuum hose, late ignition timing or incorrect camshaft timing.

8   If the reading is three to eight inches below normal and it fluctuates at that low reading, suspect an intake manifold gasket leak at an intake port or a faulty fuel injector.

9   If the needle has regular drops of about two-to-four inches at a steady rate, the valves are probably leaking. Perform a compression check or leak-down test to confirm this.

10   An irregular drop or down-flick of the needle can be caused by a sticking valve or an ignition misfire. Perform a compression check or leak-down test and read the spark plugs **(see inside rear cover)**.

11   A rapid vibration of about four in-Hg vibration at idle combined with exhaust smoke indicates worn valve guides. Perform a leak-down test to confirm this. If the rapid vibration occurs with an increase in engine speed, check for a leaking intake manifold gasket or head gasket, weak valve springs, burned valves or ignition misfire.

12   A slight fluctuation, say one inch up and down, may mean ignition problems. Check all

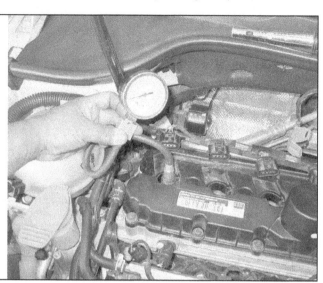

4.5 Use a compression gauge with a threaded fitting for the spark plug hole, not the type that requires hand pressure to maintain the seal (typical)

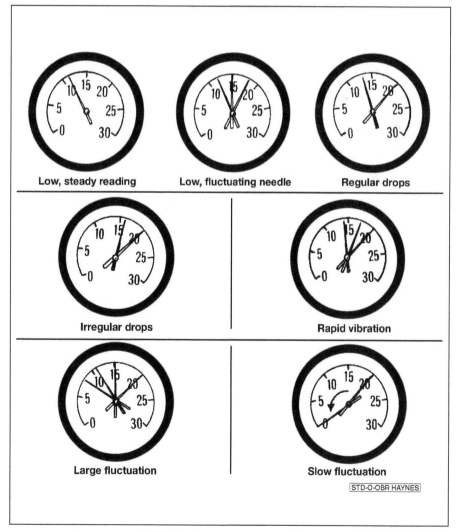

| Low, steady reading | Low, fluctuating needle | Regular drops |

| Irregular drops | Rapid vibration |

| Large fluctuation | Slow fluctuation |

STD-O-OBR HAYNES

5.6 Typical vacuum gauge readings

6.1 After tightly wrapping water-vulnerable components, use a spray cleaner on everything, with particular concentration on the greasiest areas, usually around the valve cover and lower edges of the block. If one section dries out, apply more cleaner

6.2 Depending on how dirty the engine is, let the cleaner soak in according to the directions and then hose off the grime and cleaner. Get the rinse water down into every area you can get at; then dry important components with a hair dryer or paper towels

6.3 Get an engine hoist that's strong enough to easily lift your engine in and out of the engine compartment; an adapter, like the one shown here, can be used to change the angle of the engine as it's being removed or installed

the usual tune-up items and, if necessary, run the engine on an ignition analyzer.

13   If there is a large fluctuation, perform a compression or leak-down test to look for a weak or dead cylinder or a blown head gasket.

14   If the needle moves slowly through a wide range, check for a clogged PCV system, incorrect idle fuel mixture, throttle body or intake manifold gasket leaks.

15   Check for a slow return after revving the engine by quickly snapping the throttle open until the engine reaches about 2,500 rpm and let it shut. Normally the reading should drop

to near zero, rise above normal idle reading (about 5 in-Hg over) and then return to the previous idle reading. If the vacuum returns slowly and doesn't peak when the throttle is snapped shut, the rings may be worn. If there is a long delay, look for a restricted exhaust system (often the muffler or catalytic converter). An easy way to check this is to temporarily disconnect the exhaust ahead of the suspected part and redo the test.

---

6   Engine removal - methods and precautions

Refer to illustrations 6.1, 6.2, 6.3, 6.4 and 6.5

If you've decided that an engine must be removed for overhaul or major repair work, several preliminary steps should be taken. Read all removal and installation procedures carefully prior to committing this job. Some engines are removed by lowering them to the floor, then raising the vehicle sufficiently to slide it out; this will require a vehicle hoist.

Locating a suitable place to work is extremely important. Adequate work space, along with storage space for the vehicle, will be needed. If a shop or garage isn't available, at the very least a flat, level, clean work surface made of concrete or asphalt is required. Cleaning the engine compartment and engine before beginning the removal procedure will help keep tools clean and organized **(see illustrations 6.1 and 6.2)**.

An engine hoist or A-frame will also be necessary. Make sure the equipment is rated in excess of the combined weight of the engine and transaxle. Safety is of primary importance, considering the potential hazards involved in lifting the engine out of the vehicle.

If you're a novice at engine removal, get at least one helper. One person cannot easily do all the things you need to do to lift a big heavy engine out of the engine

compartment. Also helpful is to seek advice and assistance from someone who's experienced in engine removal.

Plan the operation ahead of time. Arrange for or obtain all of the tools and equipment you'll need prior to beginning the job **(see illustrations 6.3, 6.4 and 6.5)**. some of the equipment necessary to perform engine removal and installation safely and with relative ease are (in addition to an engine hoist) a heavy duty floor jack, complete sets of wrenches and sockets as described in the front of this manual, wooden blocks, plenty of rags and cleaning solvent for mopping up spilled oil, coolant and gasoline. If the hoist must be rented, make sure that you arrange for it in advance and have everything disconnected and/or removed before bringing the hoist home. This will save you money and time.

Plan for the vehicle to be out of use for quite a while. A machine shop can do the

6.4 Get an engine stand sturdy enough to firmly support the engine while you're working on it. Stay away from three-wheeled models; they have a tendency to tip over more easily, so get a four-wheeled unit

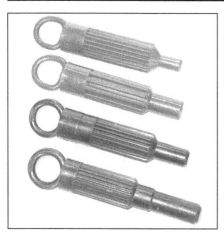

6.5 A clutch alignment tool is necessary if you plan to install a rebuilt engine mated to a manual transmission

work that is beyond the scope of the home mechanic. Machine shops often have busy schedules, so before removing the engine, consult the shop for an estimate of how long it will take to rebuild or repair the components that may need work.

7 Engine overhaul - disassembly sequence

1 It's much easier to remove the external components from the engine if it's mounted on a portable engine stand. A stand can often be rented quite cheaply from an equipment rental yard. Before the engine is mounted on a stand, the flywheel/driveplate should be removed from the engine.
2 If a stand isn't available, it's possible to remove the external engine components with it blocked up on the floor. Be extra careful not to tip or drop the engine when working without a stand.
3 If you're going to obtain a rebuilt engine, all external components must come off first, to be transferred to the replacement engine. These components include:

8.1 Before you try to remove the pistons from engines with very worn cylinders, use a ridge reamer to remove the raised material (ridge) from the top of the cylinders

Driveplate or flywheel
Ignition system components
Emissions-related components
Engine mounts and mount brackets
Fuel injection components
Intake/exhaust manifolds
Oil filter
Thermostat and housing assembly
Water pump

Note: *When removing the external components from the engine, pay close attention to details that may be helpful or important during installation. Note the installed position of gaskets, seals, spacers, pins, brackets, washers, bolts and other small items.*
4 If you're going to obtain a short block (assembled engine block, crankshaft, pistons and connecting rods), remove the timing chain, cylinder head, oil pan, oil pump pick-up tube and water pump from your engine so that you can turn in your old short block to the rebuilder as a core. See *Engine rebuilding alternatives* for additional information regarding the different possibilities to be considered.

8 Pistons and connecting rods - removal and installation

Removal

Refer to illustrations 8.1, 8.3 and 8.4
Note: *Prior to removing the piston/connecting rod assemblies, remove the cylinder head and oil pan (see Chapter 2A or 2B).*
1 Use your fingernail to feel if a ridge has formed at the upper limit of ring travel (about 1/4-inch down from the top of each cylinder). If carbon deposits or cylinder wear have produced ridges, they must be completely removed with a special tool **(see illustration)**. Follow the manufacturer's instructions provided with the tool. Failure to remove the ridges before attempting to remove the piston/connecting rod assemblies may result in piston breakage.
2 After the cylinder ridges have been removed, turn the engine so the crankshaft is facing up.
3 Before the connecting rods are removed, check the connecting rod endplay with feeler gauges. Slide them between the first connecting rod and the crankshaft throw until the play is removed **(see illustration)**. Repeat this procedure for each connecting rod. The endplay is equal to the thickness of the feeler gauge(s). Check with an automotive machine shop for the endplay service limit. If the play exceeds the service limit, new connecting rods will be required. If new rods (or a new crankshaft) are installed, the endplay may fall under the minimum allowable clearance. If it does, the rods will have to be machined to restore it. If necessary, consult an automotive machine shop for advice.
4 Check the connecting rods and caps for identification marks **(see illustration)**. If they aren't plainly marked, use paint or a marker to clearly identify each rod and cap (1, 2, 3, etc., depending on the cylinder they're associated with).
5 Loosen each of the connecting rod cap bolts 1/2-turn at a time until they can be

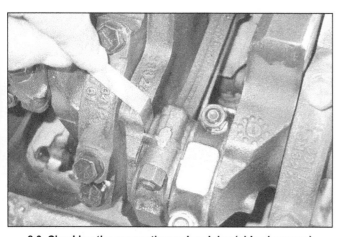

8.3 Checking the connecting rod endplay (side clearance)

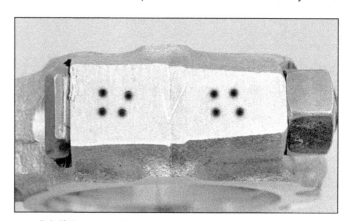

8.4 If the connecting rods and caps are not marked, use permanent ink or paint to mark the caps to the rods by cylinder number (for example, this would be the No. 4 connecting rod)

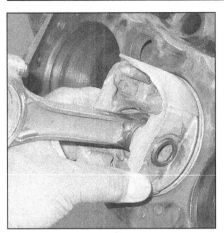

8.13 Install the piston ring into the cylinder, then push it down into position using a piston, so the ring will be square in the cylinder

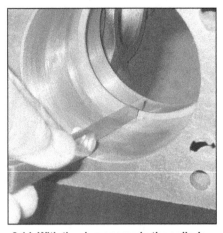

8.14 With the ring square in the cylinder, measure the ring end gap with a feeler gauge

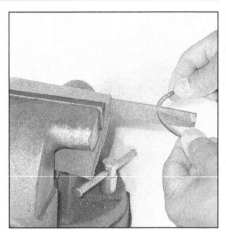

8.15 If the ring end gap is too small, clamp a file in a vise and file the piston ring ends - file from the outside of the ring inward only

removed by hand. Remove the number one connecting rod cap and bearing insert. Don't drop the bearing insert out of the cap. **Note:** *Obtain new connecting rod bolts, but save the old ones (they'll be used for checking the connecting rod bearing oil clearance during reassembly).*

6   Remove the bearing insert and push the connecting rod/piston assembly out through the top of the engine. Use a wooden or plastic hammer handle to push on the upper bearing surface in the connecting rod. If resistance is felt, double-check to make sure that all of the ridge was removed from the cylinder.

7   Repeat the procedure for the remaining cylinders.

8   The connecting rod bolts must be replaced with new ones during reassembly, but save the old ones - they will be used for checking the bearing oil clearance.

9   Reassemble the connecting rod caps and bearing inserts in their respective connecting rods and install the cap bolts finger tight. Leaving the old bearing inserts in place until reassembly will help prevent the connecting rod bearing surfaces from being accidentally nicked or gouged.

10   The pistons and connecting rods are now ready for inspection and overhaul at an automotive machine shop.

Piston ring installation

Refer to illustrations 8.13, 8.14, 8.15, 8.19a, 8.19b, 8.22a and 8.22b

11   Before installing the new piston rings, the ring end gaps must be checked. It's assumed that the piston ring side clearance has been checked and verified correct.

12   Lay out the piston/connecting rod assemblies and the new ring sets so the ring sets will be matched with the same piston and cylinder during the end gap measurement and engine assembly.

13   Insert the top (number one) ring into the first cylinder and square it up with the cylinder walls by pushing it in with the top of the piston

(see illustration). The ring should be near the bottom of the cylinder, at the lower limit of ring travel.

14   To measure the end gap, slip feeler gauges between the ends of the ring until a gauge equal to the gap width is found (see illustration). The feeler gauge should slide between the ring ends with a slight amount of drag. Check with an automotive machine shop for the correct end gap for your engine. If the gap is larger or smaller than specified, double-check to make sure you have the correct rings before proceeding.

15   If the gap is too small, it must be enlarged or the ring ends may come in contact with each other during engine operation, which can cause serious damage to the engine. The end gap can be increased by filing the ring ends very carefully with a fine file. Mount the file in a vise equipped with soft jaws, slip the ring over the file with the ends contacting the file face and slowly move the ring to remove material from the ends. When performing this operation, file only by pushing the ring

from the outside end of the file towards the vise (see illustration). Be sure to remove all raised material.

16   Excess end gap isn't critical unless it's greater than approximately 0.040-inch. Again, double-check to make sure you have the correct ring type and that you are referencing the correct section and category of specifications.

17   Repeat the procedure for each ring that will be installed in the first cylinder and for each ring in the remaining cylinders. Remember to keep rings, pistons and cylinders matched up.

18   Once the ring end gaps have been checked/corrected, the rings can be installed on the pistons.

19   The oil control ring (lowest one on the piston) is usually installed first. Some oil rings are composed of three separate components. Slip the spacer/expander into the groove (see illustration). If an anti-rotation tang is used, make sure it's inserted into the drilled hole in the ring groove. Next, install the upper side rail in the same manner (see illustration). Don't use a piston ring installation tool on the

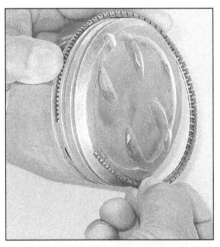

8.19a Installing the spacer/expander in the oil ring groove

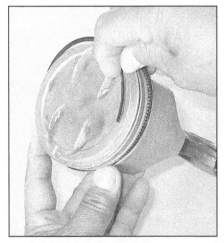

8.19b DO NOT use a piston ring installation tool when installing the oil control side rails

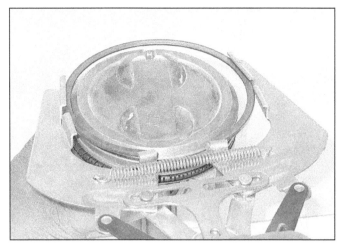

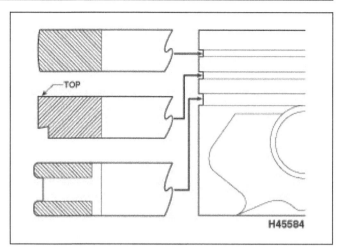

8.22a Use a piston ring installation tool to install the number 2 and the number 1 (top) rings - be sure the directional mark on the piston ring(s) is facing toward the top of the piston

8.22b Install the second compression ring with the dot, or the word TOP, facing up

oil ring side rails, as they may be damaged. Instead, place one end of the side rail into the groove between the spacer/expander and the ring land, hold it firmly in place and slide a finger around the piston while pushing the rail into the groove. Finally, install the lower side rail.

20   After the three oil ring components have been installed, check to make sure that both the upper and lower side rails can be rotated smoothly inside the ring grooves.

21   The number two (middle) ring is installed next. It's usually stamped with a mark that must face up, toward the top of the piston. Do not mix up the top and middle rings, as they have different cross-sections. **Note:** *Always follow the instructions printed on the ring package or box - different manufacturers may require different approaches.*

22   Use a piston ring installation tool and make sure the identification mark is facing the top of the piston, then slip the ring into the middle groove on the piston **(see illustration)**. Don't expand the ring any more than necessary to slide it over the piston.

23   Install the number one (top) ring in the same manner. Make sure the mark is facing up. Be careful not to confuse the number one and number two rings.

24   Repeat the procedure for the remaining pistons and rings.

Installation

25   Before installing the piston/connecting rod assemblies, the cylinder walls must be perfectly clean, the top edge of each cylinder bore must be chamfered, and the crankshaft must be in place.

26   Remove the cap from the end of the number one connecting rod (refer to the marks made during removal). Remove the original bearing inserts and wipe the bearing surfaces of the connecting rod and cap with a clean, lint-free cloth. They must be kept spotlessly clean.

Connecting rod bearing oil clearance check

Refer to illustrations 8.29, 8.34, 8.37, 8.38 and 8.41

27   Clean the back side of the new upper bearing insert, then lay it in place in the connecting rod. Make sure the tab on the bearing fits into the recess in the rod. Don't hammer the bearing insert into place and be very careful not to nick or gouge the bearing face. Don't lubricate the bearing at this time.

28   Clean the back side of the other bearing insert and install it in the rod cap. Again, make sure the tab on the bearing fits into the recess in the cap, and don't apply any lubricant. It's critically important that the mating surfaces of the bearing and connecting rod are perfectly clean and oil free when they're assembled.

29   Position the piston ring gaps at 90-degree intervals around the piston as shown **(see illustration)**.

30   Lubricate the piston and rings with clean engine oil and attach a piston ring compressor to the piston. Leave the skirt protruding about

1/4-inch to guide the piston into the cylinder. The rings must be compressed until they're flush with the piston.

31   Rotate the crankshaft until the number one connecting rod journal is at BDC (bottom dead center) and apply a liberal coat of engine oil to the cylinder walls.

32   With the mark (cavity) on top of the piston facing the front (timing chain end) of the engine, gently insert the piston/connecting rod assembly into the number one cylinder bore and rest the bottom edge of the ring compressor on the engine block.

33   Tap the top edge of the ring compressor to make sure it's contacting the block around its entire circumference.

34   Gently tap on the top of the piston with the end of a wooden or plastic hammer handle **(see illustration)** while guiding the end of the connecting rod into place on the crankshaft journal.

35   The piston rings may try to pop out of the ring compressor just before entering the cylinder bore, so keep some downward pressure on the ring compressor. Work slowly, and if any resistance is felt as the piston enters the cylin-

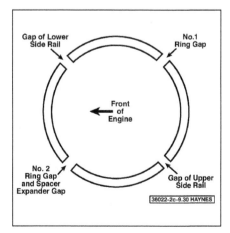

8.29 Position the piston ring end gaps as shown

8.34 Use a plastic or wooden hammer handle to push the piston into the cylinder

8.37 Place Plastigage on each connecting rod bearing journal parallel to the crankshaft centerline

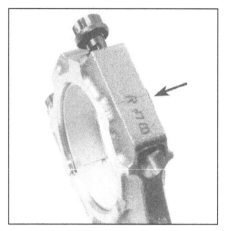

8.38 Install the connecting rod cap, making sure the cap and rod identification numbers match

8.41 Use the scale on the Plastigage package to determine the bearing oil clearance - be sure to measure the widest part of the Plastigage and use the correct scale; it comes with both standard and metric scales

der, stop immediately. Find out what's hanging up and fix it before proceeding. Do not, for any reason, force the piston into the cylinder - you might break a ring and/or the piston.

36   Once the piston/connecting rod assembly is installed, the connecting rod bearing oil clearance must be checked before the rod cap is permanently installed.

37   Cut a piece of the appropriate size Plastigage slightly shorter than the width of the connecting rod bearing and lay it in place on the number one connecting rod journal, parallel with the journal axis **(see illustration)**.

38   Clean the connecting rod cap bearing face and install the rod cap. Make sure the mating mark on the cap is on the same side as the mark on the connecting rod **(see illustration)**.

39   Install the OLD rod bolts and tighten them to the torque listed in this Chapter's Specifications. **Note:** *Use a thin-wall socket to avoid erroneous torque readings that can result if the socket is wedged between the rod cap and the bolt. If the socket tends to wedge itself between the fastener and the cap, lift up on it slightly until it no longer contacts the cap. DO NOT rotate the crankshaft at any time during this operation.*

40   Remove the fasteners and detach the rod cap, being very careful not to disturb the Plastigage.

41   Compare the width of the crushed Plastigage to the scale printed on the Plastigage envelope to obtain the oil clearance **(see illustration)**. The connecting rod oil clearance is usually about 0.002 inch (0.05 mm). Consult an automotive machine shop for the clearance specified for the rod bearings on your engine.

42   If the clearance is not as specified, the bearing inserts may be the wrong size (which means different ones will be required). Before deciding that different inserts are needed, make sure that no dirt or oil was between the bearing inserts and the connecting rod or cap when the clearance was measured. Also, recheck the journal diameter. If the Plastigage was wider at one end than the other, the journal may be tapered. If the clearance still exceeds the limit specified, the bearing will

have to be replaced with an undersize bearing. **Caution:** *When installing a new crankshaft, always use a standard size bearing.*

Final installation

43   Carefully scrape all traces of the Plastigage material off the rod journal and/or bearing face. Be very careful not to scratch the bearing - use your fingernail or the edge of a plastic card.

44   Make sure the bearing faces are perfectly clean, then apply a uniform layer of clean moly-base grease or engine assembly lube to both of them. You'll have to push the piston into the cylinder to expose the face of the bearing insert in the connecting rod.

45   Slide the connecting rod back into place on the journal, install the rod cap, install the NEW bolts and tighten them to the torque listed in this Chapter's Specifications.

46   Repeat the entire procedure for the remaining pistons/connecting rods.

47   The important points to remember are:

a) *Keep the back sides of the bearing inserts and the insides of the connecting rods and caps perfectly clean when assembling them.*

b) *Make sure you have the correct piston/ rod assembly for each cylinder.*

c) *The mark on the piston must face the front of the engine.*

d) *Lubricate the cylinder walls lightly with clean oil.*

e) *Lubricate the bearing faces when installing the rod caps after the oil clearance has been checked.*

48   After all the piston/connecting rod assemblies have been correctly installed, rotate the crankshaft a number of times by hand to check for any obvious binding.

49   As a final step, check the connecting rod endplay again. If it was correct before disassembly and the original crankshaft and rods were reinstalled, it should still be correct. If new rods or a new crankshaft were installed, the endplay may be inadequate. If so, the rods will have to be removed and taken to an automotive machine shop for resizing.

9   Crankshaft - removal and installation

Removal

Refer to illustrations 9.1 and 9.3

Note: *The crankshaft can be removed only after the engine has been removed from the vehicle. It's assumed that the flywheel or driveplate, crankshaft pulley, timing chain, oil pan, oil pump body, and piston/connecting rod assemblies have already been removed. The rear main oil seal retainer must be unbolted and separated from the block before proceeding with crankshaft removal.*

1   Before the crankshaft is removed, measure the endplay. Mount a dial indicator with the indicator in line with the crankshaft and touching the end of the crankshaft **(see illustration)**.

2   Pry the crankshaft all the way to the rear and zero the dial indicator. Next, pry the crankshaft to the front as far as possible and check the reading on the dial indicator. The distance traveled is the endplay. A typical crankshaft

9.1 Checking crankshaft endplay with a dial indicator

endplay will fall between 0.003 to 0.010-inch (0.076 to 0.25 mm). If it's greater than that, check the crankshaft thrust surfaces for wear after it's removed. If no wear is evident, new main bearings should correct the endplay.

3 If a dial indicator isn't available, feeler gauges can be used. Gently pry the crankshaft all the way to the front of the engine. Slip feeler gauges between the crankshaft and the front face of the thrust bearing or washer to determine the clearance (see illustration).

4 Loosen the main bearing cap bedplate bolts 1/4-turn at a time each, until they can be removed by hand. **Note:** *Obtain new main bearing bedplate bolts, but save the old ones (they'll be used for checking the main bearing oil clearance during reassembly).*

5 Pull the main bearing cap bedplate straight up and off the cylinder block. Try not to drop the bearing inserts if they come out with the assembly.

6 Carefully lift the crankshaft out of the engine. It may be a good idea to have an assistant available, since the crankshaft is quite heavy and awkward to handle. With the bearing inserts in place inside the engine block and main bearing caps, reinstall the main bearing cap assembly onto the engine block and tighten the bolts finger tight. Make sure you install the main bearing cap(s) with the arrow facing the front end of the engine.

Installation

7 Crankshaft installation is the first step in engine reassembly. It's assumed at this point that the engine block and crankshaft have been cleaned, inspected and repaired or reconditioned.

8 Position the engine block with the bottom facing up.

9 Remove the mounting bolts and lift off the main bearing caps.

10 If they're still in place, remove the original bearing inserts from the block and from the main bearing cap. Wipe the bearing surfaces of the block and main bearing caps with a clean, lint-free cloth. They must be kept spotlessly clean. This is critical for determining the correct bearing oil clearance.

9.3 Checking crankshaft endplay with feeler gauges at the thrust bearing journal

Main bearing oil clearance check

Refer to illustrations 9.11, 9.17, 9.19a, 9.19b and 9.21

11 Without mixing them up, clean the back sides of the new upper main bearing inserts (with grooves and oil holes) and lay one in each main bearing saddle in the block. Each upper bearing has an oil groove and oil hole in it. **Caution:** *The oil holes in the block must line up with the oil holes in the upper bearing inserts.* Install the thrust washers in the block with the grooved side facing out (see illustration). **Note:** *On Mk I models, the thrust washers are installed in the number 3 (from the rear) main bearing saddle. On Mk II models, the thrust washers are installed in the number 2 main bearing saddle.* Clean the back sides of the lower main bearing inserts (without grooves) and lay them in the corresponding main bearing cap. Make sure the tab on the bearing insert fits into the recess in the block or main bearing cap. **Caution:** *Do not hammer the bearing insert into place and don't nick or gouge the bearing faces. DO NOT apply any lubrication at this time.*

9.11 Install the thrust bearing shells with their grooves facing outward

12 Clean the faces of the bearing inserts in the block and the crankshaft main bearing journals with a clean, lint-free cloth.

13 Check or clean the oil holes in the crankshaft, as any dirt here can go only one way - straight through the new bearings.

14 Once you're certain the crankshaft is clean, carefully lay it in position in the cylinder block.

15 Before the crankshaft can be permanently installed, the main bearing oil clearance must be checked.

16 Cut several strips of the appropriate size of Plastigage (they must be slightly shorter than the width of the main bearing journal).

17 Place one piece on each crankshaft main bearing journal, parallel with the journal axis (see illustration).

18 Clean the faces of the bearing inserts in the main bearing bedplate. Hold the bearing inserts in place and install the bedplate onto the crankshaft and cylinder block. DO NOT disturb the Plastigage.

19 Apply clean engine oil to the OLD bolt threads prior to installation, then install all bolts finger-tight. Tighten the bolts (in the sequence shown; **see illustration**) progressing in two steps, to the torque listed in this Chapter's

9.17 Place the Plastigage onto the crankshaft bearing journal as shown

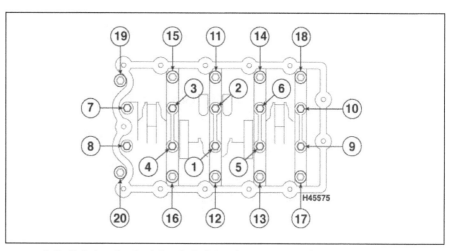

9.19 Main bearing bedplate bolt tightening sequence - Mk I models shown, Mk II models similar

ENGINE BEARING ANALYSIS

Debris

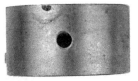

Babbitt bearing embedded with debris from machinings

Microscopic detail of debris

Microscopic detail of gouges

Overplated copper alloy bearing gouged by cast iron debris

Aluminum bearing embedded with glass beads

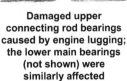

Microscopic detail of glass beads

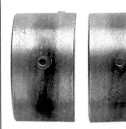

Damaged lining caused by dirt left on the bearing back

Misassembly

Result of a lower half assembled as an upper - blocking the oil flow

Excessive oil clearance is indicated by a short contact arc

Polished and oil-stained backs are a result of a poor fit in the housing bore

Result of a wrong, reversed, or shifted cap

Overloading

Damage from excessive idling which resulted in an oil film unable to support the load imposed

Damaged upper connecting rod bearings caused by engine lugging; the lower main bearings (not shown) were similarly affected

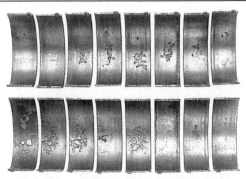

The damage shown in these upper and lower connecting rod bearings was caused by engine operation at a higher-than-rated speed under load

Misalignment

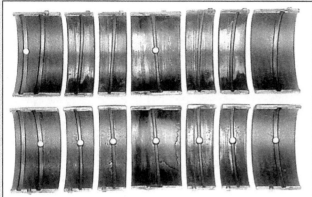

A poorly finished crankshaft caused the equally spaced scoring shown

A tapered housing bore caused the damage along one edge of this pair

A warped crankshaft caused this pattern of severe wear in the center, diminishing toward the ends

A bent connecting rod led to the damage in the "V" pattern

Lubrication

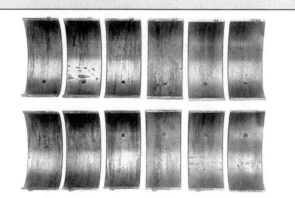

Result of dry start: The bearings on the left, farthest from the oil pump, show more damage

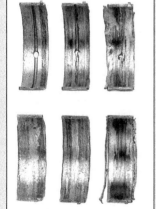

Result of a low oil supply or oil starvation

Severe wear as a result of inadequate oil clearance

Corrosion

Microscopic detail of corrosion

Corrosion is an acid attack on the bearing lining generally caused by inadequate maintenance, extremely hot or cold operation, or inferior oils or fuels

Microscopic detail of cavitation

Example of cavitation - a surface erosion caused by pressure changes in the oil film

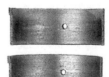

Damage from excessive thrust or insufficient axial clearance

Bearing affected by oil dilution caused by excessive blow-by or a rich mixture

9.21 Use the scale on the Plastigage package to determine the bearing oil clearance - be sure to measure the widest part of the Plastigage and use the correct scale; it comes with both standard and metric scales

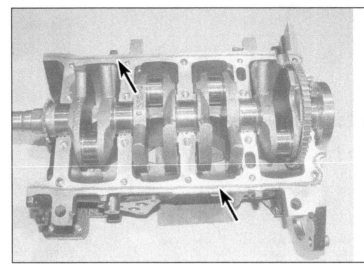

9.28 Apply a bead of sealant to the cylinder block mating surfaces

Specifications. DO NOT rotate the crankshaft at any time during this operation.

20   Remove the bolts a little at a time (and in the *reverse* order of the tightening sequence) and carefully lift the bedplate straight up and off the block. Do not disturb the Plastigage or rotate the crankshaft.

21   Compare the width of the crushed Plasti-gage on each journal to the scale printed on the Plastigage envelope to determine the main bearing oil clearance **(see illustration)**. A typical main bearing oil clearance should fall between 0.0015 to 0.0023-inch (0.038 to 0.060 mm). Check with an automotive machine shop for the clearance specified for your engine.

22   If the clearance is not as specified, the bearing inserts might be the wrong size (which means different ones will be required). Before deciding if different inserts are needed, make sure that no dirt or oil was between the bearing inserts and the cap assembly or block when the clearance was measured. If the Plastigage was wider at one end than the other, the crankshaft journal may be tapered. If the clearance still exceeds the limit specified, the bearing insert(s) will have to be replaced with an undersize bearing insert(s). **Caution:** *When installing a new crankshaft, always install a standard bearing insert set.*

23   Carefully scrape all traces of the Plasti-gage material off the main bearing journals and/or the bearing insert faces. Be sure to remove all residue from the oil holes. Use your fingernail or the edge of a plastic card - don't nick or scratch the bearing faces.

Final installation

Refer to illustration 9.28

24   Carefully lift the crankshaft out of the cyl-inder block.

25   Clean the bearing insert faces in the cyl-inder block, then apply a thin, uniform layer of moly-base grease or engine assembly lube to each of the bearing surfaces. Be sure to coat the thrust faces as well as the journal face of the thrust bearing.

26   Make sure the crankshaft journals are clean, then lay the crankshaft back in place in the cylinder block.

27   Clean the bearing insert faces and apply the same lubricant to them.

28   Apply a thin bead of sealant (Loctite 518 or Loctite 5970) to the cylinder block mating surface **(see illustration)**.

29   Hold the bearing inserts in place and install the main bearing bedplate on the crankshaft and cylinder block.

30   Apply clean engine oil to the NEW bolt threads, wipe off any excess oil, then install the bolts finger-tight.

31   Push the crankshaft forward using a screwdriver or prybar to seat the thrust bear-ing. Once the crankshaft is pushed fully for-ward to seat the thrust bearing, leave the screwdriver in position so that force stays on the crankshaft until after all of the bolts have been tightened.

32   Tighten the main bearing bedplate bolts to the torque and angle listed in this Chapter's Specifications.

33   Recheck crankshaft endplay with a feeler gauge or a dial indicator. The endplay should be correct if the crankshaft thrust faces aren't worn or damaged and if new bearings have been installed.

34   Rotate the crankshaft a number of times by hand to check for any obvious binding. It should rotate with a running torque of 50 in-lbs or less. If the running torque is too high, correct the problem at this time.

35   Install the new rear main oil seal (see Chapter 2A or 2B).

10   Engine overhaul - reassembly sequence

1   Before beginning engine reassembly, make sure you have all the necessary new parts, gaskets and seals as well as the follow-ing items on hand:

Common hand tools
A 1/2-inch drive torque wrench
New engine oil
Gasket sealant
Thread locking compound

2   If you obtained a short block it will be necessary to install the cylinder head, the timing chain, the oil pump, pick-up tube, oil pan, the water pump and the valve cover (see Chapter 2A or 2B). In order to save time and avoid problems, the external components should be installed in the following general order:

Water pump
Intake and exhaust manifolds
Fuel injection components
Emission control components
Spark plugs
Ignition coils
Oil filter
Engine mounts and mount brackets
Driveplate (automatic transaxle)
Flywheel (manual transaxle)

11   Engine - removal and installation

Warning: *Wait until the engine is completely cool before beginning this procedure.*

Note: *This is an involved operation. Read through the procedure thoroughly before start-ing work, and ensure that adequate lifting tackle and jacking/support equipment is avail-able. Make notes during disassembly to ensure that all wiring/hoses and brackets are correctly repositioned and routed on installation.*

Mk I models

Removal

Refer to illustrations 11.17, 11.23, 11.24a, 11.24b, 11.29 and 11.34

1   Relieve the fuel system pressure (see Chapter 4).

2   On Cooper models, remove the battery and battery tray (see Chapter 5). On Cooper S models, disconnect the negative battery

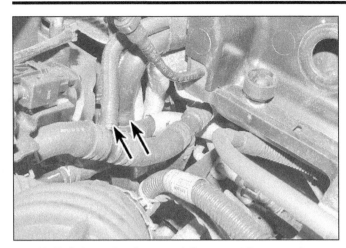

11.17 Release the clamps and disconnect the expansion tank hoses

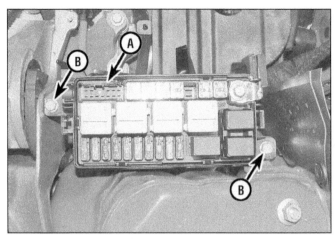

11.23 Disconnect the electrical connector (A, already unplugged) then remove the two fuse box retaining bolts (B)

cable (see Chapter 5).

3   Remove the air filter assembly (see Chapter 4).

4   Remove the intake manifold (see Chapter 4).

5   Remove the hood (see Chapter 11).

6   Loosen the front wheel bolts, raise the front of the vehicle and support it securely on jackstands. Remove the front wheels.

7   Remove both front wheelwell liners (see Chapter 11).

8   Drain the engine oil and cooling system (see Chapter 1).

9   Remove the exhaust manifold (see Chapter 4).

10   Remove the driveaxles (see Chapter 8).

11   Remove the drivebelt (see Chapter 1).

12   Disconnect and remove the upper and lower radiator hoses.

13   Place the Modular Front End (MFE) in the service position (see Chapter 11).

14   Remove the crush tubes on each side.

Cooper models

15   Release the clamps and disconnect the heater hoses from the thermostat housing

and the coolant distribution pipe.

16   Disconnect the overflow hose from the coolant filler neck.

Cooper S models

17   Release the clamps and disconnect the heater hoses from the engine compartment firewall, and the two expansion tank hoses at the pipe junction **(see illustration)**.

18   Disconnect the coolant hose from the rear of the thermostat housing.

All models

19   Unclip the fuel supply hose from the engine mounting bracket, plug the end of the hose and position it to one side.

20   Working underneath the vehicle, remove the retaining bolt, then slide the heat shield over the starter motor to the right-hand side (disengaging the shield from the rubber mounting grommets) and remove the shield.

21   Noting their installed positions, disconnect the starter motor wiring plugs. Remove the top starter motor mounting bolt to release the wiring harness bracket.

22   Remove the two retaining screws, and remove the clutch release cylinder from the

transaxle casing. There is no need to disconnect the fluid line. Release the line from its retaining bracket on the transaxle casing and position the cylinder to one side.

23   Open the fuse box cover, disconnect the engine compartment fuse box electrical connector, then remove the two screws and move the fuse box to one side **(see illustration)**.

24   Disconnect the ground strap from the left-hand chassis member in the engine compartment, then unscrew the collars and unplug the harness connector **(see illustrations)**.

25   Disconnect the shift cable(s) from the lever(s) on the transaxle (see Chapter 7A or 7B). On 2005 and earlier manual transaxle models, remove the three retaining bolts and remove the shift cable bracket from the top of the transaxle.

26   Working underneath the vehicle, remove the two bolts and remove the stabilizer bracket from the right-hand rear of the engine.

27   Disconnect the power steering pump electrical connector, and the pump cooling fan electrical connector (if equipped).

28   Disconnect the alternator, oil pressure, and crankshaft position sensor wiring plugs.

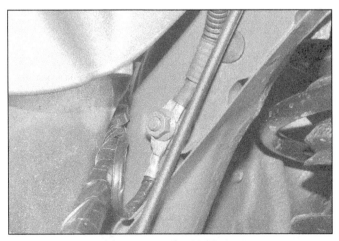

11.24a Disconnect the ground strap from the left-hand chassis member . . .

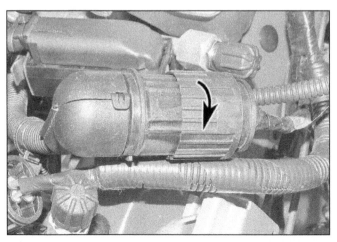

11.24b . . . then rotate the collar counterclockwise and disconnect the harness connector

**11.29 Use cable ties to suspend the compressor from the MFE -
there's no need to disconnect the refrigerant lines**

**11.34 Attach lifting chains to the right-hand end of the engine and
the lifting eye on the transaxle (Mk I models shown,
Mk II models similar)**

29   On models with air conditioning, disconnect the compressor wiring plug (the plug simply pulls apart), then remove the three retaining bolts and detach the compressor from the engine. Use cable ties (or similar) to secure the compressor to the MFE **(see illustration)**; do not detach the refrigerant lines from the compressor.

30   Make a final check to ensure that all relevant hoses, pipes and wiring have been disconnected from the engine and moved clear to allow the engine to be lifted out.

31   Place a floor jack under the transaxle casing, and take up the weight of the transaxle with the jack.

32   Remove the four retaining bolts, and the single mount-to-bracket bolt, then remove the rubber mount assembly from the transaxle casing.

33   Remove the screws securing the left-hand engine mounting bracket to the vehicle body, and remove the bracket.

34   Attach lifting chains/straps to the right-hand end of the engine, and the lifting eye on the top of the transaxle casing **(see illustration)**. Take up the weight of the engine/transaxle with a suitable hoist.

35   On models up to 12/2003, remove the engine carrier bracket from the right-hand engine mount.

36   Remove the nut securing the right-hand upper engine mounting bracket to the mount. Remove the floor jack from under the transaxle, then lift the assembly and maneuver it from the engine compartment.

37   Lower the assembly to the floor, supporting it on blocks. Disconnect the sling or chain from the transaxle, then reconnect it to the engine.

38   Support the transaxle with a jack (preferably with a transaxle adapter), remove the transaxle mounting fasteners, then separate the transaxle from the engine.

39   Remove the flywheel/driveplate and mount the engine on a stand.

Installation

40   Installation is the reverse of removal, noting the following points:

a) *Check the engine/transaxle mounts, replacing them as necessary.*

b) *Tighten all fasteners to the specified torque, where given.*

c) *Ensure that all wiring, hoses and brackets are positioned and routed as noted before removal.*

d) *Refill the engine and transaxle with oil, and refill the cooling system (see Chapter 1).*

Mk II models

Removal

Refer to illustration 11.47

41   Have the air conditioning system discharged by an automotive air conditioning technician.

42   Relieve the fuel system pressure (see Chapter 4).

43   Disconnect the cable from the negative terminal of the battery (see Chapter 5).

44   Remove the air filter housing and ducts (see Chapter 4).

45   Remove the hood (see Chapter 11).

46   Disconnect the fuel feed line from the fuel rail (see Chapter 4).

47   Follow the wiring harnesses from the engine and disconnect all electrical connectors between the engine/transaxle and the chassis **(see illustration)**.

48   Detach the power brake booster vacuum hose.

49   Detach the shift cable(s) from the transaxle (see Chapter 7A or 7B).

50   Apply the parking brake and chock the rear wheels. Loosen the front wheel bolts, then raise the front of the vehicle and support it securely on jackstands. Remove the wheels.

51   If you're working on a manual transaxle model, remove the clutch release cylinder (see Chapter 8).

52   Remove the exhaust system (see Chapter 4).

53   Drain the cooling system and engine oil (see Chapter 1).

54   Loosen the clamps and detach the heater hoses and radiator hoses from the engine.

55   Remove the cooling fans and radiator (see Chapter 3). If you're working on an S model, also remove the intercooler. Remove the radia-

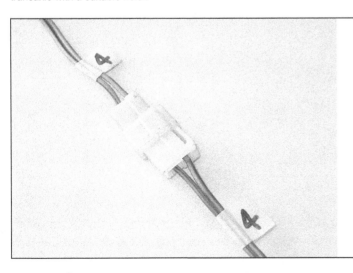

**11.47 Label
both ends of
each wire and
hose before
disconnecting it**

tor support assembly (see Chapter 11). **Note:** *These can be removed as a unit if care is taken.*

56 Remove the driveaxles (see Chapter 8).

57 If you're working on a model with an automatic transaxle, remove the starter, then remove the torque converter nuts, one at a time, rotating the crankshaft to bring each fastener into view.

58 Attach a lifting sling or chain to the lifting brackets on the engine and transaxle **(see illustration 11.34)**. Position an engine hoist and connect the sling or chain to it. If no lifting hooks or brackets are present, you'll have to fasten the chains or slings to some substantial parts of the engine/transaxle - ones that are strong enough to take the weight, but in locations that will provide good balance. Position the chain on the hoist so it balances the engine and the transaxle level with the vehicle. Take up the slack until there is slight tension on the sling or chain.

59 Detach the engine and transaxle mounts, and also the pendulum mount, or engine stabilizer link from under the engine (see Chapter 2A or 2B).

60 Recheck to be sure that nothing is still connected between the engine/transaxle and the chassis. Label and disconnect anything still remaining.

61 Slowly and carefully raise the engine/transaxle assembly and guide it out the front of the engine compartment.

62 Lower the assembly to the floor, supporting it on blocks. Disconnect the sling or chain from the transaxle, then reconnect it to the engine.

63 Support the transaxle with a jack (pref-erably with a transaxle adapter), remove the transaxle mounting fasteners, then separate the transaxle from the engine.

64 Remove the flywheel/driveplate and mount the engine on a stand.

Installation

65 Installation is the reverse of removal, noting the following points:

a) *Check the engine/transaxle mounts. If they're worn or damaged, replace them.*
b) *Attach the transaxle to the engine following the procedure described in Chapter 7A or 7B.*
c) *Add coolant, oil and transaxle fluids as needed (see Chapter 1).*
d) *Tighten all fasteners to the specified torque, where given.*
e) *Reconnect the negative battery cable (see Chapter 5).*
f) *Run the engine and check for proper operation and leaks. Shut off the engine and recheck fluid levels.*
g) *Have the air conditioning system re-charged and leak tested by the shop that discharged it.*

12 Initial start-up and break-in after overhaul

Warning: *Have a fire extinguisher handy when starting the engine for the first time.*

1 Once the engine has been installed in the vehicle, double-check the engine oil and coolant levels.

2 With the spark plugs out of the engine, and the primary (low voltage) wiring to the ignition coil(s) unplugged (see Chapter 5) and the fuel pump disabled (see Chapter 4, Section 2), crank the engine until oil pressure registers on the gauge or the light goes out.

3 Install the spark plugs and restore the ignition system and fuel pump functions.

4 Start the engine. It may take a few moments for the fuel system to build up pressure, but the engine should start without a great deal of effort.

5 After the engine starts, it should be allowed to warm up to normal operating temperature. While the engine is warming up, make a thorough check for fuel, oil and coolant leaks.

6 Shut the engine off and recheck the engine oil and coolant levels.

7 Drive the vehicle to an area with minimum traffic, accelerate from 30 to 50 mph, then allow the vehicle to slow to 30 mph with the throttle closed. Repeat the procedure 10 or 12 times. This will load the piston rings and cause them to seat properly against the cylinder walls. Check again for oil and coolant leaks.

8 Drive the vehicle gently for the first 500 miles (no sustained high speeds) and keep a constant check on the oil level. It is not unusual for an engine to use oil during the break-in period.

9 At approximately 500 to 600 miles, change the oil and filter.

10 For the next few hundred miles, drive the vehicle normally. Do not pamper it or abuse it.

11 After 2000 miles, change the oil and filter again and consider the engine broken in.

COMMON ENGINE OVERHAUL TERMS

B

Backlash - The amount of play between two parts. Usually refers to how much one gear can be moved back and forth without moving the gear with which it's meshed.

Bearing Caps - The caps held in place by nuts or bolts which, in turn, hold the bearing surface. This space is for lubricating oil to enter.

Bearing clearance - The amount of space left between shaft and bearing surface. This space is for lubricating oil to enter.

Bearing crush - The additional height which is purposely manufactured into each bearing half to ensure complete contact of the bearing back with the housing bore when the engine is assembled.

Bearing knock - The noise created by movement of a part in a loose or worn bearing.

Blueprinting - Dismantling an engine and reassembling it to EXACT specifications.

Bore - An engine cylinder, or any cylindrical hole; also used to describe the process of enlarging or accurately refinishing a hole with a cutting tool, as to bore an engine cylinder. The bore size is the diameter of the hole.

Boring - Renewing the cylinders by cutting them out to a specified size. A boring bar is used to make the cut.

Bottom end - A term which refers collectively to the engine block, crankshaft, main bearings and the big ends of the connecting rods.

Break-in - The period of operation between installation of new or rebuilt parts and time in which parts are worn to the correct fit. Driving at reduced and varying speed for a specified mileage to permit parts to wear to the correct fit.

Bushing - A one-piece sleeve placed in a bore to serve as a bearing surface for shaft, piston pin, etc. Usually replaceable.

C

Camshaft - The shaft in the engine, on which a series of lobes are located for operating the valve mechanisms. The camshaft is driven by gears or sprockets and a timing chain. Usually referred to simply as the cam.

Carbon - Hard, or soft, black deposits found in combustion chamber, on plugs, under rings, on and under valve heads.

Cast iron - An alloy of iron and more than two percent carbon, used for engine blocks and heads because it's relatively inexpensive and easy to mold into complex shapes.

Chamfer - To bevel across (or a bevel on) the sharp edge of an object.

Chase - To repair damaged threads with a tap or die.

Combustion chamber - The space between the piston and the cylinder head, with the piston at top dead center, in which air-fuel mixture is burned.

Compression ratio - The relationship between cylinder volume (clearance volume) when the piston is at top dead center and cylinder volume when the piston is at bottom dead center.

Connecting rod - The rod that connects the crank on the crankshaft with the piston. Sometimes called a con rod.

Connecting rod cap - The part of the connecting rod assembly that attaches the rod to the crankpin.

Core plug - Soft metal plug used to plug the casting holes for the coolant passages in the block.

Crankcase - The lower part of the engine in which the crankshaft rotates; includes the lower section of the cylinder block and the oil pan.

Crank kit - A reground or reconditioned crankshaft and new main and connecting rod bearings.

Crankpin - The part of a crankshaft to which a connecting rod is attached.

Crankshaft - The main rotating member, or shaft, running the length of the crankcase, with offset throws to which the connecting rods are attached; changes the reciprocating motion of the pistons into rotating motion.

Cylinder sleeve - A replaceable sleeve, or liner, pressed into the cylinder block to form the cylinder bore.

D

Deburring - Removing the burrs (rough edges or areas) from a bearing.

Deglazer - A tool, rotated by an electric motor, used to remove glaze from cylinder walls so a new set of rings will seat.

E

Endplay - The amount of lengthwise movement between two parts. As applied to a crankshaft, the distance that the crankshaft can move forward and back in the cylinder block.

F

Face - A machinist's term that refers to removing metal from the end of a shaft or the face of a larger part, such as a flywheel.

Fatigue - A breakdown of material through a large number of loading and unloading cycles. The first signs are cracks followed shortly by breaks.

Feeler gauge - A thin strip of hardened steel, ground to an exact thickness, used to check clearances between parts.

Free height - The unloaded length or height of a spring.

Freeplay - The looseness in a linkage, or an assembly of parts, between the initial application of force and actual movement. Usually perceived as slop or slight delay.

Freeze plug - See Core plug.

G

Gallery - A large passage in the block that forms a reservoir for engine oil pressure.

Glaze - The very smooth, glassy finish that develops on cylinder walls while an engine is in service.

H

Heli-Coil - A rethreading device used when threads are worn or damaged. The device is installed in a retapped hole to reduce the thread size to the original size.

I

Installed height - The spring's measured length or height, as installed on the cylinder head. Installed height is measured from the spring seat to the underside of the spring retainer.

J

Journal - The surface of a rotating shaft which turns in a bearing.

K

Keeper - The split lock that holds the valve spring retainer in position on the valve stem.

Key - A small piece of metal inserted into matching grooves machined into two parts fitted together - such as a gear pressed onto a shaft - which prevents slippage between the two parts.

Knock - The heavy metallic engine sound, produced in the combustion chamber as a result of abnormal combustion - usually detonation. Knock is usually caused by a loose or worn bearing. Also referred to as detonation, pinging and spark knock. Connecting rod or main bearing knocks are created by too much oil clearance or insufficient lubrication.

L

Lands - The portions of metal between the piston ring grooves.

Lapping the valves - Grinding a valve face and its seat together with lapping compound.

Lash - The amount of free motion in a gear train, between gears, or in a mechanical assembly, that occurs before movement can begin. Usually refers to the lash in a valve train.

Lifter - The part that rides against the cam to transfer motion to the rest of the valve train.

M

Machining - The process of using a machine to remove metal from a metal part.

Main bearings - The plain, or babbit, bearings that support the crankshaft.

Main bearing caps - The cast iron caps, bolted to the bottom of the block, that support the main bearings.

O

O.D. - Outside diameter.

Oil gallery - A pipe or drilled passageway in the engine used to carry engine oil from one area to another.

Oil ring - The lower ring, or rings, of a piston; designed to prevent excessive amounts of oil from working up the cylinder walls and into the combustion chamber. Also called an oil-control ring.

Oil seal - A seal which keeps oil from leaking out of a compartment. Usually refers to a dynamic seal around a rotating shaft or other moving part.

O-ring - A type of sealing ring made of a special rubberlike material; in use, the O-ring is compressed into a groove to provide the sealing action.

Overhaul - To completely disassemble a unit, clean and inspect all parts, reassemble it with the original or new parts and make all adjustments necessary for proper operation.

P

Pilot bearing - A small bearing installed in the center of the flywheel (or the rear end of the crankshaft) to support the front end of the input shaft of the transmission.

Pip mark - A little dot or indentation which indicates the top side of a compression ring.

Piston - The cylindrical part, attached to the connecting rod, that moves up and down in the cylinder as the crankshaft rotates. When the fuel charge is fired, the piston transfers the force of the explosion to the connecting rod, then to the crankshaft.

Piston pin (or wrist pin) - The cylindrical and usually hollow steel pin that passes through the piston. The piston pin fastens the piston to the upper end of the connecting rod.

Piston ring - The split ring fitted to the groove in a piston. The ring contacts the sides of the ring groove and also rubs against the cylinder wall, thus sealing space between piston and wall. There are two types of rings: Compression rings seal the compression pressure in the combustion chamber; oil rings scrape excessive oil off the cylinder wall.

Piston ring groove - The slots or grooves cut in piston heads to hold piston rings in position.

Piston skirt - The portion of the piston below the rings and the piston pin hole.

Plastigage - A thin strip of plastic thread, available in different sizes, used for measuring clearances. For example, a strip of plastigage is laid across a bearing journal and mashed as parts are assembled. Then parts are disassembled and the width of the strip is measured to determine clearance between journal and bearing. Commonly used to measure crankshaft main-bearing and connecting rod bearing clearances.

Press-fit - A tight fit between two parts that requires pressure to force the parts together. Also referred to as drive, or force, fit.

Prussian blue - A blue pigment; in solution, useful in determining the area of contact between two surfaces. Prussian blue is commonly used to determine the width and location of the contact area between the valve face and the valve seat.

R

Race (bearing) - The inner or outer ring that provides a contact surface for balls or rollers in bearing.

Ream - To size, enlarge or smooth a hole by using a round cutting tool with fluted edges.

Ring job - The process of reconditioning the cylinders and installing new rings.

Runout - Wobble. The amount a shaft rotates out-of-true.

S

Saddle - The upper main bearing seat.

Scored - Scratched or grooved, as a cylinder wall may be scored by abrasive particles moved up and down by the piston rings.

Scuffing - A type of wear in which there's a transfer of material between parts moving against each other; shows up as pits or grooves in the mating surfaces.

Seat - The surface upon which another part rests or seats. For example, the valve seat is the matched surface upon which the valve face rests. Also used to refer to wearing into a good fit; for example, piston rings seat after a few miles of driving.

Short block - An engine block complete with crankshaft and piston and, usually, camshaft assemblies.

Static balance - The balance of an object while it's stationary.

Step - The wear on the lower portion of a ring land caused by excessive side and back-clearance. The height of the step indicates the ring's extra side clearance and the length of the step projecting from the back wall of the groove represents the ring's back clearance.

Stroke - The distance the piston moves when traveling from top dead center to bottom dead center, or from bottom dead center to top dead center.

Stud - A metal rod with threads on both ends.

T

Tang - A lip on the end of a plain bearing used to align the bearing during assembly.

Tap - To cut threads in a hole. Also refers to the fluted tool used to cut threads.

Taper - A gradual reduction in the width of a shaft or hole; in an engine cylinder, taper usually takes the form of uneven wear, more pronounced at the top than at the bottom.

Throws - The offset portions of the crankshaft to which the connecting rods are affixed.

Thrust bearing - The main bearing that has thrust faces to prevent excessive endplay, or forward and backward movement of the crankshaft.

Thrust washer - A bronze or hardened steel washer placed between two moving parts. The washer prevents longitudinal movement and provides a bearing surface for thrust surfaces of parts.

Tolerance - The amount of variation permitted from an exact size of measurement. Actual amount from smallest acceptable dimension to largest acceptable dimension.

U

Umbrella - An oil deflector placed near the valve tip to throw oil from the valve stem area.

Undercut - A machined groove below the normal surface.

Undersize bearings - Smaller diameter bearings used with re-ground crankshaft journals.

V

Valve grinding - Refacing a valve in a valve-refacing machine.

Valve train - The valve-operating mechanism of an engine; includes all components from the camshaft to the valve.

Vibration damper - A cylindrical weight attached to the front of the crankshaft to minimize torsional vibration (the twist-untwist actions of the crankshaft caused by the cylinder firing impulses). Also called a harmonic balancer.

W

Water jacket - The spaces around the cylinders, between the inner and outer shells of the cylinder block or head, through which coolant circulates.

Web - A supporting structure across a cavity.

Woodruff key - A key with a radiused backside (viewed from the side).

Notes

Chapter 3
Cooling, heating and air conditioning systems

Contents

Specifications

Note: *Throughout this Chapter you will find references to "Mk I" and "Mk II" models; this is done to simplify which specifications and procedures apply to which models. Mk I models include 2006 and earlier Cooper/Cooper S models, and 2008 and earlier Convertible models. Mk II models include 2007 and later Cooper/Cooper S/Clubman/Clubman S and 2009 and later Convertible models.*

General

Pressure cap opening pressure	
Mk I models...	13 to 17.5 psi (0.89 to 1.2 bar)
Mk II models..	Refer to pressure specification on cap
Cooling system capacity...	See Chapter 1

Refrigerant

Quantity	
Mk I models...	14.5 ± 0.04 oz (415 ± 10 g)
Mk II models..	15.8 ± 0.4 oz (490 ± 10 g)
Type..	R134a

Torque specifications

Note: *One foot-pound (ft-lb) of torque is equivalent to 12 inch-pounds (in-lbs) of torque. Torque values below approximately 15 foot-pounds are expressed in inch-pounds, because most foot-pound torque wrenches are not accurate at these smaller values.*

	Ft-lbs (unless otherwise indicated)	**Nm**
Air conditioning compressor mounting bolts		
Mk I models..	18	25
Mk II models...	15	20
Water pump mounting bolts		
Mk I models..	22	30
Mk II models...	79 in-lbs	9
Water pump pulley bolts (Mk II models)	71 in-lbs	8
Water pump-to-supercharger bolts (Mk I Cooper S models)................	18	25
Auxiliary water pump-to-transmission bolts (Mk II Cooper S models).....	71 in-lbs	8
Friction wheel-to-engine block bolts (Mk II models)	79 in-lbs	9
Thermostat housing bolts		
Mk I models..	108 in-lbs	12
Mk II models...	71 in-lbs	8

1   General information

Warning: *Do not allow antifreeze to come in contact with your skin or painted surfaces of the vehicle. Rinse off spills immediately with plenty of water. Antifreeze is highly toxic if ingested. Never leave antifreeze lying around in an open container or in puddles on the floor; children and pets are attracted by it's sweet smell and may drink it. Check with local authorities about disposing of used antifreeze. Many communities have collection centers which will see that antifreeze is disposed of safely. Never dump used antifreeze on the ground or pour it into drains.*

Engine cooling system

All modern vehicles employ a pressurized engine cooling system with thermostatically controlled coolant circulation. The cooling system consists of a radiator, an expansion tank or coolant reservoir, a pressure cap (located on the expansion tank or radiator), a thermostat, a cooling fan, and a water pump.

The water pump circulates coolant through the engine. The coolant flows around each cylinder and around the intake and exhaust ports, near the spark plug areas and in close proximity to the exhaust valve guides.

A thermostat controls engine coolant temperature. During warm up, the closed thermostat prevents coolant from circulating through the radiator. As the engine nears normal operating temperature, the thermostat opens and allows hot coolant to travel through the radiator, where it's cooled before returning to the engine.

Heating system

The heating system consists of a blower fan and heater core located in a housing under the dash, the hoses connecting the heater core to the engine cooling system and the heater/air conditioning control head on the dashboard. Hot engine coolant is circulated through the heater core. When the heater mode is activated, a flap door in the housing opens to expose the heater core to the passenger compartment through air ducts. A fan switch on the control head activates the blower motor, which forces air through the core, heating the air.

Air conditioning system

The air conditioning system consists of a condenser mounted in front of the radiator, an evaporator mounted adjacent to the heater core, a compressor mounted on the engine, a receiver-drier or accumulator and the plumbing connecting all of the above components.

A blower fan forces the warmer air of the passenger compartment through the evaporator core (sort of a radiator-in-reverse), transferring the heat from the air to the refrigerant. The liquid refrigerant boils off into low pressure vapor, taking the heat with it when it leaves the evaporator.

2   Troubleshooting

Coolant leaks

Refer to illustration 2.2

1   A coolant leak can develop anywhere in the cooling system, but the most common causes are:

a) *A loose or weak hose clamp*
b) *A defective hose*
c) *A faulty pressure cap*
d) *A damaged radiator*
e) *A bad heater core*
f) *A faulty water pump*
g) *A leaking gasket at any joint that carries coolant*

2   Coolant leaks aren't always easy to find. Sometimes they can only be detected when the cooling system is under pressure. Here's where a cooling system pressure tester comes in handy. After the engine has cooled completely, the tester is attached in place of the pressure cap, then pumped up to the pressure value equal to that of the pressure cap rating **(see illustration)**. Now, leaks that only exist when the engine is fully warmed up will become apparent. The tester can be left connected to locate a nagging slow leak.

Coolant level drops, but no external leaks

Refer to illustrations 2.5a and 2.5b

3   If you find it necessary to keep adding coolant, but there are no external leaks, the probable causes include:

a) *A blown head gasket*
b) *A leaking intake manifold gasket (only on engines that have coolant passages in the manifold)*
c) *A cracked cylinder head or cylinder block*

4   Any of the above problems will also usually result in contamination of the engine oil, which will cause it to take on a milkshake-like appearance. A bad head gasket or cracked head or block can also result in engine oil contaminating the cooling system.

5   Combustion leak detectors (also known as block testers) are available at most auto parts stores. These work by detecting exhaust gases in the cooling system, which indicates a compression leak from a cylinder into the coolant. The tester consists of a large bulb-type syringe and bottle of test fluid **(see illustration)**. A measured amount of the fluid is added to the syringe. The syringe is placed over the cooling system filler neck and, with the engine running, the bulb is squeezed and a sample of the gases present in the cooling system are drawn up through the test fluid **(see illustration)**. If any combustion gases are present in the sample taken, the test fluid will change color.

6   If the test indicates combustion gas is present in the cooling system, you can be sure that the engine has a blown head gasket or a crack in the cylinder head or block, and will require disassembly to repair.

Pressure cap

Refer to illustration 2.8

Warning: *Wait until the engine is completely cool before beginning this check.*

7   The cooling system is sealed by a spring-loaded cap, which raises the boiling point of the coolant. If the cap's seal or spring are worn out, the coolant can boil and escape past the cap. With the engine completely cool, remove the cap and check the seal; if it's cracked, hardened or deteriorated in any way, replace it with a new one.

8   Even if the seal is good, the spring might not be; this can be checked with a cooling system pressure tester **(see illustration)**. If the cap can't hold a pressure within approximately 1-1/2 lbs of its rated pressure (which is marked on the cap), replace it with a new one.

9   The cap is also equipped with a vacuum relief spring. When the engine cools off, a vacuum is created in the cooling system. The vacuum relief spring allows air back into the system, which will equalize the pressure and prevent damage to the radiator (the radiator tanks could collapse if the vacuum is great enough). If, after turning the engine off and

2.2 The cooling system pressure tester is connected in place of the pressure cap, then pumped up to pressurize the system

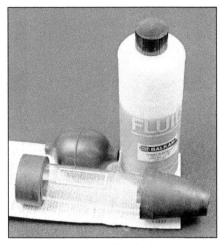

2.5a The combustion leak detector consists of a bulb, syringe and test fluid

2.5b Place the tester over the cooling system filler neck and use the bulb to draw a sample into the tester

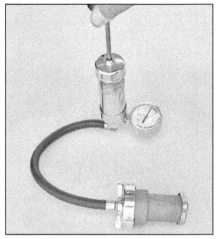

2.8 Checking the cooling system pressure cap with a cooling system pressure tester

allowing it to cool down you notice any of the cooling system hoses collapsing, replace the pressure cap with a new one.

Thermostat

Refer to illustration 2.10

10   Before assuming the thermostat **(see illustration)** is responsible for a cooling system problem, check the coolant level (see Chapter 1), drivebelt tension (see Chapter 1) and temperature gauge (or light) operation.

11   If the engine takes a long time to warm up (as indicated by the temperature gauge or heater operation), the thermostat is probably stuck open. Replace the thermostat with a new one.

12   If the engine runs hot or overheats, a thorough test of the thermostat should be performed.

13   Definitive testing of the thermostat can only be made when it is removed from the vehicle. If the thermostat is stuck in the open position at room temperature, it is faulty and must be replaced. **Caution:** *Do not drive the vehicle without a thermostat. The computer*

may stay in open loop and emissions and fuel economy will suffer.

14   To test a thermostat, suspend the (closed) thermostat on a length of string or wire in a pot of cold water.

15   Heat the water on a stove while observing thermostat. The thermostat should fully open before the water boils.

16   If the thermostat doesn't open and close as specified, or sticks in any position, replace it.

Cooling fan

Electric cooling fan

17   If the engine is overheating and the cooling fan is not coming on when the engine temperature rises to an excessive level, unplug the fan motor electrical connector(s) and connect the motor directly to the battery with fused jumper wires. If the fan motor doesn't come on, replace the motor.

18   If the radiator fan motor is okay, but it isn't coming on when the engine gets hot, the fan relay might be defective. A relay is used to control a circuit by turning it on and off in

response to a control decision by the Powertrain Control Module (PCM). These control circuits are fairly complex, and checking them should be left to a qualified automotive technician. Sometimes, the control system can be fixed by simply identifying and replacing a bad relay.

19   Locate the fan relays in the engine compartment fuse/relay box.

20   If the relay is okay, check all wiring and connections to the fan motor. Refer to the wiring diagrams at the end of Chapter 12.

21   If no obvious problems are found, the problem could be the Engine Coolant Temperature (ECT) sensor or the Powertrain Control Module (PCM). Have the cooling fan system and circuit diagnosed by a dealer service department or repair shop with the proper diagnostic equipment.

Belt-driven cooling fan

22   Disconnect the cable from the negative terminal of the battery and rock the fan back and forth by hand to check for excessive bearing play.

23   With the engine cold (and not running), turn the fan blades by hand. The fan should turn freely.

24   Visually inspect for substantial fluid leakage from the clutch assembly. If problems are noted, replace the clutch assembly.

25   With the engine completely warmed up, turn off the ignition switch and disconnect the negative battery cable from the battery. Turn the fan by hand. Some drag should be evident. If the fan turns easily, replace the fan clutch.

Water pump

26   A failure in the water pump can cause serious engine damage due to overheating.

Drivebelt-driven water pump

Refer to illustration 2.28

27   There are two ways to check the operation of the water pump while it's installed on the

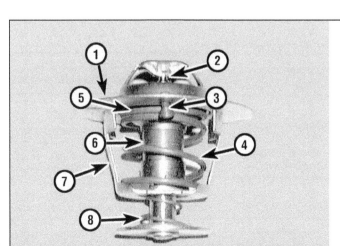

2.10 Typical thermostat:

1   Flange
2   Piston
3   Jiggle valve
4   Main coil spring
5   Valve seat
6   Valve
7   Frame
8   Secondary coil spring

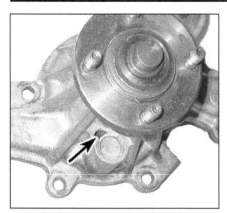

2.28 The water pump weep hole is generally located on the underside of the pump

engine. If the pump is found to be defective, it should be replaced with a new or rebuilt unit.

28 Water pumps are equipped with weep (or vent) holes **(see illustration)**. If a failure occurs in the pump seal, coolant will leak from the hole.

29 If the water pump shaft bearings fail, there may be a howling sound at the pump while it's running. Shaft wear can be felt with the drivebelt removed if the water pump pulley is rocked up and down (with the engine off). Don't mistake drivebelt slippage, which causes a squealing sound, for water pump bearing failure.

Timing chain or timing belt-driven water pump

30 Water pumps driven by the timing chain or timing belt are located underneath the timing chain or timing belt cover.

31 Checking the water pump is limited because of where it is located. However, some basic checks can be made before deciding to remove the water pump. If the pump is found to be defective, it should be replaced with a new or rebuilt unit.

32 One sign that the water pump may be failing is that the heater (climate control) may not work well. Warm the engine to normal operating temperature, confirm that the coolant level is correct, then run the heater and check for hot air coming from the ducts.

33 Check for noises coming from the water pump area. If the water pump impeller shaft or bearings are failing, there may be a howling sound at the pump while the engine is running. **Note:** *Be careful not to mistake drivebelt noise (squealing) for water pump bearing or shaft failure.*

34 It you suspect water pump failure due to noise, wear can be confirmed by feeling for play at the pump shaft. This can be done by rocking the drive sprocket on the pump shaft up and down. To do this you will need to remove the tension on the timing chain or belt as well as access the water pump.

All water pumps

35 In rare cases or on high-mileage vehicles,

another sign of water pump failure may be the presence of coolant in the engine oil. This condition will adversely affect the engine in varying degrees. **Note:** *Finding coolant in the engine oil could indicate other serious issues besides a failed water pump, such as a blown head gasket or a cracked cylinder head or block.*

36 Even a pump that exhibits no outward signs of a problem, such as noise or leakage, can still be due for replacement. Removal for close examination is the only sure way to tell. Sometimes the fins on the back of the impeller can corrode to the point that cooling efficiency is diminished significantly.

Heater system

37 Little can go wrong with a heater. If the fan motor will run at all speeds, the electrical part of the system is okay. The three basic heater problems fall into the following general categories:

a) Not enough heat
b) Heat all the time
c) No heat

38 If there's not enough heat, the control valve or door is stuck in a partially open position, the coolant coming from the engine isn't hot enough, or the heater core is restricted. If the coolant isn't hot enough, the thermostat in the engine cooling system is stuck open, allowing coolant to pass through the engine so rapidly that it doesn't heat up quickly enough. If the vehicle is equipped with a temperature gauge instead of a warning light, watch to see if the engine temperature rises to the normal operating range after driving for a reasonable distance.

39 If there's heat all the time, the control valve or the door is stuck wide open.

40 If there's no heat, coolant is probably not reaching the heater core, or the heater core is plugged. The likely cause is a collapsed or plugged hose, core, or a frozen heater control valve. If the heater is the type that flows coolant all the time, the cause is a stuck door or a broken or kinked control cable.

Air conditioning system

41 If the cool air output is inadequate:

a) Inspect the condenser coils and fins to make sure they're clear
b) Check the compressor clutch for slippage.
c) Check the blower motor for proper operation.
d) Inspect the blower discharge passage for obstructions.
e) Check the system air intake filter for clogging.

42 If the system provides intermittent cooling air:

a) Check the circuit breaker, blower switch and blower motor for a malfunction.
b) Make sure the compressor clutch isn't slipping.
c) Inspect the plenum door to make sure it's operating properly.
d) Inspect the evaporator to make sure it isn't clogged.

e) If the unit is icing up, it may be caused by excessive moisture in the system, incorrect super heat switch adjustment or low thermostat adjustment.

43 If the system provides no cooling air:

a) Inspect the compressor drivebelt. Make sure it's not loose or broken.
b) Make sure the compressor clutch engages. If it doesn't, check for a blown fuse.
c) Inspect the wire harness for broken or disconnected wires.
d) If the compressor clutch doesn't engage, bridge the terminals of the A/C pressure switch(es) with a jumper wire; if the clutch now engages, and the system is properly charged, the pressure switch is bad.
e) Make sure the blower motor is not disconnected or burned out.
f) Make sure the compressor isn't partially or completely seized.
g) Inspect the refrigerant lines for leaks.
h) Check the components for leaks.
i) Inspect the receiver-drier/accumulator or expansion valve/tube for clogged screens.

44 If the system is noisy:

a) Look for loose panels in the passenger compartment.
b) Inspect the compressor drivebelt. It may be loose or worn.
c) Check the compressor mounting bolts. They should be tight.
d) Listen carefully to the compressor. It may be worn out.
e) Listen to the idler pulley and bearing and the clutch. Either may be defective.
f) The winding in the compressor clutch coil or solenoid may be defective.
g) The compressor oil level may be low.
h) The blower motor fan bushing or the motor itself may be worn out.
i) If there is an excessive charge in the system, you'll hear a rumbling noise in the high pressure line, a thumping noise in the compressor, or see bubbles or cloudiness in the sight glass.
j) If there's a low charge in the system, you might hear hissing in the evaporator case at the expansion valve, or see bubbles or cloudiness in the sight glass.

3 Air conditioning and heating system - check and maintenance

Air conditioning system

Warning: *The air conditioning system is under high pressure. Do not loosen any hose fittings or remove any components until after the system has been discharged. Air conditioning refrigerant should be properly discharged into an EPA-approved recovery/recycling unit at a dealer service department or an automotive air conditioning repair facility. Always wear eye protection when disconnecting air conditioning system fittings.*

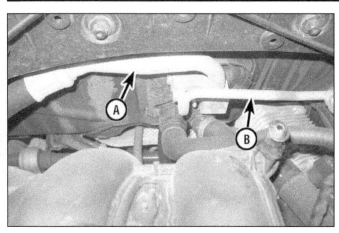

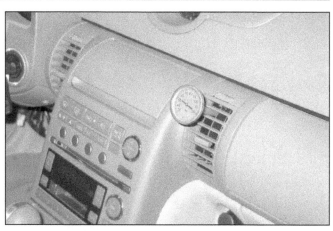

3.8 With the air conditioning system operating, the larger pipe (A) coming from the expansion valve outlet should be cold, while the high-pressure pipe (B) should be warm if the refrigerant charge is satisfactory

3.9 Insert a thermometer in the center vent, turn on the air conditioning system and wait for it to cool down; depending on the humidity, the output air should be 35 to 40 degrees cooler than the ambient air temperature

Caution 1: *All models covered by this manual use environmentally friendly R-134a. This refrigerant (and its appropriate refrigerant oils) are not compatible with R-12 refrigerant system components and must never be mixed or the components will be damaged.*

Caution 2: *When replacing entire components, additional refrigerant oil should be added equal to the amount that is removed with the component being replaced. Be sure to read the can before adding any oil to the system, to make sure it is compatible with the R-134a system.*

1   The following maintenance checks should be performed on a regular basis to ensure that the air conditioning continues to operate at peak efficiency.

a) *Inspect the condition of the compressor drivebelt. If it is worn or deteriorated, replace it (see Chapter 1).*

b) *Check the drivebelt tension (see Chapter 1).*

c) *Inspect the system hoses. Look for cracks, bubbles, hardening and deterioration. Inspect the hoses and all fittings for oil bubbles or seepage. If there is any evidence of wear, damage or leakage, replace the hose(s).*

d) *Inspect the condenser fins for leaves, bugs and any other foreign material that may have embedded itself in the fins. Use a fin comb or compressed air to remove debris from the condenser.*

e) *Make sure the system has the correct refrigerant charge.*

f) *If you hear water sloshing around in the dash area or have water dripping on the carpet, the evaporator housing drain tube is probably clogged. On these vehicles the drain tube is difficult to access. It is in the center of the longitudinal hump, directly underneath the HVAC housing, underneath the instrument panel, and exits under the vehicle, above the exhaust system heat shield (to check it requires removal of the exhaust system [see Chapter 4] and the heat shield [see Chapter 7A, Section 3]).*

2   It's a good idea to operate the system for about ten minutes at least once a month. This is particularly important during the winter months because long term non-use can cause hardening, and subsequent failure, of the seals. Note that using the Defrost function operates the compressor.

3   If the air conditioning system is not working properly, proceed to Step 6 and perform the general checks outlined below.

4   Because of the complexity of the air conditioning system and the special equipment necessary to service it, in-depth troubleshooting and repairs beyond checking the refrigerant charge and the compressor clutch operation are not included in this manual. However, simple checks and component replacement procedures are provided in this Chapter. For more complete information on the air conditioning system, refer to the *Haynes Automotive Heating and Air Conditioning Manual.*

5   The most common cause of poor cooling is simply a low system refrigerant charge. If a noticeable drop in system cooling ability occurs, one of the following quick checks will help you determine if the refrigerant level is low.

Checking the refrigerant charge

Refer to illustrations 3.8 and 3.9

6   Warm the engine up to normal operating temperature.

7   Place the air conditioning temperature selector at the coldest setting and put the blower at the highest setting.

8   After the system reaches operating temperature, feel the larger pipe exiting the evaporator at the firewall. The outlet pipe should be cold (the tubing that leads back to the compressor) **(see illustration)**. If the evaporator outlet pipe is warm, the system probably needs a charge.

9   Insert a thermometer in the center air distribution duct **(see illustration)** while operating the air conditioning system at its maximum setting - the temperature of the output air should be 35 to 40 degrees F below the ambient air temperature (down to approximately 40 degrees F). If the ambient (outside) air temperature is very high, say 110 degrees F, the duct air temperature may be as high as 60 degrees F, but generally the air conditioning is 35 to 40 degrees F cooler than ambient air.

10   Further inspection or testing of the system requires special tools and techniques and is beyond the scope of the home mechanic.

Adding refrigerant

Refer to illustrations 3.11 and 3.13

Caution: *Make sure any refrigerant, refrigerant oil or replacement component you purchase is designated as compatible with R-134a systems.*

11   Purchase an R-134a automotive charging kit at an auto parts store **(see illustration)**. A charging kit includes a can of refrigerant, a tap valve and a short section of hose that can be attached between the tap valve and the system low side service valve. **Caution:** *Never add more than one can of refrigerant to the system. If more refrigerant than that is required, the system should be evacuated and leak tested.*

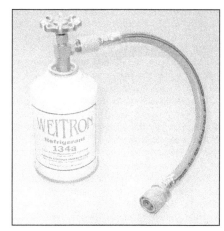

3.11 R-134a automotive air conditioning charging kit

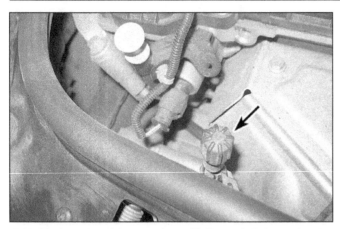

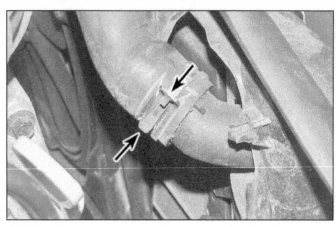

3.13 The low-side charging port on Mk II models is located in the right-front corner of the engine compartment, near the oil dipstick

4.1 Use pliers to squeeze together the two tabs of the hose clamp

12   Back off the valve handle on the charging kit and screw the kit onto the refrigerant can, making sure first that the O-ring or rubber seal inside the threaded portion of the kit is in place. **Warning:** *Wear protective eyewear when dealing with pressurized refrigerant cans.*

13   Remove the dust cap from the low-side charging port and attach the hose's quick-connect fitting to the port **(see illustration)**. **Warning:** *DO NOT hook the charging kit hose to the system high side!* The fittings on the charging kit are designed to fit **only** on the low side of the system.

14   Warm up the engine and turn On the air conditioning. Keep the charging kit hose away from the fan and other moving parts. **Note:** *The charging process requires the compressor to be running. If the clutch cycles off, you can put the air conditioning switch on High and leave the car doors open to keep the clutch on and compressor working. The compressor can be kept on during the charging by removing the connector from the pressure switch and bridging it with a paper clip or jumper wire during the procedure.*

15   Turn the valve handle on the kit until the stem pierces the can, then back the handle out to release the refrigerant. You should be able to hear the rush of gas. Keep the can upright at all times, but shake it occasionally. Allow stabilization time between each addition. **Note:** *The charging process will go faster if you wrap the can with a hot-water-soaked rag to keep the can from freezing up.*

16   If you have an accurate thermometer, you can place it in the center air conditioning duct inside the vehicle and keep track of the output air temperature. A charged system that is working properly should cool down to approximately 40 degrees F. If the ambient (outside) air temperature is very high, say 110 degrees F, the duct air temperature may be as high as 60 degrees F, but generally the air conditioning is 35 to 40 degrees F cooler than the ambient air.

17   When the can is empty, turn the valve handle to the closed position and release the connection from the low-side port. Reinstall the dust cap.

18   Remove the charging kit from the can and store the kit for future use with the piercing valve in the UP position, to prevent inad-

vertently piercing the can on the next use.

Heating systems

19   If the carpet under the heater core is damp, or if antifreeze vapor or steam is coming through the vents, the heater core is leaking. Remove it (see Section 12) and install a new unit (most radiator shops will not repair a leaking heater core).

20   If the air coming out of the heater vents isn't hot, the problem could stem from any of the following causes:

a) *The thermostat is stuck open, preventing the engine coolant from warming up enough to carry heat to the heater core. Replace the thermostat (see Section 7).*

b) *There is a blockage in the system, preventing the flow of coolant through the heater core. Feel both heater hoses at the firewall. They should be hot. If one of them is cold, there is an obstruction in one of the hoses or in the heater core, or the heater control valve is shut. Detach the hoses and back flush the heater core with a water hose. If the heater core is clear but circulation is impeded, remove the two hoses and flush them out with a water hose.*

c) *If flushing fails to remove the blockage from the heater core, the core must be replaced (see Section 12).*

Eliminating air conditioning odors

21   Unpleasant odors that often develop in air conditioning systems are caused by the growth of a fungus, usually on the surface of the evaporator core. The warm, humid environment there is a perfect breeding ground for mildew to develop.

22   The evaporator core on most vehicles is difficult to access, and factory dealerships have a lengthy, expensive process for eliminating the fungus by opening up the evaporator case and using a powerful disinfectant and rinse on the core until the fungus is gone. You can service your own system at home, but it takes something much stronger than basic household germ-killers or deodorizers.

23   Aerosol disinfectants for automotive air

conditioning systems are available in most auto parts stores, but remember when shopping for them that the most effective treatments are also the most expensive. The basic procedure for using these sprays is to start by running the system in the RECIRC mode for ten minutes with the blower on its highest speed. Use the highest heat mode to dry out the system and keep the compressor from engaging by disconnecting the wiring connector at the compressor.

24   The disinfectant can usually comes with a long spray hose. Remove the cabin air filter element (see Chapter 1), insert the nozzle into the HVAC intake and spray according to the manufacturer's recommendations.

25   Once the evaporator has been cleaned, the best way to prevent the mildew from coming back again is to make sure your evaporator housing drain tube is clear.

Automatic heating and air conditioning systems

26   Some vehicles are equipped with an optional automatic climate control system. This system has its own computer that receives inputs from various sensors in the heating and air conditioning system. This computer, like the PCM, has self-diagnostic capabilities to help pinpoint problems or faults within the system. Vehicles equipped with automatic heating and air conditioning systems are very complex and considered beyond the scope of the home mechanic. Vehicles equipped with automatic heating and air conditioning systems should be taken to dealer service department or other qualified facility for repair.

---

4   Cooling system hoses – disconnection and replacement

Refer to illustration 4.1

Warning: *Wait until the engine is completely cool before beginning this procedure.*

Note: *If the coolant is not due for replacement, it may be re-used, providing it is collected in a clean container.*

1   Spring-type hose clamps are used

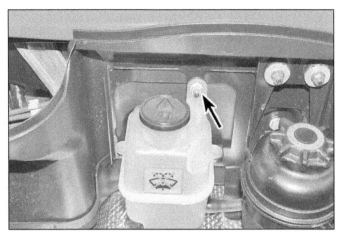

5.2a Coolant expansion tank mounting nut (Mk I models)

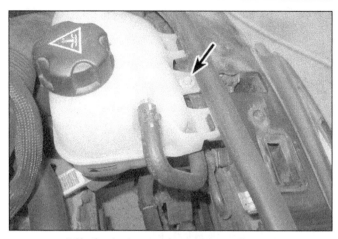

5.2b Coolant expansion tank mounting nut (Mk II models)

throughout the MINI cooling system. To release these clamps, squeeze the tags together using pliers, at the same time working the clamp away from the hose fitting **(see illustration)**.

2 On Mk II models, certain fittings use wire-type retainers. To disconnect a hose from one of these fittings, pull the wire retainer off the hose end, then detach the hose from the fitting

3 Unclip any wires, cables or other hoses which may be attached to the hose being removed. Make notes for future reference when installing, if necessary.

4 To disconnect a hose, release the retaining clamps and move them along the hose, clear of the relevant inlet/outlet. Carefully work the hose free. The hoses can be removed with relative ease when new - on an older car, they may have stuck.

5 If a hose proves to be difficult to remove, try to release it by rotating its ends before attempting to free it. Gently pry the end of the hose with a blunt instrument (such as a flat-bladed screwdriver), but do not apply too much force, and take care not to damage the fitting. Note in particular that the radiator fittings are fragile; do not use excessive force when attempting to remove the hose. If all else fails, slit the hose with a sharp knife so that it can be peeled off. Although this may prove expensive if the hose is otherwise undamaged, it is preferable to buying a new radiator. Check first, however, that a new hose is readily available.

6 When installing a hose, first slide the clamps onto the hose, then work the hose into position.

7 Work the hose into position, checking that it is correctly routed, then slide each clamp back along the hose until it passes over the flared end of the relevant inlet/outlet.

8 Refill the cooling system (see Chapter 1), then warm up the engine and check for leaks.

5 Coolant expansion tank - removal and installation

Refer to illustrations 5.2a and 5.2b

Warning: *Wait until the engine is completely*

cool before beginning this procedure.

Note: *If the coolant is not due for replacement, it may be re-used, providing it is collected in a clean container.*

1 Referring to Chapter 1, drain the cooling system sufficiently to empty the contents of the expansion tank. Do not drain any more coolant than is necessary.

2 Remove the bolt/nut securing the expansion tank to the support bracket on the engine compartment firewall **(see illustrations)**.

3 Lift the tank from place, then release the clamp(s) and disconnect the hose(s).

4 Disconnect the level sensor electrical connector, if equipped.

5 Installation is the reverse of removal, ensuring the hose(s) is/are securely reconnected. On completion, top-up the coolant level as described in Chapter 1.

6 Radiator - removal, inspection and installation

Warning: *Wait until the engine is completely cool before beginning this procedure.*

Note: *If the coolant is not due for replace-*

ment, it may be re-used, providing it is collected in a clean container.

Removal

1 If you're working on an Mk II model, have the air conditioning system discharged by a licensed air conditioning technician.

2 Drain the cooling system (see Chapter 1).

Mk I models

Refer to illustrations 6.4, 6.6 and 6.8

3 Place the Modular Front End in the service position (see Chapter 11). On models with a CVT transaxle, remove the oil cooler.

4 Release the clamps and disconnect the coolant hoses from the radiator **(see illustration)**.

5 Remove the two retaining screws and lift the air conditioning condenser from its mounting lugs. Position the condenser to one side. Do not disconnect the refrigerant lines, and take great care not to damage them.

6 Pry up the center pins, then lever out the plastic expansion rivets at the top corners of the radiator **(see illustration)**.

7 Trace the wiring back from the cooling fan, release it from the cable ties, and disconnect the electrical connector.

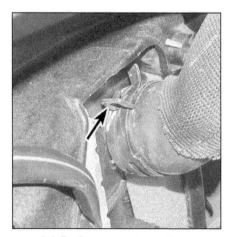

6.4 Radiator upper hose clamp

6.6 Pry up the center pins, then remove the plastic expansion rivets from the top corners of the radiator (Mk I models)

6.8 Check to be sure the radiator lower mounts are in good condition

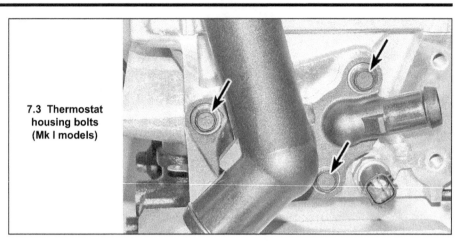

7.3 Thermostat housing bolts (Mk I models)

8   Lift the radiator from the lower mounts **(see illustration)**.

Mk II models
9   Remove the front bumper (see Chapter 11).
10   Remove the air conditioning condenser (see Section 15).
11   Squeeze the tangs and unclip the top of the radiator from the radiator support panel.
12   Pull the radiator forward to access the hose clamps, then detach the hoses.
13   Pull the radiator up and out, being careful not to damage the cooling fins.

Inspection
14   If the radiator has been removed due to suspected blockage, reverse-flush it as described in Chapter 1. Clean dirt and debris from the radiator fins, using compressed air (in which case, wear eye protection) or a soft brush. Be careful, as the fins are sharp, and easily damaged.
15   If necessary, a radiator specialist can perform a flow test on the radiator, to establish whether an internal blockage exists.
16   A leaking radiator must be referred to a

specialist for permanent repair. Do not attempt to weld or solder a leaking radiator, as damage to the plastic components may result. In most cases, however, it makes more sense to replace the radiator with a new or rebuilt unit.
17   Inspect the condition of the radiator rubber mounts, and replace them if necessary.

Installation
18   Installation is a reversal of removal, bearing in mind the following points:
  a) *Ensure that the lower lugs on the radiator are correctly engaged with the mounting rubbers.*
  b) *Make sure the cooling fan shroud is properly engaged.*
  c) *On completion, refill the cooling system as described in Chapter 1.*

---

7   Thermostat - removal, testing and installation

Warning: *Wait until the engine is completely cool before beginning this procedure.*

Note: *If the coolant is not due for replacement, it may be re-used, providing it is collected in a clean container.*

Removal
1   Drain the cooling system (see Chapter 1).

Mk I models
Refer to illustrations 7.3, 7.4 and 7.6

Cooper
2   The thermostat is installed in the coolant housing on the left-hand end of the cylinder head. To improve access, remove the battery and battery tray (see Chapter 5).
3   Disconnect the hoses from the housing, then remove the three bolts and remove the housing **(see illustration)**.
4   Pull the thermostat and sealing ring from the housing **(see illustration)**.

Cooper S
5   Remove the air filter housing (see Chapter 4).
6   Working at the left-hand end of the cylinder head, disconnect the oxygen sensor electrical connector, and the MAP sensor electrical connector, then remove the MAP sensor complete with mounting bracket **(see illustration)**. Note that the two mounting bracket bolts also secure the thermostat housing to the cylinder head
7   Release the clamps and disconnect the hoses from the thermostat housing.
8   Remove the remaining bolt and remove

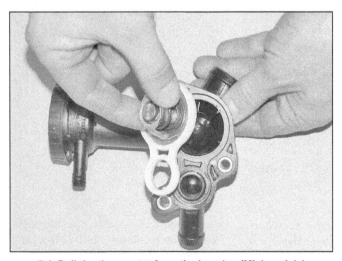

7.4 Pull the thermostat from the housing (Mk I models)

7.6 MAP sensor bracket

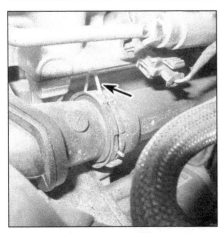

7.10 Remove this wire retainer from the coolant pipe behind the left end of the cylinder head (Mk II models)

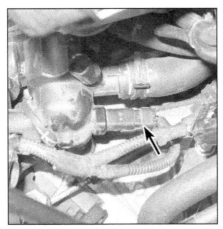

7.11 Thermostat electrical connector (Mk II models)

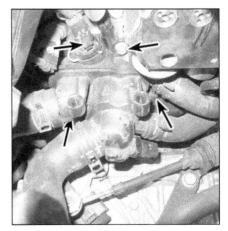

7.13 Thermostat housing mounting bolts (two not visible) and ECT electrical connector (Mk II models)

the thermostat housing from the cylinder head.
9   Pull the thermostat and sealing ring from the housing **(see illustration 7.4)**.

Mk II models

Refer to illustrations 7.10, 7.11 and 7.13

10   Pull out the wire retainer from the coolant pipe behind the left end of the cylinder head **(see illustration)**.
11   Disconnect the electrical connector from the thermostat housing **(see illustration)** and the Engine Coolant Temperature (ECT) sensor.
12   Loosen the clamps and detach the hoses from the thermostat housing.
13   Remove the thermostat mounting bolts and detach the thermostat housing from the cylinder head **(see illustration)**.

Testing

14   A rough test of the thermostat may be made by suspending it with a piece of string in a container full of water. Heat the water to bring it to a boil – the thermostat must open by the time the water boils. If not, replace it.
15   A thermostat which fails to close as the water cools must also be replaced.

Installation

Refer to illustration 7.16

16   Installation is a reversal of removal, bearing in mind the following points:
  a) *Mk I models: Examine the sealing ring for damage or deterioration, and if necessary, replace. Note that the seal is not available separately from the thermostat.*
  b) *Mk I models: Ensure that the thermostat is installed facing the correct direction, with the spring facing into the cylinder head, with the bleed valve at the top* **(see illustration)**.
  c) *Mk II models: Be sure to use a new sealing ring on the housing. Note that the thermostat is not available separately from the housing.*
  d) *Refill the cooling system (see Chapter 1).*

---

8   Engine cooling fan - removal and installation

Note: *If the coolant is not due for replacement, it may be re-used, providing it is col-*

lected in a clean container.
Note: *The engine cooling fan is controlled by the Powertrain Control Module (PCM), via inputs from the Engine Coolant Temperature (ECT) sensor. Mk I models incorporate two relays into the circuit: the low-speed relay, located in the engine compartment fuse/relay box, and the high-speed relay, located on the fan shroud, near the resistor. On Mk II models, the PCM also uses inputs from other sensors and controls fan operation/speed through a module mounted on the fan shroud, near the fan motor.*

Mk I models

Warning: *Wait until the engine is completely cool before beginning this procedure.*

Cooling fan

Refer to illustrations 8.2 and 8.3

1   Remove the radiator (see Section 6).
2   Release the retaining clips and lift the fan and shroud from the radiator **(see illustration)**.
3   If required, remove the bolts and sepa-

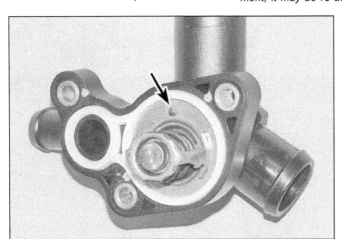

7.16 On 2006 and earlier models, install the thermostat with the spring facing the cylinder head, and the bleed valve at the top

8.2 Release the clips and remove the fan and shroud from the radiator (Mk I models)

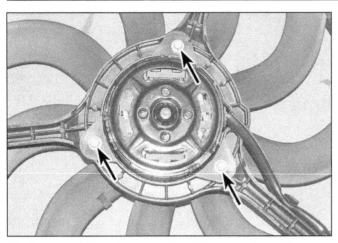

8.3 Remove the three bolts and detach the motor from the shroud (Mk I models)

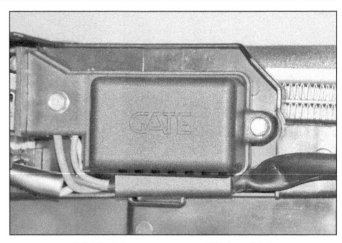

8.6 Resistor cover screws (Mk I models)

rate the fan motor from the shroud **(see illustration)**.

4   Installation is a reversal of removal.

Cooling fan resistor

Refer to illustrations 8.6 and 8.7

5   Remove the radiator (see Section 6).

6   Remove plastic cover over the resistor **(see illustration)**.

8.7 To replace the resistor, the wires must be resoldered

7   The existing resistor must be removed, and the new resistor soldered into place **(see illustration)**. Check the availability of a new resistor before unsoldering the old one.

8   The remainder of installation is a reversal of removal.

Mk II models

Refer to illustrations 8.10, 8.11a and 8.11b

9   If you're working on an S model, remove the left-side charge air duct.

10   Disconnect the cooling fan electrical connector **(see illustration)**.

11   Remove the upper and lower fan shroud mounting fasteners **(see illustrations)**.

12   Rotate the fan shroud counterclockwise (as viewed from the rear) to unlock it, then remove the fan and shroud assembly.

13   Installation is the reverse of removal. Make sure the fan shroud engages properly and clicks into place in its mounts.

---

9   Water pump - removal and installation

Warning: *Wait until the engine is completely cool before beginning this procedure.*

Note: *If the coolant is not due for replacement, it may be re-used, providing it is collected in a clean container.*

Mk I models
Removal
Cooper

Refer to illustration 9.5

1   Drain the cooling system (see Chapter 1).

2   Remove the front bumper, bumper carrier and front wheelwell liners, then place the Modular Front End (MFE) in the service position (see Chapter 11). Remove the crush tubes from each side.

3   Remove the alternator (see Chapter 5).

4   Release the clamps and disconnect the hoses from the water pump.

5   Remove the bolts and detach the water pump **(see illustration)**.

Cooper S

Refer to illustration 9.7

6   On Cooper S models, the water pump is bolted to the left-hand end of the supercharger. To remove the water pump, remove the supercharger as described in Chapter 4. Discard the water pump-to-cylinder block O-ring seal; a

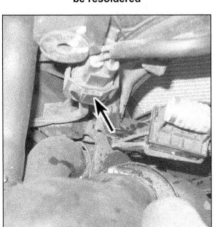

8.10 Cooling fan motor electrical connector (Mk II models)

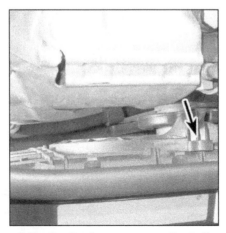

8.11a Cooling fan upper mounting bolt (Mk II models)

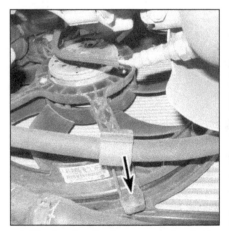

8.11b Cooling fan lower mounting bolt (Mk II models)

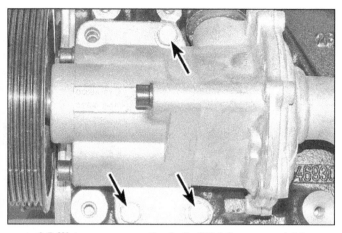

9.5 Water pump mounting bolts (Mk I Cooper models)

9.7 Water pump mounting bolts (Mk I Cooper S models)

new one must be installed.

7   Remove the retaining bolts and separate the water pump from the supercharger **(see illustration)**.

Installation

Cooper

8   Install the pump assembly to the cylinder block, tightening its retaining bolts securely.

9   The remainder of installation is the reverse of removal.

10   Refill the cooling system (see Chapter 1).

Cooper S

Refer to illustration 9.11

11   Install the water pump to the super-charger, noting how the pump drive lugs engage with the lugs on the supercharger **(see illustration)**. Tighten the bolts to the specified torque.

12   Install the supercharger (see Chapter 4).

13   Refill the cooling system (see Chapter 1).

Mk II models

Main water pump (all models)

Refer to illustrations 9.15 and 9.17

14   Drain the cooling system (see Chapter 1).

15   Loosen the water pump pulley bolts **(see illustration)**, then place the friction wheel in the service position (see Section 10, Step 3).

16   Remove the bolts and detach the pulley from the water pump, then remove the friction wheel assembly (see Section 10).

17   Remove the water pump mounting bolts and detach the pump from the engine block **(see illustration)**.

18   Clean the mating surface of the engine block (and the pump, if the same one is to be installed).

19   Installation is the reverse of removal. Be sure to use a new water pump seal. Also make sure the surface of the pulley that contacts the friction wheel is in good condition. If not, replace it.

20   Refill the cooling system (see Chapter 1).

Auxiliary water pump (S models)

Refer to illustration 9.23

Note: *The auxiliary water pump is mounted to the front of the transaxle, below the left end of the turbocharger.*

21   Drain the cooling system (see Chapter 1). Remove the mounting bolt and reposition the coolant reservoir.

22   Remove the intake hose and the charge

air hose from the turbocharger.

23   Disconnect the electrical connector from the pump **(see illustration)**.

24   Loosen the clamps and disconnect the hoses from the pump.

9.11 The dogs of the supercharger engage with the driveshaft of the water pump (Mk I Cooper S models)

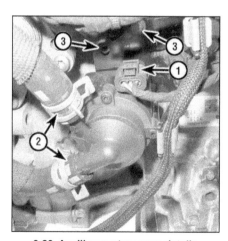

9.23 Auxiliary water pump details (Mk II S models)

1   *Electrical connector*
2   *Coolant hoses*
3   *Mounting bolts*

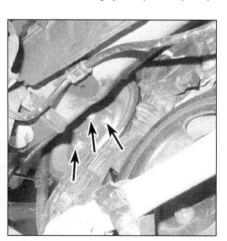

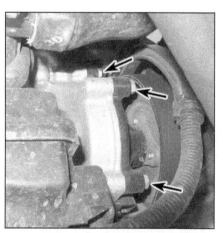

9.15 Water pump pulley bolts (Mk II models)

9.17 Water pump mounting bolts (seen from the rear, three of five bolts shown) (Mk II models)

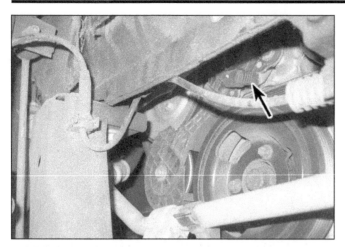

10.3 Pull this strap all the way out to place the friction wheel in the service position

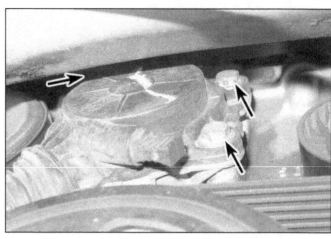

10.5 Friction wheel mounting bolts (upper rear bolt not visible) (Mk II models)

25   Remove the mounting bolts and detach the auxiliary water pump from the transmission.
26   Installation is the reverse of removal.
27   Refill the cooling system (see Chapter 1).

---

10   Friction wheel assembly (Mk II models) - removal and installation

---

Refer to illustrations 10.3 and 10.5

1   Loosen the right-front wheel bolts. Raise the front of the vehicle and support it securely on jackstands, then remove the wheel.
2   Remove the right wheelwell liner (see Chapter 11).
3   Loosen the water pump pulley bolts **(see illustration 9.15)**, then pull the strap out all the way and place the friction wheel in the service position **(see illustration)**.
4   Remove the bolts from the water pump pulley, then remove the pulley.
5   Unscrew the mounting bolts and detach the friction wheel **(see illustration)**.
6   Installation is the reverse of removal.

---

11   Heater/air conditioning control unit - removal and installation

---

Warning: *These models are equipped with a*

Supplemental Restraint System (SRS), more commonly known as airbags. Always disable the airbag system before working in the vicinity of any airbag system component to avoid the possibility of accidental deployment of the airbag(s), which could cause personal injury (see Chapter 12).
Warning: *Do not use a memory saving device to preserve the PCM or radio memory when working on or near airbag system components.*

Mk I models

Refer to illustrations 11.3, 11.4 and 11.5

1   Disconnect the cable from the negative terminal of the battery (see Chapter 5).
2   Remove the front center console as described in Chapter 11. Remove the audio unit as described in Chapter 12.
3   Remove the switch panel from below the heater control panel **(see illustration)**.
4   Remove the plastic captive nuts from each side of the panel, then depress the clips and push the heater control inwards **(see illustration)**.
5   Starting with the left-hand side first, pull the control panel from the instrument panel, then disconnect the control cables and electrical connectors **(see illustration)**. Maneuver the unit from the instrument panel.
6   Installation is the reverse of removal.

Ensure the control cables are correctly reconnected and securely held by the retaining clips; check the operation of the control knobs before securing the control panel to the instrument panel.

Mk II models

Refer to illustrations 11.8 and 11.9

7   Disconnect the cable from the negative terminal of the battery (see Chapter 5). Remove the tachometer (see Chapter 12).
8   Pry the left side trim panel, forward of the console, from the center part of the instrument panel **(see illustration)**. Also pry out the panel below the control unit at the front of the console.
9   Remove the screws from both sides of the center instrument panel bezel that surrounds the control unit **(see illustration)**.
10   Remove the center console (see Chapter 11), then pull back the center bezel/control unit far enough to disconnect the electrical connectors.
11   Remove the four screws and separate the control unit from the bezel.
12   Installation is the reverse of removal. If installing a new unit to a vehicle produced before 12/2007, you'll have to cut the recess

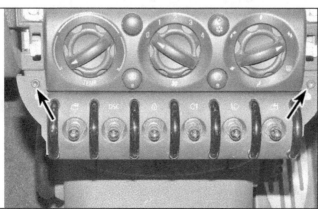

11.3 Heater/air conditioning control unit mounting screws

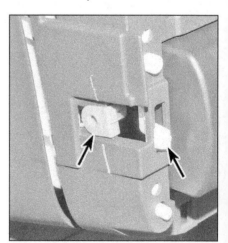

11.4 Remove the plastic nut and release the panel retaining clip

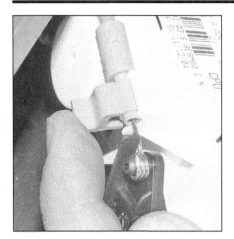

11.5 Release the retaining clip, and detach the heater cable from the control lever

11.8 Carefully pry off this trim panel from the lower part of the instrument panel

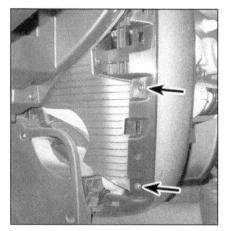

11.9 Remove the center bezel screws from each side

in the instrument panel behind the control unit a little wider (7/8-inch [23 mm]) to make room for the new, larger unit.

12 Heater core - replacement

Warning: *Wait until the engine is completely cool before beginning this procedure.*
Warning: *These models are equipped with a Supplemental Restraint System (SRS), more commonly known as airbags. Always disable the airbag system before working in the vicinity of any airbag system component to avoid the possibility of accidental deployment of the airbag(s), which could cause personal injury (see Chapter 12).*
Warning: *Do not use a memory saving device to preserve the PCM or radio memory when working on or near airbag system components.*
Note: *If the coolant is not due for replacement, it may be re-used, providing it is collected in a clean container.*
1   Drain the cooling system (see Chapter 1). Disconnect the cable from the negative terminal of the battery (see Chapter 5).

Mk I models
Refer to illustrations 12.3, 12.4, 12.6 and 12.8
2   To improve access to the heater hoses on the firewall on Cooper models, remove the battery and battery tray (see Chapter 5). On Cooper S models, remove the air filter housing (see Chapter 4).
3   Loosen the hose clamps and disconnect the hoses from the heater core pipes on the engine compartment firewall **(see illustration)**. If available, blow compressed air through the heater core to remove as much coolant from it as possible.
4   Remove the heater core cover from the driver's side footwell, below the center of the instrument panel **(see illustration)**. Note that access to the uppermost screw is extremely limited.
5   Position towels and a container beneath the heater core pipes on the left-hand side of the heating/ventilation housing to catch any spilled coolant.
6   Remove the Allen screws securing the coolant pipe clamps, and move the clamps down the pipes **(see illustration)**.
7   Free the pipes from the core, catching the

coolant in the container, then remove the two Torx screws and slide the core from the housing. Keep the open fittings angled upwards as the core is removed to minimize coolant spillage. Discard the clamps and rubber seals, as new ones must be installed. Note that new clamps are supplied with the new seals.

12.3 Loosen the clamps and disconnect the heater hoses at the firewall

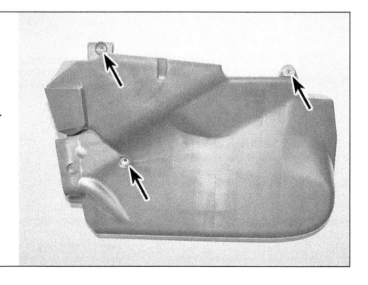

12.4 The heater core cover is secured by three screws

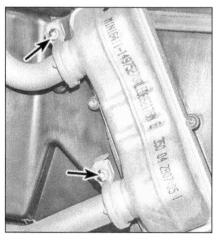

12.6 Loosen the Allen screws and slide the pipe clamps away from the heater core

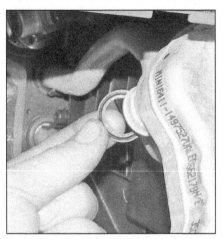

12.8 Install new seals to the heater core fittings

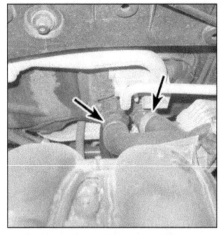

12.16 Loosen the clamps and disconnect the heater hoses at the firewall

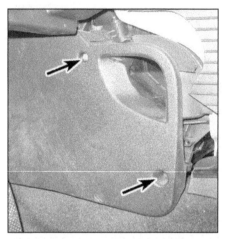

12.17 Remove these screws and slide the panel downward to detach it

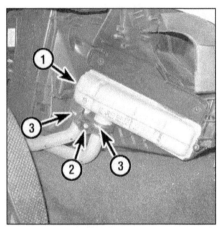

12.19 Heater core details

1   *Retaining screw*
2   *Pipe clamp screw*
3   *Protrusions on pipe flanges (must be engaged with openings in the clamp)*

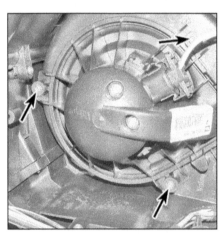

13.2 Remove the screws and withdraw the heater blower motor

13.4 Blower motor resistor screw

8   Install a new sealing ring to each of the fittings of the heater core **(see illustration)**.
9   Where necessary, replace the foam seal around the core, then ease the core into the housing, engaging the coolant pipes with the core fittings, then secure it in place with the two Torx screws. Take care not to displace the seals as the pipes are inserted.
10   Secure the pipes to the core with the new clamps. Make sure the clamps fit correctly over the flanges.
11   Install the cover and tighten the screws securely.
12   Reconnect the coolant hoses at the firewall in the engine compartment and secure the clamps.
13   Install the battery and battery tray or air filter housing as applicable.
14   Refill the cooling system (see Chapter 1).

Mk II models

Refer to illustrations 12.16, 12.17 and 12.19

15   Remove the air filter housing (see Chapter 4).

16   Loosen the hose clamps and disconnect the hoses from the heater core pipes on the engine compartment firewall **(see illustration)**. If available, blow compressed air through the heater core to remove as much coolant from it as possible.
17   Remove the screws and detach the panel below the center of the instrument panel, just inboard of the accelerator pedal **(see illustration)**.
18   Position towels and a container beneath the heater core pipes to catch any spilled coolant.
19   Remove the screw at the forward part of the heater core **(see illustration)**.
20   Remove the pipe clamp screw, then pull the core out slightly and disconnect the pipes from the heater core. Have some rags handy to catch any coolant that spills out.
21   Remove the heater core from the housing.
22   Installation is the reverse of removal. Be sure to use new seals on the pipe ends that fit into the heater core, and make sure the pipes engage the core properly, with their flange protrusions visible through the slots in the clamp.
23   Refill the cooling system (see Chapter 1).

13   Blower motor and resistor - removal and installation

Warning: *These models are equipped with a Supplemental Restraint System (SRS), more commonly known as airbags. Always disable the airbag system before working in the vicinity of any airbag system component to avoid the possibility of accidental deployment of the airbag(s), which could cause personal injury (see Chapter 12).*
Warning: *Do not use a memory saving device to preserve the PCM or radio memory when working on or near airbag system components.*

Mk I models

Blower motor

Refer to illustration 13.2

1   Remove the instrument panel (see Chapter 11).
2   Disconnect the electrical connector, then remove the three Torx screws and withdraw the motor from the heater housing **(see illustration)**.
3   Installation is a reversal of removal.

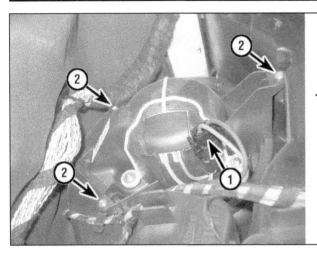

13.7 Blower motor details

1   Electrical connector
2   Mounting screws

13.9 The blower motor resistor (A) is located below the blower motor (B)

Blower motor resistor

Refer to illustration 13.4

4   Working in the passenger's side footwell, disconnect the wiring plug, remove the Torx screw and pull out the resistor **(see illustration)**.

5   Installation is a reversal of removal.

Mk II models

6   Remove the screws and detach the panel below the center of the instrument panel, just inboard of the accelerator pedal **(see illustration 12.17)**.

Blower motor

Refer to illustration 13.7

7   Disconnect the electrical connector, remove the screws and guide the blower motor out of the heater housing **(see illustration)**.

8   Installation is the reverse of removal.

Blower motor resistor

Refer to illustration 13.9

9   Push up on the retaining tang at the bottom of the resistor and remove it from the housing **(see illustration)**.

10   Disconnect the electrical connector from the resistor.

11   Installation is the reverse of removal.

---

14   Air conditioning compressor - removal and installation

Note: *If a new compressor is being installed, also install a new receiver/drier. Additionally, if the system has been opened for more than 24 hours, a new receiver/drier should be installed.*

Mk I models

Removal

Refer to illustrations 14.11a and 14.11b

1   Have the air conditioning system discharged by a licensed air conditioning technician.

2   Disconnect the cable from the negative terminal of the battery (see Chapter 5).

3   Raise the front of the vehicle and support it securely on jackstands. Block the rear wheels.

4   Remove the under-vehicle splash shield.

5   Place the Modular Front End (MFE) in the service position (see Chapter 11).

6   If the vehicle is equipped with a CVT transaxle, remove the fluid cooler and carefully position it out of the way (don't disconnect the fluid lines).

7   Remove the drivebelt (see Chapter 1).

8   Remove the air conditioning condenser (see Section 15).

9   Disconnect the compressor electrical connector from the engine harness.

10   Unscrew the nuts securing the refrigerant line retaining plates to the compressor. Separate the lines from the compressor and quickly seal the lines and compressor unions to prevent the entry of moisture into the refrigerant circuit. Discard the sealing rings, as new ones must be used on installation. **Warning:** *Failure to seal the refrigerant line unions will result in the receiver/drier becoming saturated, necessitating its replacement.*

11   Unscrew the compressor mounting bolts, then free the compressor from its mounting bracket and remove it from the engine **(see illustrations)**.

Installation

Note: *If a new compressor is being installed, follow the directions that came with the compressor regarding the draining of excess oil prior to installation.*

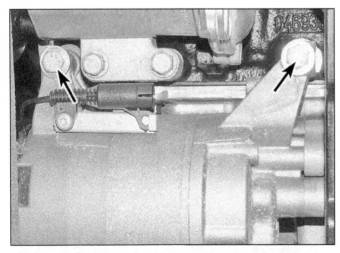

14.11a Air conditioning compressor upper mounting bolts . . .

14.11b . . . and lower mounting bolt

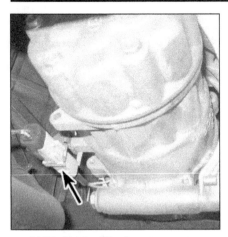

14.23 Air conditioning compressor electrical connector

12   Maneuver the compressor into position and fit the mounting bolts. Tighten the compressor front end mounting bolt to the specified torque first, followed by the rear bolt, and finally, the lower mounting bolt.

13   Lubricate the new refrigerant line sealing rings with compressor oil. Remove the plugs and install the sealing rings then quickly fit the refrigerant lines to the compressor. Ensure the refrigerant lines are correctly joined then install the retaining bolt, tightening it securely.

14   Reconnect the electrical connector, then install the drivebelt (see Chapter 1).

15   The remainder of installation is the reverse of removal.

16   Have the air conditioning system recharged with the correct type and amount of refrigerant and oil by the shop that discharged it. **Caution:** *Don't operate the system until this is done.*

17   If a new compressor has been installed, it is essential to carry out the following running-in procedure:

    a)   *Switch on the air conditioning system.*
    b)   *Set the air vents to the open position.*
    c)   *Start the engine and let the idle speed stabilize.*
    d)   *Set the blower motor to at least 75% of maximum output.*

    e)   *Allow the air conditioning system to run for at least 2 minutes at engine idle speed. Any faster could cause damage to the compressor.*

Mk II models

Removal

Refer to illustration 14.23

18   Have the air conditioning system discharged by a licensed air conditioning technician.

19   Disconnect the cable from the negative terminal of the battery (see Chapter 5).

20   Raise the front of the vehicle and support it securely on jackstands. Block the rear wheels.

21   Remove the under-vehicle splash shield.

22   Remove the drivebelt (see Chapter 1).

23   Disconnect the compressor electrical connector **(see illustrations)**.

24   Unscrew the nuts securing the refrigerant pipes retaining plates to the compressor. Separate the pipes from the compressor and quickly seal the pipe and compressor unions to prevent the entry of moisture into the refrigerant circuit. Discard the sealing rings, as new ones must be used on installation. **Warning:** *Failure to seal the refrigerant pipe unions will result in the receiver/drier becoming saturated, necessitating its replacement.*

25   Unscrew the compressor mounting bolts then free the compressor from its mounting bracket and remove it from the engine.

Installation

Note: *If a new compressor is being installed, follow the directions that came with the compressor regarding the draining of excess oil prior to installation.*

26   Maneuver the compressor into position and fit the mounting bolts, tightening the bolts to the torque listed in this Chapter's Specifications.

27   Lubricate the new refrigerant line sealing rings with compressor oil. Remove the plugs and install the sealing rings then quickly fit the refrigerant lines to the compressor. Ensure the refrigerant lines are correctly joined then install the retaining bolt, tightening it securely.

28   Reconnect the electrical connector, then install the drivebelt (see Chapter 1).

29   The remainder of installation is the reverse of removal.

30   Have the air conditioning system recharged with the correct type and amount of refrigerant and oil by the shop that discharged it. **Caution:** *Don't operate the system until this is done.*

31   If a new compressor has been installed, it is essential to carry out the following running-in procedure:

    a)   *Switch on the air conditioning system.*
    b)   *Set the air vents to the open position.*
    c)   *Start the engine and let the idle speed stabilize.*
    d)   *Set the blower motor to at least 75% of maximum output.*
    e)   *Allow the air conditioning system to run for at least 2 minutes at engine idle speed. Any faster could cause damage to the compressor.*

15   Air conditioning condenser - removal and installation

Note: *If a new condenser is being installed because the old one was damaged, also install a new receiver/drier. Additionally, if the system has been opened for more than 24 hours, a new receiver/drier should be installed.*

1   Have the air conditioning system discharged by a licensed air conditioning technician.

Mk I models

Removal

Refer to illustrations 15.3 and 15.4

2   Remove the front bumper and bumper carrier (see Chapter 11).

3   Remove the retaining bolts and disconnect the refrigerant lines from the left-hand side of the condenser. Recover the O-ring seals **(see illustration)**.

4   Remove the two retaining screws and lift the condenser from its mounting lugs **(see illustration)**.

15.3 Refrigerant line-to-condenser fitting bolts

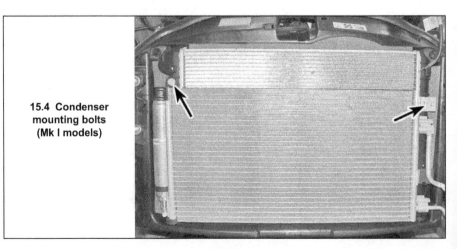

15.4 Condenser mounting bolts (Mk I models)

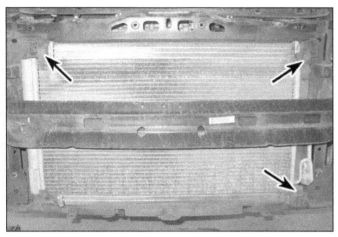

15.10 Condenser mounting bolts (Mk II models)

16.3 Receiver/drier bracket screw (Mk I models)

Installation

5   Installation is a reversal of removal. Noting the following points:

  a) *Ensure the mounting rubbers are correctly fitted then seat the condenser in position in the MFE (Modular Front End).*

  b) *Lubricate the new sealing rings with compressor oil.*

  c) *Have the air conditioning system recharged with the correct type and amount of refrigerant and oil by the shop that discharged it.* **Caution:** *Don't operate the system until this is done.*

Mk II models

Removal

Refer to illustration 15.10

6   Remove the front bumper cover (see Chapter 11).

7   Raise the front of the vehicle and support it securely on jackstands. Block the rear wheels.

8   Remove the under-vehicle splash shield.

9   From below, disconnect the refrigerant line fittings from the left side of the condenser. Also remove the fastener from above the upper refrigerant line fitting.

10   Remove the condenser mounting bolts **(see illustration).** Lift the condenser out from behind the bumper bar, being careful not to damage the fins on the condenser or radiator.

Installation

11   Installation is a reversal of removal. Noting the following points:

  a) *Ensure the mounting rubbers are correctly fitted then seat the condenser in position in the MFE (Modular Front End).*

  b) *Lubricate the new sealing rings with compressor oil.*

  c) *Have the air conditioning system recharged with the correct type and amount of refrigerant and oil by the shop that discharged it.* **Caution:** *Don't operate the system until this is done.*

16   Air conditioning receiver/drier - removal and installation

1   Have the air conditioning system discharged by a licensed air conditioning technician.

Mk I models

Refer to illustration 16.3

2   Remove the front bumper and bumper carrier (see Chapter 11).

3   Remove the screw, and remove the bracket from the top of the receiver/drier, then unscrew it from the condenser **(see illustration).**

4   Installation is a reversal of removal noting the following points:

  a) *Lubricate the seal with compressor oil.*

  b) *Have the air conditioning system recharged with the correct type and amount of refrigerant and oil by the shop that discharged it.* **Caution:** *Don't operate the system until this is done.*

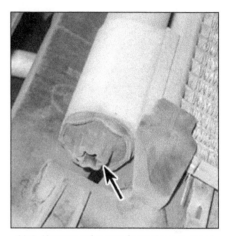

16.6 Unscrew the cap from the receiver/drier, then remove the desiccant cartridge

Mk II models

Refer to illustration 16.6

5   Remove the condenser (see Section 15).

6   Unscrew the cap from the bottom of the receiver/drier **(see illustration).** Pull out the desiccant cartridge.

7   Installation is the reverse of removal. Be sure to use new O-rings and lubricate them with clean refrigerant oil.

8   Have the air conditioning system recharged with the correct type and amount of refrigerant and oil by the shop that discharged it. ***Caution:*** *Don't operate the system until this is done.*

Notes

Chapter 4
Fuel and exhaust systems

Contents

Specifications

Note: *Throughout this Chapter you will find references to "Mk I" and "Mk II" models; this is done to simplify which specifications and procedures apply to which models. Mk I models include 2006 and earlier Cooper/Cooper S models, and 2008 and earlier Convertible models. Mk II models include 2007 and later Cooper/Cooper S/Clubman/Clubman S and 2009 and later Convertible models.*

Fuel system pressure (approximate)
Mk I models
    Cooper .... 39.5 to 47 psi (280 to 320 kPa)
    Cooper S .... 47.8 to 53.8 psi (320 to 370 kPa)
Mk II models
    Cooper .... 50.7 psi (350 kPa)
    Cooper S .... 72.5 psi (500 kPa)

Torque specifications

Note: *One foot-pound (ft-lb) of torque is equivalent to 12 inch-pounds (in-lbs) of torque. Torque values below approximately 15 foot-pounds are expressed in inch-pounds, because most foot-pound torque wrenches are not accurate at these smaller values.*

	Ft-lbs (unless otherwise specified)	Nm
Mk I models		
Air outlet duct-to-supercharger clamp bolts	18	25
Water pump-to-supercharger bolts	18	25
Fuel rail mounting bolts	18	25
Intercooler mounting bracket bolts	84 in-lbs	10
Supercharger outlet housing bolts	18	25
Supercharger mounting bolts		
M8	18	25
M10	33	45
Throttle body bolts	84 in-lbs	10

Torque specifications (continued)   **Ft-lbs** (unless otherwise specified)   **Nm**

Note: *One foot-pound (ft-lb) of torque is equivalent to 12 inch-pounds (in-lbs) of torque. Torque values below approximately 15 foot-pounds are expressed in inch-pounds, because most foot-pound torque wrenches are not accurate at these smaller values.*

Mk II models

	Ft-lbs	Nm
Fuel rail mounting bolts		
Non-turbo models	71 in-lbs	8
Turbo models	168 in-lbs	19
Fuel rail bracket-to-intake manifold stud (turbo models)	15	20
High-pressure fuel line tube nuts (turbo models)		
N14 engine		
Step 1	132 in-lbs	15
Step 2	24	33
Step 3 (after running the engine for 5 minutes)	24	33
N18 engine	19	26
High-pressure fuel pump-to-cylinder head fasteners (turbo models)		
N14 engine	96 in-lbs	11
N18 engine	84 in-lbs	10
Throttle body bolts		
To used intake manifold	61 in-lbs	7
To new intake manifold	84 in-lbs	10

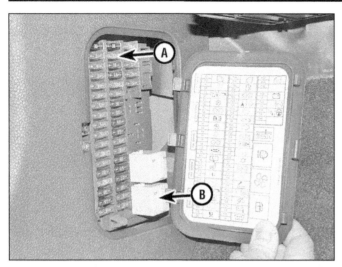

2.2a The fuel pump fuse (A) is located in the under-dash fuse/ relay box in the driver's side kick panel; (B) is the fuel pump relay (Mk I models)

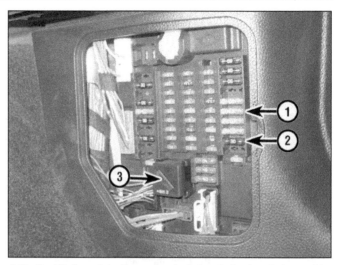

2.2b On Mk II models, the under-dash fuse/relay box is located in the passenger's kick panel

1   Fuel pump fuse - 2007 through 2010 models
2   Fuel pump fuse - 2011 models
3   Fuel pump relay

1   General information

Note: *Throughout this Chapter you will find references to "Mk I" and "Mk II" models; this is done to simplify which specifications and procedures apply to which models. Mk I models include 2006 and earlier Cooper/Cooper S models, and 2008 and earlier Convertible models. Mk II models include 2007 and later Cooper/Cooper S/Clubman/Clubman S and 2009 and later Convertible*

Fuel system warnings

Gasoline is extremely flammable and repairing fuel system components can be dangerous. Consider your automotive repair knowledge and experience before attempting repairs which may be better suited for a professional mechanic.

- Don't smoke or allow open flames or bare light bulbs near the work area
- Don't work in a garage with a gas-type appliance (water heater, clothes dryer)
- Use fuel-resistant gloves. If any fuel spills on your skin, wash it off immediately with soap and water
- Clean up spills immediately
- Do not store fuel-soaked rags where they could ignite
- Prior to disconnecting any fuel line, you must relieve the fuel pressure (see Section 3)
- Wear safety glasses
- Have a proper fire extinguisher on hand

Fuel system

The fuel system consists of the fuel tank, electric fuel pump/fuel level sending unit (located in the fuel tank), fuel rail and

fuel injectors. The fuel injection system is a multi-port system; multi-port fuel injection uses timed impulses to inject the fuel directly into the intake port of each cylinder. The Powertrain Control Module (PCM) controls the injectors. The PCM monitors various engine parameters and delivers the exact amount of fuel required into the intake ports.

Fuel is circulated from the fuel pump to the fuel rail through fuel lines running along the underside of the vehicle. Various sections of the fuel line are either rigid metal or nylon, or flexible fuel hose. The various sections of the fuel hose are connected either by quick-connect fittings or threaded metal fittings.

Exhaust system

The exhaust system consists of the exhaust manifold(s), catalytic converter(s), muffler(s), tailpipe and all connecting pipes, flanges and clamps. The catalytic converters are an emission control device added to the exhaust system to reduce pollutants.

2   Troubleshooting

Fuel pump

Refer to illustrations 2.2a and 2.2b

1   The fuel pump is located inside the fuel tank. Sit inside the vehicle with the windows closed, turn the ignition key to ON (not START) and listen for the sound of the fuel pump as it's briefly activated. You will only hear the sound for a second or two, but that sound tells you that the pump is working. Alternatively, have an assistant listen at the fuel filler cap.

2   If the pump does not come on, check the fuel pump fuse and relay **(see illustrations)**. If the fuse and relay are okay, check the wiring back to the fuel pump. If the fuse,

relay and wiring are okay, the fuel pump is probably defective. If the pump runs continuously with the ignition key in the ON position, the Powertrain Control Module (PCM) is probably defective. Have the PCM checked by a professional mechanic.

Fuel injection system

Note: *The following procedure is based on the assumption that the fuel pump is working and the fuel pressure is adequate (see Section 4).*

3   Check all electrical connectors that are related to the system. Check the ground wire connections for tightness.

4   Verify that the battery is fully charged (see Chapter 5).

5   Inspect the air filter element (see Chapter 1).

6   Check all fuses related to the fuel system (see Chapter 12).

7   Check the air induction system between the throttle body and the intake manifold for air leaks. Also inspect the condition of all vacuum hoses connected to the intake manifold and to the throttle body.

8   Remove the air intake duct from the throttle body and look for dirt, carbon, varnish, or other residue in the throttle body, particularly around the throttle plate. If it's dirty, clean it with carb cleaner, a toothbrush and a clean shop towel.

9   With the engine running and if you can access the injectors, place an automotive stethoscope against each injector, one at a time, and listen for a clicking sound that indicates operation. **Warning:** *Stay clear of the drivebelt and any rotating or hot components.*

10   If you can hear the injectors operating, but the engine is misfiring, the electrical circuits are functioning correctly, but the injectors might be dirty or clogged. Try a commercial injector cleaning product (available at auto parts stores). If cleaning the injectors doesn't help, replace the injectors.

3.4 Release the clips and slide the plastic cover up over the injectors

3.5 Remove the plastic cap on the end of the fuel rail . . .

11   If an injector is not operating (it makes no sound), disconnect the injector electrical connector and measure the resistance across the injector terminals with an ohmmeter. Compare this measurement to the other injectors. If the resistance of the non-operational injector is quite different from the other injectors, replace it.

12   If the injector is not operating, but the resistance reading is within the range of resistance of the other injectors, the PCM or the circuit between the PCM and the injector might be faulty.

---

3   Fuel pressure relief procedure

Warning: *Gasoline is extremely flammable. See* **Fuel system warnings** *in Section 1. Also, wait until the engine is completely cool before relieving the fuel pressure.*

1   Remove the fuel filler cap to relieve any pressure built up in the fuel tank, then remove the fuel pump fuse from the under-dash fuse/relay box (see Section 2). Start the engine and allow it to stall (or attempt to start it).

2   Disconnect the cable from the negative terminal of the battery (see Chapter 5). As an added precaution, continue to follow the Steps in this Section.

Mk I models

Refer to illustrations 3.4, 3.5 and 3.6

3   Cooper S models: Remove the intercooler as described in Section 15.

4   Cooper models: Release the retaining clips and pull the plastic cover from above the fuel injectors **(see illustration)**.

5   Remove the cap over the valve at the end of the fuel rail **(see illustration)**.

6   Place shop rags around and over the valve, then depress the valve core with a small screwdriver to relieve the residual pressure **(see illustration)**.

7   Properly dispose of the rags. Proceed to Step 13.

Mk II models

Cooper models

Refer to illustration 3.9

8   Remove the air filter housing (see Section 9).

9   Remove the cap from the Schrader valve at the end of the fuel rail **(see illustration)**.

10   Place shop rags around and over the valve, then depress the valve core with a small screwdriver to relieve the residual pressure.

11   Properly dispose of the rags. Proceed to Step 13.

Cooper S models

12   These models have no Schrader valve on the fuel rail, since the fuel injectors operate under extremely high pressures. The pressure in the fuel system between the fuel tank and the high pressure fuel pump on the left end of the cylinder head should already have been bled down after Step 1. If you're going to work on the high pressure side of the system (between the high pressure pump and the fuel rail), pressure can be reduced there by tapping on the injectors.

All models

13   Cover and surround any fitting to be loosened or disconnected with clean rags, and be sure to wear eye protection. In the case of Mk II S (turbocharged) models, if working on the

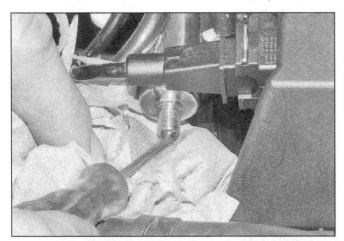

3.6 . . . then depress the valve core to relieve the fuel pressure.
Warning: *Wait until the engine is completely cool. Cover the area with shop rags, and also wear eye protection*

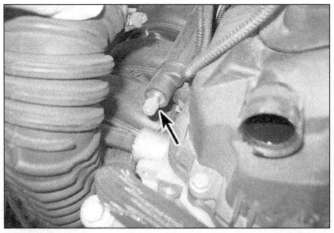

3.9 On Mk II Cooper models, the Schrader valve is on the right-end of the fuel rail

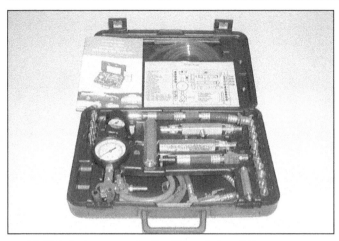

4.2 This fuel pressure testing kit contains all the necessary fittings and adapters, along with the fuel pressure gauge, to test most automotive systems

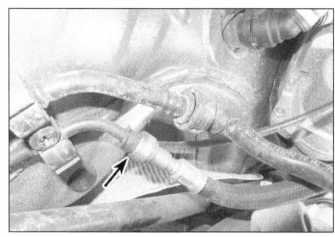

4.13 On Mk II S models, the fuel pressure gauge must be connected at this fuel line fitting, under the vehicle near the steering gear

high-pressure portion of the system, wear full-face protection and thick leather gloves, and loosen threaded fittings slowly so as to allow the pressure to seep out gradually instead of forcefully.

14 Properly dispose of fuel-soaked rags.

4 Fuel pressure - check

Warning: *Gasoline is extremely flammable. See* **Fuel system warnings** *in Section 1.*

Note: *The following procedure assumes that the fuel pump is receiving voltage and runs.*

Mk I models

Refer to illustration 4.2

Note: *Mk I models incorporate a vacuum-controlled pressure regulator on the fuel rail inlet; fuel is supplied to the rail and injectors at a predetermined pressure. When vacuum in the intake manifold drops when the throttle is opened (such as during acceleration), the fuel pressure regulator opens and allows fuel under greater pressure to enter the fuel rail to maintain the predetermined pressure, which otherwise would have been lowered due to the greater demand placed on the fuel injectors. There is no fuel return line.*

1 If you're working on a Cooper S model, remove the intercooler (see Section 15).

2 Locate the fuel pressure test port on the left end of the fuel rail **(see illustration 3.5)**. Unscrew the cap and connect a fuel pressure gauge to the Schrader valve **(see illustration)**.

3 If you're working on a Cooper S model, reinstall the intercooler.

4 Start the engine and allow it to idle. Note the reading on the gauge as soon as the pressure stabilizes, and compare it to the pressure listed in this Chapter's Specifications.

5 If the pressure is higher than specified, unplug the vacuum hose from the pressure regulator and check for the presence of vac-

uum at the hose at idle. If there is no vacuum, check the hose for a leak or restriction. If there is vacuum, replace the fuel pressure regulator (see Section 12).

6 If the pressure is lower than specified, detach the vacuum hose from the pressure regulator while watching the gauge. The pressure should increase. If it doesn't, either the pressure regulator or the fuel pump module is faulty (the filter portion of the module is not available separately).

7 Turn off the engine. Fuel pressure should not fall more than approximately 8 psi over five minutes. If it does, the problem could be a leaky fuel injector, fuel line leak, or a faulty fuel pump module.

8 Relieve the fuel system pressure as described in Section 3, Steps 1 and 2, then remove the pressure gauge. Wipe up any spilled gasoline.

Mk II models

Note: *On Mk II models, the fuel pressure regulator is part of the fuel pump module.*

Cooper models

9 Locate the fuel pressure test port on the left end of the fuel rail **(see illustration 3.9)**. Unscrew the cap and connect a fuel pressure gauge to the Schrader valve.

Cooper S models

Refer to illustration 4.13

10 These models use a direct-injection type of fuel-injection system in which the fuel is injected directly into the cylinders under extremely high pressure, not into the intake ports as on other models. This means that the fuel pressure can't be checked at the fuel rail (the high-pressure side). It can only be checked on the fuel delivery, or low-pressure, side. Because the system is not equipped with a Schrader valve, the fuel pressure gauge must be installed into the fuel line using special adapters, at the fuel line fitting under the vehicle, near the steering gear.

11 Relieve the fuel system pressure as described in Section 3, Steps 1 and 2.

12 Raise the front of the vehicle and support it securely on jackstands. Block the rear wheels to prevent the vehicle from rolling. Remove the transaxle splash shield.

13 Disconnect the fuel line fitting **(see illustration)** and attach the fuel pressure gauge using the proper adapters.

All models

14 Start the engine and allow it to idle. Note the reading on the gauge as soon as the pressure stabilizes, and compare it to the pressure listed in this Chapter's Specifications.

15 If the fuel pressure is not within specifications:

a) *Check for a restriction in the fuel line (kink, etc.). If no restrictions are found replace the fuel pump module (see Section 7).*

b) *If the fuel pressure is higher than specified, replace the fuel pump module (see Section 7).*

16 Turn off the engine. Fuel pressure should not fall more than approximately 8 psi over five minutes. If it does, the problem could be a leaky fuel injector, fuel line leak, or a faulty fuel pump module.

17 Relieve the fuel system pressure as described in Section 3, Steps 1 and 2, then remove the pressure gauge. Wipe up any spilled gasoline.

5 Fuel lines and fittings - general information and disconnection

Warning: *Gasoline is extremely flammable. See* **Fuel system warnings** *in Section 1.*

1 Relieve the fuel pressure before servicing fuel lines or fittings (see Section 3), then disconnect the cable from the negative battery terminal (see Chapter 5) before proceeding.

2 The fuel supply line connects the fuel

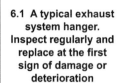

6.1 A typical exhaust system hanger. Inspect regularly and replace at the first sign of damage or deterioration

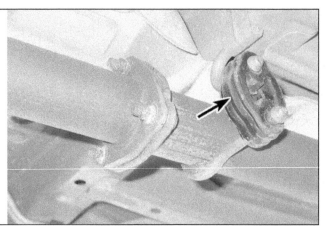

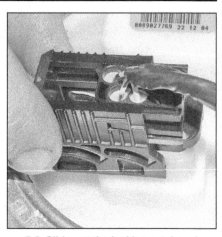

7.5 Slide out the locking catch and disconnect the fuel level sensor/pump electrical connector

pump in the fuel tank to the fuel rail on the engine. The Evaporative Emission (EVAP) system lines connect the fuel tank to the EVAP canister and connect the canister to the intake manifold.

3   Whenever you're working under the vehicle, be sure to inspect all fuel and evaporative emission lines for leaks, kinks, dents and other damage. Always replace a damaged fuel or EVAP line immediately.

4   If you find signs of dirt in the lines during disassembly, disconnect all lines and blow them out with compressed air. Inspect the fuel strainer on the fuel pump pick-up unit for damage and deterioration.

Steel tubing

5   It is critical that the fuel lines be replaced with lines of equivalent type and specification.

6   Some steel fuel lines have threaded fittings. When loosening these fittings, hold the stationary fitting with a wrench while turning the tube nut.

Plastic tubing

7   When replacing fuel system plastic tubing, use only original equipment replacement plastic tubing. **Caution:** *When removing or installing plastic fuel line tubing, be careful not to bend or twist it too much, which can damage it. Also, plastic fuel tubing is NOT heat resistant, so keep it away from excessive heat.*

Flexible hoses

8   When replacing fuel system flexible hoses, use only original equipment replacements.

9   Don't route fuel hoses (or metal lines) within four inches of the exhaust system or within ten inches of the catalytic converter. Make sure that no rubber hoses are installed directly against the vehicle, particularly in places where there is any vibration. If allowed to touch some vibrating part of the vehicle, a hose can easily become chafed and it might start leaking. A good rule of thumb is to maintain a minimum of 1/4-inch clearance around

a hose (or metal line) to prevent contact with the vehicle underbody.

6   Exhaust system servicing - general information

Refer to illustration 6.1

Warning: *Allow exhaust system components to cool before inspection or repair. Also, when working under the vehicle, make sure it is securely supported on jackstands.*

1   The exhaust system consists of the exhaust manifolds, catalytic converter, muffler, tailpipe and all connecting pipes, flanges and clamps. The exhaust system is isolated from the vehicle body and from chassis components by a series of rubber hangers **(see illustration)**. Periodically inspect these hangers for cracks or other signs of deterioration, replacing them as necessary.

2   Conduct regular inspections of the exhaust system to keep it safe and quiet. Look for any damaged or bent parts, open seams, holes, loose connections, excessive corrosion or other defects which could allow exhaust fumes to enter the vehicle. Do not repair deteriorated exhaust system components; replace them with new parts.

3   If the exhaust system components are extremely corroded, or rusted together, a cutting torch is the most convenient tool for removal. Consult a properly-equipped repair shop. If a cutting torch is not available, you can use a hacksaw, or if you have compressed air, there are special pneumatic cutting chisels that can also be used. Wear safety goggles to protect your eyes from metal chips and wear work gloves to protect your hands.

4   Here are some simple guidelines to follow when repairing the exhaust system:

a) *Work from the back to the front when removing exhaust system components.*
b) *Apply penetrating oil to the exhaust system component fasteners to make them easier to remove.*
c) *Use new gaskets, hangers and clamps.*

d) *Apply anti-seize compound to the threads of all exhaust system fasteners during reassembly.*
e) *Be sure to allow sufficient clearance between newly installed parts and all points on the underbody to avoid overheating the floor pan and possibly damaging the interior carpet and insulation. Pay particularly close attention to the catalytic converter and heat shield.*

7   Fuel pump/fuel level sensors - removal and installation

Warning: *Gasoline is extremely flammable. See* **Fuel system warnings** *in Section 1. Also, wait until the engine is completely cool before relieving the fuel pressure.*

Note: *There are two level sensors installed in the fuel tank - one in the left side of the tank, and one in the right side. The pump is integral with the left side sensor, and the filter is integral with the right side sensor.*

Removal

1   Relieve the fuel system pressure (see Section 3).

Left-hand sensor/fuel pump

Refer to illustrations 7.5, 7.6, 7.7, 7.8, 7.9a and 7.9b

2   Remove the rear seat cushion as described in Chapter 11.

3   Unscrew the five nuts, and remove the access cover from the floor.

4   Remove the gasket.

5   Slide out the locking element to disconnect the electrical connector **(see illustration)**.

6   Unscrew the fuel pump/level sensor unit locking ring and remove it from the tank. Although a MINI tool (16 1 020) is available for this task, it can be accomplished using a large pair of pliers to push on two opposite raised ribs on the locking ring. Alternatively, a homemade tool can be fabricated to engage with the raised ribs of the locking ring. Turn the ring

Disconnecting Fuel Line Fittings

Two-tab type fitting; depress both tabs with your fingers, then pull the fuel line and the fitting apart

On this type of fitting, depress the two buttons on opposite sides of the fitting, then pull it off the fuel line

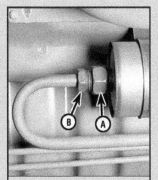

Threaded fuel line fitting; hold the stationary portion of the line or component (A) while loosening the tube nut (B) with a flare-nut wrench

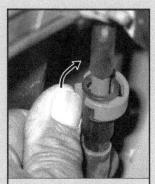

Plastic collar-type fitting; rotate the outer part of the fitting

Metal collar quick-connect fitting; pull the end of the retainer off the fuel line and disengage the other end from the female side of the fitting . . .

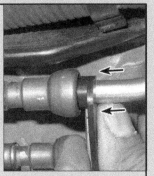

. . . insert a fuel line separator tool into the female side of the fitting, push it into the fitting and pull the fuel line off the pipe

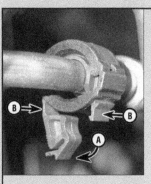

Some fittings are secured by lock tabs. Release the lock tab (A) and rotate it to the fully-opened position, squeeze the two smaller lock tabs (B) . . .

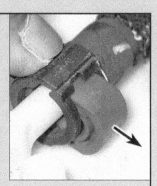

. . . then push the retainer out and pull the fuel line off the pipe

Spring-lock coupling; remove the safety cover, install a coupling release tool and close the tool around the coupling . . .

. . . push the tool into the fitting, then pull the two lines apart

Hairpin clip type fitting: push the legs of the retainer clip together, then push the clip down all the way until it stops and pull the fuel line off the pipe

7.6 Using a homemade tool to unscrew the locking ring

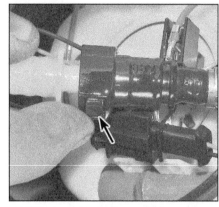

7.7 Depress the button and disconnect the fuel hose

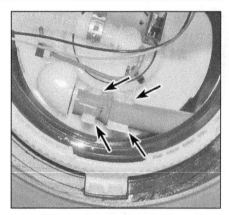

7.8 Release the transfer pipe from the retaining clips

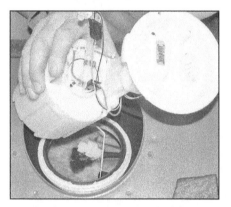

7.9a Take care not to bend the float arm when removing the sensor/pump unit

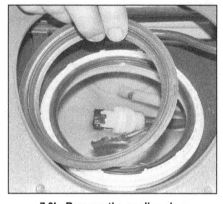

7.9b Remove the sealing ring

7.12 Squeeze together the sides of the retaining clip and disconnect the hose

counterclockwise until it can be unscrewed by hand **(see illustration)**.

7   Lift the sensor cover a little, then unclip the keeper from the fuel hose connection, depress the button and disconnect the fuel hose **(see illustration)**.

8   Disconnect the electrical connector, and release the transfer pipe from the clips on the top of the pump unit **(see illustration)**.

9   Carefully lift the level sensor/pump unit from the tank, taking care not to bend the sensor float arm (gently push the float arm

towards the unit if necessary) **(see illustration)**. Remove the sealing ring **(see illustration)**. No further disassembly is recommended - at the time of writing, it would appear the pump is not available separately from the level sensor.

Right-hand sensor and filter housing

Refer to illustrations 7.12 and 7.14

10   Remove the left-side sensor cover, and disconnect the fuel hoses as described in Steps 2 through 8.

11   Unscrew the five nuts and remove the access cover from over the right-side sensor.

12   Squeeze together the sides of the retaining clip and disconnect the fuel hose **(see illustration)**.

13   Unscrew the locking ring as described in Step 6.

14   Lift the sensor a little, pry out the clip and slide the fuel hose holder down and off the sensor assembly **(see illustration)**.

15   Lift the sensor assembly from the tank, complete with fuel hoses. Remove the sealing ring.

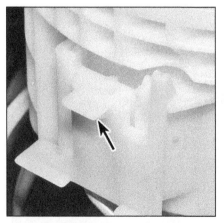

7.14 Pry out the clip and slide the hose holder down and off the sensor

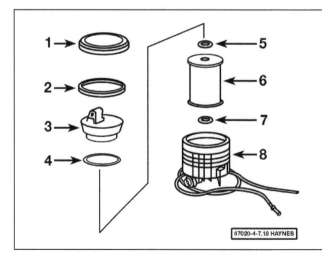

7.18 Right side fuel filter housing details

1   Retaining ring
2   Seal
3   Cap
4   O-ring
5   O-ring
6   Filter element
7   O-ring
8   Housing

67020-4-7.18 HAYNES

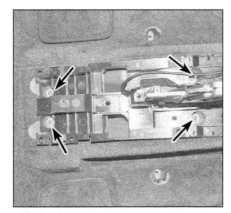

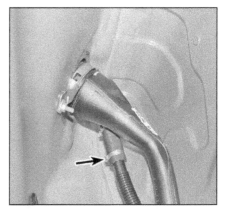

7.22 Ensure the locating lug engages correctly with the slot in the fuel tank collar

8.6 Unscrew the four nuts and remove the rear console mounting bracket (Mk I models)

8.11 Cut off the clamp and disconnect the vent hose

Filter replacement

Refer to illustration 7.18

16   Disconnect the wiring from the small hooks on the side of the housing.
17   Detach the fuel level sensor from the housing.
18   If equipped with a retaining ring at the top of the housing, unscrew it **(see illustration)**. On models without a retaining ring, turn the cap counterclockwise to unlock it. Remove the cap from the housing.
19   Remove the filter element from the housing.
20   Clean all components and replace all O-rings with new ones.
21   Reassembly is the reverse of disassembly.

Installation

Refer to illustration 7.22

22   Installation is a reversal of removal, noting the following points:

a) *Use a new sealing ring.*
b) *To allow the unit to pass through the opening in the fuel tank, insert the float arm first.*
c) *When the unit is installed, the locating lug on the unit must engage with the corresponding slot in the fuel tank collar* **(see illustration)**.
d) *Tighten the locking collar securely.*

8   Fuel tank - removal and installation

Refer to illustrations 8.6, 8.11, 8.12 and 8.14

Warning: *Gasoline is extremely flammable. See* **Fuel system warnings** *in Section 1. Also, wait until the engine is completely cool before relieving the fuel pressure.*

1   Relieve the fuel system pressure (see Section 3), then disconnect the cable from the negative terminal of the battery (see Chapter 5).
2   Before removing the fuel tank, all fuel should be drained from the tank. Since a fuel tank drain plug is not provided, it is preferable to carry out the removal operation when the tank is nearly empty. **Warning:** *If it's necessary to siphon the fuel out, use a siphoning kit, available at most auto parts stores. Never start the siphoning action by mouth.*
3   On Mk I models, loosen the left rear wheel bolts. On Mk II models, loosen the bolts on both rear wheels. Raise the rear of the vehicle and support it securely on jackstands. Remove the wheel(s).
4   On Mk I Cooper S models, detach the rear half of the battery positive cable.
5   On all Mk I models, remove the rear center console as described in Chapter 11.
6   On Mk I models, unscrew the four nuts and remove the rear console mounting

bracket **(see illustration)**. If equipped, disconnect the electrical connectors, remove the screws and remove the DSC motion sensor from the console bracket.
7   Remove the rear section of the exhaust system, and any heat shields that would interfere with removal of the fuel tank.
8   Remove the parking brake cables as described in Chapter 9.
9   Remove the rear seat cushion as described in Chapter 11. Unscrew the nuts and remove the fuel tank access covers. Disconnect the electrical connector from the left-side sensor, and the fuel hose from the right-side sensor.
10   Remove the screws, pry out the plastic rivets and, on Mk I models, remove the left-rear wheel well liner. On Mk II models, remove the wheel well liners from both sides. On convertible models, remove the chassis reinforcement supports from under the fuel tank.
11   Cut the hose clamp and disconnect the fuel vent hose from the filler neck **(see illustration)**.
12   Loosen the clamp and disconnect the fuel filler hose from the tank **(see illustration)**.
13   Support the fuel tank using a floor jack and a wood plank.
14   Remove the retaining bolts securing the tank retaining straps, and the bolt at the center-front of the tank **(see illustration)**.

8.12 Working underneath the vehicle, loosen the clamp and disconnect the fuel filler hose from the tank

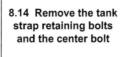

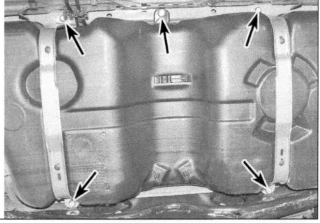

8.14 Remove the tank strap retaining bolts and the center bolt

9.1 Use a screwdriver to release the air hose clamp (Mk I Cooper models)

9.2 Air filter housing retaining bolts (Mk I Cooper models)

9.4 Slide the battery positive connection up from its bracket on the air filter housing (Mk I Cooper S models)

15   Lower the tank slightly, and disconnect the vent hose connection above the tank.

9.5 Release the clip and disconnect the outlet duct from the air cleaner cover (Mk I Cooper S models)

16   Lower the tank and maneuver it from under the vehicle.
17   Installation is the reverse of removal.

---

9   Air filter housing - removal and installation

Mk I models
Cooper
Refer to illustrations 9.1 and 9.2

1   Release the hose clamp and disconnect the air hose from the top of the air filter housing **(see illustration)**. Also release the wiring harness clip and detach the harness from the top of the housing.
2   Remove the two retaining bolts, then lift the housing from position, detaching it from the intake duct at the front of the housing **(see illustration)**.
3   Installation is the reverse of removal.

Cooper S
Refer to illustrations 9.4, 9.5, 9.6a, 9.6b, 9.6c, 9.7 and 9.8

4   Pull up the battery positive connection point from its clips, located on the side of the air filter housing, and position it to one side **(see illustration)**.
5   Release the clamp and disconnect the outlet ducting from the housing cover **(see illustration)**.
6   Remove the cover, then release the clips and slide the Powertrain Control Module (PCM) upwards from position. Slide out the locking catches and disconnect the electrical connectors, then remove the PCM **(see illustrations)**.
7   Remove the air filter housing retaining bolt, then unclip the inlet ducting from the MFE (Modular Front End), and disconnect it from the housing **(see illustration)**.
8   Slide the air filter housing forward and maneuver it from position. Note the rubber

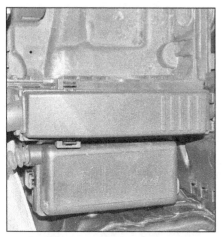

9.6a Remove the cover . . .

9.6b . . . release the retaining clips . . .

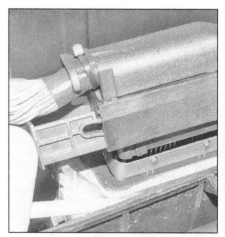

9.6c . . . then lift the PCM a little, slide out the locking catches and disconnect the electrical connectors (Mk I Cooper S models)

9.7 Air filter housing bolt (Mk I Cooper S models)

9.8 The base of the air filter housing locates over two rubber mounts on the firewall (Mk I Cooper S models)

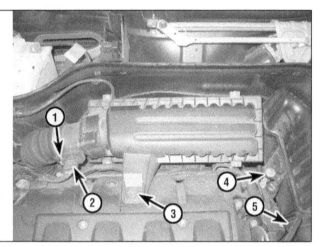

9.10 Air filter housing details (Mk II Cooper models)

1 MAF sensor hose clamp
2 MAF sensor electrical connector
3 Air filter housing mounting bolt
4 Intake air resonator mounting bolt
5 Fresh air intake duct hose clamp

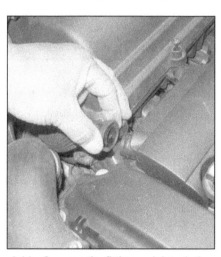

9.14a Squeeze the fitting and detach the crankcase breather hose from the valve cover. . .

mountings on the firewall panel **(see illustration)**.

9 Installation is a reversal of removal.

Mk II models

Cooper

Refer to illustration 9.10

10 Loosen the hose clamp from the Mass Airflow (MAF) sensor and the fresh air intake duct hose clamp **(see illustration)**.

11 Disconnect the electrical connector from the Mass Airflow (MAF) sensor.

12 Remove the air filter housing/intake air resonator assembly from the ducts.

13 Installation is the reverse of removal.

Cooper S

Refer to illustrations 9.14a, 9.14b and 9.15

14 Detach the crankcase breather hose from the right rear corner of the valve cover, reposition the hose, then remove the filter housing mounting screw underneath **(see illustrations)**.

15 Disconnect the electrical connector from the Mass Airflow (MAF) sensor **(see illustration)**.

16 Loosen the hose clamp and detach the air intake duct from the filter housing.

17 Detach the fresh air intake duct from the filter housing by squeezing it to disengage the tabs from the housing.

18 Remove the air filter housing.

19 Installation is the reverse of removal.

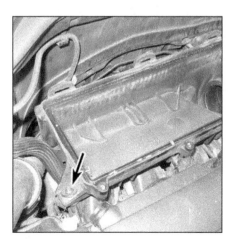

9.14b . . . then remove this filter housing fastener (Mk II Cooper S models)

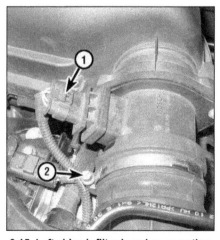

9.15 Left side air filter housing mounting details (Mk II Cooper S models)

1 MAF sensor electrical connector
2 Air intake duct hose clamp

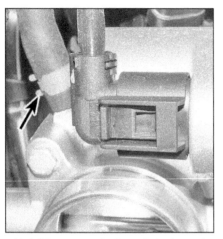

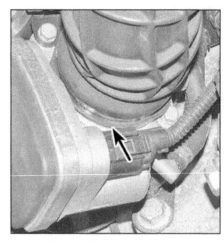

10.4 Disconnect the hose from the tank vent valve, and also disconnect the electrical connector

10.6 Throttle body mounting bolts - Cooper models

10.11 Release the clamp and disconnect the intake duct

---

10 Throttle body - removal and installation

Mk I models

Cooper

Refer to illustrations 10.4 and 10.6

1   Remove the air filter housing as described in Section 9.

2   Release the clips and pull up the plastic cover from above the fuel injectors.

3   Release the hose clamp and remove the intake hose from the throttle body.

4   Depress the clip and disconnect the hose from the tank vent valve, then release the hose from the clips along the fuel rail (**see illustration**).

5   Disconnect the electrical connectors from the throttle body.

6   Remove the four bolts and remove the throttle body from the intake manifold (**see illustration**).

7   Discard the sealing ring, as a new one must be installed.

8   Installation is a reversal of removal, but install a new sealing ring. **Note:** *If a new throttle body has been installed, the engine management PCM must be recoded and the throttle body matched using dedicated test equipment - this task must be entrusted to a MINI dealer or specialist.*

Cooper S

Refer to illustrations 10.11, 10.12, 10.13a and 10.13b

9   Remove the air filter housing as described in Section 9.

10   Disconnect the electrical connector from the throttle body.

11   Release the clamp and disconnect the air intake duct from the throttle body (**see illustration**).

12   Release the clamp and disconnect the fuel tank purge valve hose from the throttle body (**see illustration**).

13   Remove the four bolts, lift the throttle body slightly to disengage it from the locating pins on the supercharger intake duct, then slide it to the left and lift it out (**see illustra-**

tions). If necessary, loosen the bolts securing the support bracket to the transaxle bellhousing.

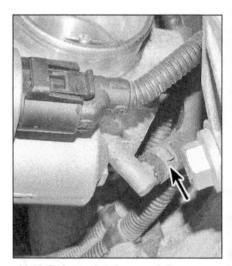

10.12 Release the clamp and disconnect the fuel tank vent valve hose

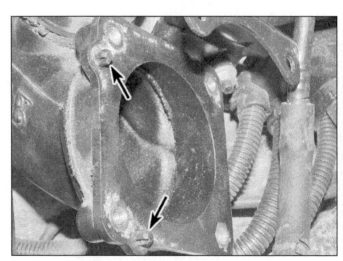

10.13a Throttle body mounting bolts - Cooper S models

10.13b Note the throttle body locating pins

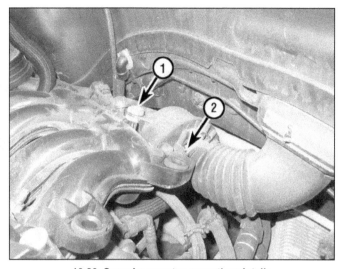

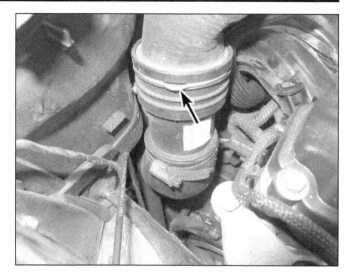

10.20 Sound generator mounting details

1   *Mounting bolt*

2   *Air duct hose clamp*
    (other clamp not visible)

10.21 Pull out the wire spring clamp and detach the air hose at the right end of the intake manifold

14   Installation is a reversal of removal, but install a new sealing ring. **Note:** *If a new throttle body has been installed, the engine management PCM must be recoded and the throttle body matched using dedicated test equipment - this task must be entrusted to a MINI dealer or specialist.*

Mk II models

Cooper

15   Remove the air filter housing (see Section 9). Also loosen the clamp and remove the duct from the throttle body.
16   Remove the throttle body mounting screws.
17   Detach the throttle body from the intake manifold, then disconnect its electrical connector.

18   Installation is a reversal of removal, but install a new sealing ring. **Note:** *If a new throttle body has been installed, the engine management PCM must be recoded and the throttle body matched using dedicated test equipment - this task must be entrusted to a MINI dealer or specialist.*

Cooper S

Refer to illustrations 10.20, 10.21 and 10.22

19   Remove the air filter housing (see Section 9).
20   Remove the sound generator **(see illustration)**.
21   Loosen the hose clamp, pry out the wire spring clamp and remove the air hose **(see illustration)**.
22   Loosen the clamp and disconnect the charge air duct from the throttle body **(see illustration)**.

23   Remove the throttle body mounting screws.
24   Detach the throttle body from the intake manifold, then disconnect its electrical connector.
25   Installation is a reversal of removal, but install a new sealing ring. **Note:** *If a new throttle body has been installed, the engine management PCM must be recoded and the throttle body matched using dedicated test equipment - this task must be entrusted to a MINI dealer or specialist.*

---

11   Fuel rail and injectors - removal and installation

Warning: *Gasoline is extremely flammable. See* **Fuel system warnings** *in Section 1. Also, wait until the engine is completely cool before relieving the fuel pressure.*
1   Relieve the fuel system pressure (see Section 3). Disconnect the cable from the negative terminal of the battery (see Chapter 5).

Mk I models

Cooper

Refer to illustration 11.3

2   Release the clips and remove the plastic cover from the over the injectors **(see illustration 3.4)**.
3   Slide up the locking catch, squeeze together the sides of the top of the connector, and disconnect the connectors from the fuel injectors **(see illustration)**. Release the clips and pull the wiring connector rail from the bracket.

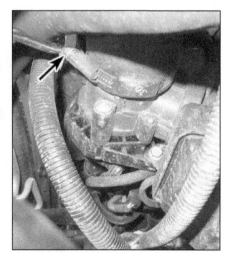

10.22 Charge air duct-to-throttle body hose clamp (viewed from below)

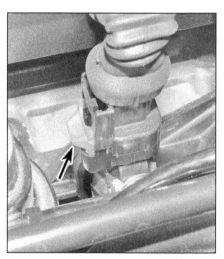

11.3 Slide up the red locking catch and squeeze together the sides of the top of the connector, then pull up to disconnect

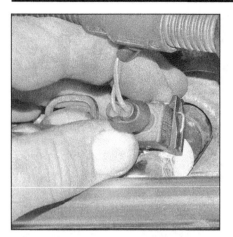

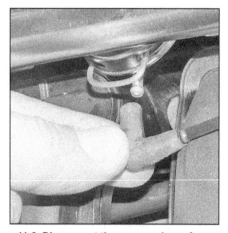

11.4 Depress the wire clip and pull the connector from the injector

11.6 Disconnect the vacuum hose from the regulator

11.7 Disconnect the fuel supply hose from the fuel rail

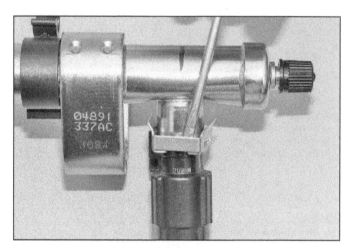

11.8 Remove the bolts and pull the fuel rail, with the injectors, from the manifold

11.9a Pry out the clip and detach the injector from the fuel rail

Cooper S

Refer to illustration 11.4

4   Press the wire locking clips in, then disconnect the electrical connectors from the top of the injectors **(see illustration)**.

11.9b The injector O-ring seals should be replaced whenever the fuel rail/injectors are removed

All models

Refer to illustrations 11.6, 11.7, 11.8, 11.9a and 11.9b

5   Release the wiring harness clips from the fuel rail.
6   Pull the vacuum hose from the regulator **(see illustration)**.
7   Disconnect the fuel supply hose from the fuel rail **(see illustration)**. Be prepared for fuel spillage.
8   Remove the two bolts and pull the fuel rail, with injectors, from the intake manifold **(see illustration)**. Unclip the vent valve hose from the fuel rail (if equipped).
9   Pry out the retaining clips and remove the injectors from the fuel rail **(see illustration)**. Replace the injector O-rings **(see illustration)**.
10   Lightly lubricate the fuel injector O-rings with a little petroleum jelly or a light film of engine oil.
11   Install the injectors to the fuel rail, and retain them in place with the clips pushed into the grooves.
12   The remainder of installation is a reversal of removal.

Mk II models

Cooper

Refer to illustrations 11.14 and 11.15

13   Remove the air filter housing (see Section 9).
14   Detach the fuel line from the fuel rail **(see illustration)**.
15   Remove the fuel rail mounting bolts **(see illustration)**.
16   Pull the fuel rail and injectors up, then disconnect the electrical connectors from the injectors.
17   Pry out the retaining clips and remove the injectors from the fuel rail **(see illustration 11.9a)**. Replace the injector O-rings **(see illustration 11.9b)**.
18   Lightly lubricate the fuel injector O-rings with a little petroleum jelly or a light film of engine oil.
19   Install the injectors to the fuel rail, and retain them in place with the clips pushed into the grooves.
20   The remainder of installation is a reversal of removal.

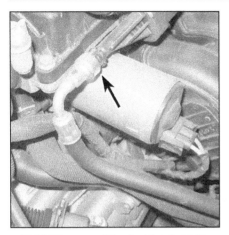

11.14 Disconnect the fuel line from the fuel rail (Mk II Cooper models)

11.15 Right-side fuel rail mounting bolt location (Mk II Cooper models)

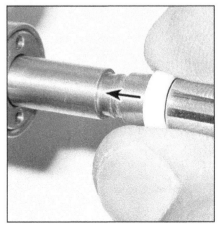

11.29 To remove the Teflon sealing ring, cut it off with a hobby knife (be careful not to scratch the injector groove)

Cooper S

Refer to illustrations 11.29, 11.34, 11.35, 11.36a, 11.36b and 11.37

Warning: *Even after relieving the fuel system pressure on the fuel delivery (low-pressure) side of the system, fuel in the fuel rail remains under extremely high pressure. Be sure to review the information in Section 3, Steps 12 and 13.*

21   Remove the intake manifold (see Chapter 2A).

22   Unscrew the tube nuts at each end of the high-pressure fuel line (between the fuel rail and the high-pressure fuel pump) (see Section 17). **Warning:** *Be sure to wear full face protection, long sleeves and thick leather gloves, and cover the fittings with shop rags while slowly loosening them so fuel under pressure in the line bleeds out slowly. Also, the manufacturer states that it is necessary to replace the high-pressure fuel line (between the fuel rail and the high-pressure fuel pump) with a new one whenever it is removed.*

23   Disconnect the electrical connector from the high-pressure fuel sensor.

24   Remove any retainers/tie-wraps and detach the wiring harness from the fuel rail.

25   Depress the tabs and disconnect the electrical connectors from the fuel injectors.

26   Remove the fuel rail mounting bolts, then pull the fuel rail off the fuel injectors.

27   Remove the spring steel retainers from the injectors. Discard the retainers - new ones must be used during installation.

28   Pull the injectors from the cylinder head. Special tools are available to remove injectors that are stuck (MINI special tool nos. 13 0 231 and 13 0 232). Sometimes improvised methods will work, too, but be careful not to damage the injectors.

29   Remove the old combustion chamber Teflon sealing ring and the upper O-ring and support ring from each injector **(see illustration)**. **Caution:** *Be extremely careful not to damage the groove for the seal or the rib in the floor of the groove. If you damage the groove or the rib, you must replace the injector.*

30   Before installing the new Teflon seal on each injector, thoroughly clean the groove for the seal and the injector shaft. Remove all combustion residue and varnish with a clean shop rag.

Teflon seal installation using the special tools

31   The manufacturer recommends that you use the tools included in the special injector tool set to install the Teflon lower seals on the injectors: Install the special seal assembly cone on the injector, install the special sleeve on the injector and use the sleeve to push on the assembly cone, which pushes the Teflon seal into place on its groove. Do NOT use any lubricants to do so.

32   Pushing the Teflon seal into place in its groove expands it slightly. There are three sizing sleeves in the special tool set with progressively smaller inside diameters. Using a clockwise rotating motion of about 180 degrees, install the slightly larger sleeve onto the injector and over the Teflon seal until the sleeve hits its stop, then carefully turn the sleeve counterclockwise as you pull it off the

11.34 Slide the new Teflon seal onto the end of a socket that's the same diameter as the end of the fuel injector . . .

injector. Use the slightly smaller sizing sleeve the same way, followed by the smallest sizing ring. The seal is now sized. Repeat this step for each injector.

Teflon seal installation without special tools

33   If you don't have the special injector tool set, the Teflon seal can be installed using this method: First, find a socket that is equal or very close in diameter to the diameter of the end of the fuel injector.

34   Work the new Teflon seal onto the end of the socket **(see illustration)**.

35   Place the socket against the end of the injector **(see illustration)** and slide the seal from the socket onto the injector. Do NOT use any lubricants to do so. Continue pushing the seal onto the injector until it seats into its mounting groove.

36   Because the inside diameter of the seal has to be stretched open to fit over the bore of the socket and the injector, its outside diameter is now slightly too large - it is no longer flush with the surface of the injector. It must be shrunk it back to its original size. To do

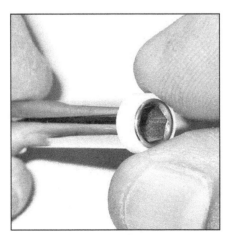

11.35 . . . align the socket with the end of the injector and slide the seal onto the injector and into its mounting groove

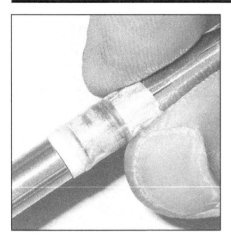

11.36a Use the socket to push a short section of plastic tubing onto the end of the injector and over the new seal . . .

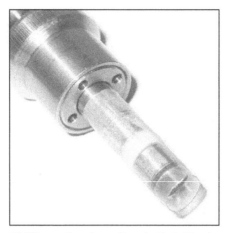

11.36b . . . then leave the plastic tubing in place for several hours to compress the new seal

11.37 Note that the upper O-ring (1) is installed *above* the support ring (2)

so, push a piece of plastic tubing with an interference fit onto the end of the socket; a plastic straw that fits tightly on the injector will work. After pushing the plastic tubing onto the socket about an inch, snip off the rest of the tubing, then use the socket to push the tubing onto the end of the injector **(see illustration)** and slide it onto the injector until it completely covers the new seal **(see illustration)**. Leave the tubing on for a few hours, then remove it. The seal should now be shrunk back its original outside diameter, or close to it.

Injector and fuel rail installation

Note: *There are two different styles of injectors: first generation and second generation. When obtaining replacement parts, you'll need the cylinder head number which is stamped on the front side of the head, just above the exhaust manifold. Cylinder heads with the first six characters "MCGU10" are of the first generation. Heads with the first six characters "MCGU15" are of the second generation. First generation injectors have a plastic top where the upper O-ring seats, and use a "compensation element" with a retaining ring at the bottom, but without a "stop choc." Second generation injectors have a metal top where the O-ring seats, and use a "decoupling" element at the bottom without a retaining ring. Be sure to obtain the correct parts, as they aren't interchangeable.*

37   Install the new support ring at the upper end of the injector. Lubricate the new upper O-ring with clean engine oil and install it on the injector. Do NOT oil the new Teflon seal. Note that the seal is installed *above* the spacer **(see illustration)**.
38   Thoroughly clean the injector bores with a small nylon brush. If any of the valves are in the way, carefully rotate the engine just enough to provide enough clearance to reach all of the bore.
39   Install the new compensation element and retaining ring, or the new decoupling element, as applicable, to the bottom of each injector.
40   Install the fuel injectors in the cylinder head (NOT in the fuel rail). You should be able to push each assembled injector into its bore in the cylinder head. The bore is tapered, so you will encounter some resistance as the Teflon seal nears the bottom of the bore. Press the injector into its bore until it stops, making sure to align the plastic projection on the injector with the hole in the cylinder head.
41   Install a new spring steel retainer on each injector, then install the fuel rail, tightening the fasteners to the torque listed in this Chapter's Specifications.
42   Install a *new* high-pressure fuel line, tightening the tube nuts to the torque listed in this Chapter's Specifications.
43   The remainder of installation is the reverse of removal.

---

12   Fuel pressure regulator - removal and installation

Warning: *Gasoline is extremely flammable. See **Fuel system warnings** in Section 1. Also, wait until the engine is completely cool before relieving the fuel pressure.*

Mk I models

Refer to illustrations 12.2a and 12.2b

1   Remove the fuel rail and injectors (see Section 11).
2   Remove the retaining clip, noting the position of the vacuum connection, then remove the regulator from the fuel rail. Note the small screen filter mounted to the regulator **(see illustrations)**. Discard the O-ring seals; new ones must be installed.
3   Installation is a reversal of removal.

Mk II models

4   The fuel pressure regulator on these models is part of the fuel pump module and is not available separately.

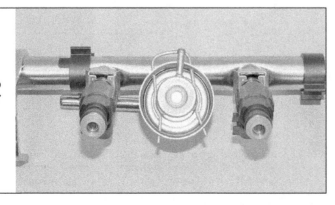

12.2a Pry out the clip and pull the regulator from the fuel rail

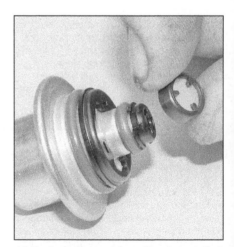

12.2b A small gauze type filter is installed on the regulator

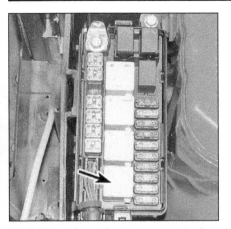

13.2 The main engine management relay is in the engine compartment fusebox

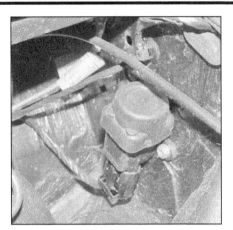

14.1 The inertia fuel cut-off switch is located in the left-rear corner of the engine compartment (on models so equipped)

15.1 Remove the four intercooler cover Torx screws

13 Main engine management relay - removal and installation

Refer to illustration 13.2

1   Ensure the ignition is switched off, then wait approximately five minutes before removing the relay to dissipate any residual electrical power. Release the clip(s) and remove the cover from the engine compartment fusebox.

2   The DME main engine management relay is the forward-most relay at the front of the fusebox **(see illustration)**. Pull the relay from the socket.

3   Installation is a reversal of removal.

14 Inertia fuel cut-off switch - removal and installation

Refer to illustration 14.1

1   The inertia fuel cut-off switch (on models so equipped) is located in the left-rear corner of the engine compartment **(see illustration)**.

2   Disconnect the electrical connector,

remove the screws and remove the switch.

3   Installation is a reversal of removal. To reset the switch, depress the button at the top of the switch.

15 Intercooler - removal and installation

Mk I models
Removal

Refer to illustrations 15.1 and 15.2

1   Remove the four Torx screws and remove the intercooler cover **(see illustration)**.

2   Remove the four retaining bolts on each side, and remove the intercooler bellows upper clamps **(see illustration)**.

3   Remove the four retaining bolts and remove the intercooler cover mounting brackets.

4   Tilt the intercooler at one end, disengage it from the bellows, and remove it from position.

Installation

Refer to illustration 15.5

5   Check the condition of the rubber bellows for splits, cracks, etc, then install the bellows to the intake and outlet ducts, and loosely install the clamps to the bellows **(see illustration)**.

6   Install the intercooler, making sure the bellows engage correctly. Tighten the clamp bolts securely. Do not lubricate the bellow sealing lips.

7   Position the cover mounting brackets, and tighten the retaining bolts securely.

8   Reinstall the intercooler cover and tighten the screws securely.

Mk II models

Refer to illustrations 15.11 and 15.12

Warning: *Wait until the engine is completely cool before beginning this procedure*

9   Apply the parking brake, then raise the front of the vehicle and support it securely on jackstands. Block the rear wheels.

15.2 Unscrew the bolts and remove the bellows clamps

15.5 Reinstall the rubber bellows to the intake and outlet ducts, then loosely reinstall the clamps

15.11 Intercooler duct hose clamps

15.12 Intercooler mounting fasteners

10   Remove the front bumper (see Chapter 11).

11   Loosen the clamps and detach the ducts from each end of the intercooler **(see illustration)**.

12   Remove the intercooler mounting fasteners **(see illustration)** and remove the intercooler.

13   Installation is the reverse of removal.

14   Refill the cooling system (see Chapter 1).

16   Supercharger - removal and installation

Warning: *Wait until the engine is completely cool before beginning this procedure.*

Removal

Refer to illustrations 16.7, 16.11, 16.12, 16.13a, 16.13b and 16.14

1   Disconnect the cable from the negative terminal of the battery (see Chapter 5).

2   Remove both wheelwell liners (see Chapter 11).

3   Remove the front bumper and bumper carrier as described in Chapter 11.

4   Release the clip, then disconnect the air intake pipe from the plastic duct attached to the MFE. Remove the bolt securing the radiator upper hose to the intercooler.

5   Place the Modular Front End (MFE)/ radiator support in the service position (see Chapter 11).

6   Remove the intake manifold as described in Chapter 2A.

7   Unscrew the bolts and remove the outlet housing from the supercharger **(see illustration)**. Discard the gasket, and cover the opening in the supercharger. Note that the new gasket should be installed with the TOP mark uppermost.

8   Remove the drivebelt and drain the engine coolant (see Chapter 1).

9   Remove the alternator (see Chapter 5).

10   Disconnect the hoses from the water pump. To improve access, release the clamp and disconnect the hose from the thermostat housing.

11   Disconnect the vacuum hoses from the supercharger intake ducting by depressing the collars at the quick-release couplings, then unscrew the bolt and remove the ducting **(see illustration)**.

12   Unscrew the retaining bolt and move the dipstick guide tube forward slightly **(see illustration)**.

13   Unscrew the five mounting bolts and remove the supercharger. Note the locating dowel in the right-hand mating face of the supercharger **(see illustrations)**. Discard the water pump-to-cylinder block rubber seal; a new one must be installed.

14   If required, remove the bolts and detach the water pump from the supercharger **(see illustration)**.

Installation

Refer to illustrations 16.15 and 16.16

15   If removed, reinstall the water pump to the supercharger, noting how the pump driveshaft engages with the supercharger shaft **(see illustration)**.

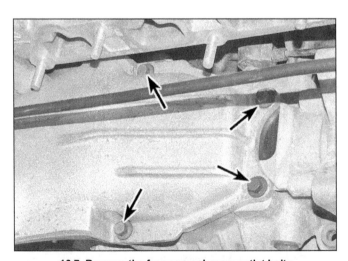

16.7 Remove the four supercharger outlet bolts

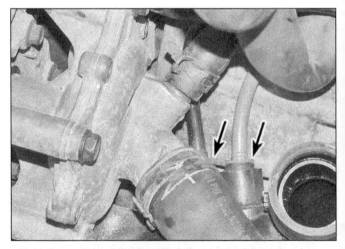

16.11 Press the collar down, and disconnect the vacuum hoses from the intake ducting

16.12 Remove the bolt securing the oil dipstick guide tube to the supercharger

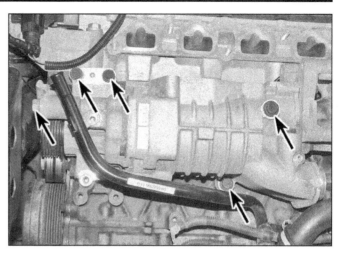

16.13a Remove the supercharger bolts

16   Reinstall the supercharger, using a new seal between the water pump and cylinder block **(see illustration)**. Tighten the retaining bolts to the torque listed in this Chapter's Specifications.

17   The remainder of installation is a reversal of removal. Top-up the coolant as described in Chapter 1.

17   High-pressure fuel pump (Mk II turbo models) - removal and installation

Warning: *Gasoline is extremely flammable. See* **Fuel system warnings** *in Section 1. Also, wait until the engine is completely cool before relieving the fuel pressure.*

Removal

Refer to illustrations 17.6a, 17.6b and 17.6c

1   Relieve the fuel system pressure (see Section 3). Disconnect the cable from the negative terminal of the battery (see Chapter 5).

16.13b Note the locating dowel

2   Remove the air filter housing (see Section 9).

3   Disconnect the fuel feed line from the high-pressure pump. This requires MINI tool

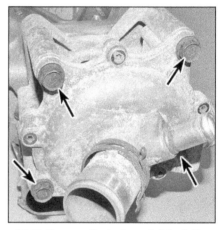

16.14 Remove the bolts and detach the water pump from the supercharger

no. 13 0 250. Install the tool over the line fitting, engaging the lugs of the tool in the openings of the fitting. Tighten the screw on the side of the tool to secure it to the fitting.

16.15 Note how the pump driveshaft engages with the supercharger

16.16 Replace the seal between the water pump and cylinder block

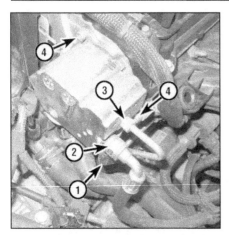

17.6a High-pressure fuel pump details

1   *Electrical connector*
2   *Fuel feed line fitting*
3   *High pressure fuel line fitting*
4   *Mounting bolts*

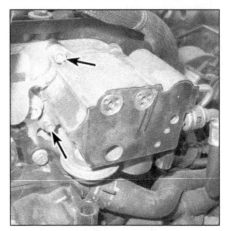

**17.6b High pressure fuel pump front
mounting bolts**

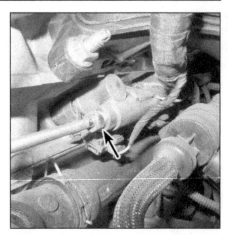

**17.6c High-pressure fuel line-to-fuel
rail fitting**

4   Cover the fitting with a shop cloth, push the fuel line into the pump, then pull the tool and line away from the pump.

5   Remove the tool from the fuel line, then plug the line and fuel pump opening to prevent the entry of dirt.

6   Unscrew the tube nuts at each end of the high-pressure fuel line (between the fuel rail and the high-pressure fuel pump) **(see illustrations)**. **Warning:** *Be sure to wear full face protection, long sleeves and thick leather gloves, and cover the fittings with shop rags while slowly loosening them so fuel under pressure in the line bleeds out slowly. Also,*

the manufacturer states that it is necessary to replace the high-pressure fuel line (between the fuel rail and the high-pressure fuel pump) with a new one whenever it is removed.

7   Disconnect the electrical connector from the pump, then remove the mounting screws and detach the pump from the cylinder head. **Warning:** *The manufacturer states that the pump mounting screws must be replaced with new ones whenever they are removed.*

Installation

8   Clean the mating surfaces of the pump and cylinder head. Install a new O-ring on the pump.

9   Turn the drive rotor on the pump so it's in the proper position to mate with the slots in

the end of the camshaft.

10   Install the pump to the cylinder head. Install the new mounting screws and tighten them hand tight at this time, so that the pump can move a little.

11   Install the **new** high-pressure fuel line, first tightening the line-to-fuel rail tube nut (hand-tight only), then the line-to-pump tube nut (also only hand-tight).

12   Tighten the pump-to-cylinder head fasteners to the torque listed in this Chapter's Specifications.

13   Tighten the high-pressure fuel line tube nuts to the torque listed in this Chapter's Specifications.

14   The remainder of installation is the reverse of removal.

Chapter 5
Engine electrical systems

Contents

1 General information and precautions

Note: *Throughout this Chapter you will find references to "Mk I" and "Mk II" models; this is done to simplify which specifications and procedures apply to which models. Mk I models include 2006 and earlier Cooper/Cooper S models, and 2008 and earlier Convertible models. Mk II models include 2007 and later Cooper/Cooper S/Clubman/Clubman S and 2009 and later Convertible models.*

General information

Ignition system

The electronic ignition system consists of the Crankshaft Position (CKP) sensor, the Camshaft Position (CMP) sensor, the Knock Sensor (KS), the Powertrain Control Module (PCM), the ignition switch, the battery, the individual ignition coils or a coil pack, and the spark plugs. For more information on the CKP, CMP and KS sensors, as well as the PCM, refer to Chapter 6.

Charging system

The charging system includes the alternator (with an integral voltage regulator), the Powertrain Control Module (PCM), the Body Control Module (BCM), a charge indicator light on the dash, the battery, a fuse or fusible link and the wiring connecting all of these components. The charging system supplies electrical power for the ignition system, the lights, the radio, etc. The alternator is driven by a drivebelt.

Starting system

The starting system consists of the battery, the ignition switch, the starter relay, the Powertrain Control Module (PCM), the Body Control Module (BCM), the Transmission Range (TR) switch, the starter motor and solenoid assembly, and the wiring connecting all of the components.

Precautions

Always observe the following precautions when working on the electrical system:

a) *Be extremely careful when servicing engine electrical components. They are easily damaged if checked, connected or handled improperly.*

b) *Never leave the ignition switched on for long periods of time when the engine is not running.*

c) *Never disconnect the battery cables while the engine is running.*

d) *Maintain correct polarity when connecting battery cables from another vehicle during jump starting - see the "Booster battery (jump) starting" Section at the front of this manual.*

e) *Always disconnect the cable from the negative battery terminal before working on the electrical system, but read the battery disconnection procedure first (see Section 3).*

It's also a good idea to review the safety-related information regarding the engine electrical systems located in the *Safety first!* Section at the front of this manual before beginning any operation included in this Chapter.

2 Troubleshooting

Ignition system

1 If a malfunction occurs in the ignition system, do not immediately assume that any particular part is causing the problem. First, check the following items:

a) *Make sure that the cable clamps at the battery terminals are clean and tight.*

b) *Test the condition of the battery (see Steps 21 through 24). If it doesn't pass all the tests, replace it.*

c) *Check the ignition coil or coil pack connections.*

d) *Check any relevant fuses in the engine compartment fuse and relay box (see Chapter 12). If they're burned, determine the cause and repair the circuit.*

Check

Refer to illustration 2.3

Warning: *Because of the high voltage generated by the ignition system, use extreme care when performing a procedure involving ignition components.*

Note 1: *The ignition system components on these vehicles are difficult to diagnose. In the event of ignition system failure that you can't diagnose, have the vehicle tested at a dealer service department or other qualified auto repair facility.*

Note 2: *You'll need a spark tester for the following test. Spark testers are available at most auto supply stores.*

2 If the engine turns over but won't start,

verify that there is sufficient ignition voltage to fire the spark plugs as follows.

3   On models with a coil-over-plug type ignition system, remove a coil and install the tester between the boot at the lower end of the coil and the spark plug **(see illustration)**. On models with spark plug wires, disconnect a spark plug wire from a spark plug and install the tester between the spark plug wire boot and the spark plug.

4   Crank the engine and note whether or not the tester flashes. **Caution:** *Do NOT crank the engine or allow it to run for more than five seconds; running the engine for more than five seconds may set a Diagnostic Trouble Code (DTC) for a cylinder misfire.*

Models with a coil-over-plug type ignition system

5   If the tester flashes during cranking, the coil is delivering sufficient voltage to the spark plug to fire it. Repeat this test for each cylinder to verify that the other coils are OK.

6   If the tester doesn't flash, remove a coil from another cylinder and swap it for the one being tested. If the tester now flashes, you know that the original coil is bad. If the tester still doesn't flash, the PCM or wiring harness is probably defective. Have the PCM checked out by a dealer service department or other qualified repair shop (testing the PCM is beyond the scope of the do-it-yourselfer because it requires expensive special tools).

7   If the tester flashes during cranking but a misfire code (related to the cylinder being tested) has been stored, the spark plug could be fouled or defective.

Models with spark plug wires

8   If the tester flashes during cranking, sufficient voltage is reaching the spark plug to fire it.

9   Repeat this test on the remaining cylinders.

10   Proceed on this basis until you have verified that there's a good spark from each spark plug wire. If there is, then you have verified that the coils in the coil pack are functioning correctly and that the spark plug wires are OK.

11   If there is no spark from a spark plug wire, then either the coil is bad, the plug wire is bad or a connection at one end of the plug wire is loose. Assuming that you're using new plug wires or known good wires, then the coil is probably defective. Also inspect the coil pack electrical connector. Make sure that it's clean, tight and in good condition.

12   If all the coils are firing correctly, but the engine misfires, then one or more of the plugs might be fouled. Remove and check the spark plugs or install new ones (see Chapter 1).

13   No further testing of the ignition system is possible without special tools. If the problem persists, have the ignition system tested by a dealer service department or other qualified repair shop.

Charging system

14   If a malfunction occurs in the charging system, do not automatically assume the alternator is causing the problem. First check the following items:

a) *Check the drivebelt tension and condition, as described in Chapter 1. Replace it if it's worn or deteriorated.*

b) *Make sure that the alternator mounting bolts are tight.*

c) *Inspect the alternator wiring harness and the connectors at the alternator and voltage regulator. They must be in good condition, tight and have no corrosion.*

d) *Check the fusible link (if equipped) or main fuse in the underhood fuse/relay box. If it is burned, determine the cause, repair the circuit and replace the link or fuse (the vehicle will not start and/or the accessories will not work if the fusible link or main fuse is blown).*

e) *Start the engine and check the alternator for abnormal noises (a shrieking or squealing sound indicates a bad bearing).*

f) *Check the battery. Make sure it's fully charged and in good condition (one bad cell in a battery can cause overcharging by the alternator).*

g) *Disconnect the battery cables (negative first, then positive). Inspect the battery posts and the cable clamps for corrosion. Clean them thoroughly if necessary (see Chapter 1). Reconnect the cables (positive first, negative last).*

Alternator - check

15   Use a voltmeter to check the battery voltage with the engine off. It should be at least 12.6 volts **(see illustration 2.21)**.

16   Start the engine and check the battery voltage again. It should now be approximately 13.5 to 15 volts.

17   If the voltage reading is more or less than the specified charging voltage, the voltage regulator is probably defective, which will require replacement of the alternator (the voltage regulator is not replaceable separately). Remove the alternator and have it bench tested (most auto parts stores will do this for you).

18   The charging system (battery) light on the instrument cluster lights up when the ignition key is turned to ON, but it should go out when the engine starts.

19   If the charging system light stays on after the engine has been started, there is a problem with the charging system. Before replacing the alternator, check the battery condition, alternator belt tension and electrical cable connections.

20   If replacing the alternator doesn't restore voltage to the specified range, have the charging system tested by a dealer service department or other qualified repair shop.

Battery - check

Refer to illustrations 2.21 and 2.23

21   Check the battery state of charge. Visually inspect the indicator eye on the top of the battery (if equipped with one); if the indicator eye is black in color, charge the battery as described in Chapter 1. Next perform an open circuit voltage test using a digital voltmeter. **Note:** *The battery's surface charge must be removed before accurate voltage measurements can be made. Turn on the high beams for ten seconds, then turn them off and let the vehicle stand for two minutes.* With the engine and all accessories Off, touch the negative probe of the voltmeter to the negative terminal of the battery and the positive probe to the positive terminal of the battery **(see illustration)**. The battery voltage should be 12.6 volts or slightly above. If the battery is less than the specified voltage, charge the battery before proceeding to the next test. Do not proceed with the battery load test unless the battery charge is correct.

22   Disconnect the negative battery cable, then the positive cable from the battery.

23   Perform a battery load test. An accurate check of the battery condition can only be performed with a load tester **(see illustration)**. This test evaluates the ability of the battery to operate the starter and other accessories during periods of high current draw. Connect the load tester to the battery terminals. Load test the battery according to the tool manufacturer's instructions. This tool increases the load demand (current draw) on the battery.

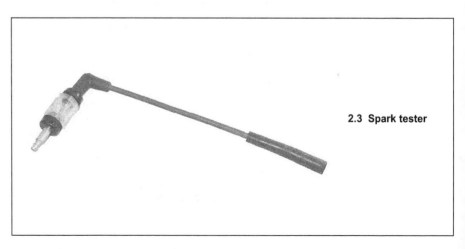

2.3 Spark tester

2.21 To test the open circuit voltage of the battery, touch the black probe of the voltmeter to the negative terminal and the red probe to the positive terminal of the battery; a fully charged battery should be at least 12.6 volts

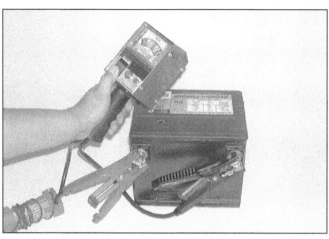

2.23 Connect a battery load tester to the battery and check the battery condition under load following the tool manufacturer's instructions

24   Maintain the load on the battery for 15 seconds and observe that the battery voltage does not drop below 9.6 volts. If the battery condition is weak or defective, the tool will indicate this condition immediately. **Note:** *Cold temperatures will cause the minimum voltage reading to drop slightly. Follow the chart given in the manufacturer's instructions to compensate for cold climates. Minimum load voltage for freezing temperatures (32 degrees F) should be approximately 9.1 volts.*

Starting system

The starter rotates, but the engine doesn't

25   Remove the starter (see Section 8). Check the overrunning clutch and bench test the starter to make sure the drive mechanism extends fully for proper engagement with the flywheel ring gear. If it doesn't, replace the starter.
26   Check the flywheel/driveplate ring gear for missing teeth and other damage. With the ignition turned off, rotate the flywheel/driveplate so you can check the entire ring gear.

The starter is noisy

27   If the solenoid is making a chattering noise, first check the battery (see Steps 21 through 24). If the battery is okay, check the cables and connections.
28   If you hear a grinding, crashing metallic sound when you turn the key to Start, check for loose starter mounting bolts. If they're tight, remove the starter and inspect the teeth on the starter pinion gear and flywheel ring gear. Look for missing or damaged teeth.
29   If the starter sounds fine when you first turn the key to Start, but then stops rotating the engine and emits a zinging sound, the problem is probably a defective starter drive

that's not staying engaged with the ring gear. Replace the starter.

The starter rotates slowly

30   Check the battery (see Steps 21 through 24).
31   If the battery is okay, verify all connections (at the battery, the starter solenoid and motor) are clean, corrosion-free and tight. Make sure the cables aren't frayed or damaged.
32   Check that the starter mounting bolts are tight so it grounds properly. Also check the pinion gear and flywheel ring gear for evidence of a mechanical bind (galling, deformed gear teeth or other damage).

The starter does not rotate at all

33   Check the battery (see Steps 21 through 24).
34   If the battery is okay, verify all connections (at the battery, the starter solenoid and motor) are clean, corrosion-free and tight. Make sure the cables aren't frayed or damaged.
35   Check all of the fuses in the underhood fuse/relay box.
36   Check that the starter mounting bolts are tight so it grounds properly.
37   Check for voltage at the starter solenoid "S" terminal when the ignition key is turned to the start position or the start button is pressed. If voltage is present, replace the starter/solenoid assembly. If no voltage is present, the problem could be the starter relay, the Transmission Range (TR) switch (see Chapter 7B) or clutch start switch, or with an electrical connector somewhere in the circuit (see the wiring diagrams at the end of Chapter 12). Also, on many modern vehicles, the Powertrain Control Module (PCM) and the Body Control Module (BCM) control the voltage signal to the starter solenoid; on such vehicles a special scan tool is required for diagnosis.

3   Battery - disconnection and reconnection

Caution: *Always disconnect the cable from the negative battery terminal FIRST and hook it up LAST or the battery may be shorted by the tool being used to loosen the cable clamps.*

Some systems on the vehicle require battery power to be available at all times, either to maintain continuous operation (alarm system, power door locks, etc.), or to maintain control unit memory (radio station presets, Powertrain Control Module and other control units). When the battery is disconnected, the power that maintains these systems is cut. So, before you disconnect the battery, please note that on a vehicle with power door locks, it's a wise precaution to remove the key from the ignition and to keep it with you, so that it does not get locked inside if the power door locks should engage accidentally when the battery is reconnected!

Devices known as "memory-savers" can be used to avoid some of these problems. Precise details vary according to the device used. The typical memory saver is plugged into the cigarette lighter and is connected to a spare battery. Then the vehicle battery can be disconnected from the electrical system. The memory saver will provide sufficient current to maintain audio unit security codes, PCM memory, etc. and will provide power to always hot circuits such as the clock and radio memory circuits. **Warning 1:** *Some memory savers deliver a considerable amount of current in order to keep vehicle systems operational after the main battery is disconnected. If you're using a memory saver, make sure that the circuit concerned is actually open before servicing it.* **Warning 2:** *If you're going to work near any of the airbag system components, the battery MUST be disconnected and a memory saver must NOT be used. If a memory saver is used, power will be supplied to the airbag, which means that it could accidentally deploy and cause serious personal injury.*

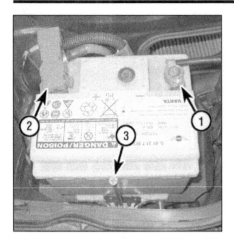

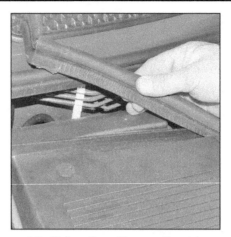

4.1 Battery mounting details - Mk II models

1   Negative terminal
2   Positive terminal
3   Hold-down bolt/clamp

4.2a Pull up the rubber weatherstrip above the battery cover . . .

4.2b . . . then push-in the retaining clips . . .

Disconnection

To disconnect the battery for service procedures requiring power to be cut from the vehicle, loosen the cable end bolt and disconnect the cable from the negative battery terminal (see Section 4). Isolate the cable end to prevent it from coming into accidental contact with the battery terminal.

Reconnection

After the battery has been reconnected, it might be necessary for the Body Control Module (BCM) to relearn the window and sunroof "reference points."

Windows, MK I models: Close the door, raise the window and hold the switch in the Up position for at least five seconds.

Windows, MK II models:

a) Close the window completely and release the switch, then hold the switch in the Up direction for one second.
b) Open the window completely and release the switch, then hold the switch in the Down direction for one second.
c) Repeat Step a).

Sunroof: Press the sunroof switch to the Tilt position until the sunroof is fully tilted, then hold the switch in that position for 20 seconds.

---

4   Battery and battery tray - removal and installation

Warning: *MK I Cooper S models incorporate a Battery Safety Terminal (BST) into the battery positive cable. This is designed to cut power to the starter and alternator (which operate on high current) in the event of a collision. The BST is a pyrotechnic (explosive) device controlled by the Multiple Restraint System (MRS). Handle it with care. Never apply voltage to the electrical connector or the contacts inside the terminal.*

Note: *When the battery is disconnected, any fault codes stored in the engine management PCM memory will be erased. If any faults are suspected (CHECK ENGINE light on), do not disconnect the battery until the fault codes have been retrieved (see Chapter 6).*

Note: *After reconnecting the battery, it may*

be necessary to carry out the window, and/or sunroof initialization procedure as described in Section 3.

Removal

Models with an engine compartment-mounted battery (all models except Mk I Cooper S models)

Refer to illustrations 4.1, 4.2a, 4.2b, 4.2c, 4.3, 4.4, 4.5, 4.6, 4.7, 4.8a, 4.8b, 4.11

1   On Mk I models, the battery is located beneath a cover on the left-hand side of the engine compartment. On Mk II models, it's located in the engine compartment on the right-hand side, under the cowl cover **(see illustration)**.

2   On MK I models, pull up the rubber weatherstrip, then release the clips and remove the plastic cover from above the battery **(see illustrations)**. On MK II models, remove the right-side cowl cover (see Chapter 11).

3   Loosen the clamp nut and disconnect the clamp from the battery negative terminal **(see illustration)**.

4.2c . . . and remove the cover

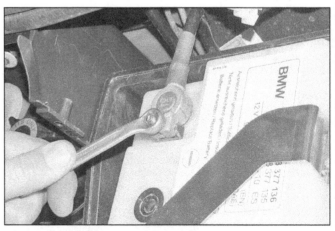

4.3 Loosen the nut and disconnect the battery negative terminal (Mk I model shown)

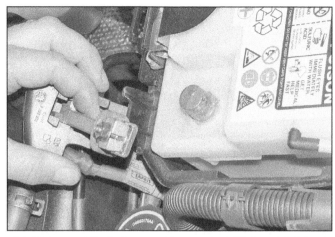

4.4 Disconnect the battery positive terminal (Mk I model shown)

4.5 Unscrew the bolts and remove the battery clamp (Mk I model shown)

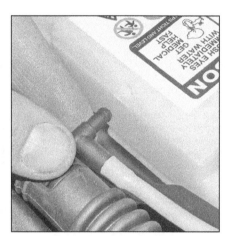

4.6 Disconnect the battery vent tube as the battery is removed (Mk I models)

4.7 Depress the clips and slide the PCM up a little (Mk I models)

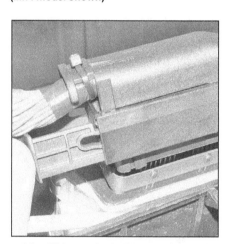

4.8a Slide out the locking catches and disconnect the PCM electrical connectors . . .

4 Disconnect the positive terminal lead in the same way **(see illustration)**.
5 Unscrew the bolts and remove the battery retaining clamp **(see illustration)**.
6 Lift the battery from its housing, disconnecting the vent hose as the battery is

removed **(see illustration)**. Take care, as the battery is heavy!

Mk I models

7 To remove the battery tray, depress the retaining clips and slide the PCM up a little

from the battery tray **(see illustration)**.
8 Slide out the locking clips and disconnect the electrical connectors from the engine management PCM **(see illustrations)**. Release the wiring harness clip and detach the harness from the battery tray, then slide the PCM up and out from the battery tray.
9 Depress the clip and slide the wiring harness clips down from the battery tray.
10 Remove the air filter housing as described in Chapter 4.

All models

11 Remove the bolts securing the battery tray **(see illustration)**.
12 The battery positive cable is clipped to the underside of the battery tray at several places. Release the clips. Lift the battery tray from place.

Models with a luggage compartment-mounted battery (Mk I Cooper S models)

Refer to illustrations 4.14, 4.15, 4.16 and 4.17

13 Open the hatch and lift the luggage compartment floor mat.

4.8b . . . then remove the PCM (Mk I models)

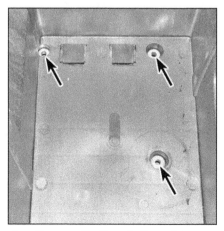

4.11 Unscrew the three bolts and remove the battery tray (Mk I models)

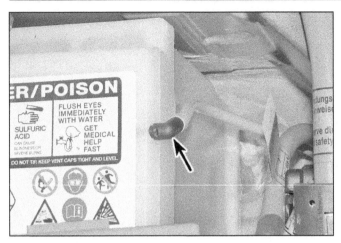

4.14 Pull the vent tube from the battery

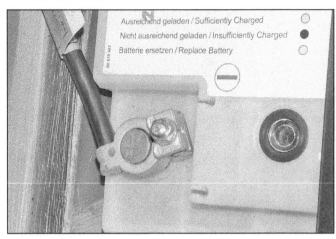

4.15 Loosen the nut and disconnect the negative cable

14   Pull the vent hose from the side of the battery **(see illustration)**.
15   Loosen the clamp nut and disconnect the battery negative cable **(see illustration)**.
16   Unclip the plastic cover from the positive terminal, then loosen the clamp nut and disconnect the cable **(see illustration)**.
17   Remove the bolt and nut, then remove battery clamping strap **(see illustration)**.
18   Lift the battery from place.

Installation

19   Installation is the reverse of removal. Smear petroleum jelly on the terminals after reconnecting the cables, to prevent corrosion. Always reconnect the positive cable first, and the negative cable last.

5   Battery cables - replacement

Warning: *Mk I Cooper S models incorporate a Battery Safety Terminal (BST) into the battery positive cable. This is designed to cut power*
to the starter and alternator (which operate on high current) in the event of a collision. The BST is a pyrotechnic (explosive) device controlled by the Multiple Restraint System (MRS). Handle it with care. Never apply voltage to the electrical connector or the contacts inside the terminal.

1   When removing the cables, always disconnect the cable from the negative battery terminal first and hook it up last, or you might accidentally short out the battery with the tool you're using to loosen the cable clamps. Even if you're only replacing the cable for the positive terminal, be sure to disconnect the negative cable from the battery first.
2   Disconnect the old cables from the battery, then trace each of them to their opposite ends and disconnect them. Be sure to note the routing of each cable before disconnecting it to ensure correct installation.
3   If you are replacing any of the old cables, take them with you when buying new cables. It is vitally important that you replace the cables with identical parts.
4   Clean the threads of the solenoid or ground connection with a wire brush to remove

rust and corrosion. Apply a light coat of battery terminal corrosion inhibitor or petroleum jelly to the threads to prevent future corrosion.
5   Attach the cable to the solenoid or ground connection and tighten the mounting nut/bolt securely.
6   Before connecting a new cable to the battery, make sure that it reaches the battery post without having to be stretched.
7   Connect the cable to the positive battery terminal first, then connect the ground cable to the negative battery terminal.

6   Ignition coils or coil pack - removal and installation

Mk I models

Refer to illustrations 6.2 and 6.3

1   Ensure the ignition is switched off.
2   The spark plug wires should be marked with the cylinder number of the spark plug they connect to. If the markings are not visible, number the wires with pieces of tape,

4.16 Lift the plastic cover and disconnect the positive cable

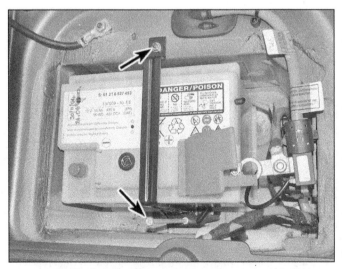

4.17 The battery clamp is secured by a nut and bolt

6.2 Pull the spark plug wires from the ignition coil

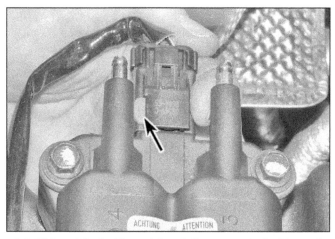

6.3 Slide out the locking catch and disconnect the ignition coil electrical connector

6.7 Remove the screws from the top of the coil cover (A), then tilt the cover forward to disconnect the clips (B) along the front edge (non-turbo models)

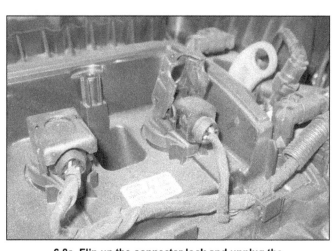

6.8a Flip up the connector lock and unplug the electrical connector

then pull the wires from the connections on the coil **(see illustration)**.

3   Slide out the red locking catch, depress the clip and disconnect the coil electrical con-nector **(see illustration)**.

4   Unscrew the four bolts, and remove the coil.

5   Installation is the reverse of removal.

6.8b Pull the coil straight out. Note: *These can be a tight fit. If necessary, carefully pry the coil out with two screwdrivers*

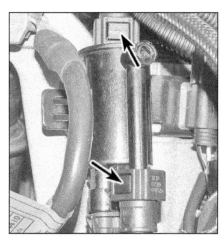

7.2 Press in the buttons and disconnect the hoses from the purge/tank vent valve

Mk II models

Refer to illustrations 6.7, 6.8a and 6.8b

6   Ensure the ignition is switched off.

7   On non-turbo models, unscrew the two bolts and remove the coil cover from the valve cover **(see illustration)**.

8   Disconnect the electrical connector and pull the coil from the valve cover **(see illustrations)**.

9   Installation is the reverse of removal.

---

7   Alternator - removal and installation

Mk I models

1   Disconnect the cable from the negative battery terminal (see Section 4).

Cooper models

Refer to illustrations 7.2, 7.5a and 7.5b

2   Disconnect the hoses, and unclip the fuel tank vent valve from the right-hand engine mount. Disconnect the electrical connector as the valve is removed **(see illustration)**.

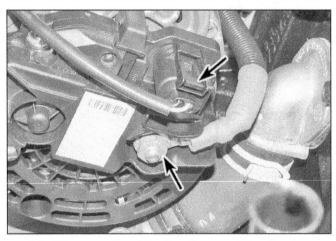

7.5a Disconnect the wiring from the rear of the alternator

7.5b Unscrew the bolts and remove the alternator (Mk I Cooper models)

7.11 Alternator mounting bolts (Mk I Cooper S models)

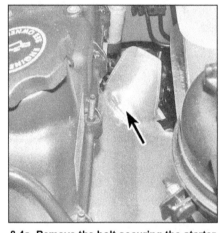

8.4a Remove the bolt securing the starter motor heat shield

Cooper S models

Refer to illustration 7.11

7 Remove the front bumper, bumper carrier and front wheel arch liners as described in Chapter 11.

8 Place the Modular Front End (MFE) in the service position (see Chapter 11).

9 Remove the drivebelt as described in Chapter 1.

10 Disconnect the electrical connections from the rear of the alternator.

11 Unscrew the mounting bolts and remove the alternator **(see illustration)**.

12 Installation is the reverse of removal.

Mk II models

13 Place the Modular Front End (MFE) in the service position (see Chapter 11).

14 Remove the drivebelt and drivebelt tensioner (see Chapter 1).

15 Cooper S models: Remove the bolt that secures the wiring harness bracket above the alternator, then position the harness aside.

16 Disconnect the electrical connections

3 Remove the intake manifold as described in Chapter 2A.

4 Remove the drivebelt as described in Chapter 1.

5 Disconnect the electrical connections

from the rear of the alternator **(see illustration)**. Unscrew the alternator mounting bolts and remove the alternator from the engine **(see illustration)**.

6 Installation is the reverse of removal.

8.4b Note how the starter motor heat shield rubber grommets locate over the lugs on the motor

8.5 Disconnect the wiring from the starter motor

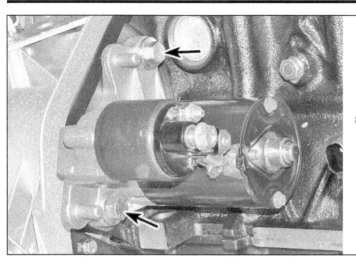

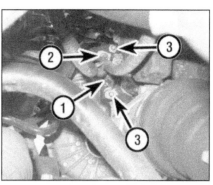

8.6 Starter motor mounting bolts (Mk I models)

8.13 Starter motor mounting details (seen from the right-front wheel opening) - Mk II models

1   Solenoid electrical connector
2   B+ cable
3   Mounting bolts

from the rear of the alternator.

17   Unscrew the alternator mounting fasteners and remove the alternator.

18   Installation is the reverse of removal.

8   Starter motor - removal and installation

1   Disconnect the cable from the negative terminal of the battery (see Section 4).

2   Apply the parking brake, then raise the front of the vehicle and support it on jackstands.

Mk I models

Refer to illustrations 8.4a, 8.4b, 8.5 and 8.6

3   Remove the exhaust manifold as described in Chapter 2A.

4   Remove the bolt and remove the heat shield from above the starter motor **(see illustrations)**.

5   Unscrew the nuts and disconnect the wiring from the rear of the starter motor **(see illustration)**. Note the positions and routing of the cables.

6   Remove the starter motor mounting bolts from the transmission bellhousing **(see illustration)**.

7   Maneuver the starter from the engine, taking care not to damage the fuel hoses.

8   Installation is a reversal of removal. Tighten the starter motor mounting bolts securely.

Mk II models

Refer to illustration 8.13

9   Remove the air filter housing (see Chapter 4).

10   Loosen the right-front wheel bolts. Raise the vehicle and support it securely on jackstands, then remove the wheel.

11   If you're working on a Cooper model, remove the vent solenoid from underneath the intake manifold.

12   If you're working on a Cooper S or JCW model, remove the vacuum tank.

13   Remove the B+ cable and the starter solenoid electrical connector, then remove the starter mounting bolts and starter **(see illustration)**.

14   Installation is a reversal of removal. Tighten the starter motor mounting bolts securely.

Notes

Chapter 6
Emissions and engine control systems

Contents

Specifications

Note: *Throughout this Chapter you will find references to "Mk I" and "Mk II" models; this is done to simplify which specifications and procedures apply to which models. Mk I models include 2006 and earlier Cooper/Cooper S models, and 2008 and earlier Convertible models. Mk II models include 2007 and later Cooper/Cooper S/Clubman/Clubman S and 2009 and later Convertible models.*

Torque specifications

Note: *One foot-pound (ft-lb) of torque is equivalent to 12 inch-pounds (in-lbs) of torque. Torque values below approximately 15 foot-pounds are expressed in inch-pounds, because most foot-pound torque wrenches are not accurate at these smaller values.*

	Ft-lbs (unless otherwise indicated)	Nm
Engine Coolant Temperature (ECT) sensor (2006 and earlier models)	84 in-lbs	17
Knock sensor mounting bolt		
Mk I models	16	22
Mk II models	15	20
Oxygen sensors	29	39
Catalytic converter (Mk II S models)		
Converter-to-turbocharger nuts		
Step 1	62 in-lbs	7
Step 2	30	40
Converter lower mounting fasteners	15	20
Converter-to-exhaust pipe clamp	18	25
Heat shield bolts	35 in-lbs	4

1   General information

To prevent pollution of the atmosphere from incompletely burned and evaporating gases, and to maintain good driveability and fuel economy, a number of emission control systems are incorporated. They include the:

Catalytic converter

A catalytic converter is an emission control device in the exhaust system that reduces certain pollutants in the exhaust gas stream. There are two types of converters: oxidation converters and reduction converters.

Oxidation converters contain a monolithic substrate (a ceramic honeycomb) coated with the semi-precious metals platinum and palladium. An oxidation catalyst reduces unburned hydrocarbons (HC) and carbon monoxide (CO) by adding oxygen to the exhaust stream as it passes through the substrate, which, in the presence of high temperature and the catalyst materials, converts the HC and CO to water vapor (H_2O) and carbon dioxide (CO_2).

Reduction converters contain a monolithic substrate coated with platinum and rhodium. A reduction catalyst reduces oxides of nitrogen (NOx) by removing oxygen, which in the presence of high temperature and the catalyst material produces nitrogen (N) and carbon dioxide (CO_2).

Catalytic converters that combine both types of catalysts in one assembly are known as "three-way catalysts" or TWCs. A TWC can reduce all three pollutants.

Evaporative Emissions Control (EVAP) system

The Evaporative Emissions Control (EVAP) system prevents fuel system vapors (which contain unburned hydrocarbons) from escaping into the atmosphere. On warm days, vapors trapped inside the fuel tank expand until the pressure reaches a certain threshold. Then the fuel vapors are routed from the fuel tank through the fuel vapor vent valve and the fuel vapor control valve to the EVAP canister, where they're stored temporarily until the next time the vehicle is operated. When the conditions are right (engine warmed up, vehicle up to speed, moderate or heavy load on the engine, etc.) the PCM opens the canister purge valve, which allows fuel vapors to be drawn from the canister into the intake manifold. Once in the intake manifold, the fuel vapors mix with incoming air before being drawn through the intake ports into the combustion chambers where they're burned up with the rest of the air/fuel mixture. The EVAP system is complex and virtually impossible to troubleshoot without the right tools and training.

Powertrain Control Module (PCM)

The Powertrain Control Module (PCM), also known as the Engine Control Module (ECM) or Engine Control Unit (ECU) is the brain of the engine management system. It also controls a wide variety of other vehicle systems. In order to program the new PCM, the dealer needs the vehicle as well as the new PCM. If you're planning to replace the PCM with a new one, there is no point in trying to do so at home because you won't be able to program it yourself.

Positive Crankcase Ventilation (PCV) system

The Positive Crankcase Ventilation (PCV) system reduces hydrocarbon emissions by scavenging crankcase vapors, which are rich in unburned hydrocarbons. A PCV valve or orifice regulates the flow of gases into the intake manifold in proportion to the amount of intake vacuum available.

The PCV system generally consists of the fresh air inlet hose, the PCV valve or orifice and the crankcase ventilation hose (or PCV hose). The fresh air inlet hose connects the air intake duct to a pipe on the valve cover. The crankcase ventilation hose (or PCV hose) connects the PCV valve or orifice in the valve cover to the intake manifold.

Information Sensors

Accelerator Pedal Position (APP) sensor - as you press the accelerator pedal, the APP sensor alters its voltage signal to the PCM in proportion to the angle of the pedal, and the PCM commands a motor inside the throttle body to open or close the throttle plate accordingly

Camshaft Position (CMP) sensor - produces a signal that the PCM uses to identify the number 1 cylinder and to time the firing sequence of the fuel injectors

Crankshaft Position (CKP) sensor - produces a signal that the PCM uses to calculate engine speed and crankshaft position, which enables it to synchronize ignition timing with fuel injector timing, and to detect misfires

Engine Coolant Temperature (ECT) sensor - a thermistor (temperature-sensitive variable resistor) that sends a voltage signal to the PCM, which uses this data to determine the temperature of the engine coolant

Fuel tank pressure sensor - measures the fuel tank pressure and controls fuel tank pressure by signaling the EVAP system to purge the fuel tank vapors when the pressure becomes excessive

Intake Air Temperature (IAT) sensor - monitors the temperature of the air entering the engine and sends a signal to the PCM to determine injector pulse-width (the duration of each injector's on-time) and to adjust spark timing (to prevent spark knock)

Knock sensor - a piezoelectric crystal that oscillates in proportion to engine vibration which produces a voltage output that is monitored by the PCM. This retards the ignition timing when the oscillation exceeds a certain threshold

Manifold Absolute Pressure (MAP) sensor - monitors the pressure or vacuum inside the intake manifold. The PCM uses this data to determine engine load so that it can alter the ignition advance and fuel enrichment

Mass Air Flow (MAF) sensor - measures the amount of intake air drawn into the engine. It uses a hot-wire sensing element to measure the amount of air entering the engine

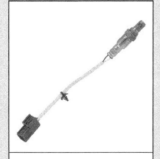

Oxygen sensors - generates a small variable voltage signal in proportion to the difference between the oxygen content in the exhaust stream and the oxygen content in the ambient air. The PCM uses this information to maintain the proper air/fuel ratio. A second oxygen sensor monitors the efficiency of the catalytic converter

Throttle Position (TP) sensor - a potentiometer that generates a voltage signal that varies in relation to the opening angle of the throttle plate inside the throttle body. Works with the PCM and other sensors to calculate injector pulse width (the duration of each injector's on-time)

Photos courtesy of Wells Manufacturing, except APP and MAF sensors.

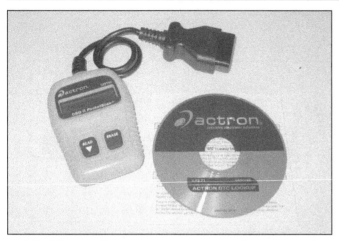 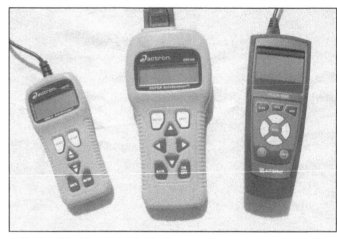

2.4a Simple code readers are an economical way to extract trouble codes when the CHECK ENGINE light comes on

2.4b Hand-held scan tools like these can extract computer codes and also perform diagnostics

2 On Board Diagnosis (OBD) system

General description

1   All models are equipped with the second generation OBD-II system. This system consists of an on-board computer known as the Powertrain Control Module (PCM), and information sensors, which monitor various functions of the engine and send data to the PCM. This system incorporates a series of diagnostic monitors that detect and identify fuel injection and emissions control system faults and store the information in the computer memory. This system also tests sensors and output actuators, diagnoses drive cycles, freezes data and clears codes.

2   The PCM is the brain of the electronically controlled fuel and emissions system. It receives data from a number of sensors and other electronic components (switches, relays, etc.). Based on the information it receives, the PCM generates output signals to control various relays, solenoids (fuel injectors) and other actuators. The PCM is specifically calibrated to optimize the emissions, fuel economy and driveability of the vehicle.

3   It isn't a good idea to attempt diagnosis or replacement of the PCM or emission control components at home while the vehicle is under warranty. Because of a federally-mandated warranty which covers the emissions system components and because any owner-induced damage to the PCM, the sensors and/or the control devices may void this warranty, take the vehicle to a dealer service department if the PCM or a system component malfunctions.

Scan tool information

Refer to illustrations 2.4a and 2.4b

4   Because extracting the Diagnostic Trouble Codes (DTCs) from an engine management system is now the first step in trouble-shooting many computer-controlled systems and components, a code reader, at the very least, will be required **(see illustration)**. More powerful scan tools can also perform many of the diagnostics once associated with expensive factory scan tools **(see illustration)**. If you're planning to obtain a generic scan tool for your vehicle, make sure that it's compatible with OBD-II systems. If you don't plan to purchase a code reader or scan tool and don't have access to one, you can have the codes extracted by a dealer service department or an independent repair shop. **Note:** *Some auto parts stores even provide this service.*

3 Obtaining and clearing Diagnostic Trouble Codes (DTCs)

All models covered by this manual are equipped with on-board diagnostics. When the PCM recognizes a malfunction in a monitored emission or engine control system, component or circuit, it turns on the Malfunction Indicator Light (MIL) on the dash. The PCM will continue to display the MIL until the problem is fixed and the Diagnostic Trouble Code (DTC) is cleared from the PCM's memory. You'll need a scan tool to access any DTCs stored in the PCM.

Before outputting any DTCs stored in the PCM, thoroughly inspect ALL electrical connectors and hoses. Make sure that all electrical connections are tight, clean and free of corrosion. And make sure that all hoses are correctly connected, fit tightly and are in good condition (no cracks or tears).

Accessing the DTCs

Refer to illustration 3.1

1   The Diagnostic Trouble Codes (DTCs) can only be accessed with a code reader or scan tool. Professional scan tools are expensive, but relatively inexpensive generic code readers or scan tools **(see illustrations 2.4a and 2.4b)** are available at most auto parts stores. Simply plug the connector of the scan tool into the diagnostic connector **(see illustration)**. Then follow the instructions included

with the scan tool to extract the DTCs.

2   Once you have outputted all of the stored DTCs, look them up on the accompanying DTC chart.

3   After troubleshooting the source of each DTC, make any necessary repairs or replace the defective component(s).

Clearing the DTCs

4   Clear the DTCs with the code reader or scan tool in accordance with the instructions provided by the tool's manufacturer.

Diagnostic Trouble Codes

5   The accompanying tables are a list of the Diagnostic Trouble Codes (DTCs) that can be accessed by a do-it-yourselfer working at home (there are many, many more DTCs available to professional mechanics with proprietary scan tools and software, but those codes cannot be accessed by a generic scan tool). If, after you have checked and repaired the connectors, wire harness and vacuum hoses (if applicable) for an emission-related system, component or circuit, the problem persists, have the vehicle checked by a dealer service department or other qualified repair shop.

3.1 The Data Link Connector (DLC) is located under the lower edge of the dash, to the left of the steering column

Diagnostic trouble codes

Code	Possible cause
P0001	Fuel volume regulator control circuit open
P0003	Fuel volume regulator control circuit low
P0004	Fuel volume regulator control circuit high
P0010	Intake camshaft position actuator circuit open (bank 1)
P0011	"A" Camshaft position - timing over-advanced (bank 1)
P0012	"A" Camshaft position - timing over-retarded (bank 1)
P0013	"B" Camshaft position - actuator circuit malfunction (bank 1)
P0014	"B" Camshaft position - timing over-advanced (bank 1)
P0015	"B" Camshaft position - timing over-retarded (bank 1)
P0016	Crankshaft position/camshaft position, bank 1, sensor A - correlation
P0017	Crankshaft position/camshaft position, bank 1, sensor B - correlation
P0019	Crankshaft position/camshaft position, bank 2, sensor B - correlation
P0030	HO_2S heater control circuit (bank 1, sensor 1)
P0031	HO_2S heater control circuit low (bank 1, sensor 1)
P0032	HO_2S heater control circuit high (bank 1, sensor 1)
P0033	Turbocharger bypass valve control circuit
P0034	Turbocharger bypass valve control circuit low
P0035	Turbocharger bypass valve control circuit high
P0036	HO_2S heater control circuit (bank 1 sensor 2)
P0037	HO_2S heater control circuit low (bank 1, sensor 2)
P0038	HO_2S heater control circuit high (bank 1, sensor 2)
P0071	Ambient air temperature sensor range/performance problem
P0072	Ambient air temperature sensor circuit low input
P0073	Ambient air temperature sensor circuit high input
P0087	Fuel rail/system pressure - too low
P0088	Fuel rail/system pressure - too high
P0100	Mass air flow or volume air flow circuit malfunction

Diagnostic trouble codes (continued)

Code	Possible cause
P0102	Mass air flow or volume air flow circuit, low input
P0103	Mass air flow or volume air flow circuit, high input
P0107	Manifold absolute pressure or barometric pressure circuit, low input
P0108	Manifold absolute pressure or barometric pressure circuit, high input
P0111	Intake air temperature circuit, range or performance problem
P0112	Intake air temperature circuit, low input
P0113	Intake air temperature circuit, high input
P0116	Engine coolant temperature circuit range/performance problem
P0117	Engine coolant temperature circuit, low input
P0118	Engine coolant temperature circuit, high input
P0120	Throttle position or pedal position sensor/switch circuit malfunction
P0121	Throttle position or pedal position sensor/switch circuit, range or performance problem
P0122	Throttle position or pedal position sensor/switch circuit, low input
P0123	Throttle position or pedal position sensor/switch circuit, high input
P0125	Insufficient coolant temperature for closed loop fuel control
P0128	Coolant thermostat (coolant temperature below thermostat regulating temperature)
P0129	Barometric pressure - too low
P0130	O_2 sensor circuit malfunction (bank 1, sensor 1)
P0131	O_2 sensor circuit, low voltage (bank 1, sensor 1)
P0132	O_2 sensor circuit, high voltage (bank 1, sensor 1)
P0133	O_2 sensor circuit, slow response (bank 1, sensor 1)
P0135	O_2 sensor heater circuit malfunction (bank 1, sensor 1)
P0136	O_2 sensor circuit malfunction (bank 1, sensor 2)
P0137	O_2 sensor circuit, low voltage (bank 1, sensor 2)
P0138	O_2 sensor circuit, high voltage (bank 1, sensor 2)
P0139	O_2 sensor circuit, slow response (bank 1, sensor 2)
P0140	O_2 sensor circuit - no activity detected (bank 1, sensor 2)
P0141	O_2 sensor heater circuit malfunction (bank 1, sensor 2)

Code	Possible cause
P0148	Fuel delivery error
P0171	System too lean (bank 1)
P0172	System too rich (bank 1)
P0190	Fuel rail pressure sensor circuit malfunction
P0192	Fuel rail pressure sensor circuit, low input
P0193	Fuel rail pressure sensor circuit, high input
P0201	Injector circuit malfunction - cylinder no. 1
P0202	Injector circuit malfunction - cylinder no. 2
P0203	Injector circuit malfunction - cylinder no. 3
P0204	Injector circuit malfunction - cylinder no. 4
P0221	Throttle position or pedal position sensor/switch B, range or performance problem
P0222	Throttle position or pedal position sensor/switch B circuit, low input
P0223	Throttle position or pedal position sensor/switch B circuit, high input
P0234	Engine overboost condition
P0237	Turbocharger boost sensor A circuit, low
P0238	Turbocharger boost sensor A circuit, high
P0243	Turbocharger wastegate solenoid A malfunction
P0245	Turbocharger wastegate solenoid A, low
P0246	Turbocharger wastegate solenoid A, high
P0261	Cylinder no. 1 injector circuit, low
P0262	Cylinder no. 1 injector circuit, high
P0264	Cylinder no. 2 injector circuit, low
P0265	Cylinder no. 2 injector circuit, high
P0267	Cylinder no. 3 injector circuit, low
P0268	Cylinder no. 3 injector circuit, high
P0270	Cylinder no. 4 injector circuit, low
P0271	Cylinder no. 4 injector circuit, high
P0299	Boost pressure control; boost pressure low
P0300	Random/multiple cylinder misfire detected

Diagnostic trouble codes (continued)

Code	Possible cause
P0301	Cylinder no. 1 misfire detected
P0302	Cylinder no. 2 misfire detected
P0303	Cylinder no. 3 misfire detected
P0304	Cylinder no. 4 misfire detected
P0313	Misfire detected with low fuel
P0324	Knock control system error
P0326	Knock sensor no. 1 circuit, range or performance problem (bank 1 or single sensor)
P0327	Knock sensor no. 1 circuit, low input (bank 1 or single sensor)
P0328	Knock sensor no. 1 circuit, high input (bank 1 or single sensor)
P0335	Crankshaft position sensor A circuit malfunction
P0336	Crankshaft position sensor A circuit - range or performance problem
P0340	Camshaft position sensor "A", circuit malfunction (bank 1)
P0341	Camshaft position sensor "A", circuit - range or performance problem
P0342	Camshaft position sensor "A", circuit - low input
P0343	Camshaft position sensor "A", circuit - high input
P0351	Ignition coil A primary or secondary circuit malfunction
P0352	Ignition coil B primary or secondary circuit malfunction
P0353	Ignition coil C primary or secondary circuit malfunction
P0354	Ignition coil D primary or secondary circuit malfunction
P0365	Camshaft position sensor "B" circuit (bank 1)
P0366	Camshaft position sensor "B" circuit range/performance (bank 1)
P0367	Camshaft position sensor "B" circuit low input (bank 1)
P0368	Camshaft position sensor "B" circuit high input (bank 1)
P0420	Catalyst system efficiency below threshold (bank 1)
P0441	Evaporative emission control system, incorrect purge flow
P0442	Evaporative emission control system, small leak detected
P0444	Evaporative emission control system, open purge control valve circuit
P0455	Evaporative emission (EVAP) control system leak detected (no purge flow or large leak)

Code	Possible cause
P0456	Evaporative emission (EVAP) control system leak detected (very small leak)
P0458	Evaporative emission control system, purge control valve - circuit low
P0459	Evaporative emission control system, purge control valve - circuit high
P0460	Fuel level sensor circuit malfunction
P0461	Fuel level sensor circuit, range or performance problem
P0462	Fuel level sensor circuit, low input
P0463	Fuel level sensor circuit, high input
P0480	Cooling fan no. 1, control circuit malfunction
P0481	Cooling fan no. 2, control circuit malfunction
P0500	Vehicle speed sensor malfunction
P0501	Vehicle speed sensor, range or performance problem
P0503	Vehicle speed sensor circuit, intermittent, erratic or high input
P0506	Idle control system, rpm lower than expected
P0507	Idle control system, rpm higher than expected
P0520	Engine oil pressure sensor/switch circuit malfunction
P0521	Engine oil pressure sensor/switch circuit, range or performance problem
P0522	Engine oil pressure sensor/switch circuit, low voltage
P0523	Engine oil pressure sensor/switch circuit, high voltage
P0524	Engine oil pressure too low
P0557	Brake booster pressure sensor - circuit - low input
P0558	Brake booster pressure sensor - circuit - high input
P0560	System voltage malfunction
P0562	System voltage low
P0563	System voltage high
P0571	Cruise control/brake switch A, circuit malfunction
P0597	Thermostat heater control system - circuit open
P0598	Thermostat heater control system - circuit low
P0599	Thermostat heater control system - circuit high
P0601	Internal control module, memory check sum error

Diagnostic trouble codes (continued)

Code	Possible cause
P0603	Internal control module, keep alive memory (KAM) error
P0604	Internal control module, random access memory (RAM) error
P0615	Starter relay - circuit malfunction
P0616	Starter relay - circuit low
P0617	Starter relay - circuit high
P0620	Generator control circuit malfunction
P0627	Fuel pump control - circuit open
P0628	Fuel pump control - circuit low
P0629	Fuel pump control - circuit high
P0634	ECM/TCM - internal temperature too high
P0641	Sensor reference voltage A - circuit open
P0651	Sensor reference voltage B - circuit open
P0686	ECM power relay control - circuit low
P0687	Engine, control relay - short to ground
P0691	Engine coolant blower motor 1 - short to ground
P0692	Engine coolant blower motor 1 - short to positive
P0694	Engine coolant blower motor 2 - short to positive
P0697	Sensor reference voltage C - circuit open
P0704	Clutch switch input circuit malfunction
P0705	Transmission range sensor, circuit malfunction (PRNDL input)
P0706	Transmission range sensor circuit, range or performance problem
P0711	Transmission fluid temperature sensor circuit, range or performance problem
P0712	Transmission fluid temperature sensor circuit, low input
P0713	Transmission fluid temperature sensor circuit, high input
P0715	Input/turbine speed sensor circuit malfunction
P0716	Input/turbine speed sensor circuit, range or performance problem
P0717	Input/turbine speed sensor circuit, no signal
P0720	Output speed sensor malfunction

Code	Possible cause
P0721	Output speed sensor circuit, range or performance problem
P0722	Output speed sensor circuit, no signal
P0729	Gear 6, incorrect ratio
P0732	Incorrect gear ratio, second gear
P0733	Incorrect gear ratio, third gear
P0734	Incorrect gear ratio, fourth gear
P0735	Incorrect gear ratio, fifth gear
P0741	Torque converter clutch, circuit performance or stuck in off position
P0815	Upshift switch circuit
P0816	Downshift switch circuit
P0961	Pressure control (PC) solenoid A - control circuit range/performance
P0962	Pressure control (PC) solenoid A - control circuit low
P0963	Pressure control (PC) solenoid A - control circuit high
P0969	Pressure control (PC) solenoid C - control circuit range/performance
P0970	Pressure control (PC) solenoid C - control circuit low
P0971	Pressure control (PC) solenoid C - control circuit high
P0973	Shift solenoid (SS) A - control circuit low
P0974	Shift solenoid (SS) A - control circuit high
P0976	Shift solenoid (SS) B - control circuit low
P0977	Shift solenoid (SS) B - control circuit high

4.1a Depress the clip and slide the pedal assembly to the left (Mk I models)

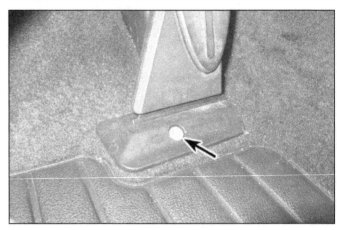

4.1b Pry out the plug and remove the screw from the base of the pedal (Mk II models)

Note: *Throughout this Chapter you will find references to "Mk I" and "Mk II" models; this is done to simplify which specifications and procedures apply to which models. Mk I models include 2006 and earlier Cooper/Cooper S models, and 2008 and earlier Convertible models. Mk II models include 2007 and later Cooper/Cooper S/Clubman/Clubman S and 2009 and later Convertible models.*

4   Accelerator Pedal Position (APP) sensor - replacement

Pedal module

Refer to illustrations 4.1a and 4.1b

1   Mk I models: Using a flat-bladed screwdriver, depress the retaining clip and slide the pedal assembly to the left **(see illustration)**. Disconnect the electrical connector as the assembly is withdrawn. Mk II models: Pry out the plug, then remove the screw at the base of the pedal **(see illustration)**.

2   To install, reconnect the electrical connector and slide the assembly into its mount, ensuring that it is correctly located. When correctly located, two clicks should be heard as

the clips engage, and the right-hand side of the assembly is flush with the side of the mounting.

Pedal mounting bracket

Refer to illustration 4.4

3   Remove the pedal module as previously described.

4   Remove the bolt, release the two clips and remove the bracket **(see illustration)**.

5   Installation is a reversal of removal.

5   Manifold Absolute Pressure/ Intake Air Temperature (MAP/IAT) sensor - replacement

Note: *This procedure applies to Mk I models only.*
Note: *New sensor seals may be required on installation.*
Note: *This sensor is also known as the T-MAP sensor (Temperature-Manifold Absolute Pressure sensor).*

Cooper models

Refer to illustration 5.2

1   Ensure the ignition is turned off, and disconnect the wiring plug from the sensor. To

improve access if required, release the clamp and disconnect the tank vent valve hose from the throttle body. The sensor is located adjacent to the throttle body.

2   Remove the two Torx screws and remove the sensor from the manifold **(see illustration)**.

3   Installation is a reversal of removal, but check the condition of the seals and replace if necessary.

Cooper S models

Refer to illustration 5.5

4   Ensure the ignition is turned off, and disconnect the sensor wiring plug.

5   Remove the two retaining screws and remove the sensor **(see illustration)**.

6   Installation is a reversal of removal, but check the condition of the seals and replace if necessary.

6   Manifold Absolute Pressure (MAP) sensor - replacement

Refer to illustrations 6.1 and 6.2
Note: *This procedure applies to Mk I Cooper S models only.*

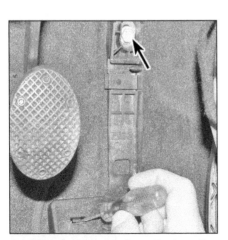

4.4 Remove the bolt, release the two clips and remove the pedal mounting bracket

5.2 Remove the two Torx screws and pull the sensor from the manifold (Cooper)

5.5 Remove the Torx screws and pull the sensor from the manifold (Cooper S)

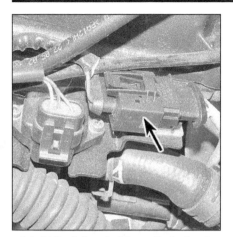

6.1 At the left end of the cylinder head, disconnect the oxygen sensor electrical connector

6.2 Remove the two screws and detach the MAP sensor

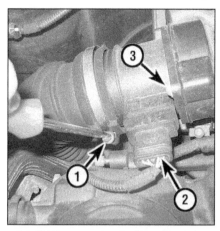

7.1 MAF sensor details (Cooper model shown, Cooper S model similar, but located at left end of the air filter housing cover)

1 Clamp
2 Electrical connector
3 Mounting screw (1 of 2)

1   Working at the left end of the cylinder head, disconnect the oxygen sensor wiring connector and detach it from its bracket **(see illustration)**.
2   Disconnect the MAP sensor wiring plug, undo the retaining screws and remove the sensor, disconnecting the hose as the sensor is withdrawn **(see illustration)**. Discard the O-ring seals; new ones must be installed.
3   Installation is a reversal of removal; replace the O-ring seals, and tighten the retaining screws securely.

2   Unplug the electrical connector from the sensor.
3   Remove the screws and detach the sensor from the air filter housing cover.
4   Check the condition of the O-ring, replacing it if necessary.
5   Installation is the reverse of removal.

3   Unscrew the retaining bolt and remove the sensor.
4   Installation is the reverse of removal. Be sure to use a new O-ring if you're installing the old sensor.

8   Intake manifold differential pressure sensor (non-turbo)/ charge air pressure sensor(s) (turbo) - replacement

Note: *This procedure applies to Mk II models only.*

Turbo models

Refer to illustrations 8.6a and 8.6b

Note: *There are two charge air pressure sensors - one on the intake manifold and one on the charge air duct.*
5   If you're removing the intake manifold charge air pressure sensor, remove the air filter housing (see Chapter 4).
6   Disconnect the electrical connector from the sensor **(see illustrations)**.
7   Unscrew the retaining bolt and remove the sensor.

7   Mass Airflow (MAF) sensor - replacement

Refer to illustration 7.1

Note: *This procedure applies to Mk II models only.*
1   Loosen the hose clamp securing the duct to the MAF sensor **(see illustration)**.

Non-turbo models

Refer to illustration 8.2
1   Remove the air filter housing (see Chapter 4).
2   Unplug the electrical connector from the sensor **(see illustration)**.

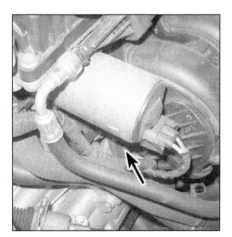

8.2 On non-turbo models, the intake manifold pressure differential sensor is located at the left rear of the engine, below the Valvetronic motor (sensor not visible in this photo; location given)

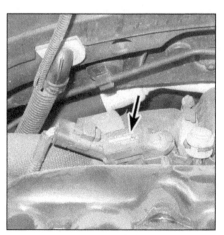

8.6a The intake manifold charge air pressure sensor is located at the center rear of the intake manifold

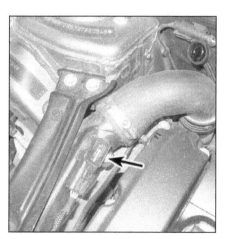

8.6b Location of the intake duct charge air pressure sensor

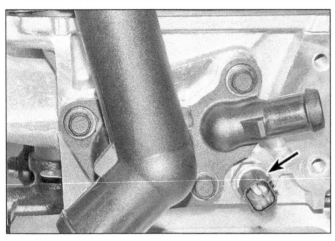

9.1 On Mk I models, the coolant temperature sensor is located at the left end of the cylinder head

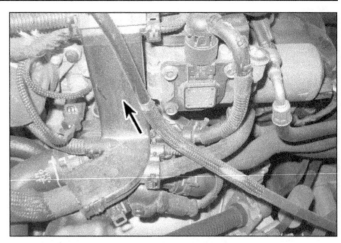

9.7 Slide the wiring harness duct up and off its bracket (Mk II models)

8   Installation is the reverse of removal. Be sure to use a new O-ring if you're installing the old sensor.

9   Engine Coolant Temperature (ECT) sensor - replacement

Warning: *Wait until the engine is completely cool before beginning this procedure.*

Mk I models

Refer to illustration 9.1

1   The sensor is located in the left-hand side of the cylinder head **(see illustration)**. Drain the cooling system (see Chapter 1).
2   On Cooper models, remove the battery and battery tray (see Chapter 5).
3   On all models, remove the air filter housing (see Chapter 4).
4   Disconnect the electrical connector, then unscrew the sensor from the cylinder head.

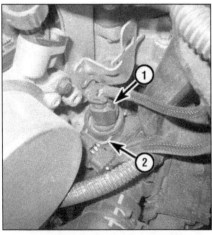

9.8 Engine Coolant Temperature (ECT) sensor details (Mk II models)

1   Electrical connector
2   Retaining clip

5   Installation is the reverse of removal. Tighten the sensor to the specified torque, and refill the cooling system as described in Chapter 1.

Mk II models

Refer to illustrations 9.7 and 9.8

6   On Cooper S models, remove the air ducts at the left end of the engine.
7   Detach the hose from the vacuum pump and slide the wiring harness duct up and off its bracket, unplugging any electrical connectors necessary to do so **(see illustration)**.
8   Unplug the electrical connector from the sensor, then pry out the clip and remove the sensor from the water passage **(see illustration)**.
9   If you're reinstalling the same sensor, replace the O-ring.
10   Installation is the reverse of removal. Refill the cooling system as described in Chapter 1.

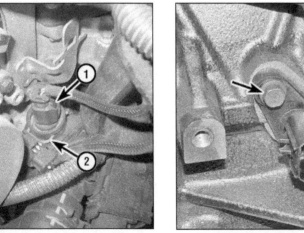

10.4 On Mk I models, the CKP sensor is located on the front face of the cylinder block adjacent to the transaxle bellhousing

10   Crankshaft Position (CKP) sensor - replacement

Mk I models

Refer to illustration 10.4

1   If you're working on a Cooper model, remove the intake manifold (see Chapter 2A).
2   If you're working on a Cooper S model, place the Modular Front End (MFE) in the service position (see Chapter 11).
3   Pry out the red catch, then disconnect the electrical connector from the sensor.
4   Remove the retaining bolt and pull the sensor from the engine block **(see illustration)**.
5   Installation is a reversal of removal. Check the condition of the O-ring and replace if necessary.

Mk II models

Refer to illustrations 10.7 and 10.8

6   Raise the vehicle and support it securely on jackstands.
7   Remove the plastic fastener, then remove the cover from over the sensor **(see illustration)**.
8   Disconnect the electrical connector from the sensor, remove the retaining bolt, then pull the sensor from the engine **(see illustration)**.
9   Installation is the reverse of removal.

11   Camshaft Position (CMP) sensor(s) - replacement

Mk I models

Refer to illustration 11.3

1   The sensor is located on the right-hand end of the cylinder head. In order to remove the sensor, remove the right-hand engine mounting bracket (see Chapter 2A).
2   Disconnect the sensor electrical connector.

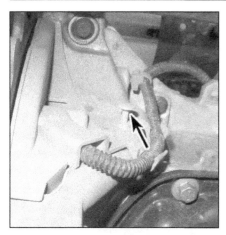

10.7 Remove the plastic fastener and cover for access to the CKP sensor (Mk II models)

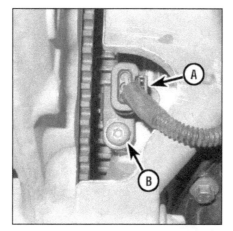

10.8 CKP sensor electrical connector (A) and mounting bolt (B) (Mk II models)

11.3 Remove the CMP sensor from the right-hand end of the cylinder head (Mk I models)

3 Remove the retaining screw and pull the sensor from the head. Discard the O-ring seal - a new one must be installed **(see illustration)**.
4 Installation is the reverse of removal.

Mk II models

Refer to illustration 11.5

Note: *These models have two CMP sensors - one for each camshaft.*
5 Disconnect the electrical connector, then remove the sensor mounting bolt **(see illustration)**.
6 Pull the sensor from the head. Discard the O-ring seal - a new one must be installed.
7 Installation is the reverse of removal.

12 Knock Sensor (KS) - replacement

Mk I models

Refer to illustration 12.6

1 The knock sensor is mounted on the front face of the cylinder block.
2 On Cooper models, remove the intake manifold (see Chapter 2A).
3 On Cooper S models, remove the supercharger (see Chapter 4).
4 Trace the wiring from the sensor and disconnect the electrical connector. Unscrew the bolt and remove the knock sensor.
5 Clean the mating surface of the cylinder block (and the sensor, too, if the original one is being installed).
6 Install the sensor, positioning it with the wiring harness at a 20-degree angle from horizontal **(see illustration)**. Tighten the mounting bolt to the torque listed in this Chapter's Specifications.

Mk II models

Note: *The knock sensor is located near the rear end of the starter motor.*
7 Raise the vehicle and support it securely

on jackstands. Because of the sensor's close proximity to the starter and the exposed battery positive cable, disconnect the cable from the negative terminal of the battery before proceeding.
8 Disconnect the electrical connector from the sensor.
9 Unscrew the bolt and remove the knock sensor.
10 Clean the mating surface of the cylinder block (and the sensor, too, if the original one is being installed).
11 Install the sensor, tightening the mounting bolt to the torque listed in this Chapter's Specifications.

13 Oxygen sensor(s) - replacement

General precautions

Note: *Because it is installed in the exhaust manifold or pipe, both of which contract when cool, an oxygen sensor might be very difficult*

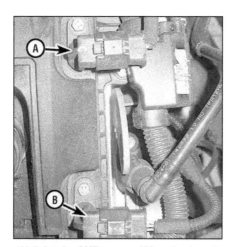

11.5 Intake CMP sensor (A) and exhaust CMP sensor (B) (Mk II models)

to loosen when the engine is cold. Rather than risk damage to the sensor or its mounting threads, start and run the engine for a minute or two, then shut it off. Be careful not to burn yourself during the following procedure.
1 Be particularly careful when servicing an oxygen sensor:

a) *Oxygen sensors have a permanently attached pigtail and an electrical connector that cannot be removed. Damaging or removing the pigtail or electrical connector will render the sensor useless.*
b) *Keep grease, dirt and other contaminants away from the electrical connector and the louvered end of the sensor.*
c) *Do not use cleaning solvents of any kind on an oxygen sensor.*
d) *Oxygen sensors are extremely delicate. Do not drop a sensor or handle it roughly.*
e) *Make sure that the silicone boot on the sensor is installed in the correct position. Otherwise, the boot might melt and it might prevent the sensor from operating correctly.*

12.6 Position the knock sensor with the wiring harness at 20-degrees from the horizontal

13.4 Note the oxygen sensor wiring harness routing

13.5 Use an oxygen sensor socket, if available, to unscrew the sensor

2   The oxygen sensors are located in the exhaust system before and after the catalytic converter. On Mk I models, the catalytic converter is part of the front section of the exhaust system. On Mk II non-turbo models, the catalytic converter is integral with the exhaust manifold. On Mk II turbo models, the catalytic converter can be unbolted from the turbocharger and the exhaust pipe.

Replacement
Mk I models
Refer to illustrations 13.4 and 13.5

3   Apply the parking brake, then raise the front of the vehicle and support it securely on jackstands.
4   Unclip the oxygen sensor cables from the retainer, and disconnect the wiring plugs **(see illustration)**.
5   Using an oxygen sensor removal socket, unscrew the sensor and remove it from the exhaust pipe **(see illustration)**.
6   Installation is the reverse of removal, noting the following points:
a) *Tighten the sensor to the specified torque.*

b) *Check that the wiring is correctly routed, and in no danger of contacting the exhaust system.*
c) *Ensure that no lubricant or dirt comes into contact with the sensor probe.*
d) *If installing the same sensor, apply a smear of copper based, high-temperature anti-seize compound to the sensor threads prior to installation (new sensors should already have anti-seize compound on their threads).*

Mk II models
Refer to illustrations 13.7a, 13.7b and 13.9

7   If you're removing an upstream sensor, trace the harness from the sensor, then unplug the electrical connector **(see illustrations)**.
8   If you're removing a downstream oxygen sensor, raise the front of the vehicle and support it securely on jackstands.
9   Trace the harness from the sensor, then unplug the electrical connector **(see illustration)**.
10   Using an oxygen sensor removal socket, unscrew the sensor and remove it.

11   Installation is the reverse of removal, noting the following points:
a) *Tighten the sensor to the specified torque.*
b) *Check that the wiring is correctly routed, and in no danger of contacting the exhaust system.*
c) *Ensure that no lubricant or dirt comes into contact with the sensor probe.*
d) *If installing the same sensor, apply a smear of copper based, high-temperature anti-seize compound to the sensor threads prior to installation (new sensors should already have anti-seize compound on their threads).*

---

14   Powertrain Control Module (PCM) - removal and installation

---

Refer to illustrations 14.3, 14.4, 14.5a and 14.5b

1   Disconnect the cable from the negative terminal of the battery (see Chapter 5). **Note:** *Disconnecting the battery will erase any fault codes stored in the PCM. It is recommended*

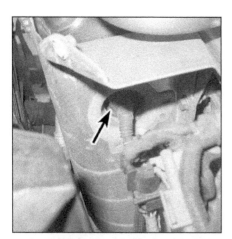

13.7a Upstream oxygen sensor location - Mk II non-turbo model

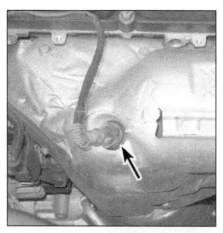

13.7b Upstream oxygen sensor location - Mk II turbo model

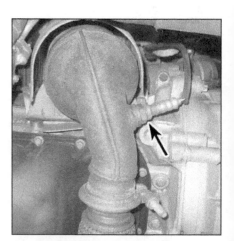

13.9 Downstream oxygen sensor location (Mk II non-turbo model shown, turbo model similar)

14.3 Remove the cover (Mk I Cooper S model shown, later S models similar) . . .

14.4 . . . and release the retaining clips

14.5a Slide out the locking catches and disconnect the electrical connectors - Mk I model shown

that the fault code memory of the module is interrogated using a code reader or scan tool prior to battery disconnection (see Section 3).
2   Mk I Cooper models: Working in the left-hand corner of the engine compartment, pull up the rubber weatherstrip, release the clips and remove the plastic cover from over the battery.
3   Cooper S models: Unclip the cover from the PCM **(see illustration)**.
4   Press back the retaining clips and slide the PCM up a little from its mountings **(see illustration)**.
5   Pull out the locking catches and disconnect the electrical connectors from the PCM, then remove it from the vehicle **(see illustrations)**.
6   Installation is a reversal of removal.
Note: *If a new module has been installed, it will have to be coded using special test equipment. Entrust this task to a MINI dealer or suitably-equipped specialist.* After reconnecting the battery, the vehicle must be driven for several miles so that the PCM can learn its basic settings. If the engine still runs erratically, the basic settings may be reinstated by a MINI dealer or specialist using special diagnostic equipment.

---

15 Evaporative emissions control (EVAP) system - component replacement

Mk I models

Charcoal canister and Leak Detection Pump (LDP) replacement

Refer to illustration 15.4

1   The canister and LDP are located above the right rear wheelwell liner.
2   Loosen the right rear wheel bolts, raise the rear of the vehicle and support it securely on jackstands. Block the front wheels to prevent the vehicle from rolling. Remove the right rear wheel.

3   Remove the screws, pry out the plastic rivets, and remove the right rear wheelwell liner.
4   Disconnect the hoses from the canister and LDP. If the hoses are secured by plastic locking clips, squeeze the sides of the clips to release them from the connection on the canister **(see illustration)**. Note the hose locations to ensure correct installation.
5   Remove the canister mounting bolt, and release the central retaining clip. Remove the canister/LDP assembly. Note the three mountings, with corresponding locating holes.
6   Installation is the reverse of removal, but ensure that the hoses are correctly reconnected as noted before removal, and make sure that the hose securing clips are correctly engaged.

Purge solenoid/tank vent valve replacement

Refer to illustration 15.9

7   The valve is located on a bracket next to the right-hand engine mounting bracket in the engine compartment.

14.5b On Mk II models, pivot the locking latches on the electrical connectors to free them

8   Ensure the ignition is switched off.
9   Squeeze together the sides of the locking collars, and disconnect the hoses from the valve **(see illustration)**.

15.4 Squeeze the sides of the collars and disconnect the hoses

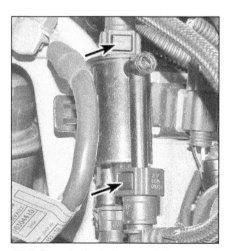

15.9 Press in the buttons and disconnect the hoses from the purge solenoid fuel tank vent valve

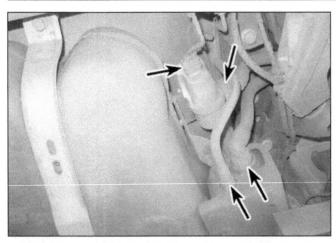

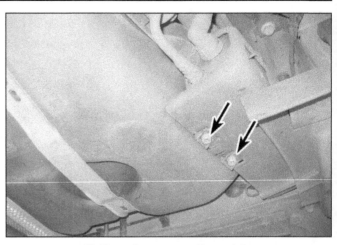

15.14 Disconnect the electrical connector and hoses from the leak detection pump and the hoses from the charcoal canister . . .

15.15 . . . then remove the canister

10   Disconnect the wiring plug from the valve.

11   Depress the clip and pull the valve from its mounting.

12   Installation is the reverse of removal, but make sure that all hoses are correctly reconnected as noted before removal.

Mk II models

Charcoal canister and Leak Detection Pump (LDP) replacement

Refer to illustrations 15.14 and 15.15

13   Raise the rear of the vehicle and support it securely on jackstands. Block the front wheels to prevent the vehicle from rolling. Remove the splash shield at the front of the right rear wheel opening.

14   Unplug the electrical connectors and disconnect the hoses from the canister and LDP **(see illustration)**.

15   Remove the mounting bolts and detach the canister/LDP from the vehicle **(see illustration)**.

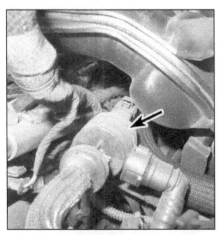

15.23 The fuel tank vent valve is located underneath the intake manifold (turbo model shown)

16   The LDP can now be separated from the canister, if necessary.

17   Installation is the reverse of removal, but make sure that all hoses are correctly reconnected as noted before removal.

Purge solenoid/tank vent valve replacement

Note: *The fuel tank vent valve is located on the back side of the engine, underneath the intake manifold.*

Non-turbo models

18   Raise the front of the vehicle and support it securely on jackstands. Block the rear wheels to prevent the vehicle from rolling. Remove the splash shield from under the engine.

19   Disconnect the electrical connector from the valve **(see illustration 15.23)**.

20   Unbolt the valve, then detach the hoses.

21   Installation is the reverse of removal.

Turbo models

Refer to illustration 15.23

22   Remove the intake manifold (see Chapter 2A).

23   Disconnect the electrical connector from the valve **(see illustration)**.

24   Unbolt the valve, then detach the hoses.

25   Installation is the reverse of removal.

---

16 Catalytic converter - testing, precautions and replacement

---

Testing

1   The performance of the catalytic converter is monitored by the oxygen sensors. When a change in readings occurs between the upstream and downstream sensors, a trouble code should be set that indicates the catalytic converter is not working efficiently.

2   An overly rich mixture, caused by a fuel system problem or an ignition misfire, can cause clogging of the converter. This is almost sure to set a trouble code related to the primary cause, but sluggish performance will result as well. A clogged converter can be verified by checking intake manifold vacuum with a vacuum gauge. A drop of approximately 3 in-Hg between idle and a steady 200 rpm generally means that there is a restriction in the exhaust system.

Precautions

3   The catalytic converter is a reliable and simple device which needs no maintenance in itself, but there are some facts an owner should be aware of if the converter is to function properly for full service life.

a) *Always keep the ignition and fuel systems well maintained in accordance with the manufacturer's instructions.*

b) *If the engine develops a misfire, do not drive the vehicle at all (or at least at little as possible) until the problem is fixed.*

c) *DO NOT push- or tow-start the vehicle - this can soak the catalytic converter with unburned fuel, causing it to overheat when the engine finally does start.*

d) *DO NOT turn off the engine at high engine speeds.*

e) *DO NOT use fuel or engine oil additives unless you read the label carefully and determine that they are safe for use with catalytic converters, as they may contain substances that are harmful to the converter.*

f) *DO NOT continue to use the vehicle if the engine burns oil to the extent of leaving a visible trail of blue smoke.*

g) *Remember that the catalytic converter operates at very high temperatures. DO NOT park the vehicle in dry undergrowth, over long grass or piles of dead leaves.*

h) *Remember that the converter is fragile - do not strike it with tools during service work.*

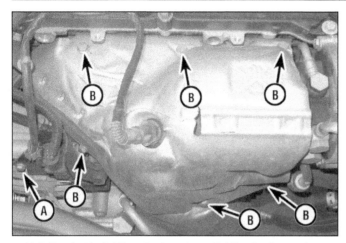

16.5 Bracket bolt (A) and exhaust manifold/turbocharger heat shield fasteners (B)

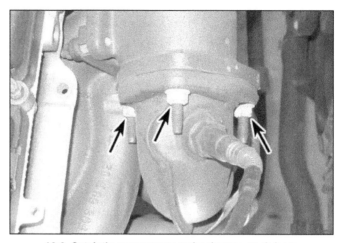

16.6 Catalytic converter-to-turbocharger studs/nuts

i) *The catalytic converter should last upwards of 100,000 miles on a well-maintained car. If the converter is defective, it must be replaced.*

Replacement

All except Mk II S models

4   The catalytic converter is an integral part of the exhaust manifold on all models except Mk II S models, and they are replaced as a unit (see Chapter 2A). On Mk II models, an additional (secondary) catalytic converter is located in the mid-section of the exhaust pipe.

Mk II S models (primary converter)

Refer to illustrations 16.5, 16.6 and 16.10

5   Remove the bracket at the right end of the exhaust manifold, then remove the exhaust manifold/turbocharger heat shield **(see illustration)**.

6   Apply penetrating oil to the catalytic converter-to-turbocharger studs/nuts **(see illustration)**.

7   Remove the upstream oxygen sensor (see Section 13).

8   Raise the front of the vehicle and support it securely on jackstands. Remove the under-vehicle splash shield.

9   Remove the downstream oxygen sensor

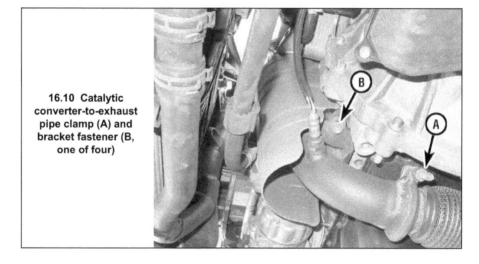

16.10 Catalytic converter-to-exhaust pipe clamp (A) and bracket fastener (B, one of four)

(see Section 13) and the catalytic converter heat shield.

10   Apply penetrating oil to the catalytic converter-to-exhaust pipe clamp fasteners and allow the oil to soak in for awhile **(see illustration)**. Remove the clamp and separate the pipe from the converter.

11   Remove the converter lower mounting fasteners.

12   Remove the catalytic converter-to-turbo-

charger studs/nuts and detach the converter from the turbocharger.

13   Installation is the reverse of removal. Thoroughly clean all mating surfaces, and use new gaskets and sealing rings. Replace any damaged fasteners, and coat all threaded fasteners with anti-seize compound. Tighten the mounting fasteners to the torque listed in this Chapter's Specifications.

Notes

Chapter 7 Part A
Manual transaxle

Contents

Specifications

Note: *Throughout this Chapter you will find references to "Mk I" and "Mk II" models; this is done to simplify which specifications and procedures apply to which models. Mk I models include 2006 and earlier Cooper/Cooper S models, and 2008 and earlier Convertible models. Mk II models include 2007 and later Cooper/Cooper S/Clubman/Clubman S and 2009 and later Convertible models.*

General
Designation
    Five-speed models
        Vehicles manufactured up to 07/2004 .... Getrag GS5-65 BH/SH
        Vehicles manufactured after 07/2004 .... Getrag GS5-52 BG
    Six-speed models .... Getrag GS6-85 BG/DG

Lubrication
Capacity .... Up to 2.0 qts (1.9 liters)
Recommended oil type .... See Chapter 1

Torque specifications

Note: *One foot-pound (ft-lb) of torque is equivalent to 12 inch-pounds (in-lbs) of torque. Torque values below approximately 15 ft-lbs are expressed in inch-pounds, since most foot-pound torque wrenches are not accurate at these smaller values.*

	Ft-lbs (unless otherwise indicated)	Nm
Transaxle-to-engine bolts		
Mk I models	63	85
Mk II models	28	38
Oil drain plug		
Five-speed transaxle	24	32
Six-speed transaxle	32	43
Oil check/fill plug		
Mk I models		
Cooper	18	25
Cooper S	32	43
Mk II models	34	43

Torque specifications (continued)

Ft-lbs (unless otherwise indicated) **Nm**

Note: One foot-pound (ft-lb) of torque is equivalent to 12 inch-pounds (in-lbs) of torque. Torque values below approximately 15 ft-lbs are expressed in inch-pounds, since most foot-pound torque wrenches are not accurate at these smaller values.

Back-up light switch		
Mk I models	18	25
Mk II models	144 in-lbs	16
Wheel bolts	81	110

1   General information

Note: *Throughout this Chapter you will find references to "Mk I" and "Mk II" models; this is done to simplify which specifications and procedures apply to which models. Mk I models include 2006 and earlier Cooper/Cooper S models, and 2008 and earlier Convertible models. Mk II models include 2007 and later Cooper/Cooper S/Clubman/Clubman S and 2009 and later Convertible models.*

1   The transaxle is contained in a cast-aluminum alloy casing bolted to the engine's left-hand end, and consists of the transmission and final drive differential.

2   Drive is transmitted from the crankshaft via the clutch to the input shaft, which has a splined extension to accept the clutch friction disc, and rotates in sealed ball-bearings. From the input shaft, drive is transmitted to the output shaft, which rotates in a roller bearing at its right-hand end, and a sealed ball-bearing at its left-hand end. From the output shaft, the drive is transmitted to the differential ring gear, which rotates with the differential case and planetary gears, thus driving the sun gears and driveaxles. The rotation of the planetary gears on their shaft allows the inner wheel to

rotate at a slower speed than the outer wheel when the car is cornering.

3   The input and output shafts are arranged side by side, parallel to the crankshaft and driveaxles, so that their gear pinion teeth are in constant mesh. In the neutral position, the output shaft gear pinions rotate freely, so that drive cannot be transmitted to the ring gear.

4   Gear selection is via a floor-mounted lever and cable mechanism. The shift cables cause the appropriate selector fork to move its respective synchro-sleeve along the shaft, to lock the gear pinion to the synchro-hub. Since the synchro-hubs are splined to the output shaft, this locks the pinion to the shaft, so that drive can be transmitted. To ensure that shifting can be made quickly and quietly, a synchromesh system is fitted to all forward gears, consisting of synchro rings and spring-loaded fingers, as well as the gear pinions and synchro-hubs.

2   Manual transaxle - draining and refilling

Refer to illustrations 2.3a and 2.3b

1   This operation is much quicker and more efficient if the vehicle has been driven a suf-

ficient amount of time to warm the engine/transaxle up to normal operating temperature.

2   Park the car on level ground, switch off the ignition and apply the parking brake firmly. For improved access, raise the front of the car and support it securely on jackstands. Note that the car must be level to ensure accuracy when refilling and checking the oil level.

3   Wipe the area around the check/fill plug. Unscrew the check/fill plug from the transaxle **(see illustrations)**. Note there are no separate sealing washers with the filler or drain plugs on 5-speed transaxles installed in vehicles manufactured after 07/2004.

4   Position a container under the drain plug and unscrew the plug.

5   Allow the oil to drain completely into the container. If the oil is hot, take precautions against scalding. Clean both the check/fill and drain plugs.

6   When the oil has finished draining, clean the drain plug threads and those of the transaxle case. Install the drain plug, tightening it to the torque listed in this Chapter's Specifications.

7   Refilling the transaxle can be an awkward operation. Hand-operated lubricant pumps are

2.3a Transaxle drain and check/fill plugs (viewed through the left-side wheelwell) - 5-speed transaxles up to 07/2004

2.3b Transaxle drain and check/fill plugs (viewed through the right-side wheelwell) - 5-speed transaxles after 07/2004 and 6-speed transaxles

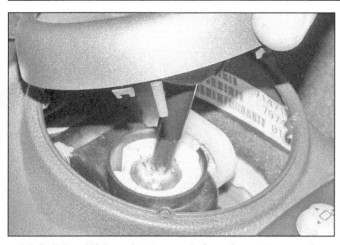

3.3 Pull the shift lever boot upwards from the center console. Note the locating pin and corresponding holes

3.5 Remove the exhaust heat shield below the shift lever housing

available at most auto parts stores and make the refilling procedure much easier. Note that the car must be supported in a level position when checking the oil level.

8   Refill the transaxle carefully until oil begins to run out of the filler plug hole, then install the check/fill plug and tighten it to the torque listed in this Chapter's Specifications. Take the car on a short drive so that the new oil is distributed fully around the transaxle components, then check the level again.

3   Shift lever and cables - removal and installation

Removal

Refer to illustrations 3.3, 3.5, 3.9a, 3.9b, 3.9c, 3.9d, 3.10a, 3.10b, 3.11, 3.12a and 3.12b

1   Firmly apply the parking brake, then raise the front of the vehicle and support it securely on jackstands.
2   Pull the shift lever knob sharply upwards and remove it.

3   Unclip the shift lever boot from the center console, release the clip and slide the boot up from the lever **(see illustration)**.
4   Remove the exhaust system as described in Chapter 4.
5   Unscrew the nuts and remove the heat shield below the shift lever housing **(see illustration)**.

Mk I models

6   On Cooper models, remove the battery and battery tray as described in Chapter 5. On Cooper S models, remove the air filter housing as described in Chapter 4.

Mk II models

7   Remove the PCM (see Chapter 6) and reposition the underhood fuse/relay panel **(see illustration 7.6)**.
8   On non-turbo models, remove the air filter housing and fresh air intake duct (see Chapter 4). On turbo models, remove the charge air ducts from above the transaxle.

All models

9   Use a pair of needle-nose pliers to pry

3.9a Use a pair of needle-nose pliers to pry the cables from the lever ballstuds (Mk I models)

the cable ends from the shift levers on the transaxle **(see illustrations)**, then detach the cable casings from the bracket on the transaxle **(see illustrations)**.

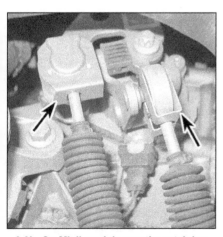

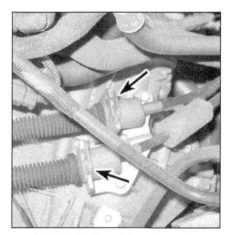

3.9b On Mk II models, pry the retaining clips back (towards the rubber boots), then pry the cable ends off the ballstuds

3.9c Squeeze the cable clip ends together and detach the cables from the bracket (Mk I models)

3.9d On Mk II models, remove the retainers and pry the cables from the bracket

3.10a Remove the four Torx bolts . . .

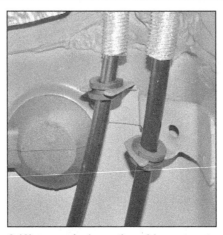

3.10b . . . and release the cable grommets from their retaining clips

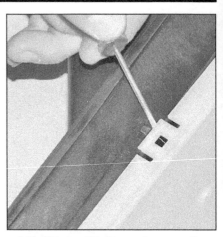

3.11 Release the clips and remove the base cover from the shift housing

3.12a Pry the cable ends from the levers . . .

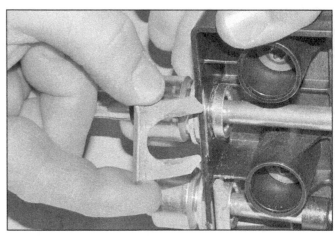

3.12b . . . then slide out the cable retainers

10   Working underneath the vehicle, remove the four Torx screws, pull the lever housing rearwards slightly, and lower the housing from position, complete with cables. Detach the cable grommets from the retaining brackets on the underside of the vehicle **(see illustrations)**.

11   With the assembly on a clean surface, carefully release the clips securing the cover to the base of the lever housing **(see illustration)**.

12   Pry the cables from the levers, then slide out the retaining clips and pull the cables from the housing **(see illustrations)**

13   The shift lever is integral with the housing and is not available separately.

Installation

14   Installation is a reversal of the removal procedure, noting the following points:
  a) *Apply grease to the ballstuds before installation.*
  b) *No adjustment of the shift cables is possible.*

---

4   Oil seals - replacement

Driveaxle oil seals

Refer to illustrations 4.2 and 4.3

1   Remove the appropriate driveaxle (see Chapter 8).

2   Carefully pry the oil seal out of the transaxle, using a large flat-bladed screwdriver **(see illustration)**.

3   Remove all traces of dirt from the area around the oil seal opening. Install the new seal into its opening, and drive it squarely into position using a seal driver or large socket which bears only on the hard outer edge of the seal, until it abuts its locating shoulder **(see illustration)**.

4   Install the driveaxle (see Chapter 8).

4.2 Carefully pry the driveaxle oil seal from the transaxle casing

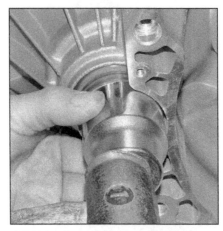

4.3 Drive the new seal into place using a seal driver or large socket

4.6 Undo the bolts and remove the clutch release bearing guide sleeve

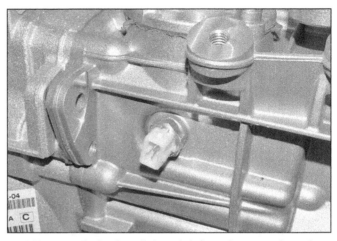

5.4 Unscrew the back-up light switch from the transaxle casing

Input shaft oil seal

Cooper models up to 07/2004, Mk II Cooper models and all Cooper S models

Refer to illustration 4.6

5   Remove the transaxle as described in Section 6 or 7, and the clutch release bearing as described in Chapter 8.

6   Remove the bolts securing the clutch release bearing guide sleeve in position, and slide the guide off the input shaft **(see illustration)**. On Mk I Cooper models, the oil seal is integral with the guide sleeve - if the seal is faulty, the guide sleeve must be replaced with the seal.

Mk I Cooper S models

7   Note the installed position of the seal and pry it from the sleeve.

8   Before installing a new guide sleeve/seal, check the input shaft's seal rubbing surface for signs of burrs, scratches or other damage, which may have caused the seal to fail in the first place. It may be possible to polish away minor faults of this sort using fine abrasive paper; however, more serious defects will require the replacement of the input shaft.

9   Use a seal driver or tubular spacer (socket) that bears only on the hard outer edge of the seal, and drive the seal into position in the guide sleeve.

10   Lubricate the lips of the new seal with clean transaxle oil, and slide the guide sleeve over the input shaft and into position.

11   Apply a little thread-locking compound to the bolts and tighten them securely.

12   The remainder of installation is a reversal of removal.

Mk II Cooper S models

13   Screw a slide hammer with a self-tapping screw adapter end into the seal, then operate the slide hammer to extract the seal. Alternatively, a self-tapping screw can be threaded into the seal and a prybar with a crow-foot end can be used to pry against the screw to dislodge the seal. **Note:** *Either method might*

require drilling a small pilot hole into the casing of the seal, to allow the screw to thread into the seal case. If so, be sure to clean up all metal fragments.

14   Before installing a new guide sleeve/seal, check the input shaft's seal rubbing surface for signs of burrs, scratches or other damage, which may have caused the seal to fail in the first place. It may be possible to polish away minor faults of this sort using fine abrasive paper; however, more serious defects will require the replacement of the input shaft.

15   Lubricate the lips of the new seal with clean transaxle oil, and slide the guide sleeve over the input shaft and into position.

16   Apply a little thread-locking compound to the bolts and tighten them securely.

17   The remainder of installation is a reversal of removal.

Mk I Cooper models from 07/2004

18   It would appear that, at the time of writing, the seal is not available separately. Consult your MINI dealer, parts supplier or transaxle specialist.

---

5   Back-up light switch - testing, removal and installation

Note: *On Mk I Cooper models and all Mk II models, the back-up light switch is located on the top of the transaxle. On Mk I Cooper S models, it's located on the firewall-side of the transaxle case.*

Testing

1   The back-up light circuit is controlled by a plunger-type switch screwed into the transaxle casing. If a fault develops, first check all fuses.

2   To test the switch, disconnect the electrical connector, and use an ohmmeter or a continuity tester to check that there is continuity between the switch terminals only when reverse gear is selected. If this is not

the case, and there are no obvious breaks or other damage to the wires, the switch is faulty, and must be replaced.

Removal

Refer to illustration 5.4

3   Where necessary, to improve access to the switch, remove the air filter housing intake duct assembly/battery and battery box (as applicable - see Chapters 4 and 5).

4   Disconnect the electrical connector, then unscrew the switch from the transaxle casing along with its sealing washer (if equipped) **(see illustration)**.

Installation

5   Install a new sealing washer to the switch (if applicable), then screw it back into position in the top of the transaxle housing and tighten it to the torque listed in this Chapter's Specifications. Install the electrical connector and test the operation of the circuit. Install any components removed for access.

---

6   Manual transaxle (Mk I models) - removal and installation

Removal

Refer to illustrations 6.6, 6.13 and 6.23

1   Disconnect the cable from the negative terminal of the battery (see Chapter 5).

2   Remove the air filter housing (see Chapter 4).

3   Loosen the front wheel bolts. Chock the rear wheels, then firmly apply the parking brake. Loosen both front wheel bolts. Raise the front of the vehicle and securely support it on jackstands. Remove both front wheels, and wheelwell liners (see Chapter 11).

4   Drain the transaxle oil as described in Section 2, then install the drain plug and tighten it to the torque listed in this Chapter's Specifications.

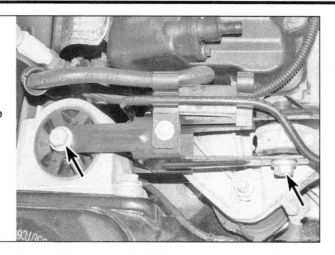

6.6 Remove the engine carrier bracket

6.13 Depress the collars and disconnect the vacuum pipes from the supercharger intake pipe

5   Remove both driveaxles (see Chapter 8).

6   On models manufactured up to 12/2003, remove the engine carrier bracket from the right-hand end of the engine **(see illustration)**.

Cooper models

7   Remove the battery and battery tray (see Chapter 5).

8   Depress the collar and disconnect the brake booster vacuum pipe from the intake manifold.

9   Remove the starter motor (see Chapter 5).

Cooper S models

10   Remove the throttle body and Manifold Absolute Pressure (MAP) sensor (see Chapter 4).

11   Remove the MAP sensor mounting bracket.

12   Remove the exhaust manifold heat shield.

13   Unscrew the retaining bolt, release the clamp and disconnect the intake pipe from the supercharger, then depress the collars and disconnect the vacuum pipes from the intake pipe **(see illustration)**.

14   Remove the retaining screw, detach the oxygen sensor wiring clip and remove the

6.23 Unscrew these two bolts and remove the plate

heat shield above the starter motor.

15   Disconnect the starter motor wiring connections, and move the wiring harness from the transaxle area (see Chapter 5).

All models

16   Install an engine support fixture to the engine, with the chain connected to the engine on the transaxle end. If no lifting hook is present, you'll either have to obtain or fabricate one (tool no. 11-8-260 or equivalent) that you can bolt to the left end of the cylinder head. Tighten the support fixture chain screw/nut to remove all slack in the chain.

17   If you don't have access to an engine support fixture, position a floor jack under the engine oil pan, with a block of wood on the jack head to protect the oil pan casing. Take up the weight of the engine with the jack.

18   Position the Modular Front End (MFE) in the service position (see Chapter 11).

19   Remove the subframe (see Chapter 2A).

20   Detach the shift cables from the transaxle (see Section 3).

21   Remove the clutch release cylinder from the transaxle (see Chapter 8). There is no need to disconnect the fluid line - free the line from its retaining clips and position the cylinder to one side.

22   Noting their locations, disconnect all electrical connectors from the transaxle. Note the harness routing and move the harness to one side.

23   Remove the semi-circular plate adjacent to the right-hand driveaxle aperture in the transaxle casing **(see illustration)**.

24   Remove the left-hand transaxle mount assembly.

25   Support the transaxle with a jack - preferably one made for this purpose. **Note:** *Transmission jack head adapters are available that replace the round head on a floor jack.* Noting their installed locations, remove the bolts securing the transaxle to the engine.

26   Using the jack, lower the transaxle approximately 1-3/5 inches (40 mm) on Cooper models, or 5-1/3 inches (135 mm) on Cooper S models and, with the help of an assistant, slide the transaxle from the engine. When lowering the engine, take great care

not to lower the assembly too far, as this may cause damage to the exhaust system or refrigerant lines.

Installation

27   Installation of the transaxle is the reverse of the removal procedure, bearing in mind the following points:

a)   *Prior to installation, check the clutch assembly and release mechanism components (see Chapter 8). Lubricate the release bearing guide with a little high-temperature grease. Do not apply too much grease, otherwise there is a possibility of the grease contaminating the clutch friction disc. Ensure no grease is applied to the input shaft/friction disc splines.*

b)   *Ensure that the locating dowels are correctly positioned prior to installation.*

c)   *Tighten all nuts and bolts to the specified torque (where given).*

d)   *Replace the driveaxle oil seals, then install the driveaxles (see Chapter 8).*

e)   *Install the release cylinder (see Chapter 8).*

f)   *On completion, refill the transaxle with the specified type and quantity of lubricant.*

---

7   Manual transaxle (Mk II models) - removal and installation

---

Removal

Refer to illustrations 7.6 and 7.9

1   Disconnect the cable from the negative terminal of the battery (see Chapter 5).

2   On Cooper models, remove the air filter housing (see Chapter 4).

3   On Cooper S models, remove the charge air ducts from above the transaxle.

4   Install an engine support fixture to the engine, with the chain connected to the engine

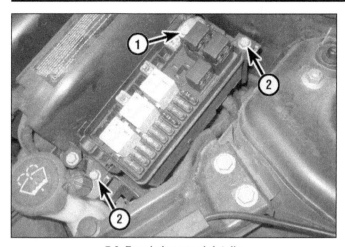

7.6 Fuse/relay panel details

1   Battery positive cable
2   Fuse/relay panel mounting bolts

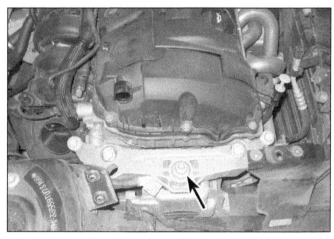

7.9 Loosen, but don't remove this nut from the right side engine mount

on the transaxle end. If no lifting hook is present, you'll either have to obtain or fabricate one (tool no. 11-8-260 or equivalent) that you can bolt to the left end of the cylinder head. Tighten the support fixture chain screw/nut to remove all slack in the chain.

5   Remove the Powertrain Control Module (PCM) (see Chapter 6).

6   Disconnect the battery cable from the underhood fuse/relay panel, then remove the fuse/relay panel bolts and move the panel out of the way **(see illustration)**.

7   Remove the transaxle-to-inner fender panel support bracket, followed by the transaxle mount (see Chapter 2A).

8   Detach the shift cables from the levers on the transaxle, then unbolt the cable bracket (see Section 3).

9   Loosen, but don't remove, the center nut of the right-side engine mount **(see illustration)**.

10   Loosen the front wheel bolts. Chock the rear wheels, then firmly apply the parking brake. Loosen both front wheel bolts. Raise the front of the vehicle and securely support it on jackstands. Remove both front wheels, and wheelwell liners (see Chapter 11).

11   Drain the transaxle oil as described in Section 2, then install the drain plug and tighten it to the torque listed in this Chapter's Specifications.

12   Remove both driveaxles (see Chapter 8).

13   Tighten the screw/nut on the support fixture to raise the engine approximately 7/16-inch (10 mm). Remove the lower engine stabilizer mount (see Chapter 2A).

14   Remove the exhaust system (see Chapter 4).

15   Remove the front subframe (see Chapter 2A).

16   Remove the clutch release cylinder from the transaxle (see Chapter 8). There is no need to disconnect the fluid line - free the line from its retaining clips and position the cylinder to one side.

17   Remove the starter (see Chapter 5).

18   Noting their locations, disconnect all electrical connectors from the transaxle. Note the harness routing and move the harness to one side.

19   Support the transaxle with a jack - preferably one made for this purpose. **Note:** *Transmission jack head adapters are available that replace the round head on a floor jack.* Noting their installed locations, remove the bolts securing the transaxle to the engine.

20   Using the jack, and with the help of an assistant, slide the transaxle from the engine. **Note:** *It may be necessary to lower the engine, using the support fixture, to facilitate transaxle removal.*

Installation

21   Installation of the transaxle is the reverse of the removal procedure, bearing in mind the following points:

a) *Prior to installation, check the clutch assembly and release mechanism components (see Chapter 8). Lubricate the release bearing guide with a little high temperature grease. Do not apply too much grease, otherwise there is a possibility of the grease contaminating the clutch friction disc. Ensure no grease is applied to the input shaft/friction disc splines.*

b) *Ensure that the locating dowels are correctly positioned prior to installation.*

c) *Replace all self-locking nuts/bolts with new ones of the same part number or design. Tighten all nuts and bolts to the specified torque (where given).*

d) *Replace the driveaxle oil seals, then install the driveaxles (see Chapter 8).*

e) *Install the release cylinder (see Chapter 8).*

f) *On completion, refill the transaxle with the specified type and quantity of lubricant.*

8   Manual transaxle overhaul - general information

1   Overhauling a manual transaxle is a difficult and involved job for the home mechanic. In addition to disassembling and reassembling many small parts, clearances must be precisely measured and, if necessary, changed by selecting shims and spacers. Internal transaxle components are also often difficult to obtain, and in many instances, extremely expensive. Because of this, if the transaxle develops a fault or becomes noisy, the best course of action is to have the unit overhauled by a transmission specialist, or to obtain an exchange reconditioned unit.

2   Nevertheless, it is not impossible for the more experienced mechanic to overhaul the transaxle, provided the special tools are available, and the job is done in a deliberate step-by-step manner, so that nothing is overlooked.

3   The tools necessary for an overhaul include internal and external snap-ring pliers, bearing pullers, a slide hammer, a set of pin punches, a dial indicator, and possibly a hydraulic press. In addition, a large, sturdy workbench and a vise will be required.

4   During transaxle disassembly, make careful notes of how each component is installed, to make reassembly easier and more accurate.

5   Before disassembling the transaxle, it will help if you have some idea what area is malfunctioning. Certain problems can be closely related to specific areas in the transaxle, which can make component examination and replacement easier. Refer to the *Troubleshooting* Section for more information.

Notes

Chapter 7 Part B
Automatic transaxle

Contents

Specifications

Note: Throughout this Chapter you will find references to "Mk I" and "Mk II" models; this is done to simplify which specifications and procedures apply to which models. Mk I models include 2006 and earlier Cooper/Cooper S models, and 2008 and earlier Convertible models. Mk II models include 2007 and later Cooper/Cooper S/Clubman/Clubman S and 2009 and later Convertible models.

General

Designations
Mk I (except S or JCW model convertibles) models .... 5-speed Continuously Variable Transmission (CVT)
Mk II
2011 and earlier models .... 6-speed Agitronic
2012 and later models .... AISIN 6-speed CVT
Automatic transaxle fluid and type.................. See Chapter 1
Torque converter stud base-to-transaxle bellhousing surface
Mk I models.................. 0.689 inch (17.5 mm)
Mk II models
Non-turbo.................. 0.465 inch (11.8 mm)
Turbo .... 0.689 inch (17.5 mm)

Torque specifications

Note: One foot-pound (ft-lb) of torque is equivalent to 12 inch-pounds (in-lbs) of torque. Torque values below approximately 15 foot-pounds are expressed in inch-pounds, because most foot-pound torque wrenches are not accurate at these smaller values.

	Ft-lbs (unless otherwise indicated)	Nm
CVT transaxle		
2011 and earlier models		
CVT switch	108 in-lbs	12
Transaxle drain plug	30	40
Transaxle fill plug	16	21
Transaxle-to-engine bolts	63	85
2012 and later models		
Transaxle drain plug		
M10 plug	20	27
M18 plug	19	25
Transaxle fill plug	19	25
Transaxle-to-engine bolts		
M8 bolts	15	66
M10 bolts	28	38
M12 bolts	49	19
6-speed Agitronic transaxle		
Transaxle fill plug	20	27
Inspection plug	20	27
Transmission Range switch mounting bolts	48 in-lbs	5.5
Transmission Range switch shaft nut (below shift lever)	62 in-lbs	7
Transmission fluid cooler bolt	26	35
Transaxle-to-engine bolts		
Mk I models	60	82
Mk II models	28	38
Torque converter-to-driveplate nuts	42	57

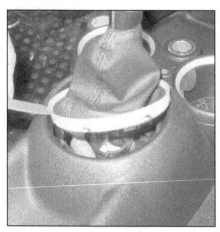

3.2 Pry up the shifter bezel and disconnect the electrical connector (Mk II models)

1   General information

Note: *Throughout this Chapter you will find references to "Mk I" and "Mk II" models; this is done to simplify which specifications and procedures apply to which models. Mk I models include 2006 and earlier Cooper/Cooper S models, and 2008 and earlier Convertible models. Mk II models include 2007 and later Cooper/Cooper S/Clubman/Clubman S and 2009 and later Convertible models.*

The automatic transaxles on the models covered in this manual include either a five-speed Continuously Variable Transmission (CVT) or a six-speed automatic with Agitronic technology (paddle shifters on the steering wheel).

The overall operation of the transaxle is managed by the Powertrain Control Module (PCM) and the Transmission Control Module (TCM). Most comprehensive troubleshooting can therefore only be carried out using dedicated electronic test equipment such as a factory scan tool.

Due to the complexity of the transaxle and its control system, major repairs and overhaul operations should be left to a dealer service department or other qualified repair facility, who will be equipped to carry out troubleshooting and repair.

2   Diagnosis - general

Automatic transaxle malfunctions may be caused by five general conditions:

a) *Poor engine performance*
b) *Improper adjustments*
c) *Hydraulic malfunctions*
d) *Mechanical malfunctions*
e) *Malfunctions in the computer or its signal network*

Diagnosis of these problems should always begin with a check of the easily repaired items: fluid level and condition (see Chapter 1) and shift cable adjustment. Next,

perform a road test to determine if the problem has been corrected or if more diagnosis is necessary. If the problem persists after the preliminary tests and corrections are completed, additional diagnosis should be done by a dealer service department or transmission repair shop. Refer to the *Troubleshooting* Section at the front of this manual for information on symptoms of transaxle problems.

Preliminary checks

1   Drive the vehicle to warm the transaxle to normal operating temperature.
2   Check the fluid level as described in Section 8.

a) *If the fluid level is unusually low, add enough fluid to bring it up to the proper level, then check for external leaks (see below).*
b) *If the fluid level is abnormally high, drain off the excess, then check the drained fluid for contamination by coolant. The presence of engine coolant in the automatic transmission fluid indicates that a failure has occurred in the transmission fluid cooler.*
c) *If the fluid is foaming, drain it and refill the transaxle, then check for coolant in the fluid.*

3   Look for a CHECK ENGINE light glowing on the instrument panel (see Chapter 6 for information). **Note:** *If the engine or its control network is malfunctioning, do not proceed with the preliminary checks until it has been repaired and runs normally.*
4   Inspect the shift cable (see Section 4). Make sure that it's properly adjusted and that it operates smoothly.

Fluid leak diagnosis

5   Most fluid leaks are easy to locate visually. Repair usually consists of replacing a seal or gasket. If a leak is difficult to find, the following procedure may help.
6   Identify the fluid. Make sure it's transmission fluid and not engine oil or brake fluid.
7   Try to pinpoint the source of the leak. Drive the vehicle several miles, then park it over a large sheet of cardboard. After a minute or two, you should be able to locate the leak by determining the source of the fluid dripping onto the cardboard.
8   Make a careful visual inspection of the suspected component and the area immediately around it. Pay particular attention to gasket mating surfaces. A mirror is often helpful for finding leaks in areas that are hard to see.
9   If the leak still cannot be found, clean the suspected area thoroughly with a degreaser or solvent, then dry it.
10   Drive the vehicle for several miles at normal operating temperature and varying speeds. After driving the vehicle, visually inspect the suspected component again.
11   Once the leak has been located, the cause must be determined before it can be properly repaired. If a gasket is replaced but the sealing flange is bent, the new gasket will

not stop the leak. The bent flange must be straightened.
12   Before attempting to repair a leak, check to make sure that the following conditions are corrected or they may cause another leak. **Note:** *Some of the following conditions cannot be fixed without highly specialized tools and expertise. Such problems must be referred to a transmission shop or a dealer service department.*

Gasket leaks

13   Check the pan periodically. Make sure the bolts are tight, no bolts are missing, the gasket is in good condition and the pan is flat (dents in the pan may indicate damage to the valve body inside).
14   If the pan gasket is leaking, the fluid level may be too high, the vent may be plugged, the pan bolts may be too tight, the pan sealing flange may be warped, the sealing surface of the housing may be damaged, the gasket may be damaged or the transaxle casting may be cracked or porous. If sealant instead of gasket material has been used to form a seal between the pan and the housing, it may be the wrong sealant.

Seal leaks

15   If a seal is leaking, the fluid level may be too high, the vent may be plugged, the seal bore may be damaged, the seal itself may be damaged or improperly installed, the surface of the shaft protruding through the seal may be damaged or a loose bearing may be causing excessive shaft movement.
16   Make sure the fluid drain/check plug in the fluid pan is in good condition. If leaking fluid is evident, replace the O-ring on the drain plug.

Case leaks

17   If the case itself appears to be leaking, the casting is porous and will have to be repaired or replaced.
18   Make sure the oil cooler hose fittings are tight and in good condition.

3   Shift lever - removal and installation

Refer to illustration 3.2

Warning: *These models are equipped with airbags. Always disable the airbag system before working in the vicinity of any airbag system component to avoid the possibility of accidental deployment of the airbag(s), which could cause personal injury (see Chapter 12).*
Warning: *Do not use a memory saving device to preserve the PCM's memory when working on or near airbag system components.*

1   Remove the shift knob by pulling it straight up and off the stalk without twisting it.
2   Pull the base of the shifter boot from the console, carefully prying if necessary. On Mk II models, disconnect the electrical connector **(see illustration)**. On Mk I models, cut the zip-tie holding the boot to the stalk and remove the boot.

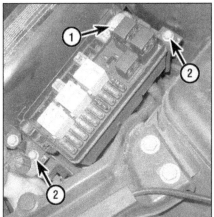
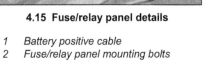

4.15 Fuse/relay panel details

1   *Battery positive cable*
2   *Fuse/relay panel mounting bolts*

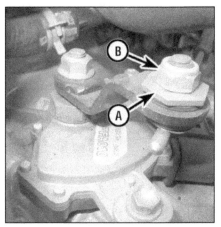

4.17a Hold the clamping sleeve (A) and unscrew the locknut (B) . . .

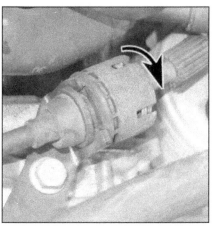

4.17b . . . . then twist the cable casing to free it from its bracket (Mk II model shown; on Mk I models the cable casing is freed by squeezing its retaining clips)

3   On models with a CVT transaxle, disconnect the shift interlock cable from the shifter.
4   On models with a CVT transaxle, disconnect the shift position switch electrical connector.
5   Remove the nut that secures the shift cable to the shift lever on the transaxle, then detach the cable casing from its bracket (see Section 4).
6   Raise the vehicle and support it securely on jackstands. Remove the rear portion of the exhaust system (see Chapter 4).
7   Remove the heat shield (see Chapter 7A, **illustration 3.5**).
8   Remove the fasteners securing the shifter base to the floorpan (see Chapter 7A, **illustration 3.10a**).
9   Working inside the vehicle, dislodge the retainer at the front of the shifter base (CVT models only) and lower the shifter down through the floorpan.
10   Installation is the reverse of removal. Be sure the gasket is positioned correctly between the shifter base and the floorpan. Adjust the shift cable and, on CVT models, the shift interlock cable (see Sections 4 and 5).

4   Shift cable - replacement

Warning: *These models are equipped with airbags. Always disable the airbag system before working in the vicinity of any airbag system component to avoid the possibility of accidental deployment of the airbag(s), which could cause personal injury (see Chapter 12).*
Warning: *Do not use a memory saving device to preserve the PCM's memory when working on or near airbag system components.*

CVT transaxle (Mk I models, except Cooper S)

Removal
1   Remove the nut that secures the shift cable to the shift lever on the transaxle (see

illustration **4.17a**), then detach the cable casing from its bracket. Remove the shift lever assembly (see Section 3).
2   Remove the clip holding the cable casing to the shifter base.
3   Pry the cable end off the ballstud and separate the cable from the shifter.

Installation
4   Installation is a reversal of removal, but lightly grease the cable end fittings, not the cable.
5   Before lowering the vehicle and before reconnecting the cable to the shift lever, adjust the cable (see Steps 6 through 11).

Adjustment
6   Apply the parking brake, then raise the front of the vehicle and support it securely on jackstands.
7   Place the shift lever in the P position.
8   Loosen the nut that secures the shift cable to the shift lever on the transaxle.
9   Make sure the selector shaft lever is in the P position. Tighten the selector lever nut.
10   Move the shift lever through the gear ranges and verify that the shifter operates correctly.
11   On CVT models, adjust the shift interlock cable (see Section 5).

6-speed Agitronic transaxle

Refer to illustrations 4.15 , 4.17a and 4.17b
Note: *On these models, the shift cable cannot be separated from the shift lever assembly and must be replaced as a unit.*
12   If you're working on a Cooper model, remove the air filter housing and air intake duct (see Chapter 4).
13   If you're working on a Cooper S model, remove the charge air ducts above the transaxle (see Chapter 4).
14   Remove the Powertrain Control Module (PCM) (see Chapter 6).
15   Disconnect the battery cable from the

underhood fuse/relay panel, then remove the fuse/relay panel bolts and move the panel out of the way **(see illustration)**.
16   Unbolt the fuse/relay panel bracket.
17   Remove the nut that secures the shift cable to the shift lever on the transaxle, then twist the cable casing and detach it from its bracket **(see illustrations)**.
18   Remove the shift lever and cable assembly (see Section 3).
19   Installation is the reverse of removal. When tightening the shift cable-to-transaxle shift lever, hold the clamping sheath with a wrench to prevent it from turning.

5   Shift interlock system - description and cable adjustment

Warning: *These models are equipped with airbags. Always disable the airbag system before working in the vicinity of any airbag system component to avoid the possibility of accidental deployment of the airbag(s), which could cause personal injury (see Chapter 12).*
Warning: *Do not use a memory saving device to preserve the ECM's memory when working on or near airbag system components.*

Description
1   The shift lock system prevents the shift lever from being shifted out of Park or Neutral until the brake pedal is applied; it achieves this via a PCM-activated solenoid. It also prevents the ignition key (on models so equipped) from being turned to the Lock position until the shift lever has been placed in the Park position; this is achieved by a shift lock cable.

Cable adjustment (CVT models)
2   Remove the center console (see Chapter 11).
3   Place the shift lever in the Park position, then turn the ignition key to the Off position and remove it.

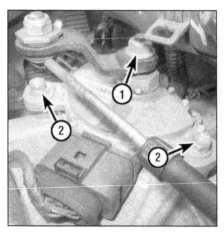

7.2 Transmission Range (TR) switch details

1   *Shift lever nut*
2   *Switch mounting bolts*

4   Loosen the park lock cable clamping bolt so the cable is free to move.
5   Move the park lock lever to its full-down position, then tighten the clamping bolt.
6   Verify proper operation of the shift lock system.

---

6   CVT switch - replacement

Note: *This switch incorporates the back-up light switch and the Park/Neutral switch.*
1   Raise the vehicle and support it securely on jackstands.
2   Locate the switch on the firewall-side of the transaxle, above the fluid pan. Disconnect the electrical connector from the switch.
3   Unscrew the switch from the case. Have a rag handy, as some fluid might drip out.
4   Installation is the reverse of removal. Tighten the switch to the torque listed in this Chapter's Specifications.

---

7   Transmission Range (TR) switch (6-speed Agitronic transaxle) - replacement

Refer to illustration 7.2
Note: *This switch is also referred to as the gear position switch.*
1   Disconnect the shift cable from the shift lever on the transaxle (see Section 4).
2   Unscrew the shift lever nut and detach the shift lever from the shaft (**see illustration**).
3   Remove the nut, retaining ring and shim from the shaft, then remove the mounting bolts and detach the switch from the transaxle.
4   Installation is the reverse of removal. Tighten the shift shaft nut to the torque listed in this Chapter's Specifications.
5   A special tool (MINI tool no. 24-4-270) is available to properly adjust the switch to the neutral position, but with a little patience an

alternative method works:
   a) *Connect the shift cable, then place the shift lever on the transaxle in the Park position.*
   b) *Move the shifter to the Reverse position and verify that the back-up lights come on. If they don't, loosen the switch mounting screws and rotate the switch one way or the other until they do, then tighten the bolts. Move the shifter to the Neutral position and make sure the back-up lights go off. Also move the shifter back to the Park position and verify that the back-up lights are off.*
   c) *With the brake pedal depressed, make sure the engine starts only in Park and Neutral.*
6   When the proper adjustment is achieved, tighten the switch mounting bolts to the torque listed in this Chapter's Specifications.

---

8   Automatic transaxle - fluid level check and replacement

Note: *The following fluid replacement procedures do not include replacing the fluid that remains in the torque converter. For a complete fluid change, inquire at a dealer service department or other properly equipped repair shop for a transaxle fluid flush/replacement.*

CVT transaxle (Mk I models, except Cooper S)
Caution: *To ensure proper refilling/obtaining proper fluid level, a scan tool capable of measuring transaxle fluid temperature is required.*

Fluid level check
Note: *Routine fluid level checks are not necessary unless there is a leak. The fluid level checking procedure consists of removing the check/fill plug and seeing if fluid runs out of the hole. If not, the fluid level is low and some must be added as described in this Section.*

Replacement
Note: *Although not required for a routine fluid change, there is a strainer in the transaxle that can be removed and cleaned, or replaced. To do so requires suspending the engine/transaxle assembly from above, lowering the subframe, then removing the fluid pan and strainer.*
1   Loosen the front wheel bolts. Raise the vehicle and support it securely on jackstands.
Note: *The front AND rear of the vehicle must be raised an equal amount.*
2   Remove the under-vehicle splash shield.
3   Reinstall the wheel bolts with washers to secure the brake discs against their hub flanges. Tighten the bolts securely.
4   Position a drain pan under the transaxle. Using an Allen wrench or hex bit and ratchet, remove the drain plug from the bottom transaxle fluid pan.
5   Install a new sealing washer, then reinstall the drain plug and tighten it to the torque listed in this Chapter's Specifications.

6   Using an Allen wrench or hex bit and ratchet, remove the check/fill plug from the front side of the transaxle. **Note:** *Some fluid might flow out when the plug is removed.*
7   Using tool no. 24-8-100 (2011 and earlier models) or tool no. 24-4-240 (2012 and later models) or equivalent fitting attached to the check/fill plug hole, pump in 4.7 quarts (4.5L) of the proper fluid into the transaxle (see Chapter 1, *Recommended fluids and lubricants*). If the fitting/adapter is equipped with a shut-off valve, turn it to the Off position. If not, remove the fitting from the transaxle and install the check/fill plug. **Note:** *On 2012 and later models, the tool must be shortened to 1.4 inches (36 mm) to fit between the body and the transaxle.*
8   Depress the brake pedal and start the engine, allowing it to run for ten seconds, then with the brake pedal still depressed, shift the lever through all gear ranges, pausing five seconds in each range, then move the shifter into the Drive range.
9   Take your foot off the brake pedal and depress the accelerator pedal to allow the transaxle to shift up through all gears, but do not exceed 2500 rpm. When the transaxle shifts into top gear, wait three seconds, then repeat two more times.
10   Gently apply the brake, bringing the driveaxles to a stop.
11   Move the shifter to the Reverse range, then take your foot off the brake pedal, depress the accelerator pedal and raise the engine speed, but not greater than 2500 rpm. Repeat two more times.
12   Gently apply the brake, bringing the driveaxles to a stop, then shift to the Park position and turn off the engine.
13   Connect a scan tool to the OBD-II diagnostic connector (see Chapter 6, Section 3).
14   Start the engine and, using the scan tool, monitor the transaxle fluid temperature.
15   When the fluid temperature increases to 85 to 120-degrees F (29 to 48-degrees C), remove the pump/adapter fitting from the transaxle, or, if you reinstalled the check/fill plug, remove the plug. **Note:** *Make sure the drain pan is in position, because fluid will run out.*
16   Wait until the fluid stops flowing from the hole. If no fluid runs out, or if not much runs out, reattach the fitting/pump and add a little more fluid, then detach the fitting and allow the excess fluid to drain out. When the fluid stops dripping, install the check/fill plug (using a new sealing washer), tightening it to the torque listed in this Chapter's Specifications.
17   Install the under-vehicle splash shield.
18   Install the wheels, then lower the vehicle and tighten the wheel bolts to the torque listed in the Chapter 1 Specifications.

6-speed Agitronic transaxle
Fluid level check
Refer to illustrations 8.21 and 8.22
Caution: *To ensure proper refilling/obtaining proper fluid level, a scan tool capable of measuring transaxle fluid temperature is required.*
19   Raise the vehicle and support it securely on jackstands. **Note:** *The front AND rear of*

the vehicle must be raised an equal amount.

20   Remove the under-vehicle splash shield. Place a drain pan underneath the transaxle.

21   Use a Torx T55 bit and remove the fill plug from the firewall side of the transaxle case (see illustration).

22   Remove the inspection plug from the transaxle fluid pan (see illustration).

23   Connect a scan tool to the OBD-II diagnostic connector (see Chapter 6, Section 3).

24   Using a fluid pump from below, or a hose and funnel from above, add the proper type of fluid (see Chapter 1, *Recommended fluids and lubricants*) to the filler hole until fluid flows from the inspection hole in the fluid pan.

25   Start the engine and allow it to idle. A little fluid should flow from the inspection hole. If not, add some until it does.

26   Depress the brake pedal and move the shifter from Park to Drive.

27   Repeat Step 26, then return the shifter to Park.

28   Monitor the transaxle fluid temperature with the scan tool. When the temperature reaches 95 to 133-degrees F (35 to 45-degrees C), check to see if any fluid is flowing from the inspection plug hole. If not, add some until it does.

29   Wait until the fluid stops flowing, then install the inspection plug (with a new sealing washer), tightening it to the torque listed in this Chapter's Specifications.

Replacement

Refer to illustration 8.34

Note: *Although not required for a routine fluid change, there is a strainer in the transaxle that can be removed and cleaned, or replaced. To do so requires suspending the engine/ transaxle assembly from above and lowering the subframe, then removing the fluid pan and strainer.*

30   Raise the vehicle and support it securely on jackstands. **Note:** *The front AND rear of the vehicle must be raised an equal amount.*

31   Remove the under-vehicle splash shield. Place a drain pan underneath the transaxle.

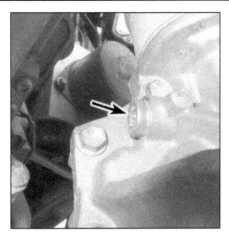

8.21 Transaxle fill plug (6-speed Agitronic transaxle)

8.22 Transaxle inspection plug (6-speed Agitronic transaxle)

32   Use a Torx T55 bit and remove the fill plug from the firewall side of the transaxle case (see illustration 8.21).

33   Remove the inspection plug from the transaxle fluid pan (see illustration 8.22).

34   Unscrew the fluid overflow tube from the fluid pan (see illustration) and allow the fluid to drain from the transaxle fluid pan.

35   Reinstall the overflow tube and tighten it securely.

36   Using a fluid pump from below, or a hose and funnel from above, add the proper type of fluid (see Chapter 1, *Recommended fluids and lubricants*) to the filler hole until fluid flows from the inspection hole in the fluid pan.

37   Connect a scan tool to the OBD-II diagnostic connector (see Chapter 6, Section 3).

38   Perform Steps 25 through 29 of the fluid level inspection procedure.

9   Transaxle fluid cooler - removal and installation

Warning: *Wait until the engine is completely cool before beginning this procedure.*

CVT transaxle (Mk I models, except Cooper S)

1   Remove the bumper cover (see Chapter 11).

2   Disconnect the cooler hoses.

3   Remove the cooler mounting screw from the right (passenger) side of the cooler, then push the cooler to the right to disengage it from its mounting bracket.

4   Installation is the reverse of removal. Check the transaxle fluid, adding as necessary (see Section 8).

6-speed Agitronic transaxle

Refer to illustration 9.8

5   Drain the engine coolant (see Chapter 1).

6   Cooper models: Remove the air filter housing (see Chapter 4).

7   Cooper S models: Remove the charge air ducts from above the transaxle.

8   Loosen the clamps and detach the hoses from the cooler (see illustration).

9   Unscrew the mounting bolt and detach the cooler from the transaxle.

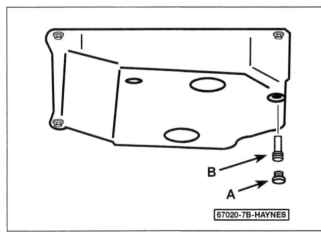

8.34 Transaxle inspection plug (A) and fluid overflow tube (B)

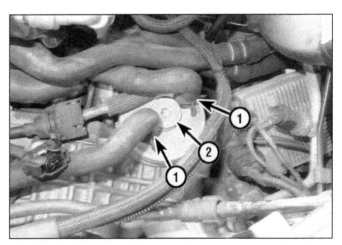

9.8 Transaxle fluid cooler details

1 Coolant hoses	2 Mounting bolt

11.10 Loosen, but don't remove this nut from the right side engine mount

11.20 Transaxle lower bolt locations (rear bolts not visible)

10   Clean the mating surface on the transaxle, then install new O-rings on the cooler.

11   Installation is the reverse of removal. Tighten the cooler mounting bolt to the torque listed in this Chapter's Specifications.

12   Refill the cooling system (see Chapter 1).

13   Check the transaxle fluid level, adding as necessary (see Section 8).

10   Automatic transaxle (Mk I models) - removal and installation

On these models, the engine and transaxle are removed as a unit, then the transaxle can be separated from the engine (see Chapter 2B). If you're working on a model with a 6-speed Agitronic transaxle, be sure to remove the torque converter-to-driveplate bolts before separating the transaxle from the engine (see Section 11, Step 19).

11   Automatic transaxle (Mk II models) - removal and installation

Refer to illustrations 11.10 and 11.20

Warning: *Wait until the engine is completely cool before beginning this procedure.*

Removal

1   Disconnect the cable from the negative terminal of the battery (see Chapter 5).

2   On Cooper models, remove the air filter housing. On Cooper S models, remove the charge air ducts from above the transaxle (see Chapter 4).

3   Install an engine support fixture to the engine, with the chain connected to the engine on the transaxle end. If no lifting hook is present, you'll either have to obtain or fabricate one (tool no. 11-8-260 or equivalent) that you can bolt to the left end of the cylinder head.

Tighten the support fixture chain screw/nut to remove all slack in the chain.

4   Remove the Powertrain Control Module (PCM) (see Chapter 6).

5   Disconnect the battery cable from the underhood fuse/relay panel, then remove the fuse/relay panel bolts and move the panel out of the way **(see illustration 4.15)**.

6   Unbolt the fuse/relay panel bracket.

7   Detach the coolant hoses from the transaxle fluid cooler (see Section 9).

8   Detach the shift cable from the lever on the transaxle, then unbolt the cable bracket (see Section 4).

9   Remove the transaxle-to-inner fender panel support bracket, followed by the transaxle mount (see Chapter 2A).

10   Loosen, but don't remove, the center nut of the right-side engine mount **(see illustration)**.

11   Loosen the front wheel bolts. Chock the rear wheels, then firmly apply the parking brake. Loosen both front wheel bolts. Raise the front of the vehicle and securely support it on jackstands. Remove both front wheels, and wheelwell liners (see Chapter 11).

12   Drain the transaxle fluid as described in Section 8.

13   Remove both driveaxles (see Chapter 8).

14   Remove the starter motor (see Chapter 5).

15   Tighten the screw/nut on the support fixture to raise the engine approximately 3/8-inch (10 mm). Remove the lower engine stabilizer mount (see Chapter 2A).

16   Remove the exhaust system (see Chapter 4).

17   Remove the front subframe (see Chapter 2A).

18   Noting their locations, disconnect all electrical connectors from the transaxle. Note the harness routing and move the harness to one side.

19   Using a socket and breaker bar on the crankshaft pulley bolt, rotate the engine (clockwise while looking at the pulley) until one of the torque converter nuts is visible through the starter opening, then remove

the nut. Continue rotating the crankshaft and removing the nuts until all six are removed. When the last nut has been removed, mark the stud and driveplate so balance will be preserved when the transaxle is reinstalled.

20   Support the transaxle with a jack - preferably one made for this purpose. **Note:** *Transmission jack head adapters are available that replace the round head on a floor jack.* Noting their installed locations, remove the bolts securing the transaxle to the engine **(see illustration)**.

21   Using the jack, and with the help of an assistant, slide the transaxle from the engine. **Note:** *It may be necessary to lower the engine, using the support fixture, to facilitate transaxle removal.*

Installation

22   Installation of the transaxle is a reversal of the removal procedure, but note the following points:

a)   *As the torque converter is reinstalled, ensure that the drive hub at the center of the torque converter hub engages properly with the recesses in the transaxle fluid pump.*

b)   *Measure the distance between the base of the torque converter studs and the transaxle bellhousing-to-engine mating surface, comparing your measurement to the dimension listed in this Chapter's Specifications.*

c)   *When installing the transaxle, make sure the marks on the torque converter and driveplate are in alignment.*

d)   *Tighten the transaxle-to-engine bolts and torque converter-to-driveplate nuts to the torque listed in this Chapter's Specifications.*

e)   *Refill the cooling system (see Chapter 1).*

f)   *Refill the transaxle (see Section 8).*

g)   *Check the shift cable adjustment (see Section 4).*

12   Automatic transaxle overhaul - general information

In the event of a fault occurring, it will be necessary to establish whether the fault is electrical, mechanical or hydraulic in nature, before repair work can be contemplated. Diagnosis requires detailed knowledge of the transaxle's operation and construction, as well as access to specialized test equipment, and so is deemed to be beyond the scope of this manual. It is therefore essential that problems with the automatic transaxle are referred to a dealer service department or other qualified repair facility for assessment.

Note that a faulty transaxle should not be removed before the vehicle has been assessed by a knowledgeable technician equipped with the proper tools, as troubleshooting must be performed with the transaxle installed in the vehicle.

Chapter 8
Clutch and driveaxles

Contents

Specifications

Note: *Throughout this Chapter you will find references to "Mk I" and "Mk II" models; this is done to simplify which specifications and procedures apply to which models. Mk I models include 2006 and earlier Cooper/Cooper S models, and 2008 and earlier Convertible models. Mk II models include 2007 and later Cooper/Cooper S/Clubman/Clubman S and 2009 and later Convertible models.*

General

Lining minimum thickness above rivet on transaxle side
    Mk I models
        Cooper .... 0.040-inch (1.0 mm)
        Cooper S.. 0.008-inch (0.2 mm)
    Mk II models (all)... 0.040-inch (1.0 mm)

Torque specifications

Note: *One foot-pound (ft-lb) of torque is equivalent to 12 inch-pounds (in-lbs) of torque. Torque values below approximately 15 ft-lbs are expressed in inch-pounds, since most foot-pound torque wrenches are not accurate at these smaller values.*

	Ft-lbs	Nm
Clutch cover/pressure plate to flywheel bolts*		
Mk I models		
M7	15	20
M8	17	21
M9	21	28
Mk II models		
Models without a self-adjusting clutch pressure plate	21	28
Models with a self-adjusting clutch pressure plate	17	23

* Use new fasteners

Torque specifications (continued)   **Ft-lbs** (unless otherwise indicated)   **Nm**

Note: *One foot-pound (ft-lb) of torque is equivalent to 12 inch-pounds (in-lbs) of torque. Torque values below approximately 15 ft-lbs are expressed in inch-pounds, since most foot-pound torque wrenches are not accurate at these smaller values.*

Clutch master cylinder mounting bolts/nuts		
Mk I models	18	24
Mk II models*		
Cylinder-to-pedal assembly	84 in-lbs	9
Pedal assembly-to-firewall	17	21
Clutch master/release cylinder mounting bolts/nuts	18	24
Driveaxle/hub nut*	134	182
Right-hand intermediate shaft bearing housing bolts		
Mk I models	18	25
Mk II models	28	38
Wheel bolts	See Chapter 1	

* Use new fasteners

1   General information

Note: *Throughout this Chapter you will find references to "Mk I" and "Mk II" models; this is done to simplify which specifications and procedures apply to which models. Mk I models include 2006 and earlier Cooper/Cooper S models, and 2008 and earlier Convertible models. Mk II models include 2007 and later Cooper/Cooper S/Clubman/Clubman S and 2009 and later Convertible models.*

Clutch

All vehicles with a manual transaxle have a single dry plate, diaphragm spring-type clutch. The cover assembly consists of a steel cover (doweled and bolted to the flywheel face), the pressure plate, and a diaphragm spring.

The clutch disc has a splined hub which allows it to slide along the splines of the transaxle input shaft. The clutch and pressure plate are held in contact by spring pressure exerted by the diaphragm in the pressure plate. Friction lining material is riveted to the friction disc (driven plate), which on Cooper models, has a spring-cushioned hub, to absorb shocks and help ensure a smooth take-up of the drive.

The clutch release bearing contacts the fingers of the diaphragm spring. Depressing the clutch pedal pushes the release bearing against the diaphragm fingers, so moving the center of the diaphragm spring inwards. As the center of the spring is pushed inwards, the outside of the spring pivots outwards, so moving the pressure plate backwards and disengaging its grip on the friction disc.

When the pedal is released, the diaphragm spring forces the pressure plate back into contact with the friction linings on the friction disc. The disc is now firmly held between the pressure plate and the flywheel, thus transmitting engine power to the transaxle.

All MINI models have a hydraulically operated clutch. A master cylinder is mounted below the clutch pedal, and takes its hydraulic fluid supply from a separate chamber in the brake fluid reservoir. Depressing the clutch pedal operates the master cylinder pushrod, and the fluid pressure is transferred along the fluid lines to a release cylinder mounted on the transaxle casing. The release cylinder pushrod operates against the release lever, onto which the release bearing is mounted - when the release cylinder operates, the release bearing moves against the diaphragm spring fingers and disengages the clutch.

The hydraulic clutch offers several advantages over a cable-operated clutch - it is completely self-adjusting, requires less pedal effort, and is less subject to wear problems.

Since many of the procedures covered in this Chapter involve working under the vehicle, make sure that it is securely supported on jackstands placed on a firm, level floor (see *Jacking and towing*).

Warning: *Brake fluid is poisonous. Take care to keep it off bare skin, and in particular out of your eyes. Brake fluid will damage paint; be careful not to spill fluid. Wash off any spilled fluid immediately with cold water. Finally, brake fluid is highly flammable, and should be handled with the same care as gasoline.*

Driveaxles

Drive is transmitted from the transaxle differential to the front wheels by means of two driveaxles. The right-hand driveaxle is in two sections, and incorporates a support bearing. The inner driveaxle boots are made of rubber, and the outer driveaxle boots are made of thermoplastic.

Each driveaxle consists of three main components: the sliding (tripod type) inner joint, the driveaxle itself, and the outer CV (constant velocity) joint. The inner end of the left-hand tripod joint is secured in the differential side gear by the engagement of a snap-ring. The inner tripod of the right-hand driveaxle is located in the intermediate shaft tripod housing. The intermediate shaft is held in the transaxle by the support bearing, which in turn is supported by a bracket bolted to the rear of the cylinder block. The outer CV joint on both driveaxles is of ball-bearing type, and is secured in the front hub by the driveaxle/hub nut.

3.4 Release the clip securing the pushrod to the clutch pedal

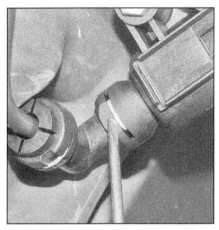

3.5 Pry out the wire clip and disconnect the pressure line from the master cylinder

2   Clutch - check

Unless you're replacing components with obvious damage, do these preliminary checks to diagnose clutch problems:

a) *The first check should be of the fluid level in the clutch master cylinder. If the fluid level is low, add fluid as necessary and inspect the hydraulic system for leaks. If the master cylinder reservoir is dry, bleed the system as described in Section 8 and recheck the clutch operation.*

b) *To check clutch spin-down time, run the engine at normal idle speed with the transaxle in Neutral (clutch pedal up - engaged). Disengage the clutch (pedal down), wait several seconds and shift the transaxle into Reverse. No grinding noise should be heard. A grinding noise would most likely indicate a bad pressure plate or clutch disc.*

c) *To check for complete clutch release, run the engine (with the parking brake applied to prevent vehicle movement) and hold the clutch pedal approximately 1/2-inch from the floor. Shift the transaxle between 1st gear and Reverse several times. If the shift is rough, component failure is indicated.*

d) *Visually inspect the pivot bushing at the top of the clutch pedal to make sure there's no binding or excessive play.*

3   Clutch master cylinder - removal and installation

Warning: *Brake fluid is poisonous. Take care to keep it off bare skin, and in particular out of your eyes. Brake fluid will damage paint; be careful not to spill fluid. Wash off any spilled fluid immediately with cold water. Finally, brake fluid is highly flammable, and should be handled with the same care as gasoline.*

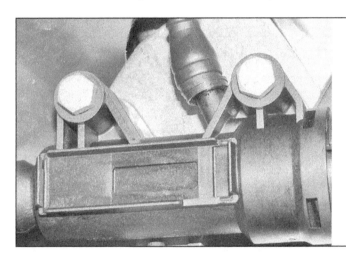

3.6 Remove the master cylinder retaining bolts

Removal

1   Working inside the vehicle, move the driver's seat fully to the rear, to allow maximum working area. Unclip the driver's side lower fascia trim panel, and remove the panel from the vehicle.

2   Have rags handy to catch spilled fluid. **Caution:** *Don't allow brake fluid to come into contact with the paint - it will damage the finish.* Also have a container ready in the engine compartment.

3   Remove the brake fluid reservoir cap, then tighten it down over a piece of plastic wrap to obtain an airtight seal. This may help to reduce the spillage of fluid when the lines are disconnected.

Mk I models

Refer to illustrations 3.4, 3.5 and 3.6

4   Release the clip securing the master cylinder pushrod to the pedal, and detach the pushrod from the pedal pin **(see illustration)**.

5   Pry out the clip and disconnect the hydraulic line from the base of the master cylinder **(see illustration)**.

6   Remove the two retaining bolts, and detach the master cylinder from the bracket **(see illustration)**. Disconnect the supply hose from the cylinder as it's withdrawn. Be pre-

pared for fluid spillage; have a plug ready and immediately plug the line to prevent leakage.

Mk II models

7   Remove the clutch pedal assembly (see Section 4).

8   Release the retainer and remove the pin securing the master cylinder pushrod to the clutch pedal, then remove the master cylinder mounting bolts.

Installation

9   Installation is the reverse of removal, noting the following points:

a) *Replace self-locking nuts with new ones. Also replace any damaged retaining clips.*

b) *Remove the plastic wrap from under the fluid reservoir cap, and top-up the fluid level (see Chapter 1).*

c) *Bleed the clutch hydraulic system (see Section 8).*

d) *On completion, operate the clutch a few times without starting the engine, then check for signs of fluid leakage at all hydraulic connections in the engine compartment.*

e) *Start the engine and check for correct clutch operation.*

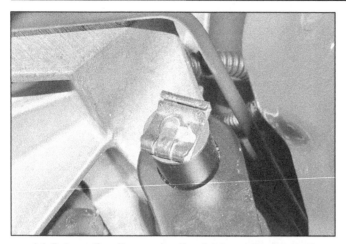

4.4 Release the clip securing the clutch pedal to the shaft

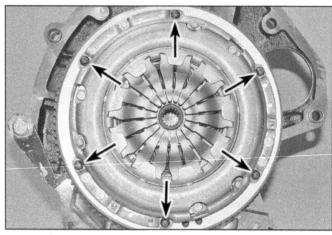

5.3 Remove the bolts securing the clutch cover to the flywheel

4 Clutch pedal - removal and installation

Removal
Mk I models

Refer to illustration 4.4

1 Working inside the vehicle, move the driver's seat fully to the rear, then remove the driver's side lower trim panel (see Chapter 11).

2 Release the clip and detach the clutch master cylinder pushrod from the pedal **(see illustration 3.4)**.

3 Unhook the over-center spring from the clutch pedal.

4 Release the clip securing the pedal to the shaft, and slide the pedal from place **(see illustration)**.

Mk II models

5 Disconnect the electrical connector from the master cylinder. Also disconnect the brake light switch electrical connector.

6 Disconnect the fluid line fitting near the rear cylinder mounting bolt. Be prepared for fluid spillage; have a plug ready and immediately plug the line to prevent leakage.

7 Remove the upper pedal cluster mounting bolt.

8 Remove the brake pedal pivot pin retaining clip, then remove the pin.

9 Remove the brake/clutch pedal assembly mounting nuts, then remove the pedal assembly and set it on a workbench.

Installation

10 Prior to installing the pedal, apply a little grease to the pivot shaft, sleeve and pedal bushings.

11 Installation is the reverse of removal.

12 Check the clutch operation, as described in Section 2.

5 Clutch components - removal, inspection and installation

Warning: *Dust produced by clutch wear is hazardous to your health. DO NOT blow it out with compressed air and DO NOT inhale it. DO NOT use gasoline or petroleum-based solvents to remove the dust. Brake system cleaner should be used to flush the dust into a drain pan. After the clutch components are wiped clean with a rag, dispose of the contaminated rags and cleaner in a covered, marked container.*

Note: *Mk II models with a self-adjusting clutch require special tools to complete this procedure.*

Removal

1 Access to the clutch may be gained in one of two ways. The engine/transaxle unit can be removed, as described in Chapter 2B, then the transaxle can be separated from the engine. Alternatively, the engine may be left in the vehicle and the transaxle removed independently, as described in Chapter 7A.

2 Once the transaxle is separated from the engine, check if there are any marks identifying the relation of the clutch cover to the flywheel. If not, make your own marks using a dab of paint or a scriber. These marks will be used if the original cover is installed, and will help to maintain the balance of the unit. A new cover may be installed in any position allowed by the locating dowels.

Models without a self-adjusting clutch

Refer to illustration 5.3

3 Unscrew the six clutch cover retaining bolts, working in a diagonal sequence, and loosening the bolts only a turn at a time **(see illustration)**. If necessary, the flywheel may be held stationary using a wide-bladed screwdriver, inserted in the teeth of the starter ring gear and resting against part of the cylinder block. The manufacturer states that new cover bolts must be used when installing.

4 Ease the clutch cover off its locating dowels. Be prepared to catch the clutch friction disc, which will drop out as the cover is removed. Note which way the disc is installed.

Models with a self-adjusting clutch

5 If a new pressure plate is to be installed, follow Steps 3 and 4 and remove it.

6 If the original pressure plate is to be used, Install MINI special tool no. 21-2-170 (or equivalent) with the arms of the tool engaged with the cutouts in the side of the clutch cover (near where the diaphragm spring is riveted to the clutch cover). Tighten the knob of the tool until the diaphragm spring is depressed about 1/4-inch (6 mm).

7 Install special tool no. 21-0-010 into the cutout near the curved portion of the cover to relieve the adjustment function.

Inspection

8 Ordinarily, when a problem occurs in the clutch, it can be attributed to wear of the clutch driven plate assembly (clutch disc). However, all components should be inspected at this time. **Note:** *If the clutch components are contaminated with oil, there will be shiny, black glazed spots on the clutch disc lining, which will cause the clutch to slip. Replacing clutch components won't completely solve the problem - be sure to check the crankshaft rear oil seal and the transaxle input shaft seal for leaks. If it looks like a seal is leaking, be sure to install a new one to avoid the same problem with the new clutch. On models up to 07/2004, the input shaft seal can be replaced as described in Chapter 7A. On models after this date, it would appear a new seal is not available - check with your MINI dealer or specialist.*

9 Check the lining on the clutch disc. There should be at least the minimum amount of lining, listed in this Chapter's Specifications, above the rivet heads. Check for loose rivets, distortion, cracks, broken springs and other obvious damage. As mentioned above, ordi-

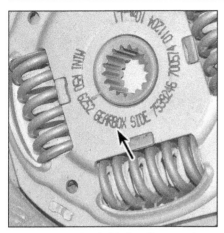

5.13a The clutch friction disc may be marked "flywheel side" or "Getriebeseite" (which mean "transmission side")

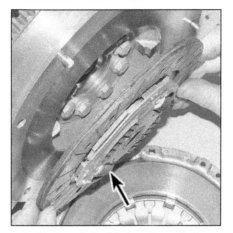

5.13b Install the friction disc with the spring hub on the gearbox side

5.16 Ensure the clutch cover locates over the dowels in the flywheel

narily the clutch disc is routinely replaced, so if you're in doubt about its condition, replace it.

10   Check the machined surfaces and the diaphragm spring fingers of the pressure plate. If the surface is grooved or otherwise damaged, replace the pressure plate. Also check for obvious damage, distortion, cracks, etc. Light glazing can be removed with medium-grit emery cloth. If a new pressure plate is required, new and factory-rebuilt units are available.

11   The release bearing should also be replaced along with the clutch disc (see Section 6).

Installation

Refer to illustrations 5.13a, 5.13b and 5.16

12   Clean the machined surfaces of the flywheel and pressure plate with brake cleaner. It's important that no oil or grease is on these surfaces or the lining of the clutch disc. Handle the parts only with clean hands.

13   Position the clutch disc and pressure plate against the flywheel with the clutch held in place with an alignment tool. Make sure it's

installed properly (most replacement clutch plates will be marked "flywheel side" or something similar - if it's not marked, install the clutch disc with the damper springs toward the transaxle) **(see illustrations)**.

14   Center the clutch disc using the type of tool that clamps the friction disc to the clutch pressure plate/cover, as there is no suitable recess in the end of the crankshaft to enable the use of the rod-and-spacer type alignment tool, unless special MINI tools (or equivalent) are available: For Mk I models, tool no. 21 6 100 (Cooper) or 21 2 210 (Cooper S). On Mk II models, tool no. 21 2 290 (Cooper) or 21 6 110 (Cooper S).

15   If a new self-adjusting clutch is being installed, install the special tools as described in Step 6, but turn the knob on the tool until the diaphragm spring fingers are depressed 3/8 to 1/2-inch (10 to 12 mm), but no further.

16   Place the clutch cover over the dowels **(see illustration)**. Install the new retaining bolts, and tighten them finger-tight so that the friction disc is gripped lightly, but can still be moved.

17   Check that the friction disc is still aligned

in relation to the pressure plate/cover assembly.

18   Once the clutch is aligned, progressively tighten the cover bolts in a diagonal sequence to the torque listed in this Chapter's Specifications. On models with a self-adjusting clutch, remove the special tools.

19   Ensure that the input shaft splines and friction disc splines are clean. Apply a thin smear of clutch assembly grease to the input shaft splines - do not apply excessively, however, or it may end up on the friction disc, causing the new clutch to slip.

20   Install the transaxle to the engine.

---

6   Clutch release bearing - removal, inspection and installation

---

Removal

Refer to illustrations 6.3 and 6.4

1   Separate the engine and transaxle as described in the previous Section.

2   On models manufactured 07/2004 and earlier, release the clips and pull the bearing from the release lever.

3   On models manufactured from 07/2004, push the end of the retaining clip back through the release lever, and maneuver the lever and bearing assembly over the end of the input shaft **(see illustration)**.

4   Turn the lever over, and release the clips securing the release bearing to the lever **(see illustration)**.

Inspection

5   Check the bearing for smoothness of operation, and replace it if there is any sign of harshness or roughness as the bearing is spun. Do not attempt to disassemble, clean or lubricate the bearing.

6   It is worth replacing the release bearing as a matter of course, unless it is known to be in perfect condition.

6.3 Push the end of the release lever retaining clip back through the lever

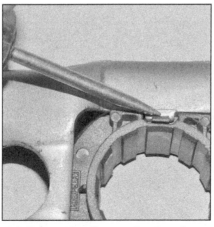

6.4 Release the clip securing the release bearing to the lever

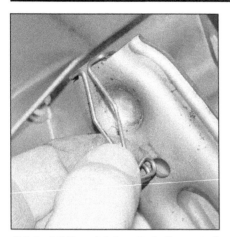

6.9 Install the retaining clip to the rear of the release lever

7.2 Remove the bolts and detach the release cylinder from the transaxle casing

7.3 Pry out the clip and pull the line from the cylinder

Installation

Models manufactured 07/2004 and earlier

7   Installation is the reverse of removal. Apply a little high-temperature grease to the ends of the release fork which bears on the back of the bearing.

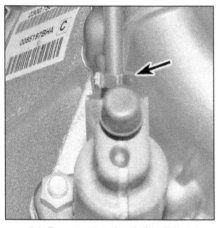

7.6 Remove the clutch line fitting

Models manufactured from 07/2004

Refer to illustration 6.9

8   Install the new release bearing to the lever, ensuring the retaining clips engage fully.
9   Pull the lever retaining clip from the mounting on the transaxle casing, and install it to the rear of the lever **(see illustration)**.
10   Position the lever/bearing assembly over the input shaft, and push the lever retaining clip over the ballstud on the transaxle.
11   The remainder of installation is the reverse of removal.

---

7   Clutch release cylinder - removal and installation

Cooper models manufactured 07/2004 and earlier

Refer to illustrations 7.2 and 7.3

1   Remove the battery and battery tray (see Chapter 5).
2   Remove the bolts securing the release cylinder to the transaxle casing **(see illustration)**.

3   Pry out the retaining clip and disconnect the fluid supply pipe from the cylinder **(see illustration)**. Be prepared for fluid spillage.
4   Installation is the reverse of removal. Bleed the system (see Section 8).

Cooper models manufactured after 07/2004 and Cooper S models

Refer to illustrations 7.6, 7.8a and 7.8b

5   Raise the front of the vehicle and support it securely on jackstands. Remove the screws and remove the engine undershield.
6   On Cooper models, unscrew the fitting and disconnect the fluid supply line from the end of the release cylinder **(see illustration)**. Be prepared for fluid spillage.
7   On Cooper S models, the fluid supply line may be clipped into place, or be retained by a fitting (see Step 6). Pry out the clip, or remove the union, and disconnect the pipe. Be prepared for fluid spillage.
8   Remove the bolts/nuts and pull the release cylinder from the transaxle housing **(see illustrations)**.
9   Installation is the reverse of removal. Bleed the system (see Section 8).

7.8a Clutch release cylinder nuts - Cooper after 07/2004

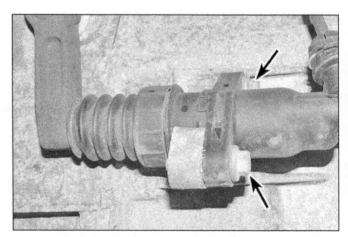

7.8b Clutch release cylinder bolts - Cooper S

8 Clutch hydraulic system - bleeding

Refer to illustration 8.3

1   The hydraulic system must be bled to remove all air anytime some part of the system has been removed or the fluid level has been allowed to fall so low that air has been drawn into the master cylinder.

2   Air in the clutch system could be the result of a leak. Do not overlook the possibility of a leak in the system, for the following reasons:

a) *Leaking fluid will damage paint and/or carpets.*

b) *The clutch system shares its fluid with the braking system, so a clutch system leak could result in brake failure from low fluid level.*

c) *Equally, fluid loss affecting the clutch system could be the result of a braking system fluid leak.*

3   The system bleed screw is located on the end of the release cylinder **(see illustration)**.

4   On Cooper models manufactured 07/2004 and earlier, remove the battery and battery tray (see Chapter 5).

5   On Cooper models manufactured after 07/2004 and Cooper S models, raise the front of the vehicle and support is securely on jackstands. Remove the screws and remove the engine undershield.

6   Remove the bleed screw cap, and attach a clear hose to the bleed nipple, and insert the other end of the hose into a suitable container.

7   Before bleeding the clutch system, it is recommended that the braking system is bled first (see Chapter 9).

8   Bleeding the clutch is much the same as bleeding the brakes - see Chapter 9 for the procedure. Ensure that the level in the brake fluid reservoir is maintained well above the MIN mark at all times, otherwise the clutch and brake hydraulic systems will both need bleeding.

9   On completion, tighten the bleed screw securely, and top-up the brake fluid level to the MAX mark. If possible, test the operation

8.3 Clutch release cylinder bleed screw

of the clutch before installing all the components removed for access.

10   Failure to bleed correctly may point to a leak in the system, or to a worn master or release cylinder. At the time of writing, it appears that the master and release cylinders are only available as complete assemblies - overhaul is not possible.

9 Driveaxles - removal and installation

Removal

Refer to illustrations 9.2, 9.4 and 9.10

1   Loosen the relevant front wheel bolts, then raise the front of the vehicle and support it securely on jackstands. Remove the relevant wheel.

2   Remove the wheel speed sensor from the steering knuckle **(see illustration)**. Release the wiring from the bracket on the suspension strut and move to one side.

3   Place a drain pan under the transaxle to catch the lubricant when the driveaxle has been removed.

4   Loosen and remove the nut securing the driveaxle to the hub **(see illustration)**. Have an assistant depress the brake pedal to prevent the hub from rotating. Discard the nut; a

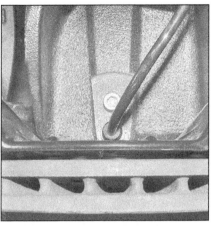

9.2 Remove the bolt and withdraw the wheel speed sensor

new one must be used.

5   Remove the nut and use a universal balljoint separator to detach the steering tie-rod end balljoint and control arm from the steering knuckle (see Chapter 10). Discard the nuts; new ones must be installed.

6   Pull the steering knuckle outwards and push the driveaxle outer joint back through the hub at the same time, disengaging it from the hub.

7   On the left-hand driveaxle, use a large screwdriver or prybar to pry the inner joint from the transaxle. Take care not to damage the casing.

8   On the right-hand side driveaxle, remove the power steering pump cooling fan (Mk I models, if equipped) (see Chapter 10), then remove the three bolts securing the driveaxle intermediate shaft bearing housing to the cylinder block.

9   Carefully pull the right-hand inner driveaxle joint from the transaxle.

10   Discard the snap-ring on the inner end of the left-hand driveaxle; a new one must be installed **(see illustration)**.

11   Check the condition of the differential oil seals, and replace them if necessary (see

9.4 Use a chisel or punch to unstake the driveaxle nut before trying to remove it

9.10 Discard the snap-ring from the inner end of the left-hand driveaxle; a new one must be installed

9.18 Use a hammer and punch to stake the driveaxle/hub nut

10.2 Cut the old boot clamps off

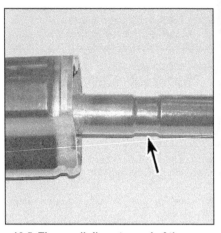

10.5 The small diameter end of the new boot must locate in the groove on the driveaxle

10.7a Tighten the large clamp with thin-nosed pliers . . .

10.7b . . . and the small clamp with boot clamp crimping pliers

Chapter 7). Check the intermediate shaft bearing. At the time of writing, the bearing is not available separately from the driveaxle - check with your dealer parts department for availability.

Installation

Refer to illustration 9.18

12   Install a new snap-ring to the groove on the inner end of the left-hand driveaxle. **Note:** *Manual and automatic transaxles use different types of snap-rings; they aren't interchangeable.* There is no snap-ring on the right-hand driveaxle.
13   Install the driveaxle(s) into the transaxle. Use a special sleeve to protect the differential oil seal as the driveaxle is inserted. If the sleeve is not used, take great care to avoid damaging the seal. **Note:** *Installation sleeves are supplied with new oil seals, where required.* Turn the driveaxle until it engages the splines on the differential gears. Make sure the snap-ring (left-hand driveaxle only) is fully engaged. **Note:** *When installing the left driveaxle, position the snap-ring with the*

gap in the 6 o'clock position (more of the clip will rest in the groove of the driveaxle, easing insertion).
14   If reinstalling the right-hand driveaxle, tighten the bolts securing the support bearing to the bracket on the cylinder block to the specified torque. On models so equipped, install the power steering pump cooling fan assembly.
15   Slide the steering knuckle/hub assembly over the end of the driveaxle, ensuring the hub splines engage correctly with the splines of the driveaxle.
16   Screw on the new driveaxle/hub nut finger-tight.
17   Locate the steering knuckle balljoint studs into the corresponding holes in the control arm and steering tie-rod end, and tighten the new nuts to the specified torque.
18   Tighten the driveaxle/hub nut to the specified torque while an assistant depresses the brake pedal to prevent the hub from rotating. With the nut correctly tightened, use a hammer and punch to stake the nut in position **(see illustration)**.

19   Install the wheel speed sensor (see Chapter 9).
20   Fill the transaxle with oil, and check the level (see Chapter 7).
21   Install the wheel, and lower the vehicle to the ground. Tighten the wheel retaining bolts to the specified torque.

10   Driveaxle inner joint boot - replacement

Caution: *On Mk I models, don't attempt to separate the inner CV joint from the shaft.*

Mk I models

Refer to illustrations 10.2, 10.5, 10.7a and 10.7b

1   Remove the driveaxle, then remove the *outer* CV joint boot as described in Section 11.
2   Note the fitted location of both of the inner joint boot retaining clamps. Cut the clamps from the boot, and slide the boot off the outer end of the driveaxle **(see illustration)**. Ensure the driveaxle is not pulled out from the inner joint.
3   Clean the driveaxle.
4   Scoop out all of the old grease from the joint housing, then pack the joint with the new grease supplied in the kit.
5   Slide the new boot along the driveaxle, and locate it on the inner joint housing. The small-diameter end of the boot must be located in the groove on the driveaxle **(see illustration)**.
6   Position the joint mid-way through its travel. Ensure that the boot is not twisted or distorted, then insert a small screwdriver under the lip of the boot at the housing end. This will allow trapped air to escape.
7   Install the new retaining clamps and tighten them securely **(see illustrations)**.
8   Install the outer CV joint and boot as described in Section 11.

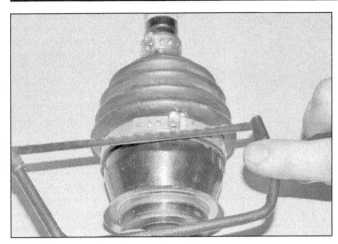

11.2 Cut the old clamps from the boot

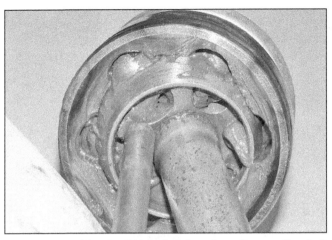

11.3 Carefully drive the CV joint hub from the splines on the shaft (Mk I models *only!*)

Mk II models

9   Remove the driveaxle, then cut off the boot clamps and remove the inner CV joint housing.

10   Remove the snap-ring from the end of the shaft, then remove the tripod from the shaft, noting how it's oriented (the flat side faces the end of the shaft). **Note:** *If the tripod is stuck on the shaft, use a three-jaw puller to remove it.*

11   Remove the boot from the shaft and clean the components.

12   Slide the new small clamp and boot onto the shaft, then install the tripod with the flat side of the spider (center portion) facing the end of the shaft. Secure the tripod with a new snap-ring.

13   Fill the housing with the grease supplied in the kit, then install it onto the tripod. Make sure the boot sealing area on the housing is clean, then install the boot over the housing, making sure it seats properly. Install the new large boot clamp.

14   Position the joint mid-way through its travel. Ensure that the boot is not twisted or distorted, then insert a small screwdriver under the small end of the boot. This will allow trapped air to escape.

15   Install the new retaining clamps and tighten them securely **(see illustrations 10.7a and 10.7b).**

11   Driveaxle outer joint boot - replacement

Caution: *On Mk II models, don't attempt to separate the outer CV joint from the shaft.*

1   Remove the driveaxle (see Section 9).

Mk I models

Refer to illustrations 11.2, 11.3, 11.4, 11.9, 11.11, 11.13a and 11.13b

2   Note the installed location of both of the outer joint boot retaining clamps, then cut the clamps from the boot, and slide the boot back along the driveaxle a little way **(see illustration)**.

3   Using a brass drift and hammer, carefully drive the outer CV joint hub from the splines on the driveaxle **(see illustration)**. Initial resistance will be felt until the internal snap-ring is released. Take care not to damage the bearing cage.

4   Extract the snap-ring from the end of the driveaxle **(see illustration)**.

5   Slide the boot, and remove it from the driveaxle together with the clamps.

6   Clean the driveaxle, and obtain a new joint retaining snap-ring. The boot retaining clamps must also be replaced. All these parts are normally supplied by MINI as a boot replacement kit.

7   Slide the new boot (together with new clamps) onto the driveaxle.

8   Install a new snap-ring to the groove in the end of the driveaxle.

9   Scoop out all of the old grease, then pack the joint with new grease supplied in the kit **(see illustration)**. Take care that the fresh grease does not become contaminated with dirt or grit as it is being applied.

10   Locate the CV joint on the driveaxle so that the splines are aligned, then push the joint until the internal snap-ring is fully engaged.

11   Move the boot along the driveaxle, and locate it over the joint and onto the outer CV joint housing. Make sure the rib in the small-diameter end of the boot is properly seated in the grooves on the driveaxle **(see illustration)**.

11.4 Pry the old circlip from the end of the driveaxle

11.9 Pack the CV joint with the grease supplied in the boot kit

11.11 The rib in the small-diameter end of the boot must be properly seated in the grooves on the driveaxle

11.13a Install the new outer clamp and tighten it with boot clamp crimping pliers . . .

11.13b . . . followed by the inner clamp

18   Make sure the small-diameter end of the boot is seated properly in its groove, then insert a small screwdriver under the lip of the boot at the housing end to allow any trapped air to escape.

19   Remove the screwdriver, install the retaining clamps in their previously noted positions, and tighten them **(see illustrations 11.13a and 11.13b)**.

12   Driveaxles - inspection

1   If excessive wear or play is evident in any driveaxle joint, remove the wheel cover (or center cover), and check that the drive-axle nut is tightened to the specified torque. Repeat this check on the other side of the vehicle.

2   Road test the vehicle, and listen for a metallic clicking from the front as the vehicle is driven slowly in a circle on full-lock. If a clicking noise is heard, this indicates wear in the outer constant velocity joint.

3   If vibration, consistent with road speed, is felt through the car when accelerating, there is a possibility of wear in the inner tripod joints.

4   Continual noise from the right-hand drive-axle, increasing with road speed, may indicate wear in the support bearing. To replace this bearing, the driveaxle and intermediate shaft must be removed, and the bearing extracted using special tools. Entrust this task to a MINI dealer or other qualified repair shop.

12   Ensure that the boot is not twisted or dis-torted, then insert a small screwdriver under the lip of the boot at the housing end to allow any trapped air to escape.

13   Remove the screwdriver, install the retaining clamps in their previously noted posi-tions, and tighten them **(see illustrations)**.

Mk II models

14   Remove the inner CV joint (see Sec-tion 10).

15   Cut the clamps from the boot, and slide the boot off the inner end of the driveaxle.

16   Clean the inner and outer joints. The outer joint will be more difficult to clean since it can't be removed, but with an ample supply of solvent or brake system cleaner, and flex-ing the joint through its range of motion, you should be able to get all of the old grease out. Allow the joint to dry thoroughly (used com-pressed air, if possible).

17   Slide the new outer boot and clamps onto the axleshaft. Fill the boot with the grease supplied in the kit. Make sure the outer joint housing is clean where the boot fits, then pull the boot onto the housing.

Chapter 9   Brakes

Contents

Specifications

Note: *Throughout this Chapter you will find references to "Mk I" and "Mk II" models; this is done to simplify which specifications and procedures apply to which models. Mk I models include 2006 and earlier Cooper/Cooper S models, and 2008 and earlier Convertible models. Mk II models include 2007 and later Cooper/Cooper S/Clubman/Clubman S and 2009 and later Convertible models.*

Front brakes

Disc minimum thickness	
Mk I models	0.748 inch (19 mm)
Mk II models	0.803 inch (20.4 mm)
Maximum disc run-out	0.002 inch (0.05 mm)
Brake pad friction material minimum thickness	1/8 inch (3.0 mm)

Rear brakes

Disc minimum thickness	0.330 inch (8.4 mm)
Maximum disc run-out	0.002 inch (0.05 mm)
Brake pad friction material minimum thickness	1/8 inch (3.0 mm)

Torque specifications

Ft-lbs (unless otherwise indicated) **Nm**

Note: *One foot-pound (ft-lb) of torque is equivalent to 12 inch-pounds (in-lbs) of torque. Torque values below approximately 15 foot-pounds are expressed in inch-pounds, because most foot-pound torque wrenches are not accurate at these smaller values.*

	Ft-lbs	Nm
Disc retaining screws		
All except 2006 and earlier JCW models	20	27
2006 and earlier JCW models	72 in-lbs	8
DSC motion sensor	72 in-lbs	8

Torque specifications (continued)

Ft-lbs (unless otherwise indicated) **Nm**

Note: One foot-pound (ft-lb) of torque is equivalent to 12 inch-pounds (in-lbs) of torque. Torque values below approximately 15 foot-pounds are expressed in inch-pounds, because most foot-pound torque wrenches are not accurate at these smaller values.

	Ft-lbs	Nm
Brake disc-to-hub Torx screw	20	27
Front brake caliper (ATE)		
Guide pin bolts		
Mk I models	22	30
Mk II models	26	35
Mounting bracket bolts	81	110
Front brake caliper mounting bolts (Brembo)	81	110
Master cylinder retaining nuts*		
Mk I models	15	20
Mk II models	17	23
Rear brake caliper		
Guide pin bolts*		
Mk I models	22	30
Mk II models	26	35
Mounting bracket bolts*	48	65
Steering column lower universal joint bolt*	22	30
Power brake booster mounting nuts*	16	21
Wheel speed sensor retaining bolts	72 in-lbs	8
Wheel bolts	See Chapter 1	

* Do not re-use - replace with a new one whenever removed.

1   General information

The braking system is of the power-assisted, dual-circuit hydraulic type.

All models are equipped with disc brakes on all wheels. ABS is standard.

The disc brakes are actuated by single-piston sliding type calipers (ATE) or four-piston fixed calipers (Brembo), both of which ensure that equal pressure is applied to each disc pad.

Hydraulic system

The hydraulic system consists of two separate circuits. The master cylinder has separate reservoirs for the two circuits, and, in the event of a leak or failure in one hydraulic circuit, the other circuit will remain operative.

Power brake booster

The power brake booster, utilizing engine manifold vacuum and atmospheric pressure to provide assistance to the hydraulically operated brakes, is mounted on the firewall in the engine compartment.

Parking brake

On all models, the parking brake provides an independent mechanical means of rear brake application. All models are equipped with rear brake calipers with an integral parking brake function. The parking brake cable operates a lever on the caliper which forces the piston to press the pad against the disc surface. A self-adjust mechanism is incorporated, to automatically compensate for brake pad wear.

Service

After completing any operation involving disassembly of any part of the brake system, always test drive the vehicle to check for proper braking performance before resuming normal driving. When testing the brakes, perform the tests on a clean, dry, flat surface. Conditions other than these can lead to inaccurate test results.

Test the brakes at various speeds with both light and heavy pedal pressure. The vehicle should stop evenly without pulling to one side or the other.

Tires, vehicle load and wheel alignment are factors which also affect braking performance.

Precautions

There are some general cautions and warnings involving the brake system on this vehicle:

a) *Use only brake fluid conforming to DOT 4 specifications.*

b) *The brake pads contain fibers which are hazardous to your health if inhaled. Whenever you work on brake system components, clean all parts with brake system cleaner. Do not allow the fine dust to become airborne. Also, wear an approved filtering mask.*

c) *Safety should be paramount whenever any servicing of the brake components is performed. Do not use parts or fasteners which are not in perfect condition, and be sure that all clearances and torque specifications are adhered to. If you are at all unsure about a certain procedure, seek professional advice. Upon completion of any brake system work, test the brakes carefully in a controlled area before putting the vehicle into normal service. If a problem is suspected in the brake system, don't drive the vehicle until it's fixed.*

d) *Used brake fluid is considered a hazardous waste and it must be disposed of in accordance with federal, state and local laws. DO NOT pour it down the sink, into septic tanks or storm drains, or on the ground.*

e) *Clean up any spilled brake fluid immediately and then wash the area with large amounts of water. This is especially true for any finished or painted surfaces.*

2   Troubleshooting

PROBABLE CAUSE	CORRECTIVE ACTION

No brakes - pedal travels to floor

1 Low fluid level 2 Air in system	1 and 2 Low fluid level and air in the system are symptoms of another problem - a leak somewhere in the hydraulic system. Locate and repair the leak
3 Defective seals in master cylinder	3 Replace master cylinder
4 Fluid overheated and vaporized due to heavy braking	4 Bleed hydraulic system (temporary fix). Replace brake fluid (proper fix)

Brake pedal slowly travels to floor under braking or at a stop

1 Defective seals in master cylinder	1 Replace master cylinder
2 Leak in a hose, line, caliper or wheel cylinder	2 Locate and repair leak
3 Air in hydraulic system	3 Bleed the system, inspect system for a leak

Brake pedal feels spongy when depressed

1 Air in hydraulic system	1 Bleed the system, inspect system for a leak
2 Master cylinder or power booster loose	2 Tighten fasteners
3 Brake fluid overheated (beginning to boil)	3 Bleed the system (temporary fix). Replace the brake fluid (proper fix)
4 Deteriorated brake hoses (ballooning under pressure)	4 Inspect hoses, replace as necessary (it's a good idea to replace all of them if one hose shows signs of deterioration)

Brake pedal feels hard when depressed and/or excessive effort required to stop vehicle

1 Power booster faulty	1 Replace booster
2 Engine not producing sufficient vacuum, or hose to booster clogged, collapsed or cracked	2 Check vacuum to booster with a vacuum gauge. Replace hose if cracked or clogged, repair engine if vacuum is extremely low
3 Brake linings contaminated by grease or brake fluid	3 Locate and repair source of contamination, replace brake pads
4 Brake linings glazed	4 Replace brake pads or shoes, check discs for glazing, service as necessary
5 Caliper piston(s) or wheel cylinder(s) binding or frozen	5 Replace calipers or wheel cylinders
6 Brakes wet	6 Apply pedal to boil-off water (this should only be a momentary problem)
7 Kinked, clogged or internally split brake hose or line	7 Inspect lines and hoses, replace as necessary

Excessive brake pedal travel (but will pump up)

1 Air in hydraulic system	1 Bleed system, inspect system for a leak

Excessive brake pedal travel (but will not pump up)

1 Master cylinder pushrod misadjusted	1 Adjust pushrod
2 Master cylinder seals defective	2 Replace master cylinder
3 Brake linings worn out	3 Inspect brakes, replace pads
4 Hydraulic system leak	4 Locate and repair leak

Brake pedal doesn't return

1 Brake pedal binding	1 Inspect pivot bushing and pushrod, repair or lubricate
2 Defective master cylinder	2 Replace master cylinder

Troubleshooting (continued)

PROBABLE CAUSE	CORRECTIVE ACTION

Brake pedal pulsates during brake application

1 Excessive brake disc runout or disc surfaces out-of-parallel	1 Have discs machined by an automotive machine shop
2 Loose or worn wheel bearings	2 Adjust or replace wheel bearings
3 Loose lug nuts	3 Tighten lug nuts

Brakes slow to release

1 Malfunctioning power booster	1 Replace booster
2 Pedal linkage binding	2 Inspect pedal pivot bushing and pushrod, repair/lubricate
3 Malfunctioning proportioning valve	3 Replace proportioning valve
4 Sticking caliper or wheel cylinder	4 Repair or replace calipers or wheel cylinders
5 Kinked or internally split brake hose	5 Locate and replace faulty brake hose

Brakes grab (one or more wheels)

1 Grease or brake fluid on brake lining	1 Locate and repair cause of contamination, replace lining
2 Brake lining glazed	2 Replace lining, deglaze disc

Vehicle pulls to one side during braking

1 Grease or brake fluid on brake lining	1 Locate and repair cause of contamination, replace lining
2 Brake lining glazed	2 Deglaze or replace lining, deglaze disc
3 Restricted brake line or hose	3 Repair line or replace hose
4 Tire pressures incorrect	4 Adjust tire pressures
5 Caliper or wheel cylinder sticking	5 Repair or replace calipers or wheel cylinders
6 Wheels out of alignment	6 Have wheels aligned
7 Weak suspension spring	7 Replace springs
8 Weak or broken shock absorber	8 Replace shock absorbers

Brakes drag (indicated by sluggish engine performance or wheels being very hot after driving)

1 Brake pedal pushrod incorrectly adjusted	1 Adjust pushrod
2 Master cylinder pushrod (between booster and master cylinder) incorrectly adjusted	2 Adjust pushrod
3 Obstructed compensating port in master cylinder	3 Replace master cylinder
4 Master cylinder piston seized in bore	4 Replace master cylinder
5 Contaminated fluid causing swollen seals throughout system	5 Flush system, replace all hydraulic components
6 Clogged brake lines or internally split brake hose(s)	6 Flush hydraulic system, replace defective hose(s)
7 Sticking caliper(s) or wheel cylinder(s)	7 Replace calipers or wheel cylinders
8 Parking brake not releasing	8 Inspect parking brake linkage and parking brake mechanism, repair as required
9 Faulty proportioning valve	9 Replace proportioning valve

Troubleshooting (continued)

PROBABLE CAUSE	CORRECTIVE ACTION

Brakes fade (due to excessive heat)

1 Brake linings excessively worn or glazed	1 Deglaze or replace brake pads
2 Excessive use of brakes	2 Downshift into a lower gear, maintain a constant slower speed (going down hills)
3 Vehicle overloaded	3 Reduce load
4 Brake discs worn too thin	4 Measure disc thickness, replace discs as required
5 Contaminated brake fluid	5 Flush system, replace fluid
6 Brakes drag	6 Repair cause of dragging brakes
7 Driver resting left foot on brake pedal	7 Don't ride the brakes

Brakes noisy (high-pitched squeal)

1 Glazed lining	1 Deglaze or replace lining
2 Contaminated lining (brake fluid, grease, etc.)	2 Repair source of contamination, replace linings
3 Weak or broken brake shoe hold-down or return spring	3 Replace springs
4 Rivets securing lining to backing plate loose	4 Replace pads
5 Excessive dust buildup on brake linings	5 Wash brakes off with brake system cleaner
6 Wear indicator on disc brake pads contacting disc	6 Replace brake pads
7 Anti-squeal shims missing or installed improperly	7 Install shims correctly

Note: *Other remedies for quieting squealing brakes include the application of an anti-squeal compound to the backing plates of the brake pads, and lightly chamfering the edges of the brake pads with a file. The latter method should only be performed with the brake pads thoroughly wetted with brake system cleaner, so as not to allow any brake dust to become airborne.*

Brakes noisy (scraping sound)

1 Brake pads worn out; rivets or backing plate metal contacting disc	1 Replace linings, have discs machined (or replace)

Brakes chatter

1 Worn brake lining	1 Inspect brakes, replace pads as necessary
2 Glazed or scored discs	2 Deglaze discs with sandpaper (if glazing is severe, machining will be required)
3 Discs heat checked	3 Check discs for hard spots, heat checking, etc. Have discs machined or replace them
4 Disc runout excessive	4 Measure disc runout, have discs machined or replace them
5 Loose or worn wheel bearings	5 Adjust or replace wheel bearings
6 Grooves worn in discs	6 Have discs machined, if within limits (if not, replace them)
7 Brake linings contaminated (brake fluid, grease, etc.)	7 Locate and repair source of contamination, replace pads
8 Excessive dust buildup on linings	8 Wash brakes with brake system cleaner
9 Surface finish on discs too rough after machining (especially on vehicles with sliding calipers)	9 Have discs properly machined
10 Brake pads glazed	10 Deglaze or replace brake pads

Troubleshooting (continued)

PROBABLE CAUSE	CORRECTIVE ACTION

Brake pads or shoes click

1 Shoe support pads on brake backing plate grooved or excessively worn	1 Replace brake backing plate
2 Brake pads loose in caliper	2 Loose pad retainers or anti-rattle clips
3 Also see items listed under Brakes chatter	

Brakes make groaning noise at end of stop

1 Brake pads worn out	1 Replace pads
2 Brake linings contaminated (brake fluid, grease, etc.)	2 Locate and repair cause of contamination, replace brake pads
3 Brake linings glazed	3 Deglaze or replace brake pads
4 Excessive dust buildup on linings	4 Wash brakes with brake system cleaner
5 Scored or heat-checked discs	5 Inspect discs, have machined if within limits (if not, replace discs)

Rear brakes lock up under light brake application

1 Tire pressures too high	1 Adjust tire pressures
2 Tires excessively worn	2 Replace tires
3 Defective proportioning valve	3 Replace proportioning valve

Brake warning light on instrument panel comes on (or stays on)

1 Low fluid level in master cylinder reservoir (reservoirs with fluid level sensor)	1 Add fluid, inspect system for leak, check the thickness of the brake pads
2 Failure in one half of the hydraulic system	2 Inspect hydraulic system for a leak
3 Piston in pressure differential warning valve not centered	3 Center piston by bleeding one circuit or the other (close bleeder valve as soon as the light goes out)
4 Defective pressure differential valve or warning switch	4 Replace valve or switch
5 Air in the hydraulic system	5 Bleed the system, check for leaks
6 Brake pads worn out (vehicles with electric wear sensors - small probes that fit into the brake pads and ground out on the disc when the pads get thin)	6 Replace brake pads (and sensors)

Brakes do not self adjust

1 Defective caliper piston seals	1 Replace calipers. Also, possible contaminated fluid causing soft or swollen seals (flush system and fill with new fluid if in doubt)
2 Corroded caliper piston(s)	2 Same as above

Rapid brake lining wear

1 Driver resting left foot on brake pedal	1 Don't ride the brakes
2 Surface finish on discs too rough	2 Have discs properly machined
3 Also see Brakes drag	

3 Anti-lock Brake System (ABS) - general information

General information

1 The anti-lock brake system is designed to maintain vehicle steerability, directional stability and optimum deceleration under severe braking conditions on most road surfaces. It does so by monitoring the rotational speed of each wheel and controlling the brake line pressure to each wheel during braking. This prevents the wheels from locking up.

2 The ABS system has three main components - the wheel speed sensors, the electronic control unit (ECU) and the hydraulic unit. Four wheel-speed sensors - one at each wheel - send a variable voltage signal to the control unit, which monitors these signals, compares them to its program and determines whether a wheel is about to lock up. When a wheel is about to lock up, the control unit signals the hydraulic unit to reduce hydraulic pressure (or not increase it further) at that wheel's brake caliper. Pressure modulation is handled by electrically-operated solenoid valves.

3 If a problem develops within the system, an "ABS" warning light will glow on the dashboard. Sometimes, a visual inspection of the ABS system can help you locate the problem. Carefully inspect the ABS wiring harness. Pay particularly close attention to the harness and connections near each wheel. Look for signs of chafing and other damage caused by incorrectly routed wires. If a wheel sensor harness is damaged, the sensor must be replaced. **Warning:** *Do NOT try to repair an ABS wiring harness. The ABS system is sensitive to even the smallest changes in resistance. Repairing the harness could alter resistance values and cause the system to malfunction. If the ABS wiring harness is damaged in any way, it must be replaced.* **Caution:** *Make sure the ignition is turned off before unplugging or reattaching any electrical connections.*

4 The MINI is also equipped with additional safety features built around the ABS system. These systems include EBFD (Electronic Brake Force Distribution), which automatically distributes braking effort between the front and rear wheels, and DSC (Dynamic Stability Control), which monitors the vehicle's cornering forces and steering wheel angle, then applies the braking force to the appropriate wheel to enhance the stability of the vehicle. The DSC motion sensor is located beneath the rear center console, and incorporates both the Yaw rate sensor and the lateral acceleration sensor. The DSC steering angle sensor is installed on the upper steering column. If a fault does develop in any of these systems, the vehicle must be taken to a MINI dealer or suitably-equipped specialist for fault diagnosis and repair.

Diagnosis and repair

5 If a dashboard warning light comes on and stays on while the vehicle is in operation,

3.12 Remove the bolt and pull the speed sensor from the steering knuckle

the ABS system requires attention. Although special electronic ABS diagnostic testing tools are necessary to properly diagnose the system, you can perform a few preliminary checks before taking the vehicle to a dealer service department.

- a) *Check the brake fluid level in the reservoir.*
- b) *Verify that the computer electrical connectors are securely connected.*
- c) *Check the electrical connectors at the hydraulic control unit.*
- d) *Check the fuses.*
- e) *Follow the wiring harness to each wheel and verify that all connections are secure and that the wiring is undamaged.*

6 If the above preliminary checks do not rectify the problem, the vehicle should be diagnosed by a dealer service department or other qualified repair shop. Due to the complex nature of this system, all actual repair work must be done by a qualified automotive technician.

Component replacement

Regulator assembly

7 Replacement of the regulator assembly requires access to specialist diagnostic and testing equipment in order to purge air from the system, and initialize and code the ECU. Therefore, we recommend this task be entrusted to a MINI dealer or other qualified repair shop.

Electronic control unit (ECU)

8 The ECU is integral with the regulator assembly, and is not available separately.

Front wheel sensor

Removal

Refer to illustration 3.12

9 Ensure the ignition is turned off.

10 Apply the parking brake, loosen the front wheel bolts, then raise the front of the vehicle and support securely on jackstands. Remove the appropriate front wheel.

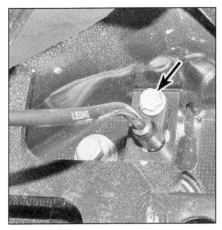

3.21 Remove the bolt and remove the rear wheel speed sensor

11 Trace the wiring back from the sensor, releasing it from any clips and ties while noting its correct routing, and disconnect the electrical connector.

12 Loosen and remove the retaining bolt and remove the sensor from the steering knuckle **(see illustration)**.

Installation

13 Ensure that the mating surfaces of the sensor and the steering knuckle are clean, and apply a dab of anti-seize compound to the hub carrier hub bore before installing.

14 Make sure the sensor tip is clean and ease it into position in the steering knuckle.

15 Clean the threads of the sensor bolt and apply a drop of thread-locking compound. Install the retaining bolt and tighten it to the specified torque.

16 Work along the sensor wiring, making sure it is correctly routed, securing it in position with all the relevant clips and ties. Reconnect the electrical connector.

17 Lower the vehicle and tighten the wheel bolts to the specified torque.

Rear wheel sensor

Removal

Refer to illustration 3.21

18 Ensure the ignition is turned off.

19 Block the front wheels, loosen the rear wheel bolts, then raise the rear of the vehicle and support it securely on jackstands (see *Jacking and towing*). Remove the appropriate wheel.

20 Trace the wiring back from the sensor, releasing it from all the relevant clips and ties while noting its correct routing, and disconnect the electrical connector.

21 Loosen and remove the retaining bolt and remove the sensor **(see illustration)**.

Installation

22 Ensure that the mating surfaces of the sensor and the hub are clean, and apply a dab of anti-seize compound to the hub bore before installing.

3.27 Remove the two Allen screws and remove the DSC motion sensor

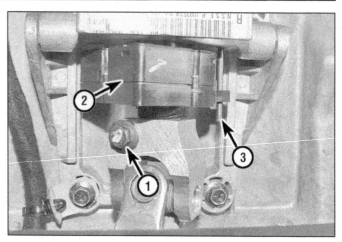

3.31 Steering column lower shaft joint pinch-bolt (1), steering angle sensor (2) and guide pin (3)

23   Make sure the sensor tip is clean and ease it into position in the steering knuckle.
24   Clean the threads of the sensor bolt and apply a drop of thread-locking compound. Install the retaining bolt and tighten it to the specified torque.
25   Work along the sensor wiring, making sure it is correctly routed, securing it in position with all the relevant clips and ties. Reconnect the electrical connector, then lower the vehicle and (where necessary) tighten the wheel bolts to the specified torque.

Dynamic stability control (DSC) motion sensor

Refer to illustration 3.27

26   Remove the rear center console (see Chapter 11).
27   Disconnect the wiring plug from the sensor, then remove the two screws and remove the sensor from the mounting bracket **(see illustration)**.
28   When installing the sensor, tighten the retaining bolts to the specified torque.

Steering angle sensor

Refer to illustration 3.31

29   Position the wheels and steering wheel in the straight-ahead position, then engage the steering lock.
30   Remove the steering column lower shroud (see Chapter 11).
31   Remove the pinch-bolt, then slide the steering column lower shaft joint from the upper shaft **(see illustration)**. Remove the wavy washer between the joint and the sensor.
32   Disconnect the sensor wiring plug, and slide the sensor down from the upper shaft.
33   When installing the sensor, ensure it locates correctly on the guide pin.

---

4   Disc brake pads - replacement

Warning: *Disc brake pads must be replaced on both front or both rear wheels at the same time - never replace the pads on only one wheel. Also, the dust created by the brake system is harmful to your health. Never blow it out with compressed air and don't inhale*
any of it. An approved filtering mask should be worn when working on the brakes. Do not, under any circumstances, use petroleum-based solvents to clean brake parts. Use brake system cleaner only!

Front pads

Refer to illustration 4.1

1   Apply the parking brake, loosen the front wheel bolts, then raise the front of the vehicle and support it securely on jackstands. Remove the front wheels. Clean the brake assembly with brake system cleaner **(see illustration)**.
2   Push the piston into its bore by pulling the caliper outwards slightly.

ATE (sliding) calipers

Mk I models

Refer to illustrations 4.3, 4.4a, 4.4b, 4.6, 4.8, 4.10, 4.11, 4.12, 4.13a, 4.13b, 4.14, 4.15a and 4.15b

3   Using a flat-bladed screwdriver, pry off the caliper retaining spring **(see illustration)**.
4   Pry off the plastic caps, then remove the

4.1 Before beginning work, thoroughly clean the caliper and disc with brake system cleaner

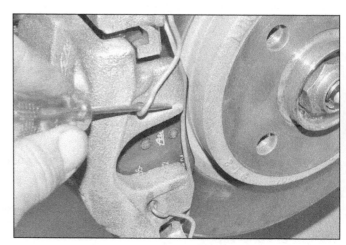

4.3 Pry off the caliper retaining spring

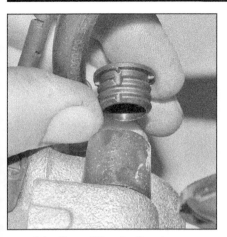

4.4a Pry out the plastic caps . . .

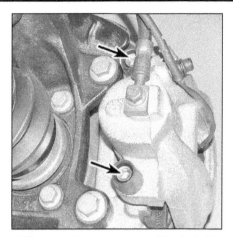

4.4b . . . and unscrew the guide pin bolts (Mk I models)

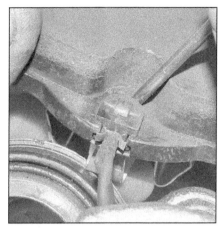

4.6 Pry the pad wear sensor from the inner brake pad

caliper guide pin bolts **(see illustrations)**.

5   Pull the caliper away, noting the inner pad is clipped to the caliper piston, while the outer pad remains in the caliper mounting bracket. Hang the caliper with a length of wire - don't let it hang by the hose **(see illustration 4.18)**.

6   Unclip the pad wear sensor wiring (if equipped), then carefully pry the sensor from the inner brake pad **(see illustration)**. Pull the inner pad from the caliper piston.

7   Withdraw the outer brake pad from the caliper mounting bracket.

8   Measure the thickness of each brake pad's friction material **(see illustration)**. If either pad is worn at any point to the specified minimum thickness or less, all four pads must be replaced. Also, the pads should be replaced if any are fouled with oil or grease; there is no satisfactory way of degreasing friction material, once contaminated. If any of the brake pads are worn unevenly, or are fouled with oil or grease, trace and rectify the cause before reassembly.

9   If the brake pads are still serviceable, carefully clean them with brake system cleaner. Clean out the grooves in the friction

material, and pick out any large embedded particles of dirt or debris. Carefully clean the pad locations in the caliper mounting bracket.

10   Prior to installing the pads, check that the guide pins slide easily in the caliper **(see illustration)**. Clean the dust and dirt from the caliper and piston with brake system cleaner. Inspect the dust boot around the piston for damage, and the piston for evidence of fluid leaks, corrosion or damage. If the boot is damaged or there is fluid leakage, it is recommended that the caliper be replaced (see Section 5).

11   If new brake pads are to be installed, the caliper piston must be pushed back into the cylinder to make room for them. Either use a C-clamp or similar tool, or use suitable pieces of wood as levers. Clamp off the flexible brake hose leading to the caliper, then connect a brake bleeding kit to the caliper bleed screw. Open the bleed screw as the piston is retracted; the surplus brake fluid will then be collected in the bleed kit vessel **(see illustration)**. Close the bleed screw just before the caliper piston is pushed fully into the caliper. This should ensure no air enters

4.8 Measure the thickness of the friction material on the brake pads

the hydraulic system. **Note:** *The ABS unit contains hydraulic components that are very sensitive to impurities in the brake fluid. Even the smallest particles can cause the system to fail through blockage. The pad retraction*

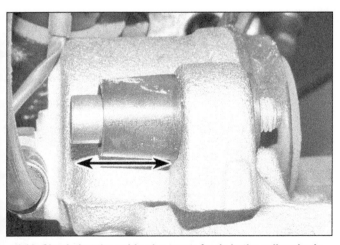

4.10 Check that the guide pins move freely in the caliper body

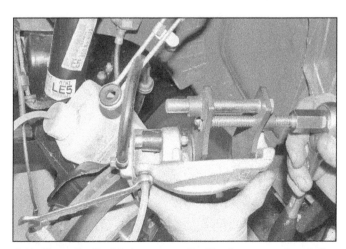

4.11 Clamp the flexible hose, open the bleed screw, and push the piston back into the caliper bore - piston retraction tool shown

4.12 Install the outer pad to the caliper mounting bracket

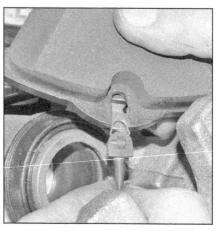

4.13a Slide the wear sensor into place . . .

4.13b . . . and install the inner pad to the caliper piston

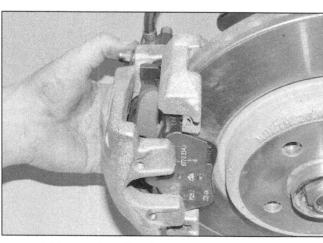

4.14 Press the caliper into position and tighten the guide pin bolts

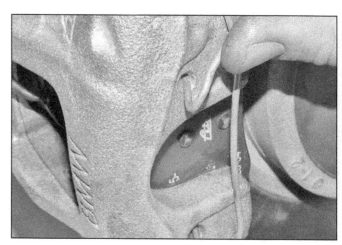

4.15a Insert the ends of the retaining spring first . . .

method described here prevents any debris in the brake fluid expelled from the caliper from being passed back to the ABS hydraulic unit.

12 Install the outer pad to the caliper mounting bracket, ensuring the friction material is against the disc **(see illustration)**.

13 Slide the wear sensor (if equipped) back into place, then fit the inner pad to the caliper piston **(see illustrations)**.

14 Press the caliper into position, then install the guide pin bolts, tightening them to the specified torque setting. Install the plastic

caps **(see illustration)**.

15 Insert the ends of the retaining spring into the holes in the caliper, then use a screwdriver to force the spring into position on the mounting bracket **(see illustrations)**. Proceed to Step 25.

Mk II models

Refer to illustrations 4.16, 4.17, 4.18, 4.19a, 4.19b and 4.20

16 Remove the pad wear sensor from the inner pad (left-front brake only) **(see illustration)**.

17 Unscrew the caliper mounting bolts, using an open-end wrench to prevent the guide pins from turning **(see illustration)**.

18 Remove the caliper from the mounting bracket and hang it with a length of wire - don't let it hang by the hose **(see illustration)**. **Note:** *Alternatively, the lower caliper mounting bolt can be removed and the caliper can be pivoted up for access to the pads.*

19 Remove the pads from the caliper mounting bracket **(see illustrations)**.

20 Remove the upper and lower pad anti-rattle clips from the caliper mounting bracket

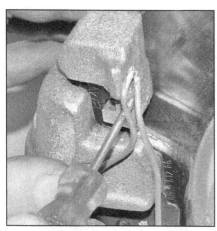

4.15b . . . then lever the spring into position

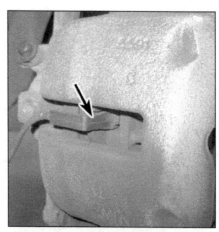

4.16 Using a pair of pliers, pull the pad wear sensor from the inner pad (left-front brake only)

4.17 On Mk II models, hold the guide pins with an open-end wrench while loosening the caliper mounting bolts

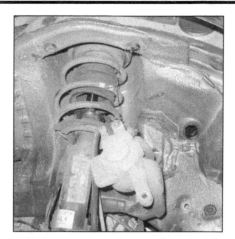

4.18 Remove the caliper and hang it with a length of wire

the outer pad against the caliper just enough to depress the pistons back a little to make room for pad removal, then slide the pad out of the caliper.

32   Refer to Step 11 and clamp-off the fluid hose, open the bleeder valve then, using two screwdrivers or pieces of wood, carefully and slowly push both of the pistons back into the caliper (at the same time). Make sure they enter their bores squarely. When the pistons have been retracted, close the bleeder valve.

33   Clean the caliper with brake system cleaner. Inspect the dust boot around the piston for damage, and the piston for evidence of fluid leaks, corrosion or damage. If the boot is damaged or there is fluid leakage, it is recommended that the caliper be replaced (see Section 5).

34   Measure the thickness of the brake pad's friction material **(see illustration 4.8)**. If the pad is worn at any point to the specified minimum thickness or less, all four pads must be replaced. Also, the pads should be replaced if any are fouled with oil or grease; there is no satisfactory way of degreasing friction material, once contaminated. If any of the brake pads are worn unevenly, or are fouled with oil or grease, trace and rectify the cause before reassembly.

35   If the brake pad is still serviceable, carefully clean it with brake system cleaner. Clean out the grooves in the friction material, and pick out any large embedded particles of dirt or debris. Carefully clean the pad locations in the caliper.

36   Apply a thin film of high-temperature brake lubricant to the pad contact surfaces of the caliper. **Note:** *The manufacturer recommends against applying any grease to the pad backing plates.*

37   Install the new outer pad, then repeat Steps 31 through 36 and replace the inner pad. **Caution:** *Don't get any brake cleaner on the side that the new pad was installed, so as not to foul the lubricant on the pad sliding surfaces.*

(see illustration). If they are cracked, worn or fit loosely on the bracket, replace them with new ones. If the old ones are to be reused, clean them thoroughly with brake system cleaner.

21   Install the anti-rattle springs onto the caliper mounting bracket, then apply a light film of high-temperature brake grease to the pad ear contact areas.

22   Install the pads into the caliper mounting bracket.

23   Perform Steps 10 and 11.

24   Install the caliper over the pads, then install the caliper mounting bolts and tighten them to the torque listed in this Chapter's Specifications. Install the pad wear sensor to the inner pad (left-front brake only), then proceed to Step 25.

All models

25   Depress the brake pedal repeatedly, until the pads are pressed into firm contact with the brake disc, and normal (non-assisted) pedal pressure is restored. Keep an eye on the fluid level in the reservoir, adding if necessary to

keep it above the MIN level.

26   Repeat the pad replacement procedure on the remaining front brake caliper.

27   Install the wheels, then lower the vehicle and tighten the wheel bolts to the torque listed in the Chapter 1 Specifications.

28   Check the brake fluid level (see Chapter 1). **Caution:** *New pads will not give full braking efficiency until they have bedded-in. Be prepared for this, and avoid hard braking as far as possible for the first hundred miles or so after pad replacement.* **Note:** *If the piston retraction method was performed properly, there should be no reason to bleed the brakes. If, however, the brake pedal feels soft or spongy after the job has been completed, bleed the brake system (see Section 9).*

Brembo (fixed) calipers

29   Remove the pad wear sensor from the inner pad (left front brake only).

30   Using a small punch, drive out the pad retaining pins and remove the spring clip.

31   Using a pair of adjustable pliers, squeeze

4.19a Remove the inner pad . . .

4.19b . . . and the outer pad from the caliper mounting bracket

4.20 Remove the anti-rattle clips from the caliper mounting bracket, then clean and inspect them

4.41 Disengage the parking brake cable end from the lever on the caliper

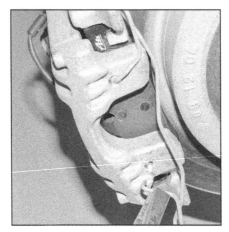

4.42 Pull the retaining spring from the caliper

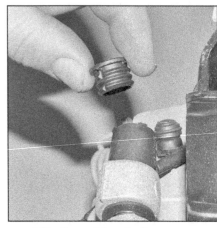

4.43a Pry out the plastic caps . . .

38   Secure the new pads with a new spring clip and new retaining pins, then install the pad wear sensor (left front brake only). If the wear sensor had made contact with the brake disc, be sure to replace it as well.

39   Perform Steps 25 through 28.

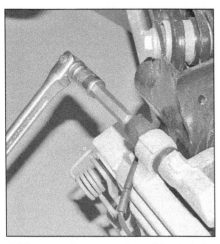

4.43b . . . and unscrew the guide pin bolts

Rear pads

Mk I models

Refer to illustrations 4.41, 4.42, 4.43a, 4.43b, 4.45a, 4.45b, 4.49, 4.50a, 4.50b, 4.51, 4.52 and 4.54

Note: *New guide pin bolts must be used on reassembly.*

40   Block the front wheels, loosen the rear wheel bolts, then raise the rear of the vehicle and support it securely on jackstands (see *Jacking and towing*). Remove the rear wheels.

41   Loosen the parking brake cable adjuster nut as described in Section 12, then using a pair of pliers, release the parking brake cable from the caliper lever **(see illustration)**.

42   Using a pair of pliers, pull the retaining spring from the caliper **(see illustration)**.

43   Pry out the rubber caps and, using a hex bit, unscrew the caliper guide pin bolts **(see illustrations)**.

44   Withdraw the caliper, leaving the outer pad in the mounting bracket, while the inner pad is removed with the caliper. Hang the caliper with a length of wire - don't let it hang by the hose **(see illustration 4.18)**.

45   Withdraw the outer pad from the caliper bracket, then unclip the pad wear sensor (if equipped) and slide the inner pad sideways from the piston **(see illustrations)**.

46   First measure the thickness of the friction material of each brake pad **(see illustration 4.8)**. If either pad is worn at any point to the specified minimum thickness or less, all four pads must be replaced. Also, the pads should be replaced if any are fouled with oil or grease; there is no satisfactory way of degreasing friction material, once contaminated. If any of the brake pads are worn unevenly, or fouled with oil or grease, trace and rectify the cause before reassembly. Examine the guide pins for signs of wear and replace if necessary.

47   If the brake pads are still serviceable, carefully clean them using a clean, fine wire brush or similar, paying particular attention to the sides and back of the metal backing. Carefully clean the pad locations in the caliper body/mounting bracket.

48   Prior to installing the pads, clean the caliper and piston with brake system cleaner. Inspect the dust seal around the piston for damage, and the piston for evidence of fluid leaks, corrosion or damage (see Section 5).

4.45a Remove the outer pad from the caliper mounting bracket . . .

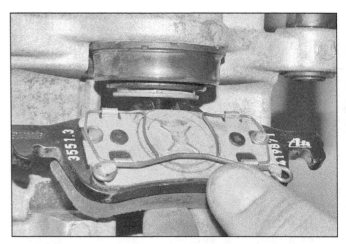

4.45b . . . and slide the inner pad from the caliper piston

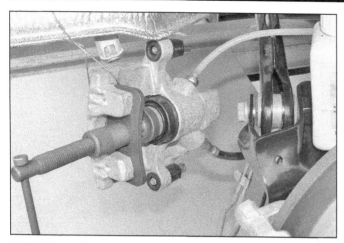

4.49 Using a piston retraction tool will rotate the piston clockwise as it's pushed back

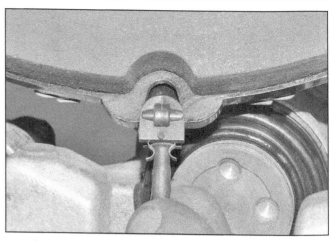

4.50a Slide the wear sensor into place on the inner pad

49   If new brake pads are to be installed, the caliper piston must be pushed back into the cylinder to make room for them. In order to retract the piston, the piston must be turned clockwise as it is pushed into the caliper. MINI tool 34 1 050 is available to retract the pistons, as are several others available from good accessory/parts retailers **(see illustration)**. Clamp off the flexible brake hose leading to the caliper, then connect a brake bleeding kit to the caliper bleed screw. Open the bleed screw as the piston is retracted; the surplus brake fluid will then be collected in the bleed kit vessel. Close the bleed screw just before the caliper piston is pushed fully into the caliper. This should ensure no air enters the hydraulic system **Note:** *The ABS unit contains hydraulic components that are very sensitive to impurities in the brake fluid. Even the smallest particles can cause the system to fail through blockage. The pad retraction method described here prevents any debris in the brake fluid expelled from the caliper from being passed back to the ABS hydraulic unit, as well as preventing any chance of damage to the master cylinder seals.*

50   Install the wear sensor (where applicable), then slide the inner brake pad into position in the caliper, ensuring the retaining clip engages correctly on the caliper piston **(see illustrations)**.
51   Install the outer pad to the caliper mounting bracket **(see illustration)**, making sure the friction material is facing the brake disc.
52   Lower the caliper into position, then insert the guide pin bolts and tighten them to the torque listed in this Chapter's Specifications **(see illustration)**.
53   Install the retaining spring to the caliper.
54   Slide the parking brake cable into the caliper bracket and reconnect the inner cable to the caliper lever **(see illustration)**.
55   Depress the brake pedal repeatedly until the pads are pressed into firm contact with the brake disc, and normal (non-assisted) pedal pressure is restored.
56   Repeat the procedure on the remaining rear brake caliper.
57   Check the operation of the parking brake, and if necessary, carry out the adjustment procedure as described in Section 12.
58   Install the wheels, then lower the vehicle

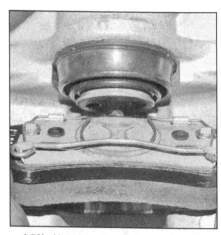

4.50b Note how the pad retaining clip must fit over the lip of the caliper piston

and tighten the wheel bolts to the torque listed in the Chapter 1 Specifications.
59   Check the brake fluid level (see Chapter 1). **Caution:** *New pads will not give full*

4.51 Install the outer pad to the caliper mounting bracket

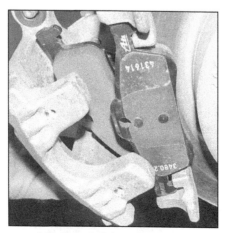

4.52 Reinstall the caliper to the mounting bracket

4.54 Slide the parking brake cable into the caliper bracket

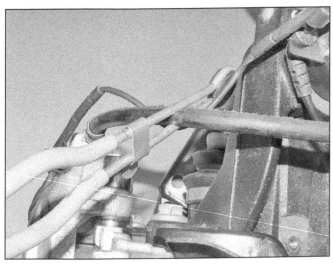

5.2 Use a clamp on the flexible brake hose

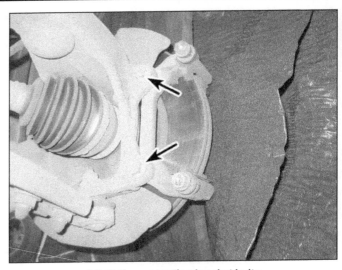

5.6 Caliper mounting bracket bolts

braking efficiency until they have bedded-in. Be prepared for this, and avoid hard braking as far as possible for the first hundred miles or so after pad replacement.

Mk II models

60 Pad replacement on these models is similar to Mk I models, with the following exceptions:

a) *The caliper mounting bolts are similar to those shown in* **illustration 4.17.**

b) *Both pads ride in the caliper mounting bracket and are supported by anti-rattle clips (refer to the Mk II model front pad replacement procedure). Be sure to lubricate the pad contact areas of the anti-rattle clips with a light film of high-temperature brake grease.*

c) *The tool used to retract the piston is MINI tool no. 34 6 300 (or an aftermarket equivalent).*

5   Disc brake caliper - removal and installation

Warning: *Dust created by the brake system is harmful to your health. Never blow it out with compressed air and don't inhale any of it. An approved filtering mask should be worn when working on the brakes. Do not, under any circumstances, use petroleum-based solvents to clean brake parts. Use brake system cleaner only.*

Warning: *Ensure the ignition is switched off before disconnecting any brake hydraulic system fittings, and do not switch it on until after the hydraulic system has been bled. Failure to do this could lead to air entering the regulator unit, requiring the unit to be bled using special MINI test equipment.*

Caution: *Brake fluid will damage paint. Cover all body parts and be careful not to spill fluid during this procedure.*

Note: *If replacement is indicated (usually*

because of fluid leakage), it is recommended that the calipers be replaced, not overhauled. New and factory rebuilt units are available on an exchange basis, which makes this job quite easy. Always replace the calipers in pairs - never replace just one of them.

Front caliper

Refer to illustration 5.2

1   Apply the parking brake, loosen the front wheel bolts, then raise the front of the vehicle and support it securely on jackstands. Remove the wheel.

2   Clamp-off the brake hose to keep contaminants out of the brake system and to prevent losing any more brake fluid than is necessary **(see illustration)**. **Note:** *If the caliper is being removed for access to another component, don't disconnect the hose.*

3   Clean the area around the caliper hose union, then loosen the union. **Caution:** *Just break it loose - don't try to unscrew it yet or the hose will be twisted.*

ATE (sliding) calipers

Removal

Refer to illustration 5.6

4   If you're working on an Mk I model, use a screwdriver to pry off the caliper retaining spring **(see illustration 4.3)**.

5   If you're working on an Mk I model, pry out the plastic caps, then loosen and remove the upper and lower caliper guide pins **(see illustrations 4.4a and 4.4b)**. If you're working on an Mk II model, remove the caliper mounting bolts **(see illustration 4.17)**. Lift the caliper away from the brake disc, then unscrew the caliper from the end of the brake hose. On Mk I models, pull the inner pad from the caliper piston.

6   If required, the caliper mounting bracket can be unbolted from the steering knuckle **(see illustration)**.

Installation

7   If previously removed, install the caliper mounting bracket to the steering knuckle, then tighten the bolts to the torque listed in this Chapter's Specifications.

8   Screw the caliper body fully onto the flexible hose union.

9   On Mk I models, install the inner brake pad to the caliper piston. Ensure that the outer brake pad (and inner pad on Mk II models) is correctly installed in the caliper mounting bracket (see Section 4), then install the caliper.

10   Install the guide pin bolts or caliper mounting bolts, then tighten them to the torque listed in this Chapter's Specifications. On Mk I models, install the plastic caps.

11   Tighten the brake hose union securely, then remove the brake hose clamp.

12   Bleed the brake circuit according to the procedure in Section 9. Make sure there are no leaks from the hose connections.

13   Install the wheel, then lower the vehicle and tighten the wheel bolts to the torque listed in the Chapter 1 Specifications. Pump the pedal several times to bring the pads into contact with the disc. Test the brakes carefully before returning the vehicle to normal service.

Brembo (fixed) calipers

Removal

14   Remove the brake pads (see Section 4).

15   Remove the bolts securing the caliper to the steering knuckle.

16   Lift the caliper away from the brake disc, then unscrew the caliper from the end of the brake hose.

Installation

17   Screw the caliper body fully onto the flexible hose union.

18   Install the caliper, tightening the mounting bolts to the torque listed in this Chapter's Specifications.

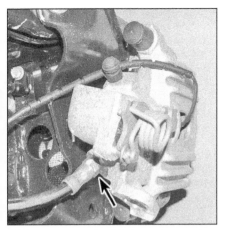

5.25 Loosen the flexible brake hose union

6.4 To check disc runout, mount a dial indicator as shown and rotate the disc

6.5 Use a micrometer to measure disc thickness

19   Tighten the brake hose union securely, then remove the brake hose clamp.
20   Install the brake pads (see Section 4).
21   Bleed the brake circuit according to the procedure in Section 9. Make sure there are no leaks from the hose connections.
22   Install the wheel, then lower the vehicle and tighten the wheel bolts to the torque listed in the Chapter 1 Specifications. Pump the pedal several times to bring the pads into contact with the disc. Test the brakes carefully before returning the vehicle to normal service.

Rear caliper

Removal

Refer to illustration 5.25

23   Block the front wheels, loosen the rear wheel bolts, then raise the rear of the vehicle and support it securely on jackstands. Remove the rear wheel.
24   Clamp-off the brake hose to keep contaminants out of the brake system and to prevent losing any more brake fluid than is necessary **(see illustration 5.2)**. **Note:** *If the caliper is being removed for access to another component, don't disconnect the hose.*
25   Wipe away all traces of dirt around the brake hose union on the caliper, then loosen the hose union **(see illustration)**. **Caution:** *Just break it loose - don't try to unscrew it yet or the hose will be twisted.*
26   Unscrew the caliper from the end of the brake hose. If required, the caliper mounting bracket can be unbolted from the trailing arm.
27   Remove the brake pads as described in Section 4.

Installation

28   If previously removed, install the caliper mounting bracket to the trailing arm, and tighten the bolts to the torque listed in this Chapter's Specifications.
29   Screw the caliper body fully onto the flexible hose union.

30   Install the brake pads as described in Section 4.
31   Install the guide pin bolts or caliper mounting bolts, then tighten them to the torque listed in this Chapter's Specifications. On Mk I models, install the plastic caps.
32   Tighten the brake hose union securely, then remove the brake hose clamp.
33   Bleed the brake circuit according to the procedure in Section 9. Make sure there are no leaks from the hose connections.
34   Install the wheel, then lower the vehicle and tighten the wheel bolts to the torque listed in the Chapter 1 Specifications. Pump the pedal several times to bring the pads into contact with the disc. Test the brakes carefully before returning the vehicle to normal service.

---

6   Brake disc - inspection, removal and installation

Warning: *Dust created by the brake system is harmful to your health. Never blow it out with compressed air and don't inhale any of it. An approved filtering mask should be worn when working on the brakes. Do not, under any circumstances, use petroleum-based solvents to clean brake parts. Use brake system cleaner only.*
Note: *This Section applies to both front and rear brake discs.*

Inspection

Refer to illustrations 6.4 and 6.5

Note: *If either disc requires replacement, BOTH should be replaced at the same time, to ensure even and consistent braking. New brake pads should also be installed.*

1   Loosen the wheel bolts, raise the vehicle and support it securely on jackstands.
2   Remove the brake caliper as outlined in Section 5. It isn't necessary to disconnect the

brake hose. After removing the caliper bolts, suspend the caliper out of the way with a piece of wire.
3   Visually inspect the disc surface for score marks and other damage. Light scratches and shallow grooves are normal after use and may not always be detrimental to brake operation, but deep scoring requires disc removal and refinishing by an automotive machine shop. Be sure to check both sides of the disc. If pulsating has been noticed during application of the brakes, suspect disc runout.
4   To check disc runout, reinstall the wheel bolts with spacers and tighten them securely. place a dial indicator at a point about 1/2-inch from the outer edge of the disc **(see illustration)**. Set the indicator to zero and turn the disc. The indicator reading should not exceed the specified allowable runout limit. If it does, the disc should be refinished by an automotive machine shop. **Note:** *The discs should be resurfaced regardless of the dial indicator reading, as this will impart a smooth finish and ensure a perfectly flat surface, eliminating any brake pedal pulsation or other undesirable symptoms related to questionable discs. At the very least, if you elect not to have the discs resurfaced, remove the glaze from the surface with emery cloth or sandpaper, using a swirling motion.*
5   It's absolutely critical that the disc not be machined to a thickness under the specified minimum thickness. The minimum (or discard) thickness is cast or stamped into the disc. The disc thickness can be checked with a micrometer **(see illustration)**.

Removal

Refer to illustration 6.7

6   Loosen and remove the two bolts securing the brake caliper mounting bracket to the steering knuckle (front) or trailing arm (rear).
7   Remove the Torx screw securing the brake disc to the hub, then remove the disc

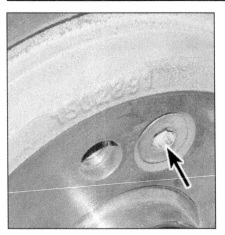

6.7 Remove the Torx screw securing the disc to the hub

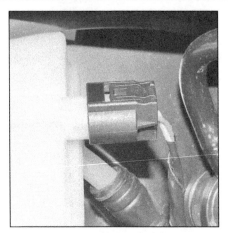

7.3 Disconnect the fluid level sensor electrical connector

(see illustration). If it is stuck, lightly tap its rear face with a rubber mallet.

Installation

8   While the disc is off, wire-brush the backside of the center portion that contacts the wheel hub. Also clean off any rust or dirt on the hub face.

9   Apply small dots of high-temperature anti-seize around the circumference of the hub, and around the raised center portion.

10   Place the disc in position and install the Torx bolt, tightening it to the torque listed in this Chapter's Specifications.

11   Install the caliper mounting bracket and caliper, tightening the bolts to the torque values listed in this Chapter's Specifications.

12   Install the wheel, then lower the vehicle. Tighten the wheel bolts to the torque listed in the Chapter 1 Specifications. Depress the brake pedal a few times to bring the brake pads into contact with the disc. Bleeding won't be necessary unless the brake hose was disconnected from the caliper. Check the operation of the brakes carefully before driving the vehicle.

13   If new or resurfaced rear discs are being installed, check the operation of the parking brake and adjust it if necessary (see Section 12).

14   Check the operation of the brakes carefully before driving the vehicle.

---

7   Master cylinder - removal and installation

Warning: *Ensure the ignition is switched off before disconnecting any brake hydraulic system fittings, and do not switch it on until after the hydraulic system has been bled. Failure to do this could lead to air entering the regulator unit, requiring the unit to be bled using special MINI test equipment.*
Caution: *Brake fluid will damage paint. Cover all body parts and be careful not to spill fluid during this procedure.*

Removal

Refer to illustrations 7.3, 7.6a and 7.6b

1   On Mk II models, remove the cowl cov-

ers (see Chapter 11).

2   Remove as much fluid as you can from the reservoir with a syringe, such as an old turkey baster. **Warning:** *If a baster is used, never again use it for the preparation of food.* Alternatively, open any convenient bleed screw in the system, and gently pump the brake pedal to expel the fluid through a plastic tube connected to the screw until the reservoir is emptied (see Section 9).

3   Disconnect the fluid level sensor electrical connector **(see illustration)**. Unclip the clutch fluid feed hose from the master cylinder reservoir, if applicable.

4   Pull the clutch master cylinder supply hose from the brake reservoir, then plug the hose. Be prepared for fluid spillage.

5   Wipe clean the area around the brake pipe unions on the side of the master cylinder, and place absorbent rags beneath the pipe unions to catch any surplus fluid. Make a note of the correct installed positions of the unions, then unscrew the union nuts and carefully withdraw the pipes. Place rags under the fluid fittings and prepare caps or plastic bags to cover the ends of the lines once they are disconnected. Wash off any spilled fluid immediately with cold water.

6   Remove the two nuts securing the master cylinder to the power brake booster, then withdraw the unit from the engine compartment. If the sealing ring installed on the rear of the master cylinder shows signs of damage or deterioration, it must be replaced. If required, unscrew the retaining Torx pin and separate the reservoir from the master cylinder **(see illustrations)**.

Installation

7   Bench bleed the new master cylinder before installing it. Mount the master cylinder in a vise, with the jaws of the vise clamping on the mounting flange.

8   Attach a pair of master cylinder bleeder tubes to the outlet ports of the master cylinder.

9   Fill the reservoir with brake fluid of the

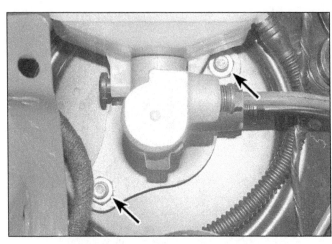

7.6a Remove the master cylinder mounting nuts

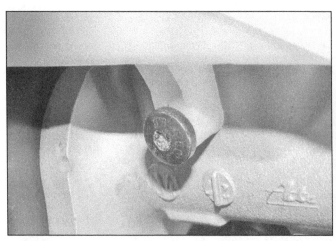

7.6b The fluid reservoir is secured to the master cylinder by a Torx screw

recommended type (see Chapter 1).

10   Slowly push the pistons into the master cylinder (a large Phillips screwdriver can be used for this) - air will be expelled from the pressure chambers and into the reservoir. Because the tubes are submerged in fluid, air can't be drawn back into the master cylinder when you release the pistons.

11   Repeat the procedure until no more air bubbles are present.

12   Remove the bleed tubes, one at a time, and install plugs in the open ports to prevent fluid leakage and air from entering. Install the reservoir cap.

13   Remove all traces of dirt from the master cylinder and servo unit mating surfaces and ensure that the sealing ring is correctly installed on the rear of the master cylinder.

14   Install the master cylinder to the booster unit. Install the new master cylinder mounting nuts, and tighten them to the torque listed in this Chapter's Specifications.

15   Wipe clean the brake pipe unions and install them to the master cylinder ports, tightening them to the torque listed in this Chapter's Specifications.

16   Press the mounting seals fully into the master cylinder ports, then carefully ease the fluid reservoir into position. Slide the reservoir retaining pin into position and tighten it securely.

17   Reconnect the clutch master cylinder supply pipe, level sensor wiring plug, and pressure sensor (if equipped).

18   Install any components removed to improve access. Fill the master cylinder reservoir with fluid, then bleed the master cylinder and the brake system as described in Section 9. To bleed the cylinder on the vehicle, have an assistant depress the brake pedal and hold the pedal to the floor. Loosen the fitting to allow air and fluid to escape. Repeat this procedure on both fittings until the fluid is clear of air bubbles. **Caution:** *Have plenty of rags on hand to catch the fluid - brake fluid will ruin painted surfaces. After the bleeding procedure is completed, rinse the area under the master cylinder with clean water.* **Note:** *The clutch system may also need to be bled (see Chapter 8).*

19   Test the operation of the brake system carefully before placing the vehicle into normal service. **Warning:** *Do not operate the vehicle if you are in doubt about the effectiveness of the brake system. It is possible for air to become trapped in the anti-lock brake system hydraulic control unit, so, if the pedal continues to feel spongy after repeated bleedings or the BRAKE or ANTI-LOCK light stays on, have the vehicle towed to a dealer service department or other qualified shop to be bled with the aid of a scan tool.*

8   Brake hoses and lines - inspection and replacement

Warning: *Ensure the ignition is switched off before disconnecting any brake hydraulic sys-*

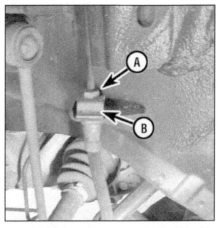

8.3 Unscrew the brake line fitting (A) with a flare-nut wrench, remove the spring clip (B), then unscrew the hose from the caliper

tem fittings, and do not switch it on until after the hydraulic system has been bled. Failure to do this could lead to air entering the regulator unit, requiring the unit to be bled using special MINI test equipment.
Caution: *Brake fluid will damage paint. Cover all body parts and be careful not to spill fluid during this procedure.*

1   About every six months, with the vehicle raised and placed securely on jackstands, the flexible hoses which connect the steel brake lines with the front and rear brake assemblies should be inspected for cracks, chafing of the outer cover, leaks, blisters and other damage. These are important and vulnerable parts of the brake system and inspection should be complete. A light and mirror will be needed for a thorough check. If a hose exhibits any of the above defects, replace it with a new one.

Flexible hoses

Refer to illustration 8.3

2   Clean all dirt away from the ends of the hose.

3   Unscrew the tube nut with a flare-nut wrench, if available, to prevent rounding-off the corners of the nut, then remove the clip securing the hose to the body (and any suspension components) **(see illustration)**.

4   Unscrew the hose from the caliper, discarding the sealing washers on either side of the fitting.

5   Attach the new brake hose to the caliper, tightening it to the torque listed in this Chapter's Specifications.

6   Pass the hose through the bracket on the chassis and thread the tube nut into the fitting. Install the spring clip, then tighten the tube nut.

7   Carefully check to make sure the suspension or steering components don't make contact with the hose. Have an assistant push down on the vehicle and also turn the steering wheel lock-to-lock during inspection.

8   Bleed the brake system (see Section 9).

Metal brake lines

9   When replacing brake lines, be sure to use the correct parts. Don't use copper tubing for any brake system components. Purchase steel brake lines from a dealer parts department or auto parts store.

10   Prefabricated brake line, with the tube ends already flared and fittings installed, is available at auto parts stores and dealer parts departments. These lines can be bent to the proper shapes using a tubing bender.

11   When installing the new line make sure it's well supported in the brackets and has plenty of clearance between moving or hot components.

12   After installation, check the master cylinder fluid level and add fluid as necessary. Bleed the brake system as outlined in Section 9 and test the brakes carefully before placing the vehicle into normal operation.

9   Brake hydraulic system - bleeding

Refer to illustration 9.8

Warning: *If air has found its way into the hydraulic control unit, the system must be bled with the use of a scan tool. If the brake pedal feels spongy even after bleeding the brakes, or the ABS light on the instrument panel does not go off, or if you have any doubts whatsoever about the effectiveness of the brake system, have the vehicle towed to a dealer service department or other repair shop equipped with the necessary tools for bleeding the system.*
Warning: *Wear eye protection when bleeding the brake system. If the fluid comes in contact with your eyes, immediately rinse them with water and seek medical attention.*
Note: *Bleeding the brake system is necessary to remove any air that's trapped in the system when it's opened during removal and installation of a hose, line, caliper, wheel cylinder or master cylinder.*
Note: *The clutch system may also need to be bled (see Chapter 8).*

1   It will probably be necessary to bleed the system at all four brakes if air has entered the system due to low fluid level, or if the brake lines have been disconnected at the master cylinder.

2   If a brake line was disconnected only at a wheel, then only that caliper or wheel cylinder must be bled.

3   If a brake line is disconnected at a fitting located between the master cylinder and any of the brakes, that part of the system served by the disconnected line must be bled.

4   Remove any residual vacuum (or hydraulic pressure) from the brake power booster by applying the brake several times with the engine off.

5   Remove the master cylinder reservoir cap and fill the reservoir with brake fluid. Rein-

9.8 When bleeding the brakes, a hose is connected to the bleed screw at the caliper and submerged in brake fluid - air will be seen as bubbles in the tube and container (all air must be expelled before moving to the next wheel)

10.8 Slide off the pushrod retaining clip

stall the cap. **Note:** *Check the fluid level often during the bleeding operation and add fluid as necessary to prevent the fluid level from falling low enough to allow air bubbles into the master cylinder.*

6   Have an assistant on hand, as well as a supply of new brake fluid, an empty clear plastic container, a length of plastic, rubber or vinyl tubing to fit over the bleeder valve and a wrench to open and close the bleeder valve.

7   Beginning at the right rear wheel, loosen the bleeder screw slightly, then tighten it to a point where it's snug but can still be loosened quickly and easily.

8   Place one end of the tubing over the bleeder screw fitting and submerge the other end in brake fluid in the container **(see illustration)**.

9   Have the assistant slowly depress the brake pedal and hold it in the depressed position.

10   While the pedal is held depressed, open the bleeder screw just enough to allow a flow of fluid to leave the valve. Watch for air bubbles to exit the submerged end of the tube. When the fluid flow slows after a couple of

seconds, tighten the screw and have your assistant release the pedal.

11   Repeat Steps 9 and 10 until no more air is seen leaving the tube, then tighten the bleeder screw and proceed to the left rear wheel, the right front wheel and the left front wheel, in that order, and perform the same procedure. Be sure to check the fluid in the master cylinder reservoir frequently.

12   Never use old brake fluid. It contains moisture which can boil, rendering the brake system inoperative.

13   Refill the master cylinder with fluid at the end of the operation.

14   Check the operation of the brakes. The pedal should feel solid when depressed, with no sponginess. If necessary, repeat the entire process. **Warning:** *Do not operate the vehicle if you are in doubt about the effectiveness of the brake system. It is possible for air to become trapped in the anti-lock brake system hydraulic control unit, so, if the pedal continues to feel spongy after repeated bleedings or the BRAKE or ANTI-LOCK light stays on, have the vehicle towed to a dealer service department or other qualified shop to be bled with the aid of a scan tool.*

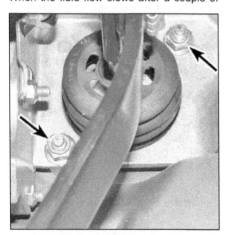

10.9 Remove the two booster retaining nuts

---

10   Power brake booster - check, removal and installation

Operating check

1   Depress the brake pedal several times with the engine off and make sure that there is no change in the pedal reserve distance.

2   Depress the pedal and start the engine. If the pedal goes down slightly, operation is normal.

Airtightness check

3   Start the engine and turn it off after one or two minutes. Depress the brake pedal several times slowly. If the pedal goes down

farther the first time but gradually rises after the second or third depression, the booster is airtight.

4   Depress the brake pedal while the engine is running, then stop the engine with the pedal depressed. If there is no change in the pedal reserve travel after holding the pedal for 30 seconds, the booster is airtight.

Removal and installation

Refer to illustrations 10.8 and 10.9

5   Remove the master cylinder (see Section 7).

6   Carefully pry the vacuum check valve from the brake booster body.

7   Pull the down the rear edge of the fascia panel above the pedals, and disengage it from the retaining clips.

8   Remove the clip and slide out the clevis pin securing the pushrod to the brake pedal **(see illustration)**.

9   Remove the two nuts securing the booster to the firewall **(see illustration)**.

10   Maneuver the unit out of position, along with its gasket (if equipped). Replace the gasket if it shows signs of damage.

11   Installation is the reverse of removal, noting the following points:

a) *Replace the brake booster mounting nuts and tighten them to the torque listed in this Chapter's Specifications.*

b) *Install the master cylinder as described in Section 7 and bleed the complete hydraulic system as described in Section 9.*

---

11   Power brake booster check valve - removal, testing and installation

---

Removal

Refer to illustration 11.1

1   Withdraw the valve from its rubber sealing grommet, using a pulling and twisting

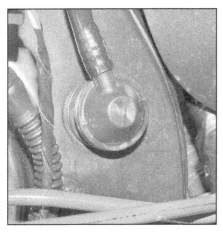

11.1 Pull and twist the brake booster check valve from the sealing grommet

12.4 Loosen the parking brake adjuster nut

4   Loosen the adjuster nut on the rod from the equalizer plate **(see illustration)**.

5   Start the engine and depress the brake pedal approximately 40 times. Stop the engine.

6   Tighten the adjuster nut just enough to eliminate any freeplay in the cables.

7   Apply the parking brake ten times. On the last application, pull the lever up and stop after the second click is emitted.

8   Tighten the adjuster nut until the rear brake pads begin to make contact with the discs.

9   Release the lever and check by hand that the rear wheels rotate freely, then check that no more than six clicks are emitted before the parking brake is fully applied.

10   Install the boot, then lower the vehicle.

motion **(see illustration)**. Remove the grommet from the brake booster.

2   The check valve is only available as an assembly with the vacuum pipe. Trace along the length of the pipe, releasing it from any retaining clips, and separate it at the connection next to the left-hand end of the cylinder head.

Testing

3   Examine the check valve for signs of damage, and replace if necessary. The valve may be tested by blowing through it in both directions. Air should flow through the valve in one direction only - when blown through from the booster unit end of the valve. Replace the valve if this is not the case.

4   Examine the rubber sealing grommet and flexible vacuum hose for signs of damage or deterioration, and replace as necessary.

Installation

5   Fit the sealing grommet into position in the brake booster.

6   Carefully ease the check valve into posi-

tion, taking great care not to displace or damage the grommet. Reconnect the vacuum hose, and install it to the retaining clips.

7   Start the engine and check for air leaks in the system.

12 Parking brake - adjustment

Refer to illustration 12.4

1   To check the parking brake adjustment, applying normal moderate pressure, pull the parking brake lever to the fully-applied position, counting the number of clicks emitted from the parking brake ratchet mechanism. If adjustment is correct, there should be 2 clicks before the brakes begins to apply, and no more than 6 before the parking brake is fully applied. If this is not the case, adjust as follows.

2   Squeeze together the sides of the parking brake lever boot, and carefully pry it from the center console.

3   Block the front wheels, then raise the rear of the vehicle and support it securely on jackstands.

13 Parking brake lever - removal and installation

Refer to illustrations 13.2 and 13.6

1   Remove the rear section of the center console (see Chapter 11).

2   Disconnect the parking brake warning switch wiring plug. Remove the four nuts and remove the center console mounting bracket **(see illustration)**. Where applicable, disconnect the wiring plug, remove the two screws and remove the DSC motion sensor.

3   Block the wheels to prevent the vehicle moving once the parking brake lever is released.

4   Referring to Section 12, release the parking brake lever and back off the adjuster nut to obtain maximum freeplay in the cable.

5   Detach both parking brake cables from the equalizer plate **(see illustration 14.5b)**.

6   Remove the three nuts and remove the lever assembly **(see illustration)**.

7   Installation is the reverse of removal. Tighten the lever retaining nuts securely, and adjust the parking brake (see Section 12).

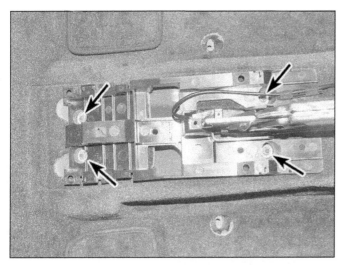

13.2 Unscrew the four nuts and remove the center console bracket

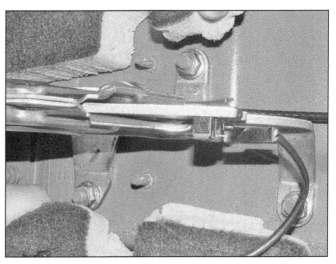

13.6 Unscrew the three nuts and remove the parking brake lever assembly

14.5a Loosen the parking brake lever adjuster nut . . .

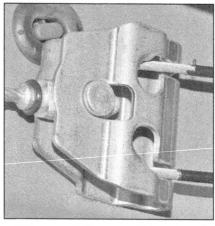

14.5b . . . and disconnect the cable end fitting from the equalizer plate

14.8 Slide a 12 mm diameter tube over the end of the cable to release the clips securing the cable to the vehicle body (a deep socket will also work)

14 Parking brake cables - removal and installation

Refer to illustrations 14.5a, 14.5b and 14.8

1   The parking brake cable consists of a left-hand section and a right-hand section connecting the rear brakes to the adjuster mechanism on the parking brake lever rod. The cables can be removed separately.
2   Firmly block the front wheels, loosen the relevant rear wheel bolts, then raise the rear of the vehicle and support it securely on jackstands. Remove the relevant rear wheel.
3   Remove the rear console (see Chapter 11).
4   Remove the four nuts and remove the console mounting bracket from the floor.
5   Loosen the parking brake adjuster nut sufficiently to be able to disengage the relevant cable end fitting from the equalizer plate **(see illustrations)**.

6   Release the cable end fitting from the lever on the brake caliper **(see illustration 4.41)**.
7   Remove the rear section of the exhaust system, then remove the heat shield from below the fuel tank.
8   Working inside the vehicle, slide a short length of 12 mm diameter tube over the end of the cable and depress the clips securing the cable outer end fitting to the vehicle body **(see illustration)**.
9   Working underneath the vehicle and noting its installed location, free the cable from the various retaining clips/brackets along its route, detach the bracket from the subframe and pull the front end of the cable from the opening in the floor. Withdraw the cable from underneath the vehicle.
10   Installation is the reverse of removal, adjusting the parking brake as described in Section 12.

15 Brake light switch - removal, installation and adjustment

Removal

Refer to illustrations 15.2a and 15.2b

1   Remove the driver's side under-dash panel (see Chapter 11).
2   Disconnect the electrical connector, then pull the switch from the pedal bracket **(see illustrations)**.

Installation and adjustment

3   Depress the brake pedal, then install the switch to the bracket, inserting it as far as possible.
4   Pull the pedal back to its stop. The switch is now correctly adjusted.
5   Reconnect the electrical connector and install the trim panel.

15.2a Disconnect the brake light switch electrical connector . . .

15.2b . . . then pull the switch from the holder

Chapter 10
Suspension and steering systems

Contents

Specifications

Note: *Throughout this Chapter you will find references to "Mk I" and "Mk II" models; this is done to simplify which specifications and procedures apply to which models. Mk I models include 2006 and earlier Cooper/Cooper S models, and 2008 and earlier Convertible models. Mk II models include 2007 and later Cooper/Cooper S/Clubman/Clubman S and 2009 and later Convertible models.*

Torque specifications

	Ft-lbs (unless otherwise indicated)	Nm
Front suspension		
Stabilizer bar clamp bolts	122	165
Stabilizer bar link nuts	41	56
Brake caliper mounting bracket bolts	See Chapter 9	
Driveaxle/hub retaining nut	See Chapter 8	
Hub and bearing assembly mounting bolts*		
Mk I models	41	56
Mk II models		
Step 1	15	20
Step 2	Tighten an additional 90-degrees	
Control arm inner balljoint		
Mk I models		
To subframe*	74	100
To control arm*	59	80
Mk II models (to subframe)		
Step 1	51	70
Step 2	Tighten an additional 90-degrees	
Control arm outer balljoint		
Mk I models		
To steering knuckle*	41	56
To control arm*	41	56
Mk II models		
To steering knuckle*		
Step 1	51	70
Step 2	Tighten an additional 90-degrees	
To control arm*		
M14 nuts	129	175
M12 nuts		
Step 1	51	70
Step 2	Tighten an additional 90-degrees	

* *Do not re-use*

Torque specifications (continued)   **Ft-lbs** (unless otherwise indicated)   **Nm**

Note: *One foot-pound (ft-lb) of torque is equivalent to 12 inch-pounds (in-lbs) of torque. Torque values below approximately 15 ft-lbs are expressed in inch-pounds, since most foot-pound torque wrenches are not accurate at these smaller values.*

Front suspension (continued)

	Ft-lbs	Nm
Control arm bracket-to-subframe bolts		
Mk I models		
Step 1	43	59
Step 2	Tighten an additional 90-degrees	
Mk II models	122	165
Suspension strut damper shaft nut*	47	64
Suspension strut-to-steering knuckle pinch-bolt		
Mk I models	60	81
Mk II models	74	100
Suspension strut upper mounting nuts*	25	34

* Do not reuse

Rear suspension

	Ft-lbs	Nm
Stabilizer bar clamp bolts	168 in-lbs	19
Stabilizer bar link-to-trailing arm*	41	56
Brake caliper bracket bolts	See Chapter 9	
Control arms-to-subframe	74	100
Control arms-to-trailing arm	74	100
Hub retaining bolts	41	56
Shock absorber lower mounting bolt		
Mk I models	103	140
Mk II models	122	165
Shock absorber damper shaft nut*	22	30
Shock absorber upper mounting bolts*	41	56
Subframe mounting bolts	74	100
Trailing arm bushing/bracke-to-trailing arm	122	165
Trailing arm bushing/bracket-to-vehicle body	74	100
Wheel speed sensor bolt	See Chapter 9	

Steering

	Ft-lbs	Nm
Power steering line union-to-pump bolt	84 in-lbs	10
Power steering pump bracket-to-subframe	168 in-lbs	19
Power steering pump cooling fan nuts*	168 in-lbs	19
Steering column mounting bolts*		
Mk I models	16	22
Mk II models	26	35
Steering column lower universal joint nut*		
Mk I models	16	22
Mk II models	21	28
Steering column upper universal joint pinch-bolt*	21	28
Steering wheel bolt		
Mk I models	33	45
Mk II models	46	63
Tie-rod end balljoint to steering knuckle*		
Mk I models	38	52
Mk II models	48	65
Tie-rod end pinch bolt (Mk II models)	21	28
Wheel bolts	See Chapter 1	

* Do not re-use

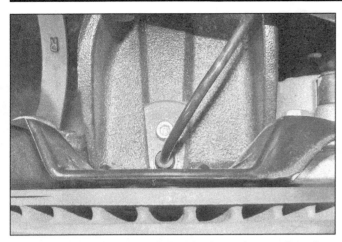

2.5 Unscrew the bolt and pull the wheel speed sensor from the steering knuckle

2.7 Note that the strut-to-steering knuckle pinch-bolts are inserted from the rear

1 General information

Note: *Throughout this Chapter you will find references to "Mk I" and "Mk II" models; this is done to simplify which specifications and procedures apply to which models. Mk I models include 2006 and earlier Cooper/Cooper S models, and 2008 and earlier Convertible models. Mk II models include 2007 and later Cooper/Cooper S/Clubman/Clubman S and 2009 and later Convertible models.*

The independent front suspension is of MacPherson strut type, incorporating coil springs and integral telescopic shock absorbers. The struts are attached to steering knuckles at their lower ends, and the knuckles are in turn attached to the lower suspension arm by balljoints. A stabilizer bar is bolted to the rear of the subframe, and is connected to the front suspension struts by links.

The multi-link rear suspension is fully independent with trailing arms attached to the vehicle body, and control arms mounted between the body-mounted subframe and the trailing arms. The gas-pressurized shock absorber/coil spring assemblies are mounted between the vehicle body and the trailing arms. A rear stabilizer bar is installed to all models.

Both front and rear wheel bearings are integral with the hubs, and no adjustment is possible.

A power-assisted rack-and-pinion steering gear is installed, together with a conventional column and telescopic coupling, incorporating two universal joints. On Mk I models, the fluid pressure for the rack is provided by an electrically operated pump mounted behind the engine on the subframe. This is an energy efficient design, as power is only provided to the pump when steering assistance is required. On some models, an electrically-operated fan is located immediately below the pump to provide a cooling airstream. On Mk II models, power assist is provided by an electric motor on the steering gear.

When working on the suspension or steering, you may come across nuts or bolts which seem impossible to loosen. These nuts and bolts on the underside of the vehicle are continually subjected to water, road grime, mud, etc, and can become rusted or seized, making them extremely difficult to remove. In order to unscrew these stubborn nuts and bolts without damaging them (or other components), use lots of penetrating oil, and allow it to soak in for a while. Using a wire brush to clean exposed threads will also ease removal of the nut or bolt, and will help to prevent damage to the threads. Sometimes a sharp blow with a hammer and punch will break the bond between a nut and bolt, but care must be taken to prevent the punch from slipping off and ruining the threads. Using a longer bar or wrench will increase leverage, but never use an extension bar/pipe on a ratchet, as the internal mechanism could be damaged. Actually *tightening* the nut or bolt slightly first may help to break it loose. Nuts or bolts which have required drastic measures to remove them should always be replaced. As a general rule, all self-locking nuts (with nylon inserts) should be replaced.

Since most of the procedures dealt with in this Chapter involve jacking up the vehicle and working underneath it, a good pair of jackstands will be needed. A hydraulic floor jack is the preferred type of jack to lift the vehicle, and it can also be used to support certain components during removal and installation operations.

Warning: *Never, under any circumstances, rely on a jack to support the vehicle while working beneath it. It is not recommended, when jacking up the rear of the vehicle, to lift beneath the rear crossmember.*

2 Steering knuckle - removal and installation

Removal

Refer to illustrations 2.5 and 2.7

1 Loosen the front wheel bolts, then jack up the front of the vehicle and support it securely on jackstands. Remove the wheel.

2 Have an assistant press the brake pedal, then loosen and remove the driveaxle nut (see Chapter 8). Discard the nut; a new one must be installed.

3 Remove the front brake disc as described in Chapter 9. If working on the left side of a model with a ride-height sensor, unbolt the sensor bracket from the control arm.

4 Unscrew the tie-rod end balljoint nut, and detach the rod from the steering knuckle using a conventional balljoint removal tool **(see illustration 23.4)**. Take care not to damage the balljoint seal.

5 Unscrew the retaining bolt, and remove the wheel speed sensor **(see illustration)**.

6 Unscrew the outer balljoint nut, then separate the balljoint from the control arm (Mk I models) or from the steering knuckle (Mk II models) using a conventional balljoint separator tool **(see illustration 8.2a)**.

7 Unscrew and remove the pinch-bolt securing the steering knuckle assembly to the front suspension strut, noting which way it is installed **(see illustration)**. Pry open the clamp a little using a wedge-shaped tool (a chisel works well), and release the steering knuckle from the strut. If necessary, tap the steering knuckle downwards with a soft-faced mallet to separate the two components.

Installation

8 Locate the assembly on the front suspension strut. Insert the pinch-bolt with its head facing the same way as removal, and tighten it to the specified torque.

9 Pull the steering knuckle assembly outwards, and insert the driveaxle to engage the splines in the hub.

10 Install the control arm balljoint into the control arm (Mk I models) or into the steering knuckle (Mk II models), then install the new nut and tighten it to the specified torque.

11 Install the brake caliper and brake disc as described in Chapter 9. Install the brake hose support bracket to the strut.

12 Install the wheel speed sensor and tighten the retaining bolt to the torque listed in the Chapter 9 Specifications.

13 Reconnect the tie-rod end balljoint to the steering knuckle, and tighten the new nut to the specified torque.

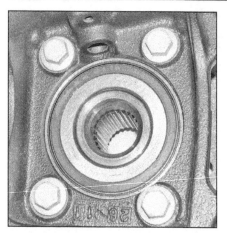

3.6 Front hub retaining bolts - shown with the driveaxle removed for clarity

14   Install the new driveaxle/hub nut and tighten it to the torque listed in Chapter 8. After tightening, stake the nut as described in Chapter 8.
15   Install the front wheel, then lower the vehicle and tighten the wheel bolts to the torque listed in Chapter 1 Specifications.

3   Hub and bearing assembly (front) - inspection and replacement

Note: *The front wheel bearings are integral with the hub assembly, and can only be replaced as a unit.*

Inspection
1   To check the bearings for excessive wear, apply the parking brake, jack up the front of the vehicle and support it securely on jackstands.
2   Grip the front wheel at the top and the bottom, and attempt to rock it. If excessive movement is noted, it may be that the hub bearings are worn. Do not confuse wear in the control arm balljoint with wear in the bearings. Hub bearing wear will show up as roughness

4.1 Pull the hose and grommet from the bracket

or vibration when the wheel is spun; it will also be noticeable as a rumbling or growling noise when driving, especially during turns. No adjustment of the wheel bearings is possible.

Replacement
Refer to illustration 3.6
3   Loosen the wheel bolts, then raise the front of the vehicle and support it securely on jackstands. Remove the driveaxle/hub nut (see Chapter 8).
4   Remove the brake disc (see Chapter 9).
5   Unscrew the retaining bolt and remove the wheel speed sensor from the steering knuckle **(see illustration 2.5)**.
6   Unscrew the four bolts securing the hub and bearing assembly to the steering knuckle, and detach the assembly **(see illustration)**. If the driveaxle sticks in the splines, a puller can be used to push it out as the bearing assembly is removed.
7   Position the new hub and bearing assembly against the steering knuckle, then install and tighten the bolts to the torque listed in this Chapter's Specifications.
8   Install the speed sensor to the steering knuckle and tighten the retaining bolt to the

torque listed in the Chapter 9 Specifications.
9   Install the brake disc and caliper as described in Chapter 9.
10   Tighten the new driveaxle/hub nut to the torque listed in the Chapter 8 Specifications, and stake it in place as described in Chapter 8.
11   Install the wheel, lower the vehicle and tighten the wheel bolts to the torque listed in the Chapter 1 Specifications.

4   Front suspension strut assembly - removal and installation

Removal
Refer to illustrations 4.1, 4.2a, 4.2b and 4.6
1   Loosen the wheel bolts, raise the front of the vehicle and support it securely on jackstands. Remove the wheel. Pull the brake hose from the bracket on the strut **(see illustration)**.
2   Remove the nut from the stabilizer bar link and disconnect it from the strut. Use an Allen key to counterhold the link ballstud or, on early models, use a second wrench **(see illustrations)**.
3   Release the wheel speed sensor and (where installed) the brake pad wear sensor wiring from the bracket on the strut.
4   Mk I models only: Unscrew the balljoint nut from the base of the steering knuckle, and use a separator tool to detach the control arm from the balljoint **(see illustration 8.2a)**.
5   Unscrew and remove the pinch-bolt securing the steering knuckle to the front suspension strut, noting which way it is installed **(see illustration 2.7)**. Lever the steering knuckle assembly down and release it from the strut. If necessary, tap the carrier downwards with a soft-faced mallet to separate the two components. **Note:** *Support the steering knuckle assembly when released from the strut, to prevent any damage to the driveaxle/wiring.*
6   Support the strut/spring assembly under the wheelwell, then remove the upper mounting nuts **(see illustration)**. Discard the nuts, new ones must be installed.

4.2a On early models, use a second wrench to counterhold the stabilizer bar link ballstud . . .

4.2b . . . on later models, use an Allen key

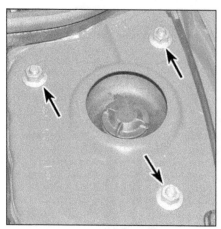

4.6 Strut upper mounting nuts

5.3 Compress the spring until the upper seat is not under tension

5.4a Pry off the plastic cap

7 Lower the suspension strut from under the wheelwell, withdrawing it from the vehicle.

Installation

8 Installation is a reversal of removal. Make sure that all fasteners are tightened to their specified torque.

5   Front suspension strut - overhaul

Refer to illustrations 5.3, 5.4a, 5.4b, 5.5, 5.6a, 5.6b, 5.6c, 5.6d, 5.6e and 5.10

Warning: *Before attempting to disassemble the front suspension strut, a tool to hold the coil spring in compression must be obtained. Do not attempt to use makeshift methods. Uncontrolled release of the spring could cause damage and personal injury. Use a high-quality spring compressor, and carefully follow the tool manufacturer's instructions provided with it. After removing the coil spring with the compressor still installed, place it in a safe, isolated area.*

5.4b We used a spark plug socket with a hexagon section on the outside and an Allen key to counterhold the damper shaft

1   If the front suspension struts exhibit signs of wear (leaking fluid, loss of damping capability, sagging or cracked coil springs), they should be disassembled and overhauled as necessary. The struts themselves cannot be serviced, and should be replaced if faulty; the springs and related components can be replaced individually. To maintain balanced characteristics on both sides of the vehicle, the components on both sides should be replaced at the same time.

2   With the strut removed from the vehicle (see Section 4), clean away all external dirt.

3   Install the coil spring compressor tools (ensuring that they are fully engaged), and compress the spring until all tension is relieved from the upper mounting **(see illustration)**.

4   Pry off the plastic cap, then hold the strut piston rod with an Allen key, and unscrew the damper shaft nut with a box-end wrench or spark plug socket with a hexagon surface on the outside **(see illustrations)**. Discard the nut, a new one must be installed.

5   Models up to 03/2002: Remove the conical washer, top mounting plate, spring seat and spring, followed by the washer bump stop and boot **(see illustration)**.

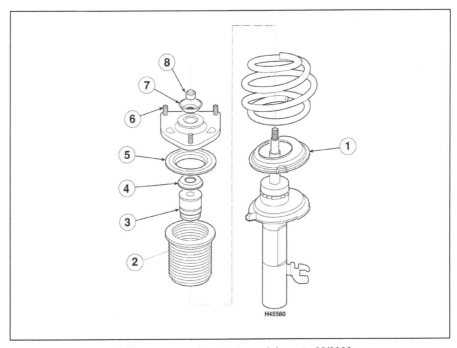

H45580

5.5 Front suspension strut - models up to 03/2002

1 *Lower spring seat*	4 *Washer*	7 *Conical washer*
2 *Boot*	5 *Upper spring seat*	8 *Damper shaft nut*
3 *Bump stop*	6 *Top mounting plate*	

5.6a Remove the top mounting plate . . .

5.6b . . . followed by the conical washer . . .

5.6c . . . flat washer . . .

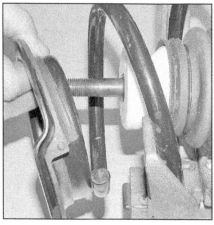

5.6d . . . upper seat and spring . . .

wear and damage, and check the bearing for smoothness of operation. Replace components as necessary.

9   Examine the strut for signs of fluid leakage. Check the damper shaft for signs of pitting along its entire length, and check the strut body for signs of damage. Test the operation of the strut, while holding it in an upright position, by attempting to move the piston. It should only be possible to move the piston a very small amount. If it's easy to move or shows little/uneven resistance, or if there is any visible sign of wear or damage to the strut, replacement is necessary.

10   Reassembly is a reversal of disassembly, noting the following points:

a) Make sure that the coil spring ends are correctly located in the upper and lower seats before releasing the compressor (see illustration).
b) Check that the bearing is correctly installed to the piston rod seat.
c) Tighten the new damper shaft nut to the specified torque.
d) The smaller diameter coil must be installed against the lower spring seat.

6   Models from 03/2002: Remove the top mount/thrust bearing, conical washer, flat washer, upper spring seat and spring, followed by the boot, bump stop and lower spring seat (see illustrations).

7   If a new spring is to be installed, the original spring must now be carefully released from the compressor. If it is to be re-used, the spring can be left in compression.

8   With the strut now completely disassembled, examine all the components for

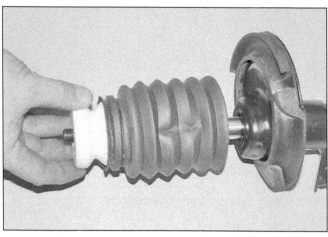

5.6e . . . and the bump stop/boot

5.10 Ensure the seat rubber fits correctly against the strut seat and the end of the spring

6.2 Unscrew the nut securing the lower end of the links to the stabilizer bar

6.4 Stabilizer bar clamp bolts - left-hand side

tighten the nuts to the torque listed in this Chapter's Specifications.

Stabilizer bar links

10   Reconnect the links to the stabilizer bar and suspension strut, then tighten the nuts to the torque listed in this Chapter's Specifications.

11   Install the wheel and lower the vehicle.

---

7   Front suspension control arm and inner balljoint - removal, overhaul and installation

---

Removal

Control arm

Refer to illustrations 7.3a, 7.3b, 7.4, 7.5, 7.6a, 7.6b and 7.6c

1   Follow the procedure for removing the front subframe as described in Chapter 2A, but only lower the subframe approximately three inches (75 mm) on Mk I models, or five inches (120 mm) on Mk II models. If you're removing a left control arm on a model with a height sensor, mark the position of the sensor link bracket to the control arm, then unbolt it.

2   On Mk I models, unbolt the outer balljoint from the steering knuckle; on Mk II models, remove the fasteners and detach the balljoint from the control arm (see Section 8).

3   On Mk I models, unscrew the bolts securing the control arm balljoint to the subframe **(see illustration)**. **Note:** *The bolts must not be re-used.* On Mk II models, remove the balljoint stud nut **(see illustration)**, attach MINI tool no. 31-1-040 to the ballstud and hit it with a hammer to free the ballstud from the subframe (an alternative tool can be fabricated).

4   Unscrew the bolts securing the control arm rear mounting bracket to the subframe,

6   Front stabilizer bar and links - removal and installation

Removal

1   Loosen the front wheel bolts, then raise the front of the vehicle and support it securely on jackstands. Remove the wheel.

Stabilizer bar

Refer to illustrations 6.2 and 6.4

2   Unscrew the nut securing the lower end of the links to the stabilizer bar. Use an Allen key to counterhold the ballstud **(see illustration)**.

3   Refer to Chapter 2A and lower the front subframe approximately five inches (120 mm).

4   Unscrew and remove the four bolts securing the stabilizer bar clamps to the subframe, then maneuver the stabilizer bar from position **(see illustration)**.

Stabilizer bar links

5   Unscrew the nuts securing the links to the stabilizer bar and suspension strut. Use an Allen key to counterhold the link's ballstuds. On early models, use a second wrench to counterhold the upper ballstud **(see illustrations 4.2a and 4.2b)**.

Installation

Stabilizer bar

6   When installing the stabilizer bar bushings make sure they are located correctly on the stabilizer bar.

7   Maneuver the stabilizer bar into position, and install the clamps and bolts. Tighten the clamp bolts to the torque listed in this Chapter's Specifications.

8   Install the front subframe as described in Chapter 2A.

9   Reconnect the stabilizer bar links and

7.3a Unscrew the two bolts securing the balljoint to the subframe - shown with the driveaxle removed for clarity (Mk I models)

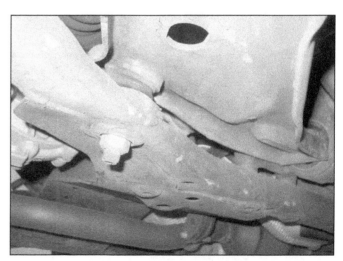

7.3b Balljoint stud-to-subframe nut (Mk II models)

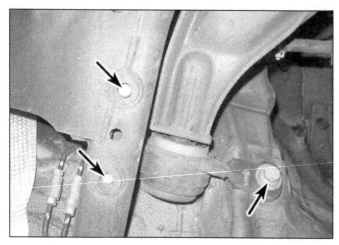

7.4 Control arm rear mounting bracket bolts - Mk II models

7.5 Use a balljoint separator to detach the balljoint from the control arm

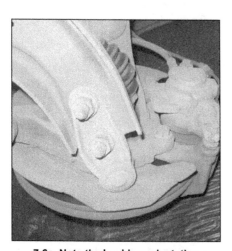

7.6a Note the bushing orientation in the arm . . .

7.6b . . . and its installed depth . . .

7.6c . . . prior to pressing it from the arm

and maneuver the arm from position **(see illustration)**.

5   On Mk I models, if required, unscrew the self-locking nut and use a balljoint separator to detach the balljoint from the control arm **(see illustration)**.

7.8 Use a balljoint separator to detach the balljoint from the control arm (Mk I models)

6   The control arm rear bushing and mounting assembly must be pressed from the arm, then the bushing must be pressed out from, and back into, the rear mounting bracket. If a press is available, note the installed depth and position of the bushing prior to removing the arm/bushing to enable installation **(see illustrations)**. If a press is not available, entrust this task to a MINI dealer or suitably-equipped specialist. **Note:** *MINI dealers have special tools which press the control arm from the bushing, and replace the bushing in the mounting bracket without disturbing the subframe.*

Inner balljoint (Mk I models)

Refer to illustration 7.8

Note: *At the time of writing, the inner balljoint was not replaceable separately.*

7   Unscrew the balljoint nut, and the bolts securing the balljoint to the front subframe **(see illustration 7.3a)**. Discard the bolts; new ones must be installed.

8   Use a separator tool to detach the balljoint from the suspension arm **(see illustration)**.

Installation

9   Installation is a reversal of removal, noting the following points:

a) *Press the bushing into the rear of the control arm, to the same depth as the original.*

b) *Replace all self-locking nuts removed.*

c) *Replace the inner balljoint-to-subframe bolts.*

d) *Tighten all fasteners to their specified torque, where given.*

e) *Have the front wheel alignment checked and, if necessary, adjusted.*

---

8   Front suspension control arm outer balljoint - replacement

---

Refer to illustrations 8.2a, 8.2b, 8.2c, 8.3a and 8.3b

1   Loosen the front wheel bolts, then jack up the front of the vehicle and support it securely on jackstands. Remove the wheel.

2   Unscrew the outer balljoint nut, then use a separator tool to detach the balljoint from the

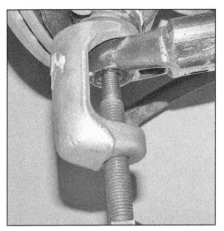

8.2a Use a balljoint separator to detach the control arm from the steering knuckle balljoint (Mk I models)

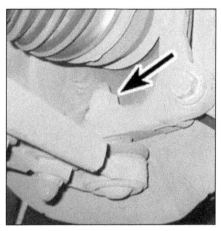

8.2b Balljoint-to-steering knuckle nut (Mk II models)

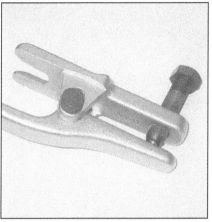

8.2c This type of balljoint separator tool, available at most auto parts stores, will be required to detach the balljoint from the steering knuckle on Mk II models

8.3a Balljoint-to-steering knuckle bolts (Mk I models)

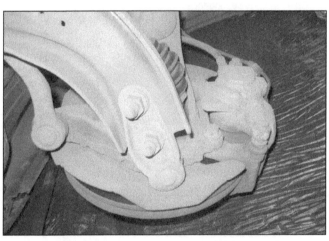

8.3b Balljoint-to-control arm nuts (Mk II models)

control arm (Mk I models, **see illustration**) or from the steering knuckle (Mk II models, **see illustrations**).

3   On Mk I models, unscrew the two bolts and remove the balljoint from the steering knuckle **(see illustration)**. On Mk II models, remove the fasteners and detach the balljoint from the control arm **(see illustration)**. Discard the fasteners - new ones must be installed.

4   Installation is the reverse of removal. Be sure to use new fasteners, tightening them to the torque listed in this Chapter's Specifications.

5   Install the wheel and lower the vehicle. Tighten the wheel bolts to the torque listed in the Chapter 1 Specifications.

9   Hub and bearing assembly (rear) - inspection and replacement

Note: *The rear wheel bearings are integral with the hub assembly, and can only be replaced as a unit.*

Inspection

1   The rear hub bearings are non-adjustable.

2   To check the bearings for excessive wear, chock the front wheels, then raise the rear of the vehicle and support it securely on jackstands. Fully release the parking brake.

3   Grip the rear wheel at the top and bottom, and attempt to rock it. If excessive movement is noted, or if there is any roughness or vibration felt when the wheel is spun, it is indicative that the hub bearings are worn.

Replacement

Refer to illustrations 9.5, 9.6a and 9.6b

4   Chock the front wheels, loosen the wheel bolts, then raise the rear of the vehicle and support it securely on jackstands. Remove the rear brake disc as described in Chapter 9.

5   Unscrew the ABS sensor retaining bolt, and pull the sensor from place **(see illustration)**.

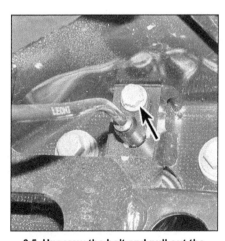

9.5 Unscrew the bolt and pull out the wheel speed sensor

6   Unscrew the four mounting bolts and remove the hub and bearing assembly from

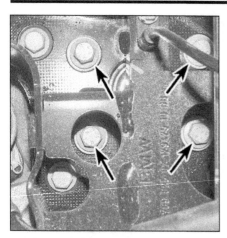

9.6a The hub assembly is retained by four bolts - Mk I model

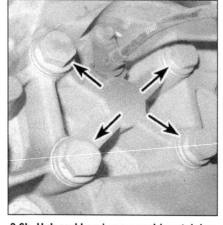

9.6b Hub and bearing assembly retaining bolts - Mk II model

10.2 Rear shock absorber lower mounting bolt

the trailing arm **(see illustrations)**.
7   Clean the hub seating face on the trailing arm.
8   Position the new hub/bearing assembly against the trailing arm, then install and tighten the bolts to the torque listed in this Chapter's Specifications.
9   Install the wheel speed sensor, and tighten the bolt to the torque listed in the Chapter 9 Specifications.
10   Install the rear brake disc as described in Chapter 9.
11   Install the wheel and lower the vehicle. Tighten the wheel bolts to the torque listed in the Chapter 1 Specifications.

---

10   Rear shock absorber - removal and installation

Removal

Refer to illustrations 10.2 and 10.3

1   Chock the front wheels, loosen the wheel bolts, then raise the rear of the vehicle and support it securely on jackstands. Remove the wheels.

2   Pry off the rubber boot (if equipped), then unscrew and remove the shock absorber lower mounting bolt **(see illustration)**.
3   Unscrew the two upper mounting bolts, and maneuver the shock absorber from place **(see illustration)**.

Installation

Note: *The final tightening of the lower mounting bolt must be carried out with the vehicle weight on the wheels, or the rear suspension raised with a floor jack to simulate normal ride height.*
4   Installation is a reversal of the removal procedure, tightening the mounting bolts to their specified torque values. **Note:** *The trailing arm will have to be pushed down in order to install the shock absorber lower mounting bolt.*

---

11   Rear shock absorber - overhaul

Refer to illustrations 11.3, 11.4, 11.5a, 11.5b, 11.9a, 11.9b and 11.9c

Warning: *Before attempting to disassemble the rear shock absorber, a tool to hold the coil spring in compression must be obtained. Do not attempt to use makeshift methods.*

Uncontrolled release of the spring could cause damage and personal injury or death. Use a high-quality spring compressor and carefully follow the tool manufacturer's instructions provided with it. After removing the coil spring with the compressor still installed, place it in a safe, isolated area.
1   If the shock absorbers exhibit signs of wear (leaking fluid, loss of damping capability, sagging or cracked coil springs) then they should be disassembled and overhauled as necessary. The shocks themselves cannot be serviced, and should be replaced if faulty; the springs and related components can be replaced individually. To maintain balanced characteristics on both sides of the vehicle, the components on both sides should be replaced at the same time.
2   With the shock absorber removed from the vehicle (see Section 10), clean away all external dirt, then mount it in a vise.
3   Install the coil spring compressor tools (ensuring that they are fully engaged), and compress the spring until all tension is relieved from the upper mount **(see illustration)**.
4   Hold the damper shaft with an Allen key, and unscrew the nut with a box-end wrench **(see illustration)**.

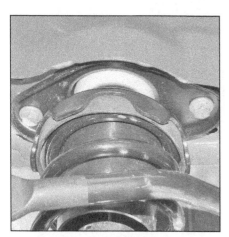

10.3 Rear shock absorber upper mounting bolts

11.3 Compress the spring until there is no pressure on the upper mounting

11.4 Use an Allen key to counterhold the damper shaft while unscrewing the nut

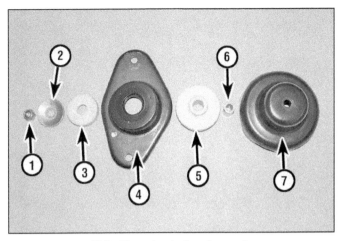

11.5a Rear shock absorber parts

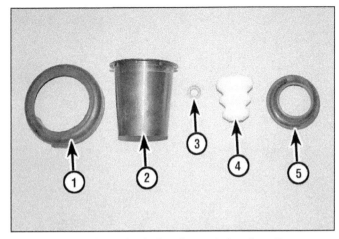

11.5b Rear shock absorber parts (continued)

1 Nut
2 Concave washer
3 Foam washer
4 Upper mounting plate
5 Foam bushing
6 Metal bushing
7 Upper spring seat

1 Spring rubber seat
2 Plastic shroud
3 Washer
4 Bump stop
5 Lower spring rubber seat

5   Withdraw the concave washer, foam washer, upper mounting plate, foam bushing, metal bushing, upper spring seat, spring rubber seat, plastic shroud, washer, bump stop, spring and lower spring rubber seat **(see illustrations)**.
6   If a new spring is to be installed, the original spring must now be carefully released from the compressor. If it is to be re-used, the spring can be left in compression.
7   With the shock now completely disassembled, examine all the components for wear and damage, and check the bearing for smoothness of operation. Replace components as necessary.
8   Examine the shock absorber for signs of fluid leakage. Check the damper shaft for signs of pitting along its entire length, and check the shock body for signs of damage. Test the operation of the shock, while holding it in an upright position, by moving the piston

through a full stroke, and then through short strokes of 2 to 4 inches (50 to 100 mm). In both cases, the resistance felt should be smooth and continuous. If the resistance is jerky, uneven, or if there is any visible sign of wear or damage to the shock, replacement is necessary.
9   Reassembly is a reversal of disassembly, noting the following points:
a) Make sure that the coil spring narrow end is correctly located in the lower seat, and the upper spring correctly locates in the upper spring seat **(see illustrations)**.
b) Reinstall the washer above the bump stop with the chamfered inner edge downwards **(see illustration)**.
c) Check that the bearing is correctly installed to the piston rod seat.
d) Tighten the damper shaft nut to the torque listed in this Chapter's Specifications.

12   Rear stabilizer bar and links - removal and installation

Removal

Refer to illustrations 12.6a, 12.6b, 12.7, 12.9a, 12.9b, 12.9c and 12.9d

1   Loosen the rear wheel bolts, then raise the rear of the vehicle, and support it securely on jackstands. Remove both rear wheels.
2   Remove the rear section of the exhaust system (see Chapter 4).
3   Unscrew the nuts and remove the heat shield above the rear muffler(s).
4   Detach the wheel speed sensor wiring harness from the retaining brackets on the subframe.
5   Release the metal brake pipes from the clips on the vehicle body upstream of the pipe connection on the underside of the subframe.

11.9a Ensure the end of the spring locates correctly with the spring seat

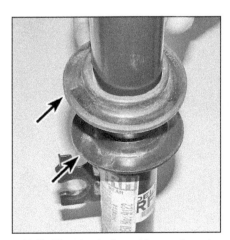

11.9b The step in the rubber seat must align with the step in the metal seat

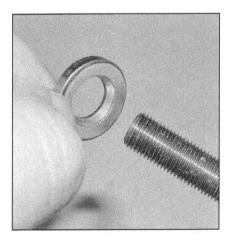

11.9c The washer above the bump stop must be installed with the chamfered inner edge downwards

12.6a Unscrew the nut securing the anti-roll link to the trailing arm - use an Allen key to counterhold the ballstud . . .

12.6b . . . then unscrew the nut securing the link to the stabilizer bar

12.7 Unscrew the two bolts on each side securing the stabilizer bar to the rear subframe

12.9a Subframe right-hand rear mounting bolt . . .

12.9b . . . right-hand front mounting bolt . . .

6   Unscrew the nuts securing the anti-roll links to the stabilizer bar and trailing arm. Use an Allen key to counterhold the ballstud as the nut is loosened **(see illustrations)**.
7   Unscrew the bolts securing the stabilizer

bar mounting clamps to the rear subframe **(see illustration)**; release the clamps (one each side).
8   Remove the fuel tank central mounting bolt.

9   Position a floor jack under the rear subframe, and remove the four subframe mounting bolts **(see illustrations)**.
10   Lower the subframe a little, and maneuver the stabilizer bar from position.

12.9c . . . left-hand rear mounting bolt . . .

12.9d . . . and left-hand front mounting bolt

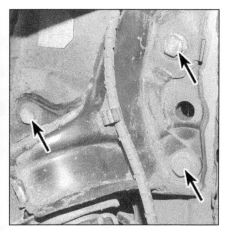

13.7a Trailing arm bolts - early Mk I models . . .

13.7b . . . and late Mk I models

13.7c Trailing arm bracket bolts - Mk II models (one bolt not visible)

11   Examine the rubber bushings for the mounting clamps and links, and replace them if necessary. The bushings and links are available individually.

Installation

12   When installing the bushings to the stabilizer bar, make sure they are located correctly with **no** lubricant (except water if required).
13   Locate the stabilizer bar on the rear subframe, then install the clamps and tighten the bolts to the torque listed in this Chapter's Specifications.
14   Lift the subframe and tighten the mounting bolts to the torque listed in the Chapter 2A Specifications.
15   The remainder of installation is a reversal of removal.

13 Rear suspension trailing arm - removal, overhaul and installation

Removal

Refer to illustrations 13.7a, 13.7b, 13.7c and 13.9

1   Remove the rear brake disc as described in Chapter 9.
2   Unscrew and remove the nut securing the stabilizer bar link to the trailing arm **(see illustration 12.6a)**. Use an Allen key to counterhold the ballstud while unscrewing the nut.
3   Unscrew the bolt and remove the wheel speed sensor from the trailing arm (see Chapter 9).
4   Mark the position of the lower control arm mounting bolt eccentric washer in relation to the trailing arm, in order to preserve the rear wheel alignment upon reassembly, then loosen and remove the bolts securing the control arms to the trailing arm **(see illustrations 14.3 and 14.4)**.
5   Loosen and remove the lower shock absorber mounting bolt.
6   Detach the wheel speed sensor and brake pad wear sensor cables from the retaining clips on the trailing arm.
7   Make alignment marks between the

trailing arm bushing mounting plate and the vehicle body to aid installment, then unscrew the bolts and maneuver the trailing arm from the vehicle **(see illustrations)**. To improve access, remove the screws, pry out the plastic rivets, and remove the plastic trim panel directly under the trailing arm mounting.
8   If required, unscrew the four bolts and detach the hub/bearing assembly from the trailing arm.

Overhaul

9   To replace the metal/rubber bushing on the trailing arm, unscrew the retaining bolt and washer **(see illustration)**.
10   At the time of writing, the bushing was not available separately from the mounting bracket. If the bushing is worn or damaged, the bracket/bushing assembly must be replaced as a compete unit.
11   If a new trailing arm is installed to models manufactured after 05/2003, a new 14 mm thread will need to be tapped where the bushing retaining bolt fits, prior to installing the bushing/mounting.
12   Position the new bracket/bushing against the trailing arm, install the bolt, but only finger-tight at this stage. The final tightening of the bolt should be carried out once the wheels are back on the ground and the vehicle normally loaded.

Installation

13   Installation is a reversal of the removal procedure, noting the following points:
a) *Delay fully tightening the control arm, and trailing arm-mounting bolts, until the vehicle wheels are on the ground, and normally loaded. Alternatively, raise the rear suspension with a floor jack to simulate normal ride height, then tighten the fasteners. Normally loaded is defined as 150 lbs (68 kg) on each front seat, 30 lbs (14 kg) in the center of the luggage compartment, and a full tank of fuel.*
b) *Align the previously-made marks prior to tightening the trailing arm bushing mounting bracket bolts*

13.9 Unscrew the bolt and slide the mounting bushing assembly from the trailing arm

c) *The wheel alignment will also require checking (see Section 24).*

14 Rear suspension control arms - removal and installation

Note: *The bushings in the suspension arms cannot be replaced separately; replace the complete arm if there is any wear or damage.*

Removal

Refer to illustrations 14.3 and 14.4

1   Loosen the relevant rear wheel bolts, then raise the rear of the vehicle and support it securely on jackstands. Remove the relevant wheel. On Mk II models, remove the rear under-vehicle splash shield.
2   On Mk I models, if removing the right-hand side control arms, remove the rear section of the exhaust system if necessary for access (see Chapter 4). On Mk II models, lower the exhaust system and remove the heat shields. On convertible models, remove the rear X-brace. On

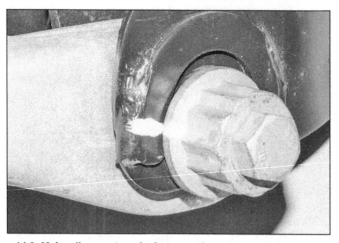

14.3 Make alignment marks between the outer control arm bolt eccentric washer and the trailing arm

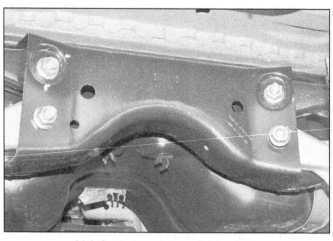

14.4 Control arm inner retaining bolts

models with a ride height sensor, unbolt the sensor bracket.

3   Prior to removing the lower control arm outer retaining bolt, make alignment marks between the bolt eccentric washer and the trailing arm, to preserve the rear wheel alignment upon installation **(see illustration)**.

4   Loosen and remove the control arm retaining bolts and maneuver the arm from under the vehicle **(see illustration)**.

Installation

5   Installation is a reversal of the removal procedure, aligning the previously-made marks (lower control arm outer bolt eccentric washer only). Delay fully tightening the control arm mounting bolts until the vehicle is back on its wheels and normally-loaded, or until the rear suspension is raised with a floor jack to simulate normal ride height.

15 Rear suspension subframe - removal and installation

Removal

1   Loosen the rear wheel bolts, then raise the rear of the vehicle and support it securely on jackstands. Remove both rear wheels. On Mk II models, remove the rear under-vehicle splash shield.

2   Remove the rear section of the exhaust system (see Chapter 4) and the heat shields.

3   Remove both parking brake cables as described in Chapter 9.

4   Remove the upper and lower control arms as described in Section 14.

5   Disengage the wheel speed sensor wiring harnesses from the retaining brackets on the subframe.

6   Loosen the union nut, and disconnect the brake hose from each side under the subframe. Be prepared for fluid spillage. Plug the ends of the pipes to prevent fluid loss and dirt/water from entering.

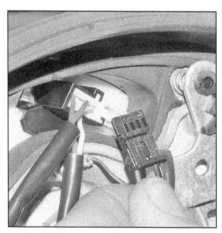

16.3 Disconnect the steering wheel switch electrical connector(s)

7   Unscrew the bolts securing the stabilizer bar clamps to the subframe **(see illustration 12.7)**.

8   Place a floor jack and length of wood under the fuel tank to support it, then unscrew the two bolts securing the fuel tank retaining straps to the subframe.

9   Unscrew the four subframe mounting bolts, lower the subframe and maneuver it from under the vehicle **(see illustrations 12.9a through 12.9d)**.

Installation

10   Installation is a reversal of the removal procedure, noting the following points:

a) Tighten all bolts to the specified torque.
b) Have the rear wheel alignment checked and, if necessary, adjusted.

16 Steering wheel - removal and installation

Warning: All models are equipped with an airbag system. Make sure that the safety

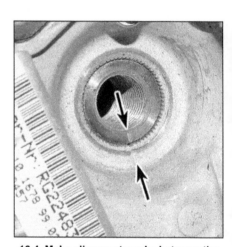

16.4 Make alignment marks between the steering wheel and column shaft

recommendations given in Chapter 12 are followed, to prevent personal injury.

Removal

Refer to illustrations 16.3 and 16.4

1   Disconnect the cable from the negative terminal of the battery (see Chapter 5). **Warning:** Before proceeding, wait a minimum of 5 minutes, as a precaution against accidental deployment of the airbag unit. This period ensures that any stored energy in the back-up capacitor is dissipated.

2   Adjust the steering wheel all the way to the rear. Remove the airbag module (see Chapter 12). **Warning:** Position the airbag module in a safe place, with the mechanism facing downwards (and the trim side facing up) as a precaution against accidental deployment.

3   Disconnect the steering wheel switch electrical connectors from the airbag system clockspring **(see illustration)**.

4   Unscrew the retaining bolt from the center of the steering wheel. If there are no marks, mark the position of the steering wheel on the steering column shaft for installation **(see illustration)**.

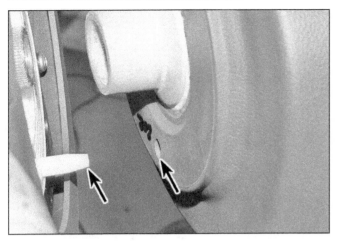

16.6 The pin on the clockspring must align with the hole in the rear of the steering wheel

16.7 Install the steering wheel bolt and tighten it to the specified torque

5   Remove the steering wheel from the top of the column.

Installation

Refer to illustrations 16.6 and 16.7

6   Make sure that the front wheels are still facing straight-ahead, then locate the steering wheel on the top of the steering column. Align the marks made on removal. Note the pin on the clockspring unit which must locate in the rear of the steering wheel **(see illustration)**.

7   Install the retaining bolt **(see illustration)**, and tighten it to the torque listed in this Chapter's Specifications while holding the steering wheel. Do not tighten the bolt with the steering lock engaged, as this may damage the lock.

8   Reconnect the electrical connector(s) for the horn and other steering wheel switches (where applicable).

9   Install the airbag as described in Chapter 12.

10   Reconnect the cable to the negative terminal of the battery (see Chapter 5).

17 Steering column - removal, inspection and installation

Warning: *All models are equipped with an airbag system. Make sure that the safety recommendations given in Chapter 12 are followed, to prevent personal injury.*

Removal

Refer to illustrations 17.3a, 17.3b, 17.5, 17.6a, 17.6b and 17.8

1   Remove the steering wheel as described in Section 16. On Mk II models, remove the tachometer (see Chapter 12).

2   Remove the upper and lower steering column shrouds as described in Chapter 11. On Mk II models, remove the instrument cluster (see Chapter 12).

3   Remove the column switch/clockspring unit from the top of the column. Disconnect the wiring plugs as the assembly is withdrawn **(see illustrations)**.

4   On Mk I models, slide out the locking

17.3a Clockspring unit/column switch retaining screws (Mk I models)

catch and disconnect the ignition switch electrical connector.

5   Remove the pinch-bolt securing the column universal joint to the steering rack pinion **(see illustration)**. Discard the nut, a new one must be installed.

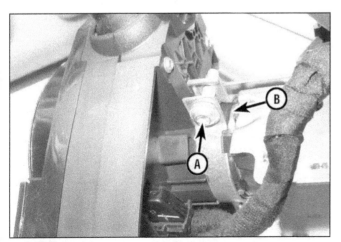

17.3b On Mk II models, loosen the clamp screw (A), carefully pry up the tab (B) on each side of the clockspring/column switch assembly, then pull the unit from the column

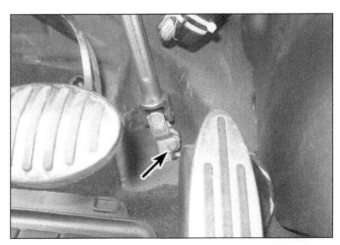

17.5 Steering column universal joint pinch bolt (Mk II models)

17.6a Slide the joint back from the pinion (Mk I models) . . .

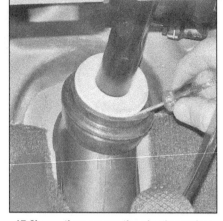

17.6b . . . then pry up the plastic washer (Mk I models)

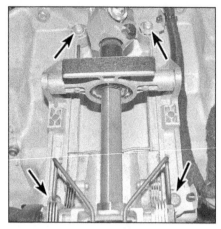

17.8 Unscrew the steering column mounting bolts (Mk I model shown, Mk II models similar)

6   On Mk I models, slide the universal joint back from the pinion, then, working in the driver's side footwell, pry the plastic washer at the base of the column from the rubber boot on the firewall. Pull the lower section of the column upwards through the firewall **(see illustrations)**.

7   On vehicles with stability assistance, disconnect the steering angle sensor electrical connector.

8   Unscrew the mounting bolts and maneuver the column assembly from the vehicle **(see illustration)**.

Inspection

9   With the steering lock disengaged, attempt to move the steering wheel up-and-down and also to the left-and-right without turning the wheel, to check for steering column bearing wear, steering column shaft joint play and steering wheel or steering column looseness. The steering column cannot be repaired, if any faults are detected, install a new column.

10   Examine the height adjustment lever mechanism for wear and damage.

11   With the steering column removed, check the universal joints for wear, and examine the column upper and lower shafts for any signs of damage or distortion. Where evident, the column should be replaced completely.

Installation

12   Installation is a reversal of removal, noting the following points:

a)   *Make sure the wheels are still in the straight-ahead position when the steering column is installed.*

b)   *Install new locking nuts to the steering column mounting bracket.*

c)   *Install a new pinch-bolt for the steering column universal joint, which joins to the steering gear.*

d)   *Tighten bolts to their specified torque.*

e)   *See Chapter 12 before installing the airbag.*

f)   *If the electronic stability program warning light (if equipped), comes on after installing the steering column, the system will need reconfiguring at a MINI dealer with dedicated test equipment, or at a suitably equipped specialist.*

18   Steering gear - removal and installation

Removal

Refer to illustrations 18.4, 18.8, 18.9, 18.10a, 18.10b and 18.10c

1   Center the steering wheel so that the front wheels are in the straight-ahead position. Remove the keys and lock the steering in position. Disconnect the cable from the negative terminal of the battery (see Chapter 5).

2   Loosen both front wheel bolts, then jack up the front of the vehicle and support it securely on jackstands. Remove both front wheels.

3   Remove the right-hand wheelwell liner

(see Chapter 11). On Mk II models, lower the subframe five inches (120 mm) (see Chapter 2A, Section 11).

4   Remove the steering column lower universal joint-to-steering gear pinion pinch-bolt, then push the joint to the rear and detach it from the pinion **(see the accompanying illustration and illustration 17.5)**.

5   Unscrew the tie-rod end nuts, and detach the rods from the steering knuckles **(see illustration 23.4)**. Take care not to damage the balljoint seals.

6   Remove the connecting links from the stabilizer bar, taking care not to damage the balljoint seals.

7   Mk I models: Siphon the fluid from the power steering fluid reservoir, or be prepared for fluid spillage when the pipes are disconnected from the rack.

8   Mk I models: Position a container beneath the steering gear, then unscrew the bolts securing the power steering lines to the gear. Identify the lines for position, then disconnect the pipes and remove the sealing washers **(see illustration)**. Allow the fluid to drain into the container. Cover the openings

18.4 Unscrew the steering column lower joint pinch-bolt (Mk I models)

18.8 Unscrew the power steering line fitting bolts (Mk I models)

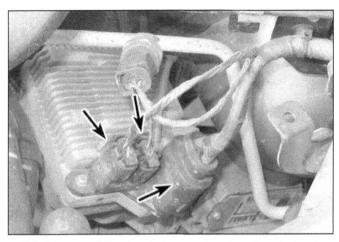

18.9 Disconnect the electrical connectors from the Electronic Power Steering module (Mk II models)

18.10a Steering gear mounting bolts - left-hand side shown

in the gear and also the ends of the fluid lines to prevent contamination of the hydraulic circuit.

9   Unbolt and remove the heat shield and support bracket from the gear. On Mk II models, disconnect the electrical connectors from the Electronic Power Steering module **(see illustration)**.

10   Remove the steering gear mounting bolts **(see illustrations)**.

11   Maneuver the steering gear out the right-hand side of the engine compartment, through the wheelwell opening.

Installation

12   Installation is a reversal of removal, noting the following points:

a) *Replace the steering fluid line sealing washers.*

b) *Replace the steering column lower universal joint pinch-bolt.*

c) *Fill and bleed the system with steering fluid as described in Section 20.*

d) *Have the front wheel alignment checked and, if necessary, adjusted.*

19   Power steering gear boots - replacement

Refer to illustrations 19.2a and 19.2b

1   Remove the tie-rod end and its locknut from the tie-rod, as described in Section 23. Make sure that a note is made of the exact position of the tie-rod end on the tie-rod, in order to retain the front wheel alignment setting on installation.

2   Release the outer and inner retaining clamps, and disconnect the boot from the steering gear housing **(see illustrations)**.

3   Slide the boot off the tie-rod.

4   Apply grease to the tie-rod inner joint. Wipe clean the seating areas on the steering gear housing and tie-rod.

5   Slide the new boot onto the tie-rod and steering gear housing.

6   Install a new inner and outer retaining clamps.

7   Install the tie-rod end as described in Section 23.

8   Have the front wheel alignment checked, and if necessary adjusted.

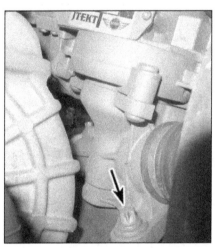

18.10b Steering gear mounting bolt - left side (Mk II models)

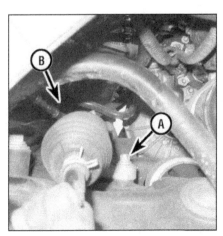

18.10c Steering gear mounting bolts - right side (A) and center (B, not visible) (Mk II models)

19.2a Boot outer clamp . . .

19.2b . . . and inner clamp

20.2 Add fluid up to the MAX mark on the dipstick

21.5 Unscrew the bolt and disconnect the fluid pressure line from the pump

ing fan (if equipped) as described in Section 22.

4   Position a container beneath the power steering pump.

5   Unscrew the bolt and disconnect the fluid pressure line from the pump, then cut through the clamp and disconnect the supply hose **(see illustration)**. Allow the fluid to drain into the container.

6   Disconnect the pump electrical connectors.

7   Unscrew the nut on the underside, and the two bolts at the rear (between the subframe and the steering gear) securing the pump bracket **(see illustrations)**. Maneuver the pump, complete with bracket, from the vehicle.

8   If required, unscrew the nuts and detach the pump from the bracket and rubber mounts.

Installation

9   Installation is a reversal of removal, noting the following points.

a) *Tighten the bolts and unions to the specified torque.*

b) *Where installed, the O-ring on the high-pressure outlet should be replaced.*

c) *Bleed the power steering hydraulic system as described in Section 20.*

---

20 Power steering hydraulic system (Mk I models) - bleeding

Refer to illustration 20.2

Warning: *Do not hold the steering wheel against the full lock stops for more than five seconds, as damage to the steering pump may result.*

1   Following any operation in which the power steering fluid lines have been disconnected, the power steering system must be bled, to remove any trapped air.

2   With the front wheels in the straight-ahead position, check the power steering fluid level in the reservoir and, if low, add fresh fluid until it reaches the MAX mark on the dipstick **(see illustration)**. Pour the fluid slowly, to prevent air bubbles forming, and use only the specified fluid (see Chapter 1).

3   Start the engine and slowly turn the steering from lock-to-lock.

4   Stop the engine, and check the hoses and connections for leaks. Check the fluid level and add more if necessary. Make sure the fluid in the reservoir does not fall below

the MIN mark on the dipstick, as air could enter the system.

5   Start the engine again, allow it to idle, then bleed the system by slowly turning the steering wheel from side-to-side several times. This should purge the system of all internal air. However, if air remains in the system (indicated by the steering operation being very noisy), leave the vehicle overnight, and repeat the procedure again the next day.

6   On completion, switch off the engine, and return the wheels to the straight-ahead position. Check for any leaks.

---

21 Power steering pump (Mk I models) - removal and installation

Removal

Refer to illustrations 21.5, 21.7a and 21.7b

1   Disconnect the cable from the negative terminal of the battery (see Chapter 5).

2   Raise the front of the vehicle and support it securely on jackstands.

3   Remove the power steering pump cool-

---

22 Power steering pump cooling fan (Mk I models) - removal and installation

Refer to illustration 22.2

1   Raise the front of the vehicle and support it securely on jackstands.

2   Unscrew the two nuts and lower the fan assembly **(see illustration)**. Discard the nuts; new ones must be installed.

3   Disconnect the electrical connector as the fan assembly is withdrawn.

4   Installation is a reversal of removal, tightening the new retaining nuts to the specified torque.

21.7a Unscrew the nut on the underside of the subframe . . .

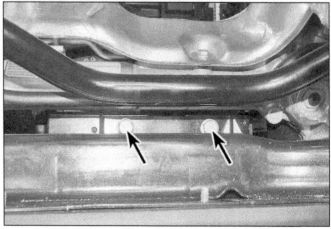

21.7b . . . and the bolts at the rear

22.2 Unscrew the power steering pump cooling fan retaining nuts

23.2a Tie-rod end locknut - early Mk I models

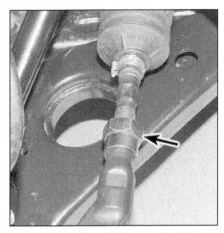

23.2b Tie-rod end locknut and collar - later Mk I models

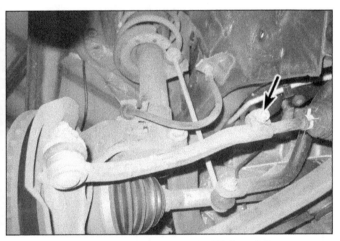

23.2c Tie-rod end pinch bolt - Mk II models

23.4 Use a balljoint separator to detach the tie-rod end balljoint from the steering knuckle

23 Tie-rod end - replacement

Removal

Refer to illustrations 23.2a, 23.2b, 23.2c and 23.4

1　Apply the parking brake, then raise the front of the vehicle and support it securely on jackstands. Remove the appropriate front wheel.

2　Count the number of threads exposed on the inner section of the tie-rod to aid installation. On Mk I models, loosen the locknut on the tie-rod by a quarter-turn **(see illustrations)**. Hold the tie-rod end stationary with another wrench engaged with the special flats while loosening the locknut. On Mk II models, loosen the pinch bolt **(see illustration)**.

3　Unscrew and remove the tie-rod end balljoint retaining nut.

4　To release the tapered shank of the balljoint from the steering knuckle, use a balljoint separator tool **(see illustration)** (if the balljoint is to be re-used, take care not to

damage the dust cover when using the separator tool).

5　Unscrew the tie-rod end from the tie-rod, counting the number of turns necessary to remove it. If necessary, hold the tie-rod stationary with grips.

Installation

Refer to illustration 23.7

6　Screw the tie-rod end onto the tie-rod by the number of turns noted during removal, until it just contacts the locknut.

7　Engage the shank of the balljoint with the steering knuckle, and install the nut. Tighten the nut to the torque listed in this Chapter's Specifications. If the ballstud turns while the nut is being tightened, use an Allen key **(see illustration)** to hold the shank or press up on the balljoint. The tapered fit of the shank will lock it, and prevent rotation as the nut is tightened.

8　On Mk I models, tighten the locknut on the tie-rod, while holding the tie-rod end as before. On Mk II models, tighten the tie-rod end pinch bolt.

9　Install the wheel and lower the vehicle. Tighten the wheel bolts to the torque listed in the Chapter 1 Specifications.

10　Have the front wheel alignment checked and, if necessary, adjusted.

23.7 If the balljoint shank turns, insert an Allen key into its end

24   Wheel alignment - general information

Refer to illustration 24.2

A wheel alignment refers to the adjustments made to the wheels so they are in proper angular relationship to the suspension and the ground. Wheels that are out of proper alignment not only affect vehicle control, but also increase tire wear.

The angles normally measured are camber, caster and toe-in **(see illustration)**. Camber and caster are not always adjustable but are usually checked to see if any suspension components are worn or damaged. The toe-in angle is commonly adjusted on all vehicles in the front and on the rear of vehicles with independent rear suspension.

Getting the proper wheel alignment is a very exacting process, one in which complicated and expensive machines are necessary to perform the job properly. Because of this, you should have a technician with the proper equipment perform these tasks. We will, however, use this space to give you a basic idea of what is involved with a wheel alignment so you can better understand the process and deal intelligently with the shop that does the work.

Toe-in is the turning in of the wheels. The purpose of a toe specification is to ensure parallel rolling of the wheels. In a vehicle with zero toe-in, the distance between the front edges of the wheels will be the same as the distance between the rear edges of the wheels. The actual amount of toe-in is normally only a fraction of an inch. Incorrect toe-in will cause the tires to wear improperly by making them scrub against the road surface.

Camber is the tilting of the wheels from vertical when viewed from one end of the vehicle. When the wheels tilt out at the top, the camber is said to be positive (+). When the wheels tilt in at the top the camber is negative (-). The amount of tilt is measured in degrees from vertical and this measurement is called the camber angle. This angle affects the amount of tire tread that contacts the road and compensates for changes in the suspension geometry when the vehicle is cornering or traveling over an undulating surface.

Caster is the tilting of the front steering axis from vertical. A tilt toward the rear is positive caster and a tilt toward the front is negative caster.

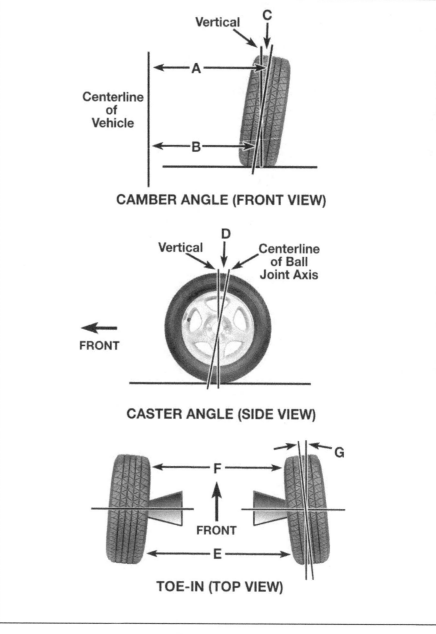

24.2 Wheel alignment details

A minus B = C (degrees camber)
D = degrees caster
E minus F = toe-in (measured in inches)
G = toe-in (expressed in degrees)

Chapter 11 Body

Contents

1   General information

Note: *Throughout this Chapter you will find references to "Mk I" and "Mk II" models; this is done to simplify which procedures apply to which models. Mk I models include 2006 and earlier Cooper/Cooper S models, and 2008 and earlier Convertible models. Mk II models include 2007 and later Cooper/Cooper S/ Clubman/Clubman S and 2009 and later Convertible models.*

The bodyshell and underframe on all models features variable thickness steel, achieved by laser-welded technology used to join steel panels of different gauges. This gives a stiffer structure, with mounting points being more rigid, which gives an improved crash performance.

An additional safety crossmember is incorporated between the A-pillars in the upper area of the firewall, and the instrument panel and steering column are secured to it. The lower firewall area is reinforced by additional systems of members connected to the front of the vehicle. The body side rocker panels (sills) have been divided along the length of the vehicle by internal reinforcement; this functions like a double tube which increases its strength. All doors are reinforced and incorporate side impact protection, which is secured in the door structure.

All sheet metal surfaces which are prone to corrosion are galvanized. The painting process includes a base color which closely matches the final topcoat, so that any stone damage is not as noticeable. The front fenders are of a bolt-on type to ease their replacement if required.

Automatic seat belts are installed on all models, and the front seat belts are equipped with a pyrotechnic pretension seat belt stack, which is attached to the seat frame of each front seat. In the event of a serious front impact, the system is triggered and pulls the stalk buckle downwards to tension the seat belt. It is not possible to reset the tensioner once fired, and it must therefore be replaced. The tensioners are fired by an explosive charge similar to that used in the airbag, and are triggered via the airbag control module.

Many of the procedures in this Chapter require the battery to be disconnected. Refer to Chapter 5.

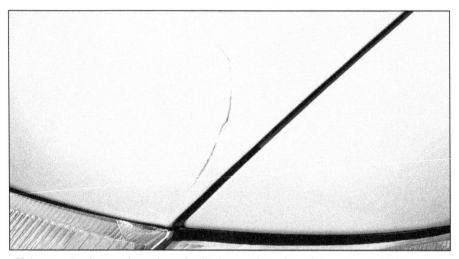

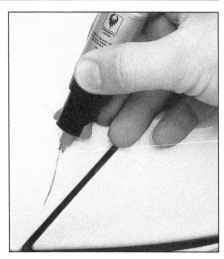

Make sure the damaged area is perfectly clean and rust free. If the touch-up kit has a wire brush, use it to clean the scratch or chip. Or use fine steel wool wrapped around the end of a pencil. Clean the scratched or chipped surface only, not the good paint surrounding it. Rinse the area with water and allow it to dry thoroughly

Thoroughly mix the paint, then apply a small amount with the touch-up kit brush or a very fine artist's brush. Brush in one direction as you fill the scratch area. Do not build up the paint higher than the surrounding paint

2   Body repair - minor damage

Repair of minor paint scratches

No matter how hard you try to keep your vehicle looking like new, it will inevitably be scratched, chipped or dented at some point. If the metal is actually dented, seek the advice of a professional. But you can fix minor scratches and chips yourself. Buy a touch-up paint kit from a dealer parts department or an auto parts store. To ensure that you get the right color, you'll need to have the specific make, model and year of your vehicle and, ideally, the paint code, which is located on a special metal plate under the hood or in the door jamb.

Plastic body panel repair

The following repair procedures are for minor scratches and gouges. Repair of more serious damage should be left to a dealer service department or qualified auto body shop. Below is a list of the equipment and materials necessary to perform the following repair procedures on plastic body panels.

> *Wax, grease and silicone removing solvent*
> *Cloth-backed body tape*
> *Sanding discs*
> *Drill motor with three-inch disc holder*
> *Hand sanding block*
> *Rubber squeegees*
> *Sandpaper*
> *Non-porous mixing palette*
> *Wood paddle or putty knife*
> *Curved-tooth body file*
> *Flexible parts repair material*

Flexible panels (bumper trim)

1   Remove the damaged panel, if necessary or desirable. In most cases, repairs can be carried out with the panel installed.
2   Clean the area(s) to be repaired with a

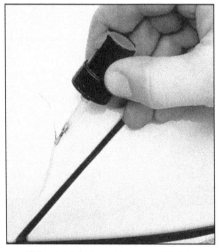

If the vehicle has a two-coat finish, apply the clear coat after the color coat has dried

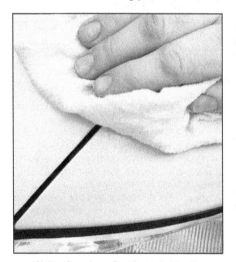

Wait a few days for the paint to dry thoroughly, then rub out the repainted area with a polishing compound to blend the new paint with the surrounding area. When you're happy with your work, wash and polish the area

wax, grease and silicone removing solvent applied with a water-dampened cloth.
3   If the damage is structural, that is, if it extends through the panel, clean the backside of the panel area to be repaired as well. Wipe dry.
4   Sand the rear surface about 1-1/2 inches beyond the break.
5   Cut two pieces of fiberglass cloth large enough to overlap the break by about 1-1/2 inches. Cut only to the required length.
6   Mix the adhesive from the repair kit according to the instructions included with the kit, and apply a layer of the mixture approximately 1/8-inch thick on the backside of the panel. Overlap the break by at least 1-1/2 inches.
7   Apply one piece of fiberglass cloth to the adhesive and cover the cloth with additional adhesive. Apply a second piece of fiberglass cloth to the adhesive and immediately cover the cloth with additional adhesive in sufficient quan-

tity to fill the weave.
8   Allow the repair to cure for 20 to 30 minutes at 60-degrees to 80-degrees F.
9   If necessary, trim the excess repair material at the edge.
10   Remove all of the paint film over and around the area(s) to be repaired. The repair material should not overlap the painted surface.
11   With a drill motor and a sanding disc (or a rotary file), cut a "V" along the break line approximately 1/2-inch wide. Remove all dust and loose particles from the repair area.
12   Mix and apply the repair material. Apply a light coat first over the damaged area; then continue applying material until it reaches a level slightly higher than the surrounding finish.
13   Cure the mixture for 20 to 30 minutes at

60-degrees to 80-degrees F.

14   Roughly establish the contour of the area being repaired with a body file. If low areas or pits remain, mix and apply additional adhesive.

15   Block sand the damaged area with sandpaper to establish the actual contour of the surrounding surface.

16   If desired, the repaired area can be temporarily protected with several light coats of primer. Because of the special paints and techniques required for flexible body panels, it is recommended that the vehicle be taken to a paint shop for completion of the body repair.

Steel body panel repair

See photo sequence

Repair of dents

17   When repairing dents, the first job is to pull the dent out until the affected area is as close as possible to its original shape. There is no point in trying to restore the original shape completely as the metal in the damaged area will have stretched on impact and cannot be restored to its original contours. It is better to bring the level of the dent up to a point that is about 1/8-inch below the level of the surrounding metal. In cases where the dent is very shallow, it is not worth trying to pull it out at all.

18   If the backside of the dent is accessible, it can be hammered out gently from behind using a soft-face hammer. While doing this, hold a block of wood firmly against the opposite side of the metal to absorb the hammer blows and prevent the metal from being stretched.

19   If the dent is in a section of the body which has double layers, or some other factor makes it inaccessible from behind, a different technique is required. Drill several small holes through the metal inside the damaged area, particularly in the deeper sections. Screw long, self-tapping screws into the holes just enough for them to get a good grip in the metal. Now pulling on the protruding heads of the screws with locking pliers can pull out the dent.

20   The next stage of repair is the removal of paint from the damaged area and from an inch or so of the surrounding metal. This is easily done with a wire brush or sanding disk in a drill motor, although it can be done just as effectively by hand with sandpaper. To complete the preparation for filling, score the surface of the bare metal with a screwdriver or the tang of a file or drill small holes in the affected area. This will provide a good grip for the filler material. To complete the repair, see the Section on filling and painting.

Repair of rust holes or gashes

21   Remove all paint from the affected area and from an inch or so of the surrounding metal using a sanding disk or wire brush mounted in a drill motor. If these are not available, a few sheets of sandpaper will do the job just as effectively.

22   With the paint removed, you will be able to determine the severity of the corrosion and decide whether to replace the whole panel, if possible, or repair the affected area. New body panels are not as expensive as most people think and it is often quicker to install a new panel than to repair large areas of rust.

23   Remove all trim pieces from the affected area except those which will act as a guide to the original shape of the damaged body, such as headlight shells, etc. Using metal snips or a hacksaw blade, remove all loose metal and any other metal that is badly affected by rust. Hammer the edges of the hole in to create a slight depression for the filler material.

24   Wire-brush the affected area to remove the powdery rust from the surface of the metal. If the back of the rusted area is accessible, treat it with rust inhibiting paint.

25   Before filling is done, block the hole in some way. This can be done with sheet metal riveted or screwed into place, or by stuffing the hole with wire mesh.

26   Once the hole is blocked off, the affected area can be filled and painted. See the following subsection on filling and painting.

Filling and painting

27   Many types of body fillers are available, but generally speaking, body repair kits which contain filler paste and a tube of resin hardener are best for this type of repair work. A wide, flexible plastic or nylon applicator will be necessary for imparting a smooth and contoured finish to the surface of the filler material. Mix up a small amount of filler on a clean piece of wood or cardboard (use the hardener sparingly). Follow the manufacturer's instructions on the package, otherwise the filler will set incorrectly.

28   Using the applicator, apply the filler paste to the prepared area. Draw the applicator across the surface of the filler to achieve the desired contour and to level the filler surface. As soon as a contour that approximates the original one is achieved, stop working the paste. If you continue, the paste will begin to stick to the applicator. Continue to add thin layers of paste at 20-minute intervals until the level of the filler is just above the surrounding metal.

29   Once the filler has hardened, the excess can be removed with a body file. From then on, progressively finer grades of sandpaper should be used, starting with a 180-grit paper and finishing with 600-grit wet-or-dry paper. Always wrap the sandpaper around a flat rubber or wooden block, otherwise the surface of the filler will not be completely flat. During the sanding of the filler surface, the wet-or-dry paper should be periodically rinsed in water. This will ensure that a very smooth finish is produced in the final stage.

30   At this point, the repair area should be surrounded by a ring of bare metal, which in turn should be encircled by the finely feathered edge of good paint. Rinse the repair area with clean water until all of the dust produced by the sanding operation is gone.

31   Spray the entire area with a light coat of primer. This will reveal any imperfections in the surface of the filler. Repair the imperfections with fresh filler paste or glaze filler and once more smooth the surface with sandpaper. Repeat this spray-and-repair procedure until you are satisfied that the surface of the filler and the feathered edge of the paint are perfect. Rinse the area with clean water and allow it to dry completely.

32   The repair area is now ready for painting. Spray painting must be carried out in a warm, dry, windless and dust free atmosphere. These conditions can be created if you have access to a large indoor work area, but if you are forced to work in the open, you will have to pick the day very carefully. If you are working indoors, dousing the floor in the work area with water will help settle the dust that would otherwise be in the air. If the repair area is confined to one body panel, mask off the surrounding panels. This will help minimize the effects of a slight mismatch in paint color. Trim pieces such as chrome strips, door handles, etc., will also need to be masked off or removed. Use masking tape and several thickness of newspaper for the masking operations.

33   Before spraying, shake the paint can thoroughly, then spray a test area until the spray painting technique is mastered. Cover the repair area with a thick coat of primer. The thickness should be built up using several thin layers of primer rather than one thick one. Using 600-grit wet-or-dry sandpaper, rub down the surface of the primer until it is very smooth. While doing this, the work area should be thoroughly rinsed with water and the wet-or-dry sandpaper periodically rinsed as well. Allow the primer to dry before spraying additional coats.

34   Spray on the top coat, again building up the thickness by using several thin layers of paint. Begin spraying in the center of the repair area and then, using a circular motion, work out until the whole repair area and about two inches of the surrounding original paint is covered. Remove all masking material 10 to 15 minutes after spraying on the final coat of paint. Allow the new paint at least two weeks to harden, then use a very fine rubbing compound to blend the edges of the new paint into the existing paint. Finally, apply a coat of wax

---

3   Body repair - major damage

1   Major damage must be repaired by an auto body shop specifically equipped to perform body and frame repairs. These shops have the specialized equipment required to do the job properly.

2   If the damage is extensive, the frame must be checked for proper alignment or the vehicle's handling characteristics may be adversely affected and other components may wear at an accelerated rate.

3   Due to the fact that all of the major body components (hood, fenders, etc.) are separate and replaceable units, any seriously damaged components should be replaced rather than repaired. Sometimes the components can be found in a wrecking yard that specializes in used vehicle components, often at considerable savings over the cost of new parts.

These photos illustrate a method of repairing simple dents. They are intended to supplement *Body repair - minor damage* in this Chapter and should not be used as the sole instructions for body repair on these vehicles.

1 If you can't access the backside of the body panel to hammer out the dent, pull it out with a slide-hammer-type dent puller. Tap with a hammer near the edge of the dent to help 'pop' the metal back to its original shape, about 1/8-inch below the surface of the surrounding metal

2 Using coarse-grit sandpaper, remove the paint down to the bare metal. Clean the repair area with wax/silicone remover.

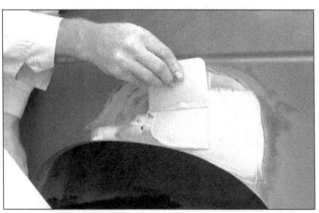

3 Following label instructions, mix up a batch of plastic filler and hardener, then quickly press it into the metal with a plastic applicator. Work the filler until it matches the original contour and is slightly above the surrounding metal

4 Let the filler harden until you can just dent it with your fingernail. File, then sand the filler down until it's smooth and even. Work down to finer grits of sandpaper - always using a board or block - ending up with 360 or 400 grit

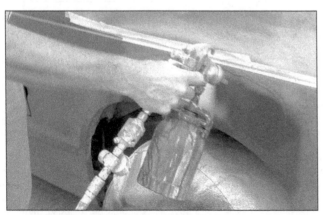

5 When the area is smooth to the touch, clean the area and mask around it. Apply several layers of primer to the area. A professional-type spray gun is being used here, but aerosol spray primer works fine

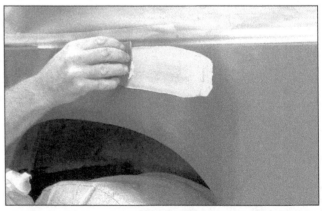

6 Fill imperfections or scratches with glazing compound. Sand with 360 or 400-grit and re-spray. Finish sand the primer with 600 grit, clean thoroughly, then apply the finish coat. Don't attempt to rub out or wax the repair area until the paint has dried completely (at least two weeks)

4   Upholstery, carpets and vinyl trim - maintenance

Upholstery and carpets

1   Every three months remove the floor-mats and clean the interior of the vehicle (more frequently if necessary). Use a stiff whiskbroom to brush the carpeting and loosen dirt and dust, then vacuum the upholstery and carpets thoroughly, especially along seams and crevices.

2   Dirt and stains can be removed from carpeting with basic household or automotive carpet shampoos available in spray cans. Follow the directions and vacuum again, then use a stiff brush to bring back the "nap" of the carpet.

3   Most interiors have cloth or vinyl upholstery, either of which can be cleaned and maintained with a number of material-specific cleaners or shampoos available in auto supply stores. Follow the directions on the product for usage, and always spot-test any upholstery cleaner on an inconspicuous area (bottom edge of a backseat cushion) to ensure that it doesn't cause a color shift in the material.

4   After cleaning, vinyl upholstery should be treated with a protectant. **Note:** *Make sure the protectant container indicates the product can be used on seats - some products may make a seat too slippery.* **Caution:** *Do not use protectant on vinyl-covered steering wheels.*

5   Leather upholstery requires special care. It should be cleaned regularly with saddle-soap or leather cleaner. Never use alcohol, gasoline, nail polish remover or thinner to clean leather upholstery.

6   After cleaning, regularly treat leather upholstery with a leather conditioner, rubbed in with a soft cotton cloth. Never use car wax on leather upholstery.

7   In areas where the interior of the vehicle is subject to bright sunlight, cover leather seating areas of the seats with a sheet if the vehicle is to be left out for any length of time.

Vinyl trim

8   Don't clean vinyl trim with detergents, caustic soap or petroleum-based cleaners. Plain soap and water works just fine, with a soft brush to clean dirt that may be ingrained. Wash the vinyl as frequently as the rest of the vehicle.

9   After cleaning, application of a high-quality rubber and vinyl protectant will help prevent oxidation and cracks. The protectant can also be applied to weather-stripping, vacuum lines and rubber hoses, which often fail as a result of chemical degradation, and to the tires.

5   Fastener and trim removal

Refer to illustration 5.4

1   There is a variety of plastic fasteners used to hold trim panels, splash shields and other parts in place in addition to typical screws, nuts and bolts. Once you are familiar with them, they can usually be removed without too much difficulty.

2   The proper tools and approach can prevent added time and expense to a project by minimizing the number of broken fasteners and/or parts.

3   The following illustration shows various types of fasteners that are typically used on most vehicles and how to remove and install them **(see illustration)**. Replacement fasteners are commonly found at most auto parts stores, if necessary.

Fasteners

This tool is designed to remove special fasteners. A small pry tool used for removing nails will also work well in place of this tool

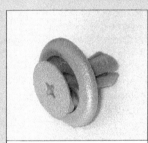

A Phillips head screwdriver can be used to release the center portion, but light pressure must be used because the plastic is easily damaged. Once the center is up, the fastener can easily be pried from its hole

Here is a view with the center portion fully released. Install the fastener as shown, then press the center in to set it

This fastener is used for exterior panels and shields. The center portion must be pried up to release the fastener. Install the fastener with the center up, then press the center in to set it

This type of fastener is used commonly for interior panels. Use a small blunt tool to press the small pin at the center in to release it . . .

. . . the pin will stay with the fastener in the released position

Reset the fastener for installation by moving the pin out. Install the fastener, then press the pin flush with the fastener to set it

This fastener is used for exterior and interior panels. It has no moving parts. Simply pry the fastener from its hole like the claw of a hammer removes a nail. Without a tool that can get under the top of the fastener, it can be very difficult to remove

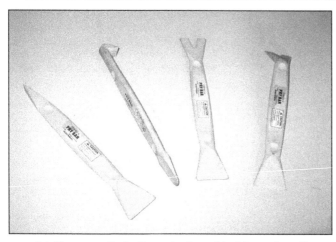

5.4 These small plastic pry tools are ideal for prying off trim panels

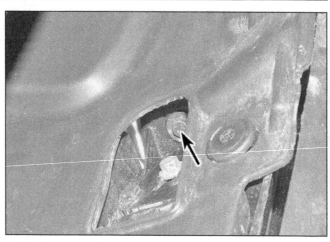

6.2a Remove the bumper bolts through the wheelwell liner ...

4   Trim panels are typically made of plastic and their flexibility can help during removal. The key to their removal is to use a tool to pry the panel near its retainers to release it without damaging surrounding areas or breaking-off any retainers. The retainers will

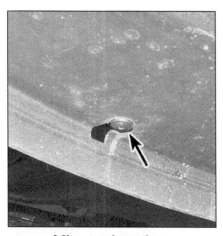

6.2b ... underneath ...

usually snap out of their designated slot or hole after force is applied to them. Stiff plastic tools designed for prying on trim panels are available at most auto parts stores **(see illustration)**. Tools that are tapered and wrapped in protective tape, such as a screwdriver or small pry tool, are also very effective when used with care.

6   Bumpers - removal and installation

Note: *Headlight washers are installed on some models; as the bumper is being removed, disconnect as required.*

Mk I models
Front bumper cover and bumper carrier

Refer to illustrations 6.2a, 6.2b, 6.2c, 6.2d, 6.3 and 6.4

1   Apply the parking brake, raise the front of the vehicle and support it securely on jackstands. Remove the under-vehicle splash shield.

2   The bumper is secured by three screws on each side and retaining clips at the top edge **(see illustrations)**. Remove the screws, release the clips and pull the bumper cover forwards. Noting their installed locations, disconnect the electrical connectors as the bumper is removed.
3   To remove the bumper carrier, feed the ambient temperature sensor wiring harness back through the hole in the bumper carrier **(see illustration)**.
4   Remove the screws/nuts from each side and remove the bumper carrier **(see illustration)**.
5   Installation is the reverse of removal. Check all electrical components that have been disconnected.

Rear bumper cover

Refer to illustrations 6.7a, 6.7b, 6.8, 6.9, 6.11, 6.12a and 6.12b

6   Loosen the rear wheel bolts, chock the front wheels, raise the rear of the vehicle and support it securely on jackstands. Open the hatch, and remove both rear wheels. Remove the rear

6.2c ... and at the top ...

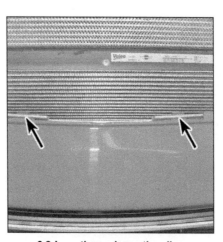

6.2d ... then release the clips (Mk I models)

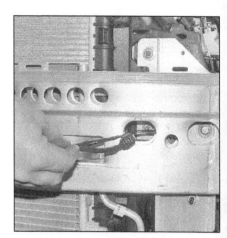

6.3 Feed the ambient temperature sensor wiring back through the bumper carrier (Mk I models)

6.4 Remove the bumper carrier nuts and bolt (Mk I models)

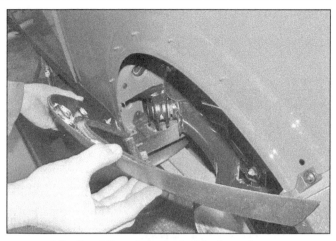

6.7a Carefully pull the fender trim from place . . .

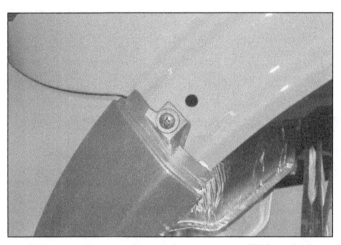

6.7b . . . and remove the rear bumper screw (Mk I models)

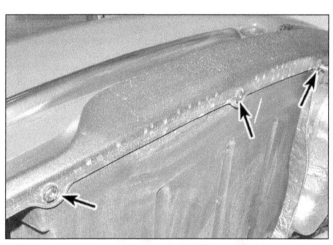

6.8 Remove the three screws securing the spare wheel carrier to the rear bumper (Mk I models)

wheelwell liners as described in Section 29.

7   Release the retaining clip, then carefully pull the rear of the fender trims from the fender to access and remove the outer bumper cover screws **(see illustrations)**.

8   On all Cooper models, remove the three screws securing the rear bumper to the spare wheel carrier **(see illustration)**.

9   On Cooper models from 07/2004 and all Cooper S models, remove the bumper central retaining screw on the underside **(see illustration)**.

10   On Cooper models up to 07/2004, pry out the central back-up light and remove the screw in the light opening.

11   Remove the two screws in the hatch opening securing the bump stops/bumper upper edge **(see illustration)**.

6.9 Central bumper screw (Mk I models)

6.11 Remove the bump stop retaining screw from each side (Mk I models)

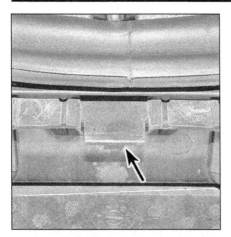

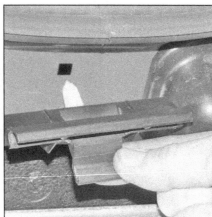

6.12a **Reach up and release the central retaining clip (shown with the bumper removed) . . .**

6.12b **. . . if it comes away with the bumper, push it into place prior to installing the bumper (Mk I models)**

12   Reach up behind and release the upper central retaining clip as the bumper is withdrawn. Disconnect any electrical connectors as the bumper is removed **(see illustration)**. If the central retaining clip came away from the vehicle body with the bumper cover, release it

from the bumper and install it to the vehicle body prior to installation **(see illustration)**.
13   Installation is a reversal of the removal procedure. Check all electrical components that have been disconnected.

Mk II models

Front bumper cover and bumper carrier

Refer to illustrations 6.16, 6.17 and 6.18

14   Remove the radiator grille (see Section 7). Detach the front of the wheelwell liners from the bumper cover (see Section 29).
15   Apply the parking brake, raise the front of the vehicle and support it securely on jackstands.
16   Remove the fasteners along the underside of the bumper cover **(see illustration)**.
17   Remove the upper mounting fasteners from each side of the bumper cover **(see illustration)**, then pull the bumper cover forwards. Note their installed locations and disconnect the electrical connectors as the cover is removed.
18   Remove the bumper carrier fasteners from each side, then remove the carrier **(see illustration)**.
19   Installation is the reverse of the removal procedure. If the absorbing foam bolster became dislodged, properly position it on the reinforcement and secure it with duct tape or industrial adhesive, then install the bumper cover.

Rear bumper cover and bumper carrier

Refer to illustrations 6.22, 6.23 and 6.24

20   Raise the rear of the vehicle and support it securely on jackstands. Chock the front wheels.
21   Carefully pry the fender trims from the fender **(see illustration 6.7a)**. Disconnect the electrical connectors for the side marker light wiring harnesses. If equipped, remove the mudguards.
22   Remove the lower bumper cover bracket screw from each side of the cover, rearward of the wheels **(see illustration)**.
23   Remove the cover upper mounting fasteners from each side **(see illustration)**, then pull the bumper away from the vehicle. Note their installed locations and disconnect the

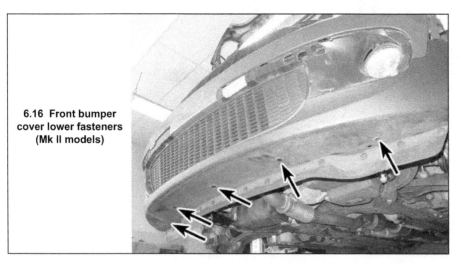

6.16 **Front bumper cover lower fasteners (Mk II models)**

6.17 **Front bumper cover upper fasteners (Mk II models)**

6.18 **Front bumper carrier fasteners (both sides similar) (Mk II models)**

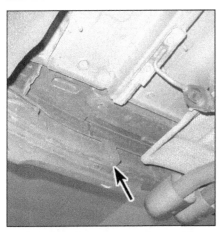

6.22 Rear bumper cover lower side mounting screw (Mk II models)

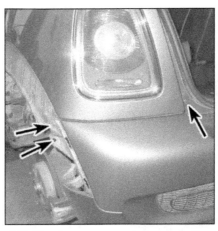

6.23 Rear bumper cover upper mounting fasteners (Mk II models)

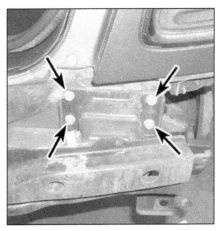

6.24 Rear bumper carrier fasteners (both sides similar) (Mk II models)

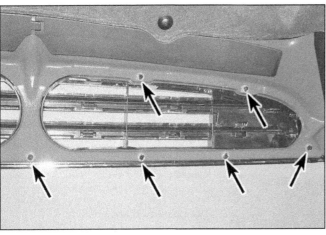

7.2 Unscrew the radiator grille nuts and remove the washers (right-side nuts shown) (Mk I models)

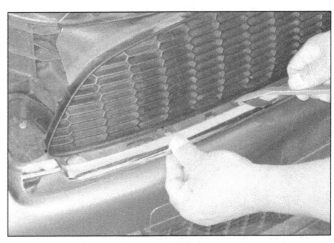

7.5 Pry off the trim strip at the base of the grille, starting at the ends (Mk II models)

electrical connectors as the cover is removed.

24   Remove the bumper carrier fasteners from each side, then remove the carrier **(see illustration)**.

25   Installation is the reverse of the removal procedure. If the absorbing foam bolster became dislodged, properly position it on the reinforcement and secure it with duct tape or industrial adhesive, then install the bumper cover.

7 Radiator grille - removal and installation

1   Support the hood in the open position.

Mk I models

Refer to illustration 7.2

2   Remove the nuts and washers **(see illustration)**.

3   Pull the grille forwards and detach it from the hood.

4   Installation is a reversal of the removal procedure.

Mk II models

Refer to illustrations 7.5, 7.6 and 7.7

5   Starting with the tabs at each end, carefully pry off the chrome trim strip at the base of the grille, using a plastic trim stick **(see illustration)**.

6   Remove the grille upper mounting fasteners **(see illustration)**.

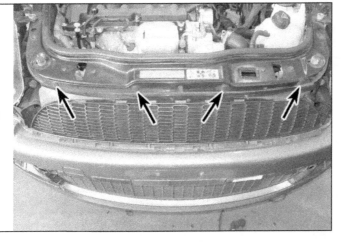

7.6 Remove the plastic upper mounting fasteners

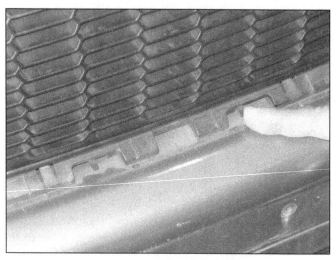

7.7 Depress the retainers along the bottom edge of the grille and pull the grille forward (Mk II models)

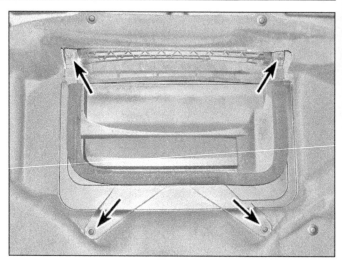

8.2 Remove the air intake duct fasteners (Cooper S models)

8.3a Remove the screws, pry out the plastic expansion rivets . . .

7   Depress the retainers at the bottom of the grille **(see illustration)**, then pull the grille forward and remove it.

8   Installation is the reverse of the removal procedure.

8   Hood - removal, installation and adjustment

Refer to illustrations 8.2, 8.3a, 8.3b, 8.4, 8.7a, 8.7b and 8.7c

1   Open the hood.

2   On Cooper S models, remove the two screws and two nuts, then remove the air intake duct from the underside of the hood **(see illustration)**.

3   Remove the screws and plastic expansion rivets, then remove the sound insulation panel from the hood **(see illustrations)**.

4   Disconnect the windshield washer wiring connector from the bottom of the jets, and dis-connect the hose from the junction **(see illustration)**.

5   On Mk I models, disconnect the electrical connectors from the rear of the headlights. Trace the wiring harness along its route and release it from the retaining clips.

6   Have an assistant support the hood, then pry out the clips and disconnect the support struts from the balljoints on the hood (see Section 18).

7   Make alignment marks between the hinges and the hood to aid installation. Remove the four bolts, disconnect the ground strap (if equipped), and lift the hood from the vehicle **(see illustrations)**.

8   Installation is a reversal of the removal procedure, noting the following points:

a) *Position the hood hinges within the out-line marks made during removal, but if necessary, alter its position to provide a uniform gap around all seams.*

b) *Adjust the front height by repositioning the latches.*

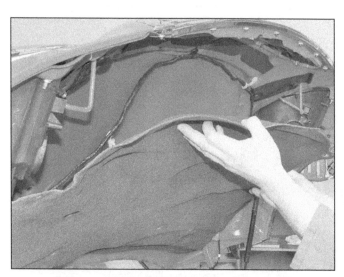

8.3b . . . and remove the sound insulation panel

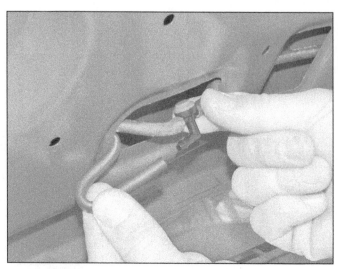

8.4 Disconnect the washer jet hose at the junction

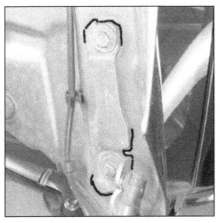

8.7a Mark the position of the hinges on the hood (Mk II model shown)

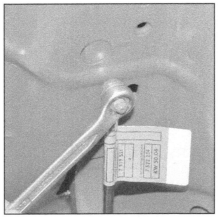

8.7b Disconnect the ground strap, if equipped . . .

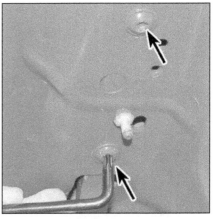

8.7c . . . then remove the two fasteners on each side securing the hood to the hinges

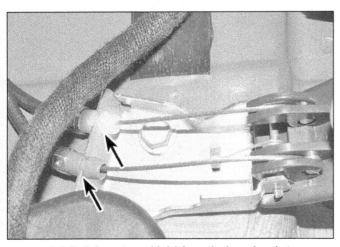

9.2 Pull the outer cable(s) from the lever bracket (Mk I model shown)

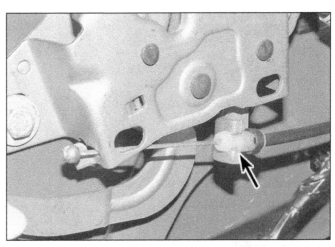

9.4 Pull the outer cable from the support bracket (Mk I model shown)

9   Hood release cable - removal and installation

Removal

Refer to illustrations 9.2 and 9.4

1   Remove the driver's side sill/footwell kick panel trim as described in Section 25.

Mk I models

2   Pull the outer cable(s) from the lever bracket, then disengage the cable end fitting from the release lever **(see illustration)**.

3   Noting its routing, trace the cable(s) forward, and release it from any retaining clips.

4   Pull the outer cable from the support bracket, then disengage the cable end fitting from the hood latch **(see illustration)**.

5   Pull the cable from place.

Mk II models

6   Working on the driver's side footwell, lift the cable housing from the lever bracket and remove the cable end from the release lever.

7   Working on the engine compartment fenderwell near the windshield washer fluid reservoir, lift the cable housing cover and disconnect the cable from the bracket and the cable connector.

8   Disconnect the forward cable from the latch and pull the cable from the front bulkhead. Pull the rear cable from the engine compartment firewall by way of the passenger compartment (driver's side footwell).

Installation

9   Install the new cable and secure it in place with the various retaining clips.

10   Engage the cable end fittings with the release lever and hood latch, then push the outer cables into place in the support brackets.

11   Check the operation of the hood latch/release lever prior to closing the hood. There is no adjustment provision for the cables.

10   Door trim panel - removal and installation

Mk I models

Refer to illustrations 10.2, 10.3a, 10.3b, 10.4, 10.6, 10.7 and 10.8

1   Completely lower the door window.

2   Carefully pry the safety reflector from the trim panel **(see illustration)**.

3   On vehicles manufactured after 09/2002, slide a flat-bladed tool into the gap at the front of the armrest to depress the retaining clip,

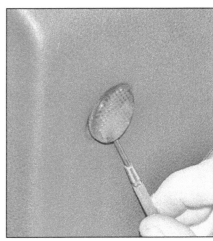

10.2 Pry the reflector from the door trim (Mk I models)

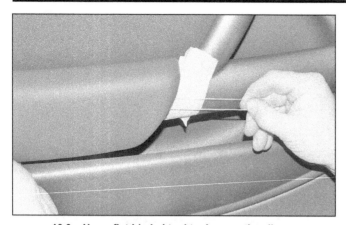

10.3a Use a flat-bladed tool to depress the clip . . .

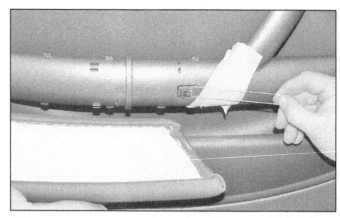

10.3b . . . then slide the armrest forward (Mk I models after 09/2002)

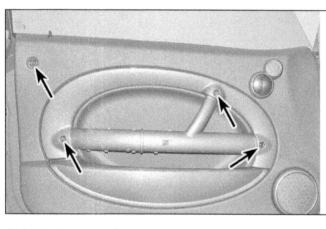

10.4 Door inner trim panel screws (Mk I models)

10.6 Use a piece of card to protect the paint when releasing the panel clips

then slide the armrest forwards to remove it **(see illustrations)**.

4   Remove the two long, and two short screws **(see illustration)**.

5   Pry out the courtesy light in the bottom of the trim panel, and disconnect the electrical connector.

6   Using a flat-bladed tool, carefully release the push-on clips around the perimeter of the trim panel **(see illustration)**.

7   Pull the panel from the door to release the upper clips. Lift it upwards and over the lock button, then guide the interior door han-

dle through the panel **(see illustration)**. Note the locations of the color-coded trim clips and replace any that are damaged.

8   Installation is a reversal of the removal procedure. To aid installation, unclip the interior door handle/speaker trim from the door trim panel, and install it once the trim panel is in place **(see illustration)**.

Mk II models

Refer to illustrations 10.9, 10.10 and 10.11

9   Pry out the retaining pin at the bottom of the inside door handle trim piece **(see illus-**

tration). **Caution:** *Don't try to remove the trim piece - it comes off with the panel.*

10   Using a trim panel tool slipped between the trim panel and the door, carefully pry out the door panel retainers from the door sheet-metal **(see illustration)** and remove the panel.

11   Remove the armrest, if necessary **(see illustration)**.

12   Installation is the reverse of removal. Be sure to replace any broken clips.

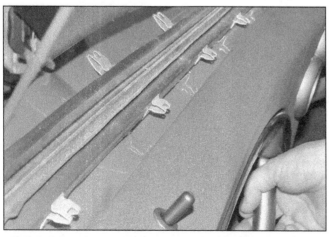

10.7 Pull the trim panel from the door to release the upper clips

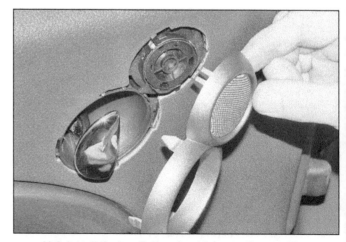

10.8 Install the handle/speaker trim once the panel has been installed

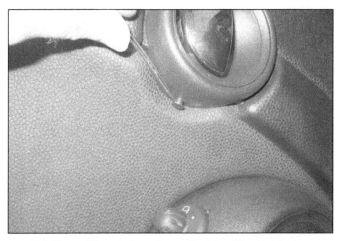

10.9 Pry out the pin from the underside of the inside door handle trim

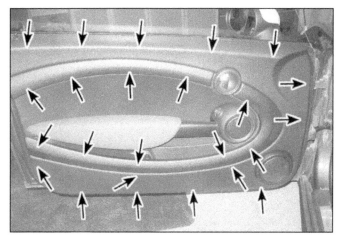

10.10 Door trim panel retaining clip locations

11 Door window glass - removal, installation, adjustment and initialization

Mk I models

Removal

Refer to illustrations 11.2, 11.3 and 11.4

1   Remove the door inner trim panel as described in Section 10.
2   Carefully pry up and remove the exterior window seal **(see illustration)**.
3   Raise the glass to a height of approximately 8-1/2 inches (22 cm) above the door upper edge, then reach between the window and the door skin and, using an Allen key, unscrew the large serrated nuts securing the window to the regulator **(see illustration)**.
4   Disengage the window from the regulator, and lift the window glass from the door **(see illustration)**.

Installation

5   Installation is a reversal of the removal procedure, making sure that the glass is correctly located in the support brackets.

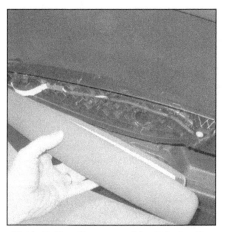

10.11 Pull on the armrest to dislodge the retaining clips

Adjustment

6   The position of the window glass may need to be adjusted to reduce wind noise.

Pretension adjustment

Refer to illustration 11.7

7   Open the door, and then slowly close

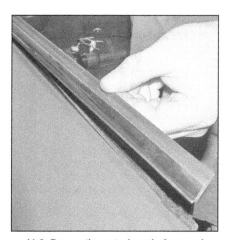

11.2 Pry up the exterior window seal

it so that the door lock clicks over the first position on the latch. There should be a gap between the door edge and the vehicle body. In this position, the top of the window should be just touching the door seal. This can be checked by placing a sheet of paper between the top of the window and the door seal. With

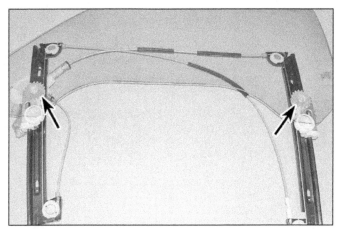

11.3 Remove the large serrated nuts securing the window to the regulator - window assembly removed for clarity

11.4 Lift the window glass from the door

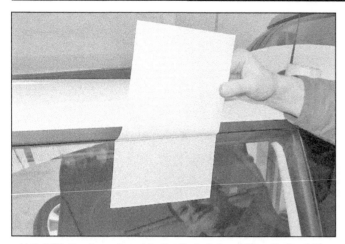

11.7 With the door closed in the first click position, it should be just possible to pull the paper from place with light resistance

11.9 Use masking tape on the window, and measure the amount of window-to-seal overlap

the door closed in the first click position, it should be possible to pull the paper from place with light resistance **(see illustration)**.

8   If adjustment is required, loosen the adjusting bolts on the underside of the door, and reposition the window. Tighten the bolts once the correct position is achieved.

Position adjustment

Refer to illustration 11.9

9   With the window fully raised, firmly close the door and measure the distance the window overlaps the door seal. At the top of the glass, the window must overlap by at least 3/16-inch (5.0 mm), and at the front of the glass by at least 9/64-inch (3.5 mm). Use masking tape on the window to measure the amount of overlap **(see illustration)**.

10   To adjust the window position, loosen the serrated nuts securing the window to the regulator, and reposition the window until the correct position is achieved **(see illustration 11.3)**. Tighten the serrated nuts and check the window operation.

Initialization

11   See Chapter 5, Section 3 for the initialization procedure.

Mk II models

Removal

Refer to illustrations 11.14 and 11.16

12   Remove the door trim panel (see Section 10).

13   Carefully pry up and remove the exterior window seal **(see illustration 11.2)**.

14   Position the window glass 3 inches (75 mm) above the door **(see illustration)**.

15   Disconnect the cable from the negative terminal of the battery (see Chapter 5).

16   Remove the access covers and loosen the window securing screws enough to free the glass (approximately 3/4-inch [19 mm]) **(see illustration)**, then lift the glass out of the door.

Installation

17   Installation is the reverse of the removal procedure. Make sure the glass fits properly

with its front and rear brackets, and tighten the screws securely. Proceed to adjust and initialize the window.

Adjustment

Refer to illustration 11.20

Caution: *Do not close the door completely until adjustment has been verified, as glass breakage may result.*

Note: *Up to three adjustments might be required to properly align the window to the window opening. One (inclination) is described here, but the other two (retraction depth and pretension) require special tools not normally available outside of a MINI dealership. If satisfactory results cannot be achieved with the inclination adjustment, take the vehicle to a MINI dealer service department or other qualified repair shop for adjustment.*

Note: *The following procedure assumes that the door is properly adjusted.*

18   Raise the window all the way, then gently close the door so that the latch clicks one notch (don't try to close the door completely).

11.14 With the window raised 3 inches (75 mm) . . .

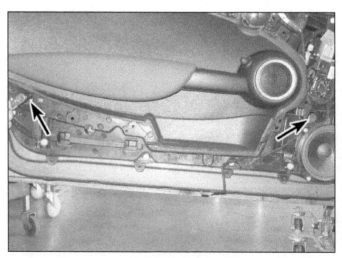

11.16 . . . access the window securing screws through these holes

11.20 Remove the covers from the holes in the bottom of the door for access to the window angle inclination screws

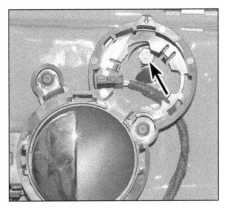

12.3a One of the window regulator bolts is hidden behind the door treble speaker

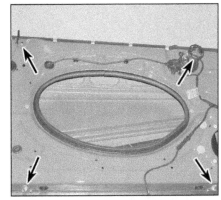

12.3b Window regulator bolts

19   The top of the window should contact the weatherstripping at the top of the door opening when the top of the glass is pushed in with a finger (it shouldn't take much force to make contact).

20   If the glass touches the weatherstrip without pushing the glass in, or if the gap between the glass and the weatherstrip is too large, remove the access covers from the bottom of the door **(see illustration)**, loosen the screws and tilt the window in-or-out as necessary, then tighten the screws securely.

Initialization

21   See Chapter 5, Section 3 for the initialization procedure.

---

12   Door window regulator and motor - removal and installation

---

Mk I models

Refer to illustrations 12.3a, 12.3b, 12.4 and 12.8

1   Remove the window glass (see Section 11).

2   Depress the retaining clips and pry the door treble speaker from position. Disconnect the electrical connector as it's withdrawn.

3   Pry out the plastic plugs (if equipped), and unscrew the regulator bolts **(see illustrations)**.

4   Remove the three bolts securing the regulator to the door frame **(see illustration)**.

5   Unclip the regulator cables from the door frame.

6   Disconnect the electrical connector from the window regulator.

7   Withdraw the window regulator mechanism downwards and out through the hole in the inner door panel.

8   If required, remove the four Torx screws and separate the motor from the regulator **(see illustration)**. A motor should be available as a separate item - check with a MINI dealer or other parts supplier prior to disassembly.

9   Installation is a reversal of the removal procedure, but carry out the adjustment procedure as described in the previous Section.

Mk II models

10   Remove the door trim panel and armrest (see Section 10).

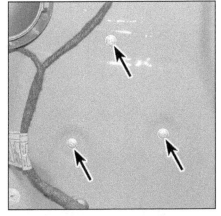

12.4 Remove the three bolts securing the regulator to the door frame

Motor

Refer to illustration 12.11

11   Disconnect the electrical connector from the motor, then remove the motor retaining screws and detach the motor **(see illustration)**.

12   Installation is the reverse of removal.

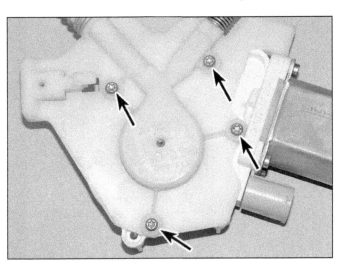

12.8 The motor is secured to the regulator by four bolts

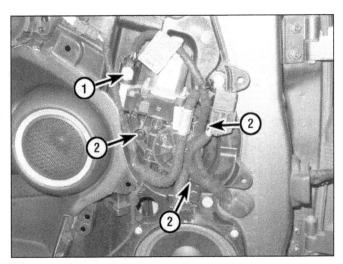

12.11 Window motor electrical connector (1) and retaining screws (2)

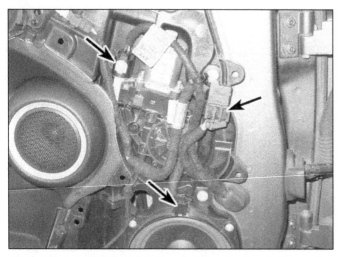

12.14 Disconnect the electrical connectors for the window motor, speaker and door harness

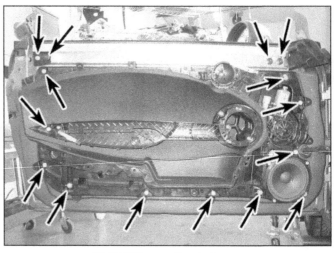

12.15 ITM mounting fasteners

12.16 Unlatch the cable retainer and detach the cable from the inner door handle

12.18 Window lift retaining screws

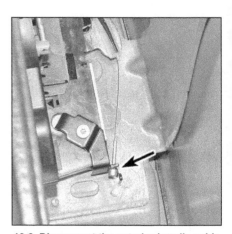

13.2 Disconnect the exterior handle cable end fitting from the lock

Integrated Trim Module (ITM) and regulator assembly

Refer to illustrations 12.14, 12.15, 12.16 and 12.18

13　Remove the window glass (see Section 11).

14　Disconnect the electrical connectors at the front of the ITM **(see illustration)**.

15　Remove the ITM retaining fasteners **(see illustration)**.

16　Pull the ITM up and off the door, then unlatch the cable retainer and detach the cable from the inner door handle **(see illustration)**.

17　Remove the motor from the ITM **(see illustration 12.11)**, then detach the motor drive from the other side of the ITM by freeing it from the plastic retaining tangs.

18　Detach the cables from any retaining clips, then remove the fasteners and slide the window lifts up and off the ITM **(see illustration)**.

19　Reassembly and installation is the reverse of removal. Adjust the window glass and initialize the motor (see Section 11).

13　Door handle and lock components - removal and installation

Mk I models

Exterior handle

Refer to illustrations 13.2 and 13.3

1　Remove the door inner trim panel as described in Section 10. Fully raise the window.

2　Disengage the cable end fitting from the lock, and pull the operating cable from the bracket on the lock **(see illustration)**.

3　Remove the two retaining Allen screws and maneuver the handle from the door frame **(see illustration)**. Installation is a reversal of the removal procedure.

Interior handle

Refer to illustrations 13.5 and 13.6

4　Remove the door inner trim panel as described in Section 10.

5　Remove the three retaining screws and

detach the handle assembly from the door frame **(see illustration)**. If required, depress the retaining clip and detach the treble speaker, disconnecting the electrical connectors as it's withdrawn.

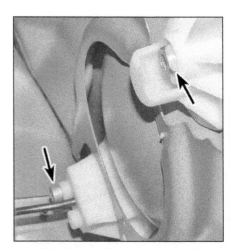

13.3 Remove the two screws and maneuver the handle from the door

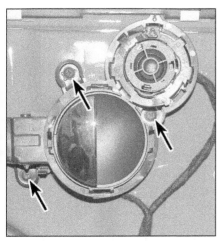

13.5 Interior handle retaining screws

13.6 Disconnect the interior handle cable

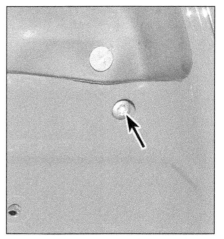

13.11 Pry out the rubber grommet and loosen the cylinder retaining screw

6   Using a thin-bladed screwdriver, unclip the outer cable locating clip, then release the inner operating cable from the handle assembly **(see illustration)**. Installation is a reversal of the removal procedure.

Lock motor/module

Refer to illustrations 13.11, 13.13, 13.14, 13.16a and 13.16b

7   Remove the door window as described in Section 11.
8   Remove the two bolts and move the rear window guide forwards to allow access to the door lock **(see illustration 12.3b)**.
9   Disconnect the electrical connector from the door lock motor.
10   Disconnect the exterior handle operating cable from the door lock **(see illustration 13.2)**.
11   If removing the driver's lock, remove the exterior handle (see Steps 1 through 3), then pry out the rubber grommet on the end of the door, and loosen the cylinder retaining screw **(see illustration)**.
12   Remove the interior door handle and detach the cable (see Steps 4 through 6). Free the cable from the retaining clips on

13.13 Pry the lock button operating rod from the lock

the door frame.
13   Unclip the lock button operating rod from the door lock **(see illustration)**.
14   Loosen and remove the three Torx bolts from the lock assembly at the rear edge of the door **(see illustration)**.
15   Maneuver the assembly out through the

13.14 Remove the three Torx bolts and maneuver the lock from the door

door inner frame, complete with the interior handle operating cable.
16   With the door lock assembly on a clean working surface, unclip the trim and disconnect the operating cable from the lever **(see illustrations)**. Installation is a reversal of the removal procedure. If installing the driver's

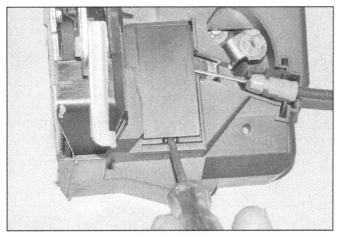

13.16a Pry up the trim . . .

13.16b . . . and disconnect the interior handle operating cable

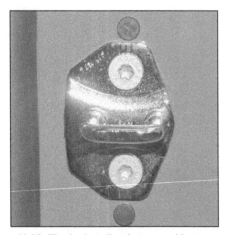

13.22 The lock striker is secured by two Torx screws

13.23 Remove the bolt securing the check strap to the door pillar . . .

13.24 . . . and the two bolts securing the check strap to the door

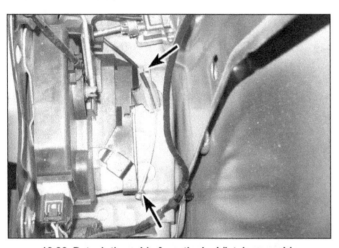

13.26 Detach the cable from the lock/latch assembly

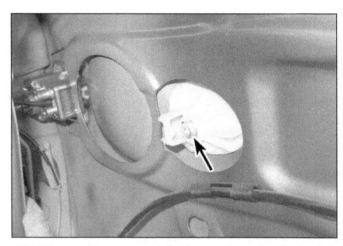

13.27 Remove the screw and plastic mount securing the front of the door handle

side lock, ensure the cylinder operating shaft engages correctly with the lock.

Lock cylinder

17   Remove the exterior handle (see Steps 1 through 3).

13.28 The door handle rear mounting screw is accessed through this hole in the door sheetmetal

18   Remove the three screws securing the door lock to the edge of the door frame **(see illustration 13.14)**.
19   Pry out the rubber grommet in the end of the door and remove the cylinder retaining Torx screw **(see illustration 13.11)**.
20   Lower the lock slightly, and remove the cylinder from the door skin. Installation is a reversal of removal.

Striker

Refer to illustration 13.22

21   Using a pencil, mark the position of the striker on the pillar.
22   Remove the mounting screws using a Torx key, then remove the striker **(see illustration)**. Installation is a reversal of the removal procedure, but check that the door lock passes over the striker centrally. If necessary, reposition the striker before fully tightening the mounting screws.

Check strap

Refer to illustrations 13.23 and 13.24

23   Using a Torx key, remove the check strap mounting bolt from the door pillar **(see illustration)**.

24   Pry the rubber grommet from the door opening, then remove the mounting screws and withdraw the check strap from the door **(see illustration)**. Installation is a reversal of the removal procedure.

Mk II models

Exterior handle

Refer to illustrations 13.26, 13.27 and 13.28

25   Remove the Integrated Trim Module (ITM) (see Section 12).
26   Detach the cable from the lock/latch assembly **(see illustration)**.
27   Remove the screw and detach the plastic mount at the front of the handle **(see illustration)**.
28   Remove the screw at the rear of the door handle **(see illustration)**, then detach the handle from the door.
29   Installation is the reverse of the removal procedure.

Lock/latch assembly

Refer to illustration 13.33

30   Remove the Integrated Trim Module (ITM) (see Section 12).

13.33 Door lock/latch mounting screws

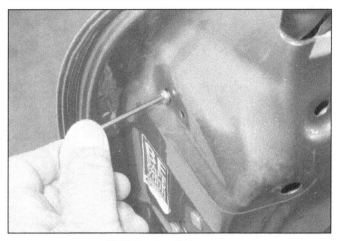

13.39 Lock cylinder screw

14.2a Remove the screw and pull the wiring connector from the pillar . . .

14.2b . . . then pull the slide lock up and disconnect the connector

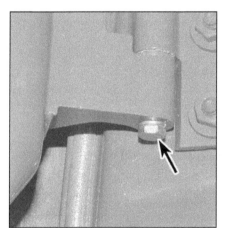

14.3 Door hinge bolt

31 Remove the exterior door handle (see Steps 25 through 28).

32 Detach the cables and inner door lock rod **(see illustration 13.26)**.

33 Remove the lock/latch assembly mounting screws from the end of the door **(see illustration)**.

34 Detach the lock/latch assembly and disconnect the electrical connector.

35 Installation is the reverse of removal. Make sure the lock cylinder actuating rod engages the lock/latch properly.

Lock cylinder

Refer to illustration 13.39

36 Remove the door exterior handle (see Steps 25 through 28).

37 Detach the lock cylinder cable from the lock/latch assembly **(see illustration 13.26)**.

38 Remove the lock/latch mounting screws from the end of the door **(see illustration 13.33)**.

39 Remove the lock cylinder screw from the end of the door **(see illustration)**, then remove the lock cylinder.

40 Installation is the reverse of removal.

Make sure the lock cylinder actuating rod engages the lock/latch properly.

Striker

41 This procedure is the same as for Mk I models. See Steps 21 and 22.

Check strap

42 Remove the door trim panel (see Section 10).

43 Remove the door speaker from the bottom of the Integrated Trim Module.

44 Pry the rubber grommet from the door opening, then remove the mounting screws and withdraw the check strap from the door **(see illustration 13.24)**. Guide the check strap out through the speaker opening.

45 Installation is the reverse of removal.

14 Door - removal and installation

Refer to illustrations 14.2a, 14.2b and 14.3

1 Remove the check strap mounting bolt from the door pillar **(see illustration 13.23)**.

2 Remove the screw, pull the wiring connector from the A-pillar, then slide out the

locking latch and disconnect the electrical connector **(see illustrations)**.

3 Remove the retaining bolts in the top and bottom hinges **(see illustration)**.

4 Carefully lift the door from the hinges.

5 Installation is a reversal of the removal procedure, but check that the door latch passes over the striker centrally. If necessary, reposition the striker.

15 Fender (front) - removal and installation

Mk I models

1 Apply the parking brake and block the rear wheels. Loosen the front wheel bolts. Raise the front of the vehicle and support it securely on jackstands, then remove the wheels.

2 Remove the side marker light (see Chapter 12).

3 Remove the wheelwell liner (see Section 29).

4 Remove the fasteners at the rear edge of the fender, the bottom of the fender by the rocker panel, mid-way up the fender, and from

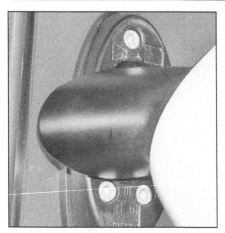

16.1 Rotate the mirror 90-degrees and remove the screws (three on Mk I models [shown], two on Mk II models)

16.4a Rotate the mirror base 90-degrees . . .

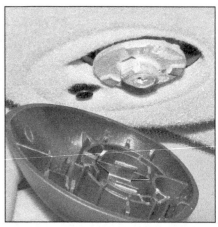

16.4b . . . and lower it from the mounting

the side marker light recess. Remove the fender panel.

5   Installation is the reverse of removal.

Mk II models

6   Apply the parking brake and block the rear wheels. Loosen the front wheel bolts. Raise the front of the vehicle and support it securely on jackstands, then remove the wheels.

7   Remove the side marker light (see Chapter 12).

8   Remove the wheelwell liner and the fender trim (see Section 29).

9   Remove the fender mounting bolts and remove the fender.

10   Installation is the reverse of removal

---

16   Mirrors - removal and installation

Exterior mirrors

Refer to illustration 16.1

1   Rotate the mirror to expose the mount-

ing screws **(see illustration)**.

2   Remove the Allen screws, then pull the mirror from the door. Disconnect the electrical connector as the mirror is withdrawn.

3   Installation is a reversal of the removal procedure. Take care not to drop the rubber grommet inside the door panel when removing the mirror, as the interior door trim will have to be removed to retrieve it.

Interior mirror

Models up to 07/2004

Refer to illustrations 16.4a and 16.4b

4   Twist the mirror base 90-degrees and lower the base from the mount **(see illustrations)**. On models with an auto-dimming mirror, disconnect the electrical connectors as the mirror is withdrawn.

5   Position the mirror base at 90-degrees to the mount, then rotate it to the normal position while holding the base against the mount. Reconnect the electrical connectors (where applicable) before installing the mirror.

Models from 07/2004

Refer to illustration 16.6 and 16.7

6   Pry apart the two halves of the mirror base cover, and disconnect the electrical connector **(see illustration)**.

7   Rotate the mirror base 90-degrees and detach it from the mount **(see illustration)**.

8   Position the mirror base at 90-degrees to the mounting, then rotate it to the normal position while holding the base against the mount. Reconnect the electrical connectors (where applicable) before installing the mirror.

9   Install the two halves of the cover.

---

17   Rear hatch - removal and installation

Mk I models

Refer to illustrations 17.4, 17.5 and 17.8

1   The hatch may be unbolted from the hinges and the hinges left in position.

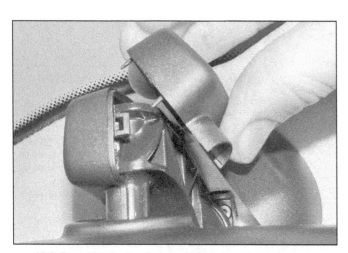

16.6 Pry apart the two halves of the mirror base cover . . .

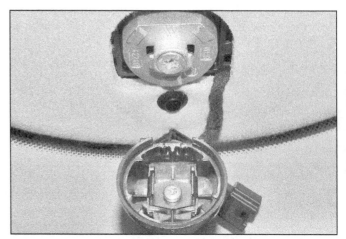

16.7 . . . then twist the mirror base 90-degrees and detach it from the mounting

17.4 Pry the rubber grommet from the hatch

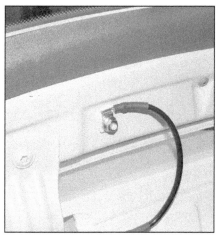

17.5 Disconnect the hatch ground wire (if equipped)

17.8 Unscrew the hatch hinge Torx bolts

2   Remove the hatch trim panels as described in Section 25.
3   Disconnect the wiring harness connectors through the hatch inner skin opening. Attach a strong fine cord to the end of the wiring harness, to act as an aid to guiding the wiring through the hatch when it is reinstalled.
4   Pry the rubber grommet from the hatch **(see illustration)**, and pull out the wiring harness. Untie the cord, leaving it in position in the hatch for guiding the wire through on installation.
5   Remove the nut securing the ground lead to the hatch **(see illustration)**.
6   Have an assistant support the hatch in its open position.
7   Using a small screwdriver, pry off the clip securing the struts to the hatch **(see illustration 18.2)**. Pull the sockets from the ballstuds, and move the struts downwards.
8   Remove the hinge Torx bolts (one on each side) from the hatch **(see illustration)**. Withdraw the hatch from the body opening, taking care not to damage the paint.
9   Installation is a reversal of the removal procedure, but check that the hatch is located centrally in the body opening, and that the striker enters the latch centrally. If necessary, loosen the mounting nuts and reposition the hatch.

Mk II models

Refer to illustrations 17.10 and 17.13

10   Open the hatch, pry out the covers and remove the fasteners **(see illustration)**. Remove the spoiler from the top of the hatch.
11   Using a small screwdriver, pry off the clip securing the struts to the hatch **(see illustration 18.2)**. Pull the sockets from the ballstuds, and move the struts downwards.
12   Detach the washer fluid hose and wiring harness connectors.
13   Close the hatch without allowing it to latch, then, with the help of an assistant and working through the openings previously covered by the spoiler, remove the hatch mount-

ing nuts **(see illustration)**.
14   Remove the hatch, being careful not to damage the paint.
15   Installation is a reversal of the removal procedure, but check that the hatch is located centrally in the body opening, and that the striker enters the latch centrally. If necessary, loosen the mounting nuts and reposition the hatch.

18   Support struts - removal and installation

Refer to illustration 18.2

1   Have an assistant support the hatch/hood in its open position.
2   Pry off the upper spring clip securing the strut to the hatch/hood, then pull the socket from the ballstud **(see illustration)**.
3   Similarly pry off the bottom clip, and pull the socket from the ballstud. Withdraw the strut.

17.13 Pry out the rubber covers and remove the hatch mounting nuts (one on each side) (Mk II models)

17.10 The spoiler mounting fasteners are accessed through these holes on the underside of the hatch (Mk II models)

4   Installation is a reversal of the removal procedure, making sure that the strut is installed the same way up as it was removed.

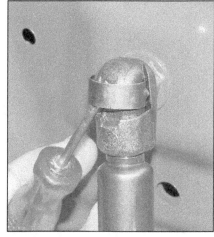

18.2 Pry off the support strut spring clip

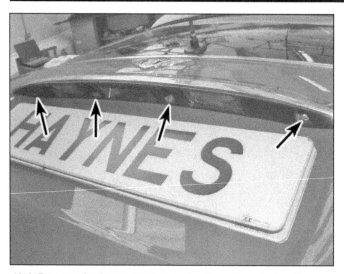

19.4 Remove the four screws securing the release button/license plate light assembly

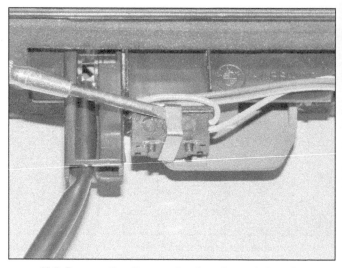

19.5 Remove the clip securing the microswitch to the release lever

19   Rear hatch lock components - removal and installation

Lock release button

Refer to illustrations 19.4 and 19.5

1   Disconnect the battery negative (ground) cable (see Chapter 5).
2   Remove the hatch inner trim panel as described in Section 25.
3   Disconnect the electrical connector from the hatch lock assembly.
4   Remove the four screws securing the release button/license plate light assembly from the outside of the hatch **(see illustration)**.
5   If required, release the clip and detach the microswitch from the release lever **(see illustration)**. To replace the microswitch, it's necessary to cut the wires, and re-solder them to the new switch.

Lock/latch assembly

Refer to illustrations 19.8 and 19.9

6   **Mk I models:** With the hatch open, pry up the center pins, then lever out the three plastic expansion rivets. Pull the hatch sill trim panel upwards to release the four retaining clips.
7   **Mk II models:** Remove the rear hatch trim panel (see Section 25).
8   Disconnect the electrical connector from the hatch lock assembly **(see illustration)**.
9   Open the emergency release cable junction box (on models so equipped), then slide the box forwards to detach it from the lock bracket **(see illustration)**.
10   Make alignment marks between the lock bracket and the vehicle body to aid installation, then remove the two Torx screws and remove the lock assembly.
11   Installation is a reversal of the removal procedure.

20   Central locking system components - removal and installation

Body control module

Refer to illustrations 20.3 and 20.4

Note: *If the module is to be replaced, the new unit must be reprogrammed/coded prior to use. This can only be carried out by a MINI dealer or specialist.*

1   To remove the module, first remove the passenger's side (Mk I models) or driver's side (Mk II models) sill/footwell kick panel as described in Section 25.
2   Disconnect the battery negative cable (see Chapter 5).
3   Remove the securing nuts, and remove the module **(see illustration)**.
4   Slide out the locking elements/lift the locking latches, and disconnect the electrical connec-

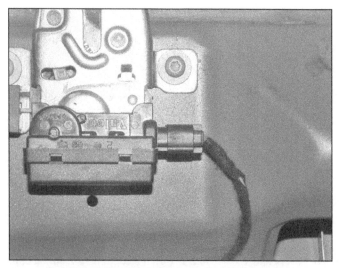

19.8 Disconnect the hatch lock electrical connector

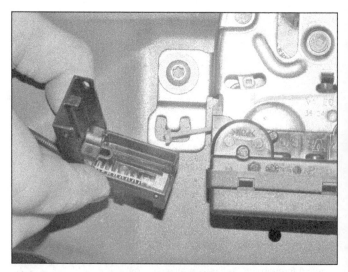

19.9 Open the junction box and slide it forward to disconnect the emergency release cable

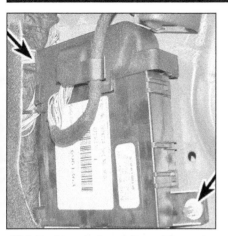

20.3 Body control module securing nuts (Mk I model)

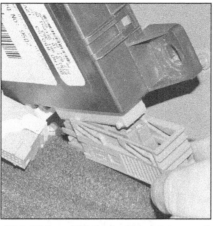

20.4 Slide out/lift the locking latches and disconnect the electrical connectors (Mk I model)

tors as the unit is withdrawn (see illustration).

5   Installation is the reverse of removal.

Door lock motors

6   The door lock motor is integral with the door lock/latch. Removal of the door lock is described in Section 13.

Rear hatch lock motor

7   The motor is integral with the hatch lock/latch. Removal of the lock/latch is described in Section 19.

---

21 Windshield and fixed glass - removal and installation

The windshield and rear window on all models are bonded in place with a special adhesive, as are the rear side windows. Special tools are required to cut free the old units and fit new ones; special cleaning solutions and primer are also required. It is therefore recommended that this work is entrusted to a MINI dealer or automotive window glass specialist.

---

22 Body side-trim moldings and adhesive emblems - removal and installation

Removal

1   Body side trims and moldings are attached either by retaining clips or adhesive bonding. On bonded moldings, insert a length of strong cord (fishing line is ideal) behind the molding or emblem concerned. With a sawing action, break the adhesive bond between the molding or emblem and the panel.

2   Thoroughly clean all traces of adhesive from the panel using rubbing alcohol and allow the location to dry.

3   On moldings with retaining clips, unclip the moldings from the panel, taking care not to damage the paint.

Installation

4   Peel back the protective paper from the rear face of the new molding or emblem.

Carefully fit it into position on the panel concerned, but take care not to touch the adhesive. When in position, apply hand pressure to the molding/emblem for a short period, to ensure maximum adhesion to the panel.

5   Replace any broken retaining clips before installing trims or moldings.

---

23 Sunroof - general information and initialization

General information

Refer to illustrations 23.2a, 23.2b and 23.2c

1   Due to the complexity of the sunroof mechanism, considerable expertise is needed to repair, replace or adjust the sunroof components successfully. Removal of the roof first requires the headliner to be removed, which is a complex and tedious operation, and not a task to be undertaken lightly. Therefore, any problems with the sunroof should be referred to a MINI dealer or other qualified repair shop.

2   On models with an electric sunroof, if the motor fails to operate, first check the relevant fuse. If the fault cannot be traced and rectified, then sunroof can be opened and closed manually using an Allen key to turn the motor spindle (a suitable key is supplied with the vehicle tool kit). To gain access to the motor, unclip the reading lights/switch panel/clock console from the headliner. Insert the Allen key into the motor spindle, and move the sunroof to the required position (see illustrations).

Initialization

3   With the battery reconnected and the ignition on, press the sunroof operating switch into the tilt position and hold it there.

4   Once the sunroof has reached the fully-tilted position, hold the switch in that position for approximately 20 seconds.

5   Close the sunroof and release the switch as normal.

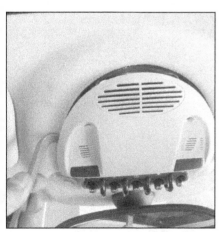

23.2a Pry the overhead console (Mk I models) or overhead console trim cover (Mk II models, shown) for access to the sunroof motor drive

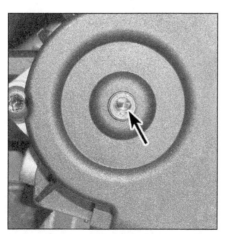

23.2b Insert the Allen key from the tool kit into the end of the motor spindle (Mk I model shown)

23.2c On Mk II models, insert the Allen key into this hole to engage the sunroof motor drive

24.4 Slide the seat rearwards, and remove the Torx bolts

24.5 Tilt the seat back and disconnect the electrical connectors

24.7 Rear seat cushion central clip (shown with the seat removed for clarity)

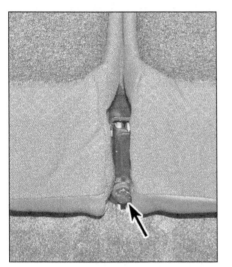

24.10 Remove the rear seat backrest retaining screw

---

24 Seats - removal and installation

Front seat

Refer to illustrations 24.4 and 24.5

1   On models with electrically operated or heated seats, or with side airbags, disconnect the battery negative cable, and position the cable away from the battery (see Chapter 5). **Warning:** *If equipped with side-impact airbags, wait a minimum of 5 minutes before proceeding, as a precaution against accidental firing of the airbag unit. This period ensures that any stored energy is dissipated.*
2   Slide the seat fully forwards.
3   Remove the two rear seat runner Torx bolts.
4   Slide the seat fully rearwards, then remove the two front seat runner Torx bolts **(see illustration)**.
5   Tilt the seat backwards, and disconnect the various seat electrical connectors from the seat base, noting their positions **(see illustration)**. Remove the seat from

the vehicle. Installation is the reverse of removal.

Rear seat cushion

Refer to illustration 24.7

6   Lift the front edge of the seat cushion to release the clips.
7   Push the rear of the seat cushion downwards to disengage the central clip **(see illustration)**.
8   Lift the cushion and disengage the four retaining clips under the child seat mount (if equipped). Remove the cushion from the vehicle. Installation is the reverse of removal.

Rear seat backrest

Refer to illustrations 24.10 and 24.11

9   Remove the parcel shelf (if equipped).
10   Fold the backrest forwards, then remove the retaining screw **(see illustration)**. Lift the rear of the bracket and disengage the backrest from the central mounting.
11   Hold the backrest at an angle of approximately 45-degrees, and pull it from the side mounting **(see illustration)**. Withdraw the backrest from inside the vehicle. Installation is the reverse of removal.

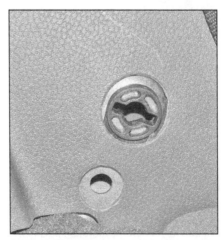

24.11 Hold the backrest at 45-degrees and pull it from the mount

---

25 Interior trim panels - removal and installation

Warning: *These models are equipped with a Supplemental Restraint System (SRS), more commonly known as airbags. Always disable the airbag system before working in the vicinity of any airbag system component to avoid the possibility of accidental deployment of the airbag(s), which could cause personal injury (see Chapter 12).*

A-pillar trim

Mk I models

Refer to illustration 25.1

1   Pull away the rubber weatherstrip from the door opening adjacent to the pillar trim, then carefully pull the trim from place, starting at the upper rear edge. Note that the trim is brittle – breakage is likely where the plastic lugs of the trim engage the metal clips. Note how the lower edge of the trim engages **(see illustration)**. On models with head curtain airbags, the pillar trim retaining straps must be cut through to release the trim, thus destroying the trim. Installation is the reverse of removal.

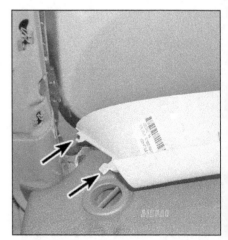

25.1 The lugs at the lower edge of the A-pillar trim clip into the instrument panel

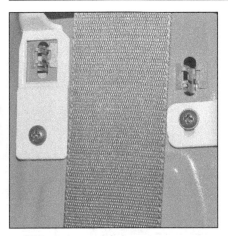

25.5a Later models have two screws securing the lower edge of the B-pillar trim, early models only have one

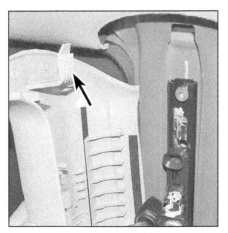

25.5b Pull the B-pillar trim downwards to disengage the upper clip

25.10 Pull the C-pillar trim towards the center of the vehicle to release the clips

Mk II models

2   Pry out the trim plug near the top of the A-pillar trim, remove the screw underneath, then carefully pry the trim from the pillar, starting from the top. If equipped, disconnect the electrical connector for the tweeter speaker. Installation is the reverse of removal, but make sure the weatherstrip and trim panel engage correctly.

B-pillar trim
Mk I models

Refer to illustrations 25.5a and 25.5b

3   Remove the luggage compartment side panel trim.
4   Remove the front seat belt lower anchorage Torx bolt. **Warning:** *The manufacturer recommends replacing seat belt bolts with new ones whenever they are removed.*
5   Remove the retaining screws from the lower trim panel, then pull the pillar trim downwards to release the retaining clips **(see illustrations)**.
6   Installation is the reverse of removal. Use thread-locking compound on the seat belt bolt.

Mk II models

7   If equipped, carefully pry off the trim cover near the top of the B-pillar trim panel and remove the screw, then pry the panel from the pillar.
8   Installation is the reverse of removal, but make sure the weatherstrip and trim panel engage correctly.

C-pillar trim
Mk I models

Refer to illustration 25.10

9   Remove the luggage compartment side panel trim, then pull away the rubber weatherstrip from the hatch opening adjacent to the pillar trim.
10   Carefully pull the pillar trim towards the center of the car to release the trim clips **(see illustration)**. The pillar trim is brittle – breakage is likely where the plastic lugs of the trim engage the metal clips. On models with side curtain airbags, the pillar trim top retaining strap must be cut through to release the trim,

thus destroying the trim. Installation is the reverse of removal.

Mk II models

11   Remove the rear seat (see Section 24).
12   Remove the rear seat belt lower anchorage Torx bolt. **Warning:** *The manufacturer recommends replacing seat belt bolts with new ones whenever they are removed.*
13   Carefully pry off the trim cover near the top of the panel, remove the screw, then carefully pry off the panel.
14   Installation is the reverse of removal, but make sure the weatherstrip and trim panel engage correctly.

Rear hatch trim
Mk I models

Refer to illustrations 25.15a, 25.15b, 25.16 and 25.18

15   Remove the six trim panel retaining screws, then use a flat-bladed tool to release the push-on clips around the panel's perimeter, and remove the panel **(see illustrations)**.

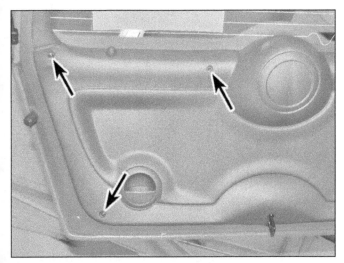

25.15a Remove the hatch trim retaining screws (right-side screws shown - left-side identical) . . .

25.15b . . . release the clips at the perimeter, and remove the panel

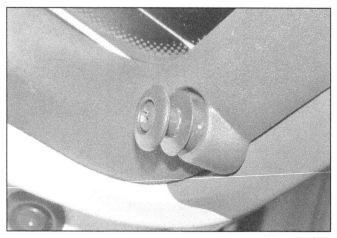

25.16 Remove the screw and detach the parcel shelf strap button on each side

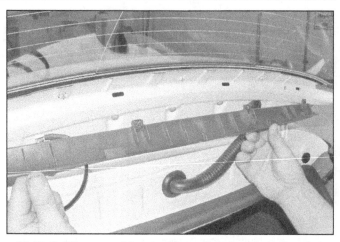

25.18 Pull the upper trim panel from place to release the clips

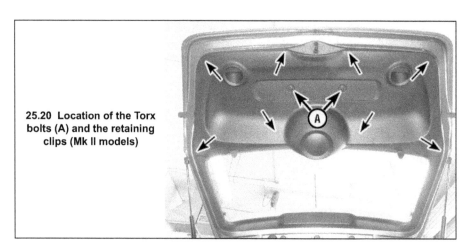

25.20 Location of the Torx bolts (A) and the retaining clips (Mk II models)

Sill/footwell kick panel trim

Refer to illustrations 25.22, 25.24 and 25.25

22   On Mk I models, pull away the rubber weatherstrip adjacent to the sill trim **(see illustration)**.

23   Remove the seat belt lower anchorage Torx bolt.

24   On the passenger's side trim of Mk I models, pry up the center pin, then lever out the complete plastic expansion rivet just in front of the fusebox cover **(see illustration)**.

25   Pull the panel trim horizontally towards the center of the vehicle to release the push-on clips **(see illustration)**.

26   Installation is the reverse of the removal procedure, but make sure the weatherstrip and trim panel engage correctly.

16   Remove the screw on each side securing the parcel shelf strap buttons **(see illustration)**.

17   Use a flat-bladed tool to release the push-on clips and remove the side trims.

18   Release the clips and pull the upper trim panel from place **(see illustration)**.

19   Installation is the reverse of removal.

Mk II models

Refer to illustration 25.20

20   Remove the two Torx screws from the center of the panel, then carefully pry the panel from the hatch in the area of the clips **(see illustration)**.

21   Installation is the reverse of the removal procedure.

Steering column shrouds
Mk I models

Refer to illustrations 25.27a, 25.27b and 25.27c

27   To release the lower shroud, pry off the rubber cover from the ignition switch, remove

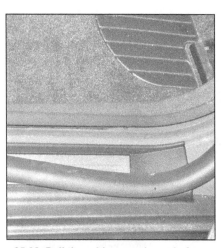

25.22 Pull the rubber weatherstrip from the door opening

25.24 Pry up the center pin, then remove the plastic expansion rivet

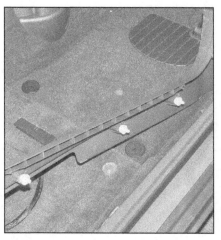

25.25 Pull the panel horizontally from the sill to release the clips

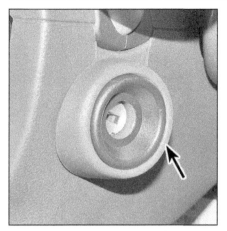

25.27a Pry the rubber cover from the ignition switch . . .

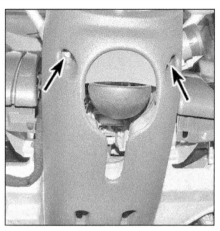

25.27b . . . then remove the two Torx screws

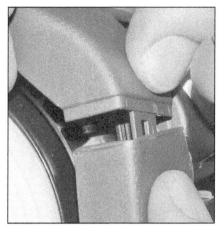

25.27c Squeeze the shrouds together where they join to release the clips

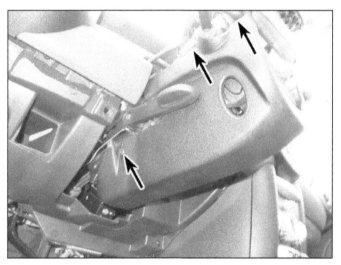

25.33 Lower steering column shroud mounting screws

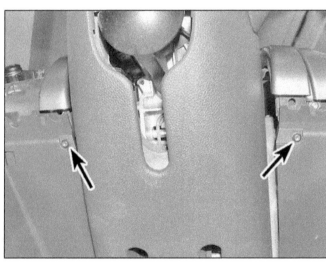

25.35 Remove the two Torx screws – models up to 07/2004

the two Torx screws, and release the two clips at the lower end of the shroud **(see illustrations)**. Squeeze together the side of the shrouds, and detach the lower shroud from the upper shroud. **Note:** *On some models we found it necessary to remove the airbag (see Chapter 12) and loosen the steering wheel bolt, allowing the steering wheel to move rearwards a little to provide sufficient clearance to remove the shroud.*

28 To remove the upper shroud, first remove the shroud mounted instrument(s), as described in Chapter 12, then remove the lower shroud as previously described. Lift the upper shroud from position.

29 Installation is the reverse of the removal procedure.

Mk II models

Refer to illustration 25.33

30 Remove the steering wheel (see Chapter 10) and the tachometer (see Chapter 12).

31 Carefully squeeze the top of the upper shroud in at the front and rear while prying the upper shroud off.

32 Remove the driver's side lower dash panel (see Step 38).

33 Remove the lower shroud mounting screws and detach the shroud from the steering column **(see illustration)**.

34 Installation is the reverse of the removal procedure.

Driver's side lower dash panel
Mk I models

Refer to illustrations 25.35 and 25.36

35 On models up to 07/2004, remove the two Torx screws and pull the panel rearwards to release it from the retaining clips **(see illustration)**. Disconnect the headlamp range control switch electrical connector as the panel is withdrawn.

36 On models from 07/2004, pull down the rear edge of the panel, and release it from the retaining clips, then pull the panel from the front hinges **(see illustration)**. Disconnect the headlight range control switch electrical connector as the panel is withdrawn.

37 Installation is the reverse of removal.

Mk II models

Refer to illustration 25.38

38 Remove the three screws from the bottom edge of the panel, then carefully pry the

25.36 On models from 07/2004, pull down the rear edge of the panel, then pull it rearwards (right-hand drive Mk I model shown)

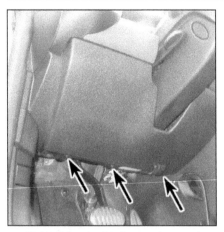

25.38 Driver's side lower trim panel
screws (Mk II models)

26.2 Release the two clips and slide the
end trim from the parking brake lever

26.3 Lift the locking tab and slide the grip
from the lever

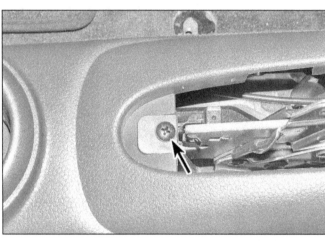

26.5a Remove the screw at the rear of the parking brake
lever opening . . .

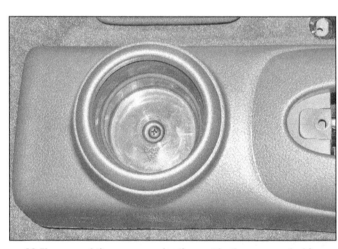

26.5b . . . and the screw under the mat in the rear cupholder

top of the panel out to dislodge the clips **(see illustration)**.

39   Installation is the reverse of the removal procedure.

---

26 Center console - removal and installation

Mk I models

Rear center console

Refer to illustrations 26.2 and 26.3

1   Release the clips securing the parking brake lever boot to the console.

2   Use two small screwdrivers to depress the clip on each side, and remove the lever end trim **(see illustration)**.

3   Lift the locking tab, then slide the lever grip and boot from the lever **(see illustration)**.

Models up to 07/2004

4   Remove the console retaining screw at the rear of the parking brake lever opening, then pull the console upwards and release it

from the four push-on clips **(see illustration 26.5a)**. Disconnect any electrical connectors as the console is removed.

Models from 07/2004

Refer to illustrations 26.5a, 26.5b and 26.6

5   Remove the console retaining screw at the rear of the parking brake lever opening, then lift out the mat in the base of the rear

cupholder and remove the exposed screw **(see illustrations)**.

6   Starting at the rear, lift the console from place to release the four retaining clips **(see illustration)**, disconnecting any electrical connectors as the console is removed. Note how the front of the rear console fits under the rear of the front console

7   Fully apply the parking brake lever to

26.6 The rear console
retaining clips locate in
two holes each side

withdraw the center console from the vehicle. Installation is the reverse of removal.

Front center console

Refer to illustrations 26.9a, 26.9b, 26.11, 26.13, 26.15a, 26.15b and 26.17

8   Pull the gear lever knob straight up to detach it from the lever.

9   Pry up the gear lever boot surround from the center console, and lift the boot over the lever. If necessary, cut the cable tie to allow the boot to pass over the lever **(see illustrations)**.

10   Remove the ashtray (if equipped) from the cupholder.

11   Remove the screw in the bottom of both cupholders **(see illustration)**.

12   Remove the mirror adjustment/heated seat switch(es) from the front center console as described in Chapter 12, Section 4.

13   Remove the two screws in the switch plate recess **(see illustration)**.

14   On models from 07/2004, pull the rear edge of the driver's side lower dash panel downwards to release the clips, then detach the panel hinges from the instrument panel **(see illustration 25.36)**. Disconnect the headlight range control switch electrical con-

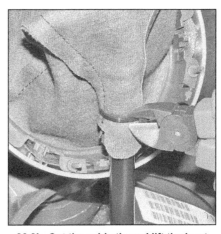

26.9a Note how the pins on the base of the gear lever boot trim locate in the holes in the center console

nector as the panel is withdrawn.

15   Open the glove box lid and remove the two screws on each side securing the center console pillars to the instrument panel, then slide the center console up the pillars a little **(see illustrations)**.

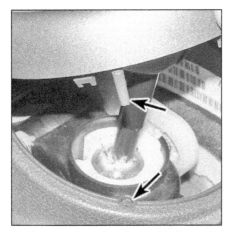

26.9b Cut the cable tie and lift the boot from the lever

16   Pull the console to the rear, and disconnect the cigarette lighter socket electrical connector as the console is removed.

17   If required, unscrew the two nuts and remove the console base from the vehicle floor **(see illustration)**. Installation is the reverse of removal.

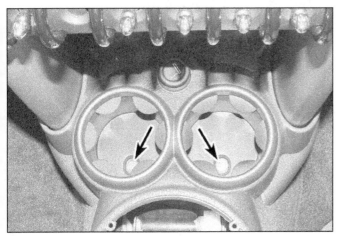

26.11 Remove the two screws in the base of the cupholders

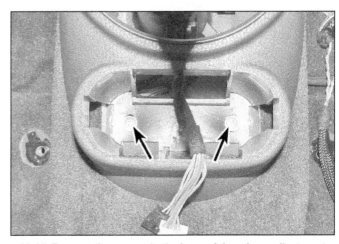

26.13 Remove the screws in the base of the mirror adjustment switch recess

26.15a Remove the two screws on each side securing the console pillars . . .

26.15b . . . then slide the center console up the pillars a little

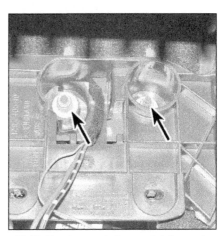

26.17 Front console base retaining nuts

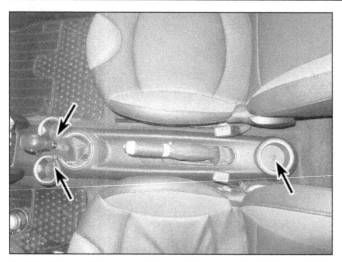

26.19a Remove these console mounting screws (Mk II models) . . .

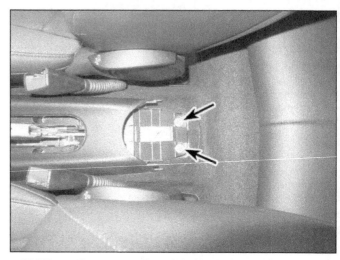

26.19b . . . then remove the cup holder and remove these two screws underneath

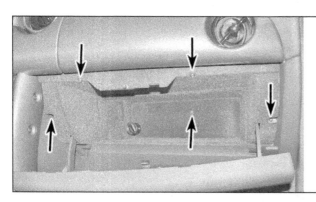

27.2 Remove the five screws and pull the glove box rearwards

2   Remove the screws, then pull the glove box from the instrument panel **(see illustration)**. Detach the cooling hose (if equipped) as the glove box is removed.

3.   Installation is the reverse of removal.

Storage tray

Refer to illustrations 27.4a and 27.4b

4   Remove the three screws, release the two retaining clips and pull the storage tray rearwards to detach it from the instrument panel **(see illustrations)**.

5   Installation is a reversal of the removal procedure.

Mk II models

Refer to illustrations 26.19a and 26.19b

18   Remove the shift knob and shifter boot (see Chapter 7A or 7B).

19   Remove the console fasteners, then pull the sides of the console near the parking brake lever outwards to detach the retaining clips **(see illustrations)**.

20   Installation is the reverse of removal.

27 Glove box - removal and installation

Glove box with lid

Refer to illustration 27.2

1   Open the glove box, and pry out the light unit. Disconnect the electrical connector as the light is withdrawn.

28 Instrument panel - removal and installation

Refer to illustrations 28.5, 28.8, 28.9a, 28.9b, 28.10, 28.16, 28.17, 28.18, 28.19a, 28.19b, 28.19c, 28.19d, 28.20, 28.21 and 28.22

Warning: *Wait until the engine is completely cool before beginning this procedure.*

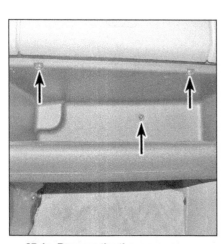

27.4a Remove the three screws . . .

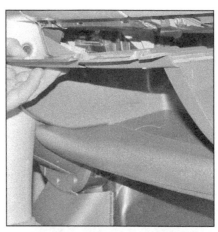

27.4b . . . and pull the storage tray rearwards to release the retaining clips

28.5 Release the clamps and disconnect the heater hoses from the engine compartment firewall

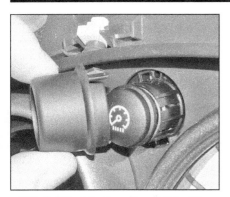

28.8 Pull the trim from the switch either side of the speedometer

28.9a Early models have three Torx screws securing the instrument panel surround . . .

28.9b . . . while later models have two

Note: *This procedure applies to Mk I models only, as on those models it is necessary to remove the instrument panel for access to the blower motor. Other than that, there are no other service procedures on either the Mk I or Mk II models that require instrument panel removal.*

1 Disconnect the cable from the negative terminal of the battery (see Chapter 5).

2 Drain the cooling system as described in Chapter 1.

3 On Cooper models, remove the battery and battery tray as described in Chapter 5.

4 On Cooper S models, remove the air cleaner housing (see Chapter 4).

5 Loosen the clamps and disconnect the heater hoses from the connections on the engine firewall **(see illustration)**.

6 On air conditioned models, have the refrigerant discharged as described in Chapter 3.

7 Remove the passenger's side airbag as described in Chapter 12.

8 Unclip the trim from the switches on either side of the speedometer **(see illustration)**.

9 Remove the screws and pull the center instrument panel trim panel from its retaining clips **(see illustrations)**.

10 Remove the four Torx screws and remove the speedometer **(see illustration)**. Disconnect the electrical connectors as the unit is withdrawn.

11 Remove the heater control panel (see Chapter 3).

12 Remove the driver's side lower dash panel as described in Section 25.

13 Remove the bolts securing the front seat belt lower anchorage points, pull the rubber weatherstrips from the lower door opening, then unclip the sill trim panel on each side by pulling the panel towards the center of the vehicle. Note that the passenger's side trim panel is also secured by a plastic expansion rivet adjacent to the fusebox at the front of the panel. Pry up the center pin, then lever out the complete rivet (see Section 25).

14 Remove the steering column shrouds as described in Section 25.

15 On air-conditioned models, remove the two bolts securing the air conditioning refrig-

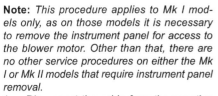

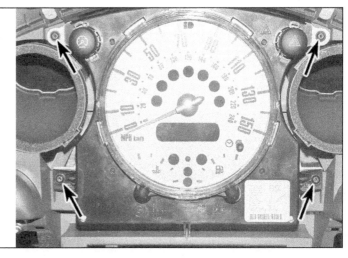

28.10 Remove the four screws and remove the speedometer

erant lines to the connection on the firewall. Discard the O-rings; new ones must be installed. Plug/seal the openings of the lines (see Chapter 3).

16 Set the steering wheel in the straight-ahead position, and engage the steering lock. Make alignment marks between the steering column universal joint and the steering gear input shaft, then remove the joint pinch-bolt/

nut **(see illustration)**. Discard the nut (a new one must be installed).

17 Slide the universal joint to the rear to detach it from the steering rack pinion, then pull the lower section of the steering column upwards, and pry the column plastic spacer from the rubber seal in the driver's footwell **(see illustration)**.

18 Remove the Torx screw and remove the

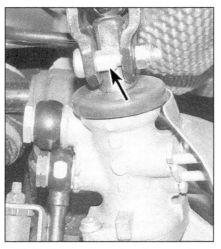

28.16 Remove the nut and remove the steering column lower joint pinch-bolt

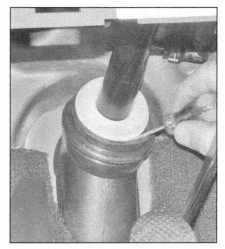

28.17 Pry the plastic spacer from the rubber boot in the driver's footwell

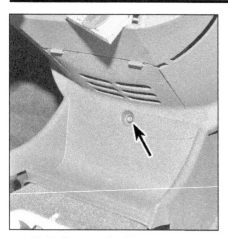

28.18 Remove the Torx screw and remove the lower panel

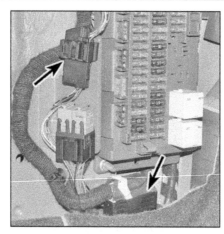

28.19a Disconnect the electrical connectors at the fusebox . . .

harness adjacent to the steering column (**see illustrations**).

20   Unscrew the instrument panel side-mounting screws (two on each side) (**see illustration**).

21   Remove the two bolts securing the instrument panel underneath the steering column (**see illustration**).

22   Working in the engine compartment, remove the nut securing the heater housing to the firewall (**see illustration**).

23   With the help of an assistant, pull the instrument panel, heater housing and steering column from the firewall, and maneuver out from the vehicle. Take care not to damage the heater pipes where they fit through the firewall.

Installation

24   Installation is a reversal of the removal procedure, noting the following points:

a) *Refill the cooling system (see Chapter 1).*
b) *Check the operation of all electrical components.*
c) *Have the air conditioning system recharged by the shop that discharged it.*

lower panel from the instrument panel center panel (**see illustration**).

19   In order for the panel to be removed complete with the steering column, etc., various electrical connectors must be unplugged. Two

connectors on the fusebox, four plugs under the steering column, one plug on the body computer in the passenger's side footwell, and one plug to the right of the heater housing. Release the cable ties securing the wiring

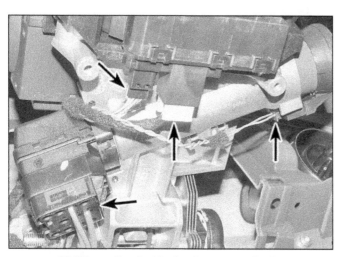

28.19b . . . the electrical connectors under the steering column . . .

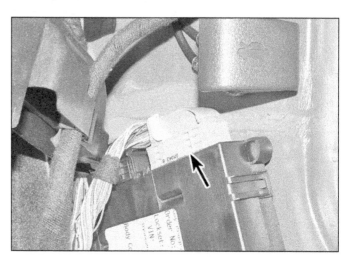

28.19c . . . one electrical connector on the body control computer . . .

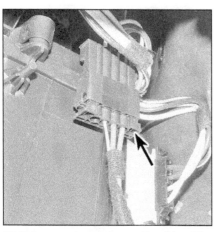

28.19d . . . and one plug on the side of the heater housing

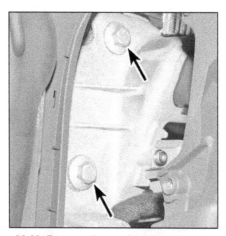

28.20 Remove the two instrument panel crossmember bolts on each side (left-hand bolts shown)

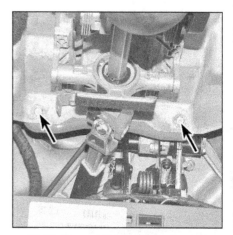

28.21 Remove the instrument panel crossmember bolts under the steering column

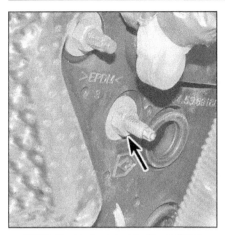

28.22 Heater housing-to-firewall nut

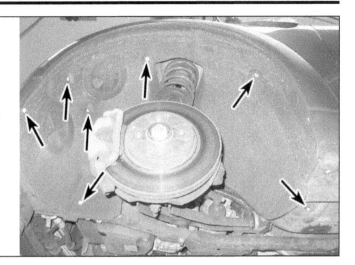

29.2 Loosen the screws, then pry out the plastic expansion rivets securing the wheelwell liner

29 Wheelwell liner and fender trim - removal and installation

Wheelwell liner

Removal

Front

Refer to illustration 29.2

1   Apply the parking brake. If the wheel is to be removed (to improve access), loosen the wheel bolts. Raise the front of the vehicle and support it securely on jackstands. Remove the front wheel.
2   Remove the screws/plastic expansion rivets securing the liner **(see illustration)**.
3   Remove the screws and clips securing the liner to the outer edge of the wheelwell and bumper. Withdraw the liner from under the vehicle.

Rear

Refer to illustrations 29.5, 29.7a and 29.7b

4   Chock the front wheels, and engage Park (automatic) or 1st gear (manual). If the wheel is to be removed (to improve access),

loosen the wheel bolts. Raise the rear of the vehicle and support it on jackstands. Remove the rear wheel.
5   Remove the screws and remove the rear mudflaps, if equipped **(see illustration)**.
6   Remove the screws securing the liner to the outer edge of the wheelwell and bumper.
7   Remove the plastic expansion rivets/nut securing the liner to the inner wheelwell, and withdraw the liner from under the vehicle **(see illustrations)**.

Installation

8   Installation is a reversal of the removal procedure. If the wheels were removed, tighten the wheel bolts to the torque listed in the Chapter 1 Specifications.

Fender trim

Refer to illustration 29.12

9   Disconnect the electrical connectors from the side marker light.
10   Remove the fasteners from the front of the wheelwell liner, pull it back to access the fastener securing the fender trim to the front bumper cover, then remove the fastener.

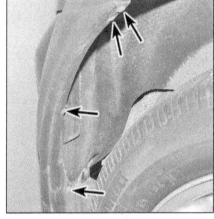

29.5 Remove the mudflap retaining screws

11   Remove the fastener from behind the side marker light.
12   Carefully pull the trim from the fender to dislodge the retaining clips **(see illustration)**.
13   Installation is the reverse of the removal procedure.

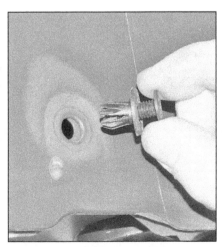

29.7a Loosen the screws, and pry out the plastic expansion rivets

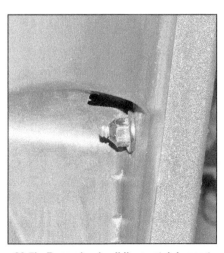

29.7b Rear wheelwell liner retaining nut

29.12 Pull the fender trim away from the fender to detach the retaining clips (if necessary, carefully pry with a plastic trim tool)

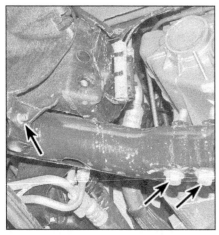

30.4 Remove the three crush tube bolts from each side

30.5a Insert the 8 mm x 100 mm bolts into the right-side . . .

---

30   Modular Front End (MFE)/radiator support panel - repositioning for service

1   Disconnect the cable from the negative terminal of the battery (see Chapter 5). Apply the parking brake and block the rear wheels. Loosen the front wheel bolts. Raise the front of the vehicle and support it securely on jackstands, then remove the wheels.

Mk I models

Refer to illustrations 30.4, 30.5a and 30.5b

2   Remove the front wheelwell liners and fender trims (see Section 29). Remove the bumper cover and bumper carrier (see Section 6).
3   If you're working on a model with a CVT automatic transaxle, remove the fluid cooler mounted to the top of the radiator and plug the line fittings.
4   Remove the three bolts from each side securing the crush tubes to the MFE and the subframe **(see illustration)**.
5   Thread two 8 mm diameter bolts of the correct thread pitch, 100 mm long, into the

bumper support members **(see illustrations)**.
6   Slide the MFE forwards, supported by the bolts, making sure nothing is still connected that impedes its travel.
7   When reassembling, tighten the crush tube-to-subframe bolts to 74 ft-lbs (100 Nm) and the crush tube-to-MFE bolts to 44 inch-lbs (5 Nm).
8   On models with a CVT transaxle, check the fluid level, adding as necessary (see Chapter 7B).

Mk II models

Refer to illustrations 30.14, 30.15, 30.16, 30.20, 30.21, 30.22, 30.23a, 30.23b, 30.23c, 30.23d, 30.24a and 30.24b

9   Remove the fasteners securing the front edges of the wheelwell liners to the bumper cover (see Section 29).
10   Detach the fender trims (see Section 29). Remove the under-vehicle splash shield.
11   Disconnect the electrical connector from the air conditioning compressor harness (see Chapter 3).
12   On S models, detach the intercooler ducts from the intercooler (see Chapter 4).
13   If equipped with headlight washers, detach the electrical connector from the headlight washer pump.

30.5b . . . and left side of the chassis members

14   Remove the subframe extension bolts from each side **(see illustration)**.
15   Remove the fender front mounting bolt from each side **(see illustration)**.
16   Detach the refrigerant line from the right side of the radiator support panel **(see illustration)**.
17   Remove the screw securing the coolant

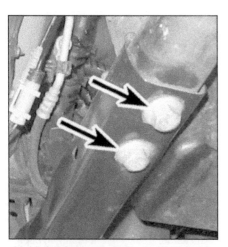

30.14 Subframe mounting bolts (left side shown, right side similar)

30.15 Remove the fender front mounting bolt from each side

30.16 Detach the refrigerant line from the right side of the radiator support panel

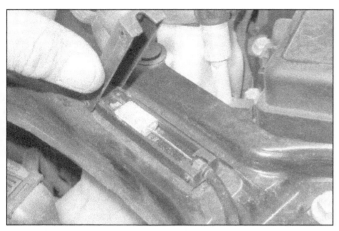

30.20 Disconnect the hood release cable at the junction at the left side of the engine compartment

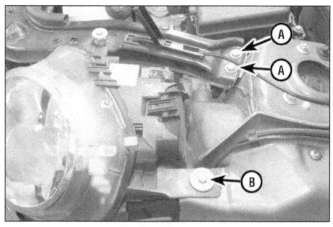

30.21 Radiator support mounting bolts (A) and headlight housing rear mounting bolt (B)

30.22 Use a plastic trim tool to pry the lower grill from the bumper cover

30.23a Insert the 8 mm x 100 mm bolts into these holes (left side shown)

30.23b Here's the right-side bolt installed (bumper cover removed for clarity)

expansion tank (see Chapter 3, if necessary).
18   Detach the duct from the air intake at the left side of the radiator support.
19   On S models, disconnect the electrical connector from the intake duct charge air pressure sensor (see Chapter 6). Also detach the intake duct from the tube directly

behind the sensor.
20   Flip up the cover and disconnect the hood release cable from the junction **(see illustration)**.
21   Remove the radiator support mounting bolts and the headlight housing rear mounting bolt **(see illustration)**.

22   Pry the lower grille from the bumper cover **(see illustration)**.
23   Thread two 8 mm diameter bolts of the correct thread pitch, 100 mm long, into the bumper support members **(see illustrations)**. Remove the bumper reinforcement fasteners from each side **(see illustrations)**.

30.23c Working through the bumper cover, remove these bumper reinforcement fasteners (not all are visible here)

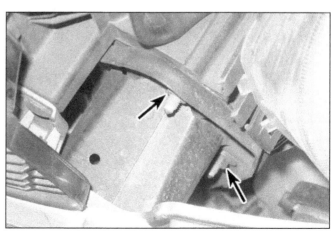

30.23d Bumper reinforcement fasteners, viewed from above (not all are visible here)

30.24a Pull the radiator support forward approximately 2-1/2 inches (6.5 cm) . . .

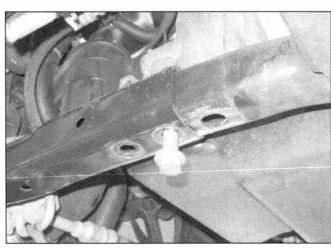

30.24b . . . then install a subframe extension bolt into the front hole of the subframe and into the rear hole of the subframe extension on each side to prevent the assembly from sliding off

24   Pull the radiator support/bumper assembly forward 2-1/2 inches (6.5 cm), making sure the subframe extensions don't slide off the subframe (see illustration). To prevent this, insert a subframe mounting bolt into the front hole of the subframe and into the rear hole of the extension, on each side (see illustration).

25   When reassembling, tighten the subframe extension/crush tube bolts to 74 ft-lbs (100 Nm) and the bumper reinforcement bolts to 16 ft-lbs (22 Nm).

31   Cowl covers - removal and installation

Mk I models

1   Refer to Chapter 12, Section 11 for the cowl cover removal and installation procedure.

Mk II models

Refer to illustrations 31.3, 34.4a and 31.4b

2   Remove the windshield wiper arms (see Chapter 12, Section 10).
3   Peel the rubber weatherstrip from the welt at the front of the cowl covers (see illustration).
4   Remove the fasteners from the cowl covers, then remove the covers (see illustrations).
5   Installation is the reverse of removal.

31.3 Peel the weatherstrip from the welt at the front of the cowl covers

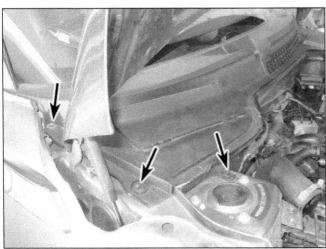

31.4a Right-side cowl cover fasteners

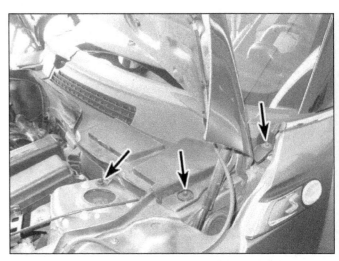

31.4b Left-side cowl cover fasteners

Chapter 12
Chassis electrical system

Contents

1 General information and electrical troubleshooting

Refer to illustrations 1.8a, 1.8b, 1.9 and 1.12

Note: *Throughout this Chapter you will find references to "Mk I" and "Mk II" models; this is done to simplify which procedures apply to which models. Mk I models include 2006 and earlier Cooper/Cooper S models, and 2008 and earlier Convertible models. Mk II models include 2007 and later Cooper/Cooper S/ Clubman/Clubman S and 2009 and later Convertible models.*

The electrical system is a 12-volt, negative ground type. Power for the lights and all electrical accessories is supplied by a lead/ acid-type battery that is charged by the alternator.

This Chapter covers repair and service procedures for the various electrical components not associated with the engine. Information on the battery, alternator, ignition system and starter motor can be found in Chapter 5.

It should be noted that when portions of the electrical system are serviced, the negative cable should be disconnected from the battery to prevent electrical shorts and/or fires.

Troubleshooting

A typical electrical circuit consists of an electrical component, any switches, relays, motors, fuses, fusible links or circuit breakers related to that component and the wiring and connectors that link the component to both the battery and the chassis. To help you pinpoint an electrical circuit problem, wiring diagrams are included at the end of this Chapter.

Before tackling any troublesome electrical circuit, first study the appropriate wiring diagrams to get a complete understanding of what makes up that individual circuit. Trouble spots, for instance, can often be narrowed down by noting if other components related to the circuit are operating properly. If several components or circuits fail at one time, chances are the problem is in a fuse or ground connection, because several circuits are often routed through the same fuse and ground connections.

Electrical problems usually stem from simple causes, such as loose or corroded connections, a blown fuse, a melted fusible link or a failed relay. Visually inspect the condition of all fuses, wires and connections in a problem circuit before troubleshooting the circuit.

If test equipment and instruments are going to be utilized, use the diagrams to plan ahead of time where you will make the necessary connections in order to accurately pinpoint the trouble spot.

The basic tools needed for electrical troubleshooting include a circuit tester or voltmeter (a 12-volt bulb with a set of test leads can also be used), a continuity tester, which includes a bulb, battery and set of test leads, and a jumper wire, preferably with a circuit breaker incorporated, which can be used to bypass electrical components **(see illustrations)**. Before attempting to locate a problem with test instruments, use the wiring diagram(s) to decide where to make the connections.

Voltage checks

Voltage checks should be performed if a circuit is not functioning properly. Connect one lead of a circuit tester to either the negative battery terminal or a known good ground. Connect the other lead to a connector in the circuit being tested, preferably nearest to the battery or

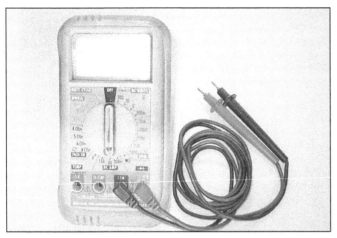

1.8a **The most useful tool for electrical troubleshooting is a digital multimeter that can check volts, amps, and test continuity**

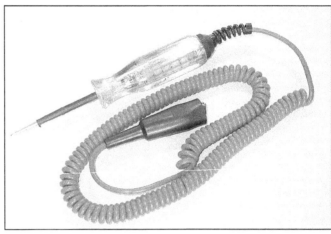

1.8b **A test light is a very handy tool for checking voltage**

fuse **(see illustration)**. If the bulb of the tester lights, voltage is present, which means that the part of the circuit between the connector and the battery is problem free. Continue checking the rest of the circuit in the same fashion. When you reach a point at which no voltage is present, the problem lies between that point and the last test point with voltage. Most of the time the problem can be traced to a loose connection. **Note:** *Keep in mind that some circuits receive voltage only when the ignition key is in the Accessory or Run position.*

Finding a short

One method of finding shorts in a circuit is to remove the fuse and connect a test light or voltmeter in place of the fuse terminals. There should be no voltage present in the circuit. Move the wiring harness from side-to-side while watching the test light. If the bulb goes on, there is a short to ground somewhere in that area, probably where the insulation has rubbed through. The same test can be performed on each component in the circuit, even a switch.

Ground check

Perform a ground test to check whether a component is properly grounded. Disconnect the battery and connect one lead of a continuity tester or multimeter (set to the ohms scale), to a known good ground. Connect the other lead to the wire or ground connection being tested. If the resistance is low (less than 5 ohms), the ground is good. If the bulb on a self-powered test light does not go on, the ground is not good.

Continuity check

A continuity check is done to determine if there are any breaks in a circuit - if it is passing electricity properly. With the circuit off (no power in the circuit), a self-powered continuity tester or multimeter can be used to check the circuit. Connect the test leads to both ends of the circuit (or to the power end and a good ground), and if the test light comes on the circuit is passing current properly **(see illustration)**. If the resistance is low (less than 5 ohms), there is continuity; if the reading is 10,000 ohms or higher, there is a break somewhere in the cir-

cuit. The same procedure can be used to test a switch, by connecting the continuity tester to the switch terminals. With the switch turned On, the test light should come on (or low resistance should be indicated on a meter).

Finding an open circuit

When diagnosing for possible open circuits, it is often difficult to locate them by sight because the connectors hide oxidation or terminal misalignment. Merely wiggling a connector on a sensor or in the wiring harness may correct the open circuit condition. Remember this when an open circuit is indicated when troubleshooting a circuit. Intermittent problems may also be caused by oxidized or loose connections.

Electrical troubleshooting is simple if you keep in mind that all electrical circuits are basically electricity running from the battery, through the wires, switches, relays, fuses and fusible links to each electrical component (light bulb, motor, etc.) and to ground, from which it is passed back to the battery. Any electrical problem is an interruption in the flow of electricity to and from the battery.

1.9 **In use, a basic test light's lead is clipped to a known good ground, then the pointed probe can test connectors, wires or electrical sockets - if the bulb lights, the part being tested has battery voltage**

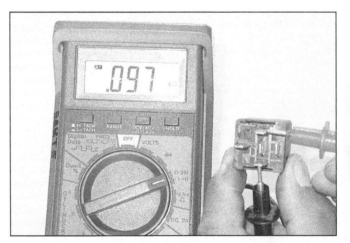

1.12 **With a multimeter set to the ohms scale, resistance can be checked across two terminals - when checking for continuity, a low reading indicates continuity, a high reading indicates lack of continuity**

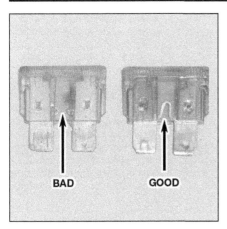

2.1a When a fuse blows, the element between the terminals melts

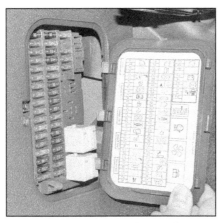

2.1b On Mk I models, the interior fusebox is located behind the driver's side kick panel

2.1c On Mk II models, the interior fusebox is located behind the passenger's side kick panel

2   Fuses, relays and Body Control Module (BCM) - check and replacement

Refer to illustrations 2.1a, 2.1b, 2.1c, 2.4 and 2.5

Note: *It is important to note that the ignition switch and the appropriate electrical circuit must always be switched off before any of the fuses (or relays) are removed and replaced. If electrical components/units have to be removed, the battery negative cable must be disconnected (see Chapter 5).*

Note: *The Body Control Module (BCM) is also referred to as the ZKE module.*

1   Fuses are designed to break a circuit when a predetermined current is reached, in order to protect components and wiring which could be damaged by excessive current flow **(see illustration)**. Any excessive current flow will be due to a fault in the circuit, usually a short-circuit (see Section 1). The main central fusebox, which also carries some relays, is located in the footwell behind the driver's side kick panel (Mk I models), or behind the passenger's side kick panel (Mk II models) **(see illustrations)**.

2   A Body Control Module is located in the behind the passenger's side kick panel trim (Mk I models) or driver's side kick panel trim (Mk II models). This module controls the following functions:

a) *Battery saver - the interior lights and chimes are automatically shut-off after a predetermined period of inactivity.*
b) *Turn signals and hazard lights.*
c) *Courtesy lighting*
d) *Heated rear window*
e) *Heated mirrors*
f) *Windshield wiper/washer*
g) *Rear wiper/washer*
h) *Seat belt warning*
i) *Lights-on warning*
j) *Door ajar warning*
k) *Chimes warnings*
l) *Central locking*
m) *Alarm systems*

3   A special scan tool is required for BCM diagnosis (most consumer-grade scan tools are not capable of that function). See Chapter 11, Section 20, for the BCM removal and installation procedure.

4   The auxiliary fusebox is located in the left side of the engine compartment **(see illustration)**, and is accessed by unclipping and removing the cover. The auxiliary fusebox also contains some relays. **Note:** *Some models also have a fusible link at the rear of the luggage compartment.*

5   Each circuit is identified by numbers on the main fusebox and on the inside of the auxiliary fusebox cover. The best way to check a fuse is with a test light: If there's power on one of the exposed terminal tips of the fuse but not the other, the fuse is blown. Plastic tweezers are attached to the auxiliary fusebox to remove and install the fuses and relays. To remove a fuse, use the tweezers provided to pull it out of the holder, then slide the fuse sideways from the tweezers **(see illustration)**.

6   Always replace a fuse with one of an identical rating. Never substitute a fuse of a higher rating, or make temporary repairs using wire or metal foil; more serious dam-

age, or even fire, could result. The fuse rating is stamped on top of the fuse. Never replace a fuse more than once without tracing the source of the trouble.

7   Relays are electrically operated switches, which are used in certain circuits. The various relays can be removed from their locations by carefully pulling them from the sockets. Some of the relays in the fuse boxes have a plastic bar on its upper surface to enable the use of the tweezers.

8   If a component controlled by a relay becomes inoperative and the relay is suspect, listen to the relay as the circuit is operated. If the relay is functioning, it should be possible to hear it click as it is energized. If the relay proves satisfactory, the fault lies with the components or wiring of the system. If the relay is not being energized, then either the relay is not receiving a switching voltage, or the relay itself is faulty. (Do not overlook the relay socket terminals when tracing faults.) Testing is by the substitution of a known good unit, but be careful; while some relays are identical in appearance and in operation, others look similar, but perform different functions.

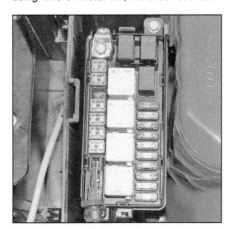

2.4 The auxiliary fusebox is located on the left-hand side of the engine compartment (Mk I model shown, Mk II similar)

2.5 Use the tweezers provided to pull the fuse from the fusebox

3   Electrical connectors - general information

Most electrical connections on these vehicles are made with multiwire plastic connectors. The mating halves of many connectors are secured with locking clips molded into the plastic connector shells. The mating halves of some large connectors, such as some of those under the instrument panel, are held together by a bolt through the center of the connector.

To separate a connector with locking clips, use a small screwdriver to pry the clips apart carefully, then separate the connector halves. Pull only on the shell, never pull on the wiring harness as you may damage the individual wires and terminals inside the connectors. Look at the connector closely before trying to separate the halves. Often the locking clips are engaged in a way that is not immediately clear. Additionally, many connectors have more than one set of clips.

Each pair of connector terminals has a male half and a female half. When you look at the end view of a connector in a diagram, be sure to understand whether the view shows the harness side or the component side of the connector. Connector halves are mirror images of each other, and a terminal shown on the right side end-view of one half will be on the left side end-view of the other half.

It is often necessary to take circuit voltage measurements with a connector connected. Whenever possible, carefully insert a small straight pin (not your meter probe) into the rear of the connector shell to contact the terminal inside, then clip your meter lead to the pin. This kind of connection is called "back-probing." When inserting a test probe into a terminal, be careful not to distort the terminal opening. Doing so can lead to a poor connection and corrosion at that terminal later. Using the small straight pin instead of a meter probe results in less chance of deforming the terminal connector.

Electrical connectors

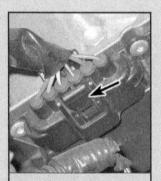

Most electrical connectors have a single release tab that you depress to release the connector

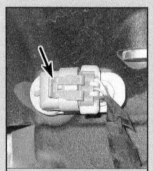

Some electrical connectors have a retaining tab which must be pried up to free the connector

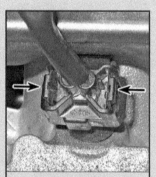

Some connectors have two release tabs that you must squeeze to release the connector

Some connectors use wire retainers that you squeeze to release the connector

Critical connectors often employ a sliding lock (1) that you must pull out before you can depress the release tab (2)

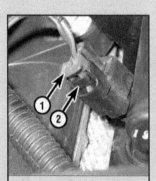

Here's another sliding-lock style connector, with the lock (1) and the release tab (2) on the side of the connector

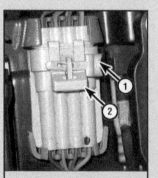

On some connectors the lock (1) must be pulled out to the side and removed before you can lift the release tab (2)

Some critical connectors, like the multi-pin connectors at the Powertrain Control Module employ pivoting locks that must be flipped open

4.2a Insert a straightened paper clip into the hole to depress the locking plunger . . .

4.2b . . . and pull the lock cylinder from place

4.3 Remove the two retaining screws on the underside of the ignition switch housing

4 Switches - removal and installation

Warning: *These models are equipped with a Supplemental Restraint System (SRS), more commonly known as airbags. Always disable the airbag system before working in the vicinity of any airbag system component to avoid the possibility of accidental deployment of the airbag(s), which could cause personal injury (see Chapter 12).*
Note: *Before removing any electrical switches, disconnect the cable from the negative terminal of the battery (see Chapter 5).*

Ignition switch, lock barrel and immobilizer transponder (Mk I models)

Refer to illustrations 4.2a, 4.2b, 4.3 and 4.4
1   Remove the steering column shrouds (see Chapter 11).
2   Insert the ignition key, and turn it to the accessory position. Insert a straightened paper clip through the hole in the end of the

lock housing to depress the locking plunger, and withdraw the lock barrel **(see illustrations).**
3   To remove the switch, which is on the opposite side of the steering column, disconnect the electrical connector, then remove the two set screws and slide the switch from position **(see illustration).**
4   To remove the immobilizer transponder, disconnect the electrical connector, then use a small screwdriver to pry the transponder ring from the end of the ignition switch **(see illustration).**
5   Installation is the reverse of the removal procedure.

Start/Stop switch (Mk II models)

Refer to illustrations 4.8, 4.10 and 4.11
6   Remove the driver's side lower dash panel (see Chapter 11).
7   Remove the tachometer (see Section 8).
8   Remove the trim strip above the Start/Stop switch **(see illustration).**
9   Remove the bezel around the speedometer/instrument panel (see Section 8).

4.4 Carefully pry the transponder ring from the ignition switch

10   Remove the screws and detach the switch panel from the instrument panel **(see illustration).**
11   Disconnect the electrical connector from the switch, then depress the retaining

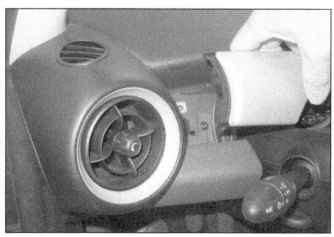

4.8 Carefully pry the trim strip from the left side of the instrument panel

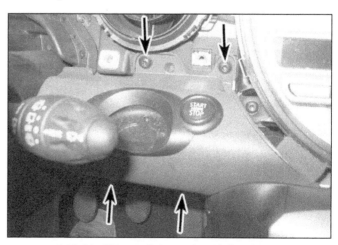

4.10 Start/Stop switch panel mounting screws

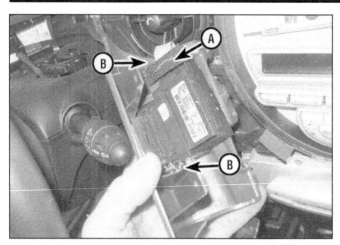

4.11 Disconnect the Start/Stop switch electrical connector (A), then release the retaining tangs (B, not visible) and detach the trim from the front of the switch

4.14 Remove the retaining screw and slide the switch from position

tangs on the back of the switch and remove the switch cover trim to expose the mounting screws **(see illustration)**. Remove the screws and detach the switch from the switch panel.

12   Installation is the reverse of removal.

Headlight, turn signal and wiper multi-function switches

Mk I models

Refer to illustration 4.14

13   Remove the steering column upper and lower shrouds (see Chapter 11).

14   Remove the retaining screw, then slide the switch from the housing. Depress the clips and disconnect the electrical connector as the switch is withdrawn **(see illustration)**.

15   Installation is the reverse of removal.

Mk II models

16   These switches are incorporated in the airbag clockspring unit and are not replaceable separately. Refer to Section 17 for the removal procedure.

Hazard warning switch

Mk I models

17   The hazard warning switch is integral with the central instrument cluster. Removal of the unit is described in Section 8.

Mk II models

Refer to illustration 4.21

18   Remove the driver's side lower dash panel (see Chapter 11).

19   Remove the tachometer (see Section 8).

20   Remove the trim strip above the Start/Stop switch **(see illustration 4.8)**. Also remove the right-side trim strip in the same manner.

21   Remove the plastic covers from the corners of the center instrument bezel, then remove the screws **(see illustration)**. Detach the bezel from the instrument panel.

22   Disconnect the electrical connector from the switch, then push the switch out through the front of the bezel to release the clips.

23   Installation is the reverse of removal.

Power window/central locking/ foglight switches

Refer to illustration 4.25

24   Remove the front center console as described in Chapter 11.

25   Remove the two Torx screws securing the switch panel to the instrument panel **(see illustration)**. Release the locking catch and disconnect the electrical connector as the panel is withdrawn.

---

5   Bulbs (exterior lights) - replacement

1   Whenever a bulb is replaced, note the following points:

a)   *Remember that if the light has just been in use, the bulb may be extremely hot.*

b)   *Always check the bulb contacts and holder, ensuring that there is clean, firm metal-to-metal contact between the bulb and its terminals.*

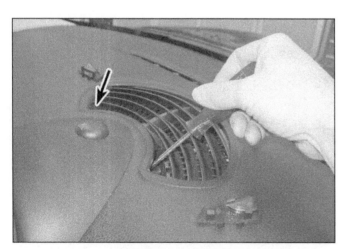

4.21 Pull out the small plastic covers, then remove the screws from underneath

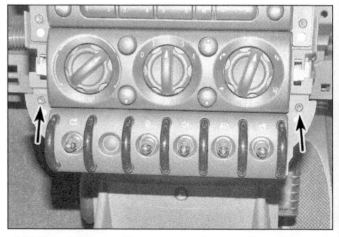

4.25 Unscrew the two Torx screws and remove the switch panel

Bulb removal

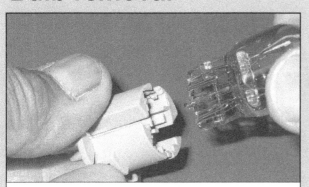

To remove many modern exterior bulbs from their holders, simply pull them out

On bulbs with a cylindrical base ("bayonet" bulbs), the socket is spring-loaded; a pair of small posts on the side of the base hold the bulb in place against spring pressure. To remove this type of bulb, push it into the holder, rotate it 1/4-turn counterclockwise, then pull it out

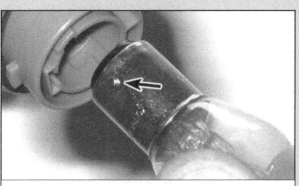

If a bayonet bulb has dual filaments, the posts are staggered, so the bulb can only be installed one way

To remove most overhead interior light bulbs, simply unclip them

---

c) Always ensure that the new bulb is of the correct rating and that it is completely clean before installing it; this applies particularly to headlight/foglight bulbs.

d) Do not touch the glass of halogen-type bulbs (headlights, front fog lights) with your fingers, as this will lead to premature failure of the new bulb due to overheating; if the glass is accidentally touched, clean it with rubbing alcohol.

Halogen headlight
Mk I models
Refer to illustration 5.2a, 5.2b, 5.3, 5.4a, 5.4b, 5.5a and 5.5b

2   Open the hood. At the rear of the headlight unit, remove the cover for access to the low beam bulb, or the circular cover for access to the high beam **(see illustrations)**.

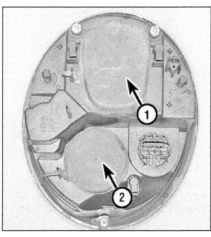

5.2a Pull off the low beam cover (1) or high beam cover (2) - models up to 07/2004

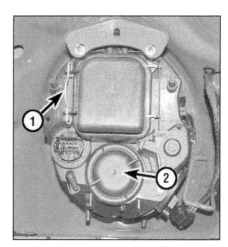

5.2b Release the clip (1) and remove the low beam cover or rotate the circular cover (2) counterclockwise for access to the high beam bulb - models from 07/2004

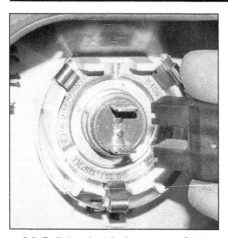

5.3 Pull the electrical connector from the bulb

5.4a Release the retaining clip and remove the bulb

5.4b Note how the tab on the bulb flange locates in the slot in the headlight

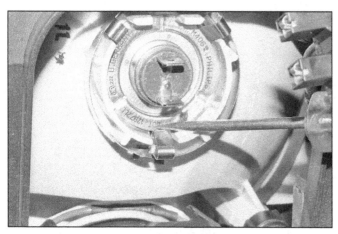

5.5a Press the middle clip outwards, and pull the bulb from the headlight

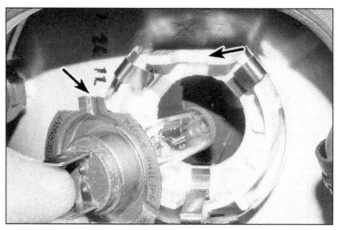

5.5b Note how the tab on the bulb flange locates in the slot in the headlight

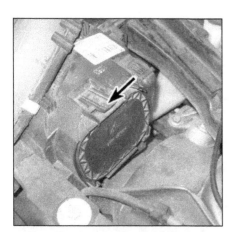

5.7 Depress this tab and open the cover for access to the headlight bulb

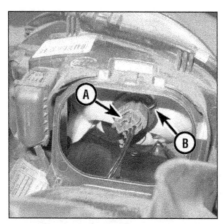

5.8 On this style headlight, depress the tab (A) and disconnect the electrical connector, then turn the bulb holder (B) counterclockwise and remove it from the housing

3   Disconnect the electrical connector from the rear of the bulb **(see illustration)**.

4   On models up to 07/2004, release the retaining clip and remove the bulb. Note how the tab fits in the slot on the rear of the headlight **(see illustrations)**.

5   On models from 07/2004, press the middle retaining clip outwards, and remove the bulb. Note how the tab fits in the slot on the rear of the headlight **(see illustrations)**.

6   Install the new bulb by pressing it into the retaining clips. The remainder of installation is a reversal of the removal procedure.

Mk II models

Refer to illustrations 5.7 and 5.8

7   Open the hood, then depress the tab and swing open the cover on the back of the headlight **(see illustration)**.

8   Disconnect the electrical connector, then remove the bulb holder from the headlight housing; there are two styles you might encounter. One is similar to the Mk I models, with a wire retaining clip (see Steps 4 and 5). The other style holder is removed by turning counterclockwise **(see illustration)**.

9   Remove the bulb/bulb holder from the housing.

10   Installation is the reverse of the removal procedure.

Xenon (HID) bulbs

Warning: *Some models use High Intensity Discharge (HID) bulbs instead of halogen bulbs. These can be identified by the high-voltage warning sticker on the headlight housing. According to the manufacturer, the high voltages produced by this system can be fatal in the event of a shock. Also, the voltage can remain in the circuit even after the headlight switch has been turned to OFF and the ignition key has been removed. Therefore, for*

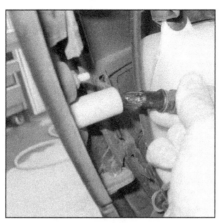

5.12 Rotate the side marker light bulb holder counterclockwise and pull it from the light unit - bumper removed for clarity

5.14 On Mk II models, the side marker light is accessible after removing the fasteners and peeling back the wheelwell liner

5.16 Rotate the turn signal bulb holder counterclockwise - shown with the bumper removed for clarity

your safety, we don't recommend that you try to remove one these headlight bulbs. Instead, have this service performed by a dealer service department or other qualified repair shop.
Warning: *Never attempt to check for voltage at the bulb socket of an HID headlight.*

Side marker light

Mk I models

Refer to illustration 5.12

11   Open the hood, reach behind the indicator/sidelight unit and disconnect the electrical connector.

12   Rotate the bulb holder counterclockwise, then pull the bulb holder out from the rear of the light unit **(see illustration)**. Pull the wedge-type bulb from the bulb holder.

13   Installation is the reverse of removal.

Mk II models

Refer to illustration 5.14

Note: *This procedure applies to the front and rear side marker lights.*

14   Remove the fasteners from the wheelwell liner in the area of the light (see Chapter 11), then peel back the liner for access to the bulb holder. Twist the holder counterclock-

wise to remove it, then pull the bulb straight out of the holder **(see illustration)**. Installation is the reverse of removal.

Turn signals

Mk I models

Refer to illustration 5.16

15   Open the hood, reach behind the indicator light unit, and disconnect the electrical connector.

16   Rotate the bulb holder counterclockwise, and withdraw it from the light unit **(see illustration)**.

17   Twist the bulb counterclockwise, and remove it from the bulb holder.

18   Installation is the reverse of removal.

Mk II models

Refer to illustrations 5.19 and 5.20

19   Turn the front wheels to gain access to the covers at the front of the wheelwell, then turn the upper cover counterclockwise to remove it **(see illustration)**.

20   Turn the bulb holder counterclockwise to remove it from the housing **(see illustration)**, then remove the bulb from the holder.

21   Installation is the reverse of the removal procedure.

5.19 The front turn signal bulb is accessible after removing the upper cover in the wheelwell liner

Side turn signals

Mk I models

Refer to illustration 5.22 and 5.23

22   Push the light housing rearwards and remove it from the front fender **(see illustration)**.

5.20 Turn the bulb holder counterclockwise to remove it from the housing

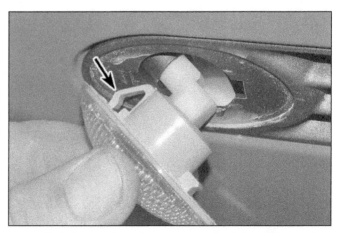

5.22 Push the side turn signal light rearwards to compress the spring clip

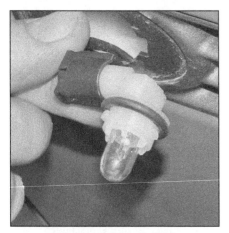

5.23 Rotate the bulb holder counterclockwise to detach it from the housing

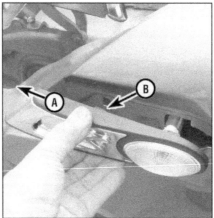

5.25 Push the side turn signal light housing forward (A) to detach the retaining tang (B) from the fender

5.26 Rotate the bulb holder counterclockwise to remove it from the housing

23 Turn the bulb holder counterclockwise, and disconnect it from the turn signal unit **(see illustration)**.
24 Pull the wedge-type bulb from the holder. Installation is the reverse of removal.

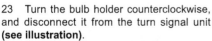

5.30 Rotate the fog light bulb holder collar counterclockwise - early Mk I models

Mk II models

Refer to illustrations 5.25 and 5.26
25 Open the hood, then push the turn signal light housing forward to detach it from the fender **(see illustration)**.
26 Turn the bulb holder counterclockwise and remove it from the housing **(see illustration)**, then pull the bulb straight out of the housing.
27 Installation is the reverse of removal.

Front fog light - Mk I models

28 Remove the relevant front wheelwell liner as described in Chapter 11.
29 Disconnect the electrical connector from the bulb holder.

Early models with H7 bulbs

Refer to illustration 5.30
30 Rotate the bulb holder collar counterclockwise and remove the bulb holder assembly **(see illustration)**.
31 Pull the bulb from the holder.
32 Installation is the reverse of removal.

Later models with H11 bulbs

Refer to illustration 5.33
33 Rotate the bulb holder counterclockwise and remove the bulb assembly **(see illustration)**. Note that the bulb is integral with the bulb holder assembly, and must be replaced as a complete unit.
34 Installation is the reverse of removal.

Front fog light and parking light - Mk II models

Refer to illustrations 5.35 and 5.36
35 Turn the front wheels to gain access to the covers at the front of the wheelwell, then turn the lower cover counterclockwise to remove it **(see illustration)**.
36 Turn the bulb holder counterclockwise to remove it from the housing **(see illustration)**. To replace the fog light bulb, disconnect the electrical connector and install the new bulb (the bulb is integral with the holder). To replace the parking light bulb, pull it straight out of its holder

5.33 Rotate the bulb holder counterclockwise - late Mk I models

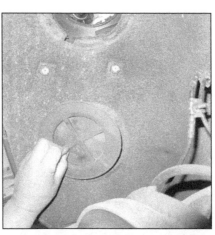

5.35 The fog light bulb and front parking light bulb are accessible after removing the lower cover in the wheelwell liner

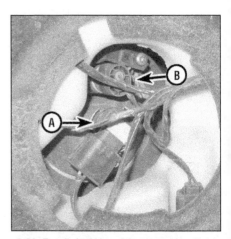

5.36 Fog light (A) and front parking light (B) bulb holders

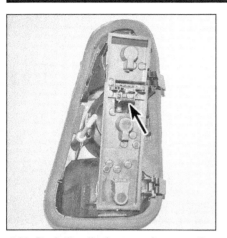

5.39 Push the clip down, and remove the bulb holder assembly - models up to 07/2004

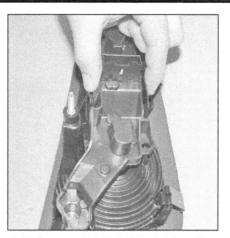

5.40 Squeeze together the retaining clips and remove the bulb holder assembly - models from 07/2004

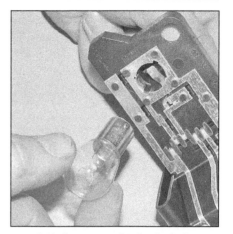

5.41 Depress and twist the bulb counterclockwise

37 Installation is the reverse of the removal procedure.

Rear turn signal/brake/tail/ back-up light

Mk I models

Refer to illustrations 5.39, 5.40 and 5.41

38 Release the clips and remove the appropriate access panel from the luggage compartment side panel trim.

39 On models up to 07/2004, push the retaining clip down and remove the bulb holder assembly from the rear of the light unit **(see illustration)**.

40 On models from 07/2004, squeeze together the retaining clips and remove the bulb holder assembly from the rear of the light unit **(see illustration)**.

41 Depress and twist the appropriate bulb counterclockwise to remove it from the bulb holder **(see illustration)**.

42 Installation is a reversal of the removal procedure. Make sure that the rear light cluster is located correctly.

Mk II models

All except Clubman models

Refer to illustration 5.44

43 Open the rear hatch/tailgate, then remove the access cover from the luggage compartment trim panel.

44 Turn the bulb holder counterclockwise and remove it from the housing **(see illustration)**.

45 Twist the bulb counterclockwise and remove it from the holder.

46 Installation is the reverse of the removal procedure.

Clubman models

47 Open the rear door, then remove the screw at the top of the taillight. Move the taillight downward, then outward and up to detach it.

48 Working on the back side of the housing, push outward on the tang and detach the bulb panel from the housing.

49 Twist the bulb counterclockwise to remove it. Refer to **illustration 5.44** for the bulb arrangement.

50 Installation is the reverse of removal, making sure the weather seal has not become displaced.

License plate light

Mk I models

Refer to illustrations 5.51 and 5.52

51 Insert a flat-bladed screwdriver in the recess on the left of the license plate light lens, lever the lens away from the tailgate center, and carefully pry out the light unit **(see illustration)**.

52 Release the festoon-type bulb from the

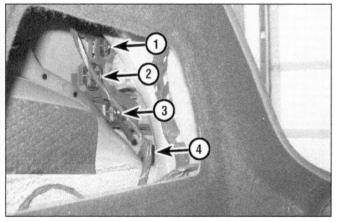

5.44 Taillight bulb arrangement (Mk II hard top and Convertible models; Clubman arrangement similar, but without no. 3)

1	Brake light/parking light	3	Brake light/parking light
2	Turn signal	4	Back-up light

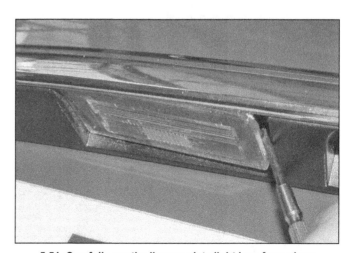

5.51 Carefully pry the license plate light lens from place

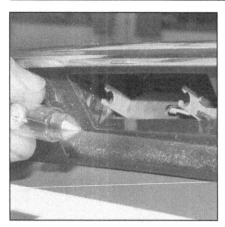

5.52 The festoon type bulb fits between the two spring contacts

5.54 When prying out a license plate light lens, pry only in the notched area

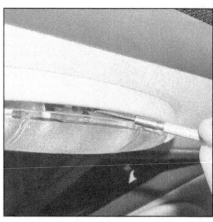

6.2 Pry the interior light lens from place

contact springs **(see illustration)**.
53   Installation is a reversal of the removal procedure. Make sure that the tension of the contact springs is sufficient to hold the bulb firmly.

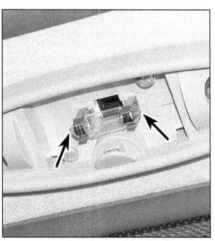

6.3 Push the contacts to one side, and remove the main festoon bulb

Mk II models

Refer to illustration 5.54
54   Insert a small screwdriver into the notch in the light lens and carefully pry the lens from the hatch or tailgate **(see illustration)**.

All except Clubman models

55   Release the festoon-type bulb from the contact springs **(see illustration 5.52)**.
56   Installation is a reversal of the removal procedure. Make sure that the tension of the contact springs is sufficient to hold the bulb firmly.

Clubman models

57   Twist the bulb holder counterclockwise to remove it from the housing, then pull the bulb from the holder.
58   Installation is the reverse of removal.

---

6   Bulbs (interior lights) - replacement

1   Whenever a bulb is replaced, note the following points:

  a)   *Remember that if the light has just been in use, the bulb may be extremely hot.*
  b)   *Always check the bulb contacts and*

    holder, ensuring that there is clean metal-to-metal contact between the bulb and its terminals. Clean off any corrosion or dirt before installing a new bulb.
  c)   *Wherever bayonet-type bulbs are used, ensure that the terminals bear firmly against the bulb contact.*
  d)   *Always ensure that the new bulb is of the correct rating and that it is completely clean before installing it.*

Interior lights
Mk I models

Refer to illustrations 6.2, 6.3, 6.4 and 6.5
2   Pry the lens from place by inserting a small flat-bladed screwdriver on the opposite side of the lens to the switch **(see illustration)**.
3   The main bulb is a festoon type. Push the contacts to one side and remove the bulb **(see illustration)**.
4   To remove the two small side bulbs, remove the two screws and remove the light unit from the headlining **(see illustration)**.
5   Twist the bulb holder counterclockwise and pull the capless bulb from the holder **(see illustration)**. Installation is a reversal of the removal procedure.

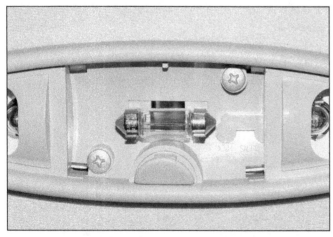

6.4 Remove the two screws and lower the light unit

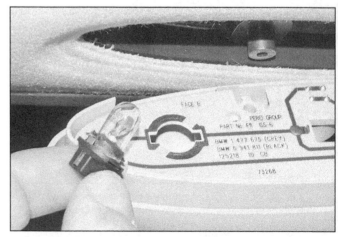

6.5 Twist the bulb holder counterclockwise

6.6a Carefully pry off the dome light lens . . .

6.6b . . . then pull the bulb straight out of the housing

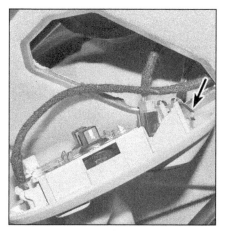

6.7 Push the map reading light unit forwards to compress the spring clip

Mk II models

Refer to illustrations 6.6a and 6.6b

6   Pry the rear dome light from the headliner. Turn the bulb holder counterclockwise, remove it from the housing, then remove the bulb **(see illustrations)**. Installation is the reverse of removal.

Map reading lights
Mk I models

Refer to illustrations 6.7, 6.8 and 6.9

7   Ensure that the interior light is switched off. Push the light unit forwards slightly, then pull the rear edge downwards to detach it from the headlining **(see illustration)**.
8   Rotate the bulb holder counterclockwise and pull it from the light **(see illustration)**.
9   The capless bulb is a push-fit in the bulb holder **(see illustration)**.
10   Installation is a reversal of the removal procedure.

Mk II models

Refer to illustrations 6.11a and 6.11b

11   Pry the lens from the overhead console, then pull the bulb straight out **(see illustrations)**. Installation is the reverse of removal.

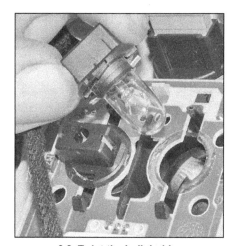

6.8 Twist the bulb holder counterclockwise

6.9 Pull the capless bulb from place

Instrument panel illumination and warning lights

12   The instrument panel and warning light are illuminated by LEDs. No provision is made for the replacement of these LEDs.

Heater control/fan switch illumination

13   The control panel and switches are illuminated by LEDs. No provision is made for the replacement of these LEDs.

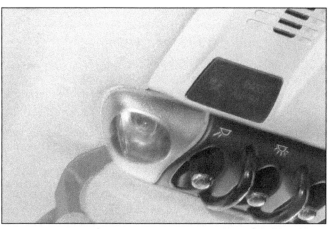

6.11a Carefully pry off the map light lens . . .

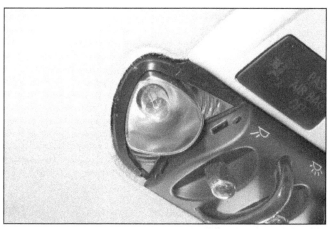

6.11b . . . then pull the bulb straight out of the housing

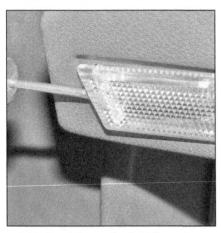

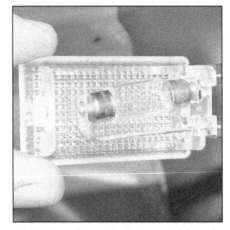

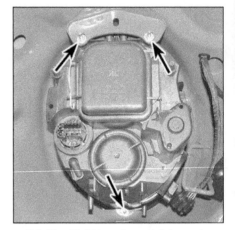

6.14 Carefully pry the light from place

6.15 Move the contacts to one side and remove the festoon bulb

7.3 Headlight housing retaining nuts (Mk I models)

Luggage compartment/ footwell/glove box lights

Refer to illustrations 6.14 and 6.15

14   With the light switched off, pry out the light using a small screwdriver **(see illustration)**.
15   Pull the festoon type bulb out from

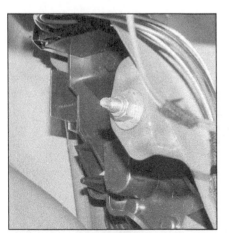

7.5 Headlight housing retaining bolts (Mk II models)

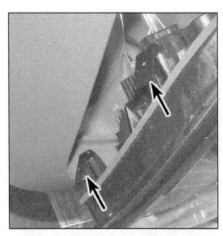

7.8a Remove the taillight housing nut . . .

between its terminals to remove **(see illustration)**. On later models the bulb pulls straight out of the socket.
16   Installation a reversal of the removal procedure.

7   Headlight and taillight housings - removal and installation

1   Before removing any light unit, note the following points:
  a) *Ensure that the light is switched off before starting work.*
  b) *Remember that if the light has just been in use, the bulb and lens may be extremely hot.*

Headlight housing
Mk I models
Refer to illustration 7.3

2   Depress the retaining clip and disconnect the electrical connector from the rear of the headlight unit.
3   Remove the three retaining nuts and carefully pull the headlight unit from the hood **(see illustration)**. No disassembly of the headlight unit is possible.

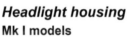

7.8b . . . and release the retaining clips

4   Installation is a reversal of removal. Have the headlight alignment checked at a qualified service center as soon as possible.

Mk II models - halogen headlights
Refer to illustration 7.5

5   Open the hood. Remove the headlight housing mounting screws, remove the assembly, and disconnect the electrical connectors **(see illustration)**.
6   Installation is a reversal of removal. Have the headlight alignment checked at a qualified service center as soon as possible.

Xenon (HID) headlights
Warning: *Some models use High Intensity Discharge (HID) bulbs instead of halogen bulbs. These can be identified by the high-voltage warning sticker on the headlight housing. According to the manufacturer, the high voltages produced by this system can be fatal in the event of a shock. Also, the voltage can remain in the circuit even after the headlight switch has been turned to OFF and the ignition key has been removed. Therefore, for your safety, we don't recommend that you try to remove one these headlight housings. Instead, have this service performed by a dealer service department or other qualified repair shop.*

Taillight housing
Mk I models
Refer to illustration 7.8a and 7.8b

7   Release the clips and remove the appropriate access panel from the luggage compartment side panel trim.
8   Remove the light unit retaining nut, then squeeze together the retaining clips and remove the light unit from the vehicle **(see illustrations)**.
9   Disconnect the wiring plug as the unit is withdrawn.
10   Installation is the reverse of removal. Make sure the weatherstripping has not become displaced.

7.11 Pry off the trim ring . . .

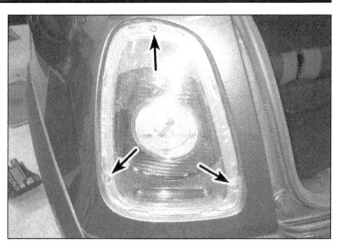

7.12 . . . for access to the taillight screws

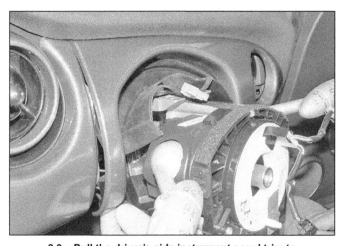

8.3a Pull the driver's side instrument panel trim to release the clips . . .

8.3b . . . followed by the passenger's side; early models . . .

Mk II models

Hardtop and Convertible models

Refer to illustrations 7.11 and 7.12

11   Carefully pry off the trim ring from the taillight lens **(see illustration)**.
12   Remove the screws **(see illustration)** and detach the housing, then disconnect the electrical connectors.
13   Installation is the reverse of removal.

Clubman models

14   Taillight removal is part of taillight bulb replacement. See Section 5 for the procedure.

---

8   Instruments - removal and installation

Caution: *The main instrument panel must be kept in the upright position to avoid silicone liquid leaking from the gauges.*
Note: *If an instrument or the instrument panel is being replaced, the stored configuration data must be loaded into a MINI diagnostic system and downloaded into the new unit once installed. Refer to your local MINI dealer service department or other qualified repair shop.*

Mk I models

Main instrument (speedometer) panel

Refer to illustrations 8.3a, 8.3b, 8.3c, 8.4, 8.5a, 8.5b, 8.6a and 8.6b

1   Disconnect the battery negative cable (see Chapter 5).

8.3c . . . and late models

2   Remove the upper steering column shroud as described in Chapter 11.
3   Carefully pull the driver's side and passenger's side instrument panel trims from their retaining clips **(see illustrations)**.
4   Pull the switch trims from position **(see illustration)**.

8.4 Pull the trims from the instrument lighting control and hazard switch

8.5a Early models have three Torx screws securing the instrument panel surround . . .

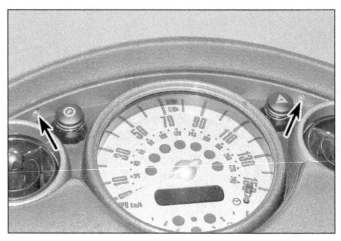

8.5b . . . while later models have two Torx screws

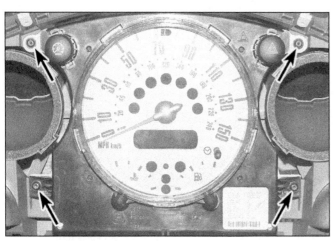

8.6a The instrument cluster is retained by four Torx screws

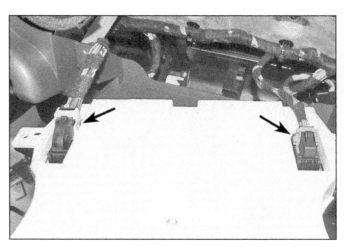

8.6b Pull over the locking catches and disconnect the electrical connectors

5   Remove the Torx retaining screws from the instrument panel surround, then unclip it from the instrument panel **(see illustrations)**.
6   Remove the four Torx screws and pull the instrument cluster from the instru-ment panel. Disconnect the electrical con-nectors as the unit is withdrawn **(see illus-trations)**. No disassembly of the instrument cluster is possible. Installation is the reverse of removal.

Steering column-mounted instruments

Refer to illustration 8.7

7   Remove the two screws and detach the instrument from the column shroud. Discon-nect the electrical connector(s) as the instru-ment is withdrawn **(see illustration)**.
8   Installation is a reversal of the removal procedure.

Mk II models
Tachometer

Refer to illustrations 8.10 and 8.11

Note: *If you're removing the tachometer to access the instrument trim panel behind the tach, remove the tachometer as described in Steps 9 through 11, then unclip the trim panel, lift the panel upwards and off of the instru-ment panel.*

9   Disconnect the battery negative cable (see Chapter 5), then remove the left lower dash panel underneath the steering column (see Chapter 11).
10   Follow the wiring harness down to the left side of the steering column, then disconnect the electrical connector **(see illustration)**.

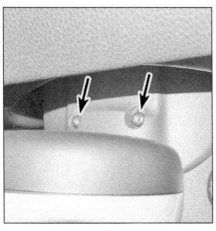

8.7 Remove the two Torx screws and slide the steering column-mounted instrument rearwards

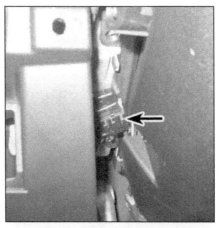

8.10 The tachometer electrical connector is near the bottom of the steering column shroud, on the left side

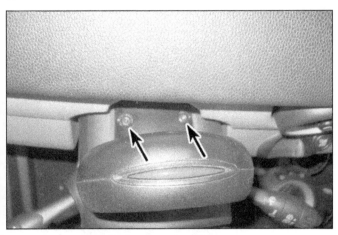

8.11 Tachometer mounting screws

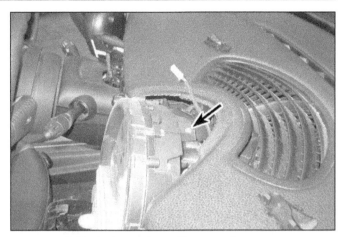

8.15a Instrument panel top mounting screw

11 Remove the screws and detach the tachometer from the steering column **(see illustration)**.

12 Installation is the reverse of removal.

Main instrument (speedometer) panel

Refer to illustrations 8.15a and 8.15b

13 Remove the tachometer (see Steps 9 through 11)

14 Remove the center instrument bezel and the Start/Stop switch (see Section 4) and the center instrument panel covers (see Chapter 11).

15 Remove the instrument panel screws **(see illustrations)**, pull the panel away from the dash, and disconnect the electrical connectors.

16 Installation is the reverse of the removal procedure.

8.15b Instrument panel lower mounting screws

---

9 Horn(s) - removal and installation

Refer to illustrations 9.3a and 9.3b

Note: *On Mk I models there are two horns - one on each side of the vehicle behind the* front bumper. Mk II models also have two horns; both are located on the right side of the vehicle behind the front bumper.

1 Remove the front bumper cover as described in Chapter 11.

2 Disconnect the electrical connector(s) from the horn terminal(s).

3 Unscrew the mounting Torx bolt(s), and withdraw the horn with its mounting bracket **(see illustrations)**.

4 Installation is a reversal of the removal procedure. Tighten the horn unit mounting bolt securely, making sure the mounting bracket is located correctly.

9.3a The horns are retained by one Torx bolt each (Mk I models)

9.3b Horn mounting bolts (Mk II models)

10.2 Loosen the wiper arm nut one or two turns - rear wiper shown

10.3a If necessary, use a puller to remove the wiper arm

10.3b Where the washer protrudes through the spindle of the rear wiper arm, use a socket as a spacer to prevent damage to the jet

10 Wiper arms - removal and installation

Refer to illustrations 10.2, 10.3a and 10.3b

1   With the wiper(s) parked (in the normal at-rest position), mark the positions of the blade(s) on the windshield, using a wax crayon or strips of masking tape.
2   Lift up the plastic cap from the bottom of the wiper arm, and loosen the nut one or two turns **(see illustration)**.
3   Lift the wiper arm, and carefully release it from the taper on the spindle by moving it from side-to-side. If the wiper arm is stubborn to remove, use a puller **(see illustration)**. On the rear wiper arm where the washer jet is in the center of the spindle, use a socket under the puller spindle to prevent any damage to the jet **(see illustration)**.
4   Completely remove the nut, and withdraw the wiper arm from the spindle.
5   Installation is a reversal of the removal procedure. Make sure that the arm is installed in the previously-noted position before tightening its nut.

11 Windshield wiper motor and linkage - removal and installation

Mk I models

Refer to illustrations 11.4, 11.5, 11.6, 11.7, 11.8, 11.10a, 11.10b, 11.11a, 11.11b, 11.12, 11.15 and 11.16

1   Remove the wiper arms (see Section 10).
2   Raise the front of the vehicle and support it securely on jackstands. Remove both front wheelwell liners as described in Chapter 11.
3   Remove the windshield washer reservoir as described in Section 13.
4   Reach behind and push out the front of the side turn signal light surround trim from the left-hand fender, then remove the trim **(see illustration)**. Disconnect the electrical connector as it's withdrawn.
5   Remove the Torx screw from each side through the side turn signal opening **(see illustration)**.
6   Remove the three screws at the front outer

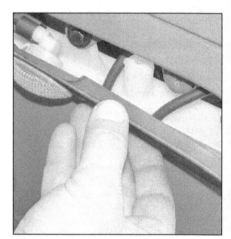

11.4 Reach behind and push out the side turn signal trim

edges of the cowl panel **(see illustration)**.
7   Remove the nut on each side securing the cowl panel to the A-panel (behind the hood hinge mountings) **(see illustration)**.

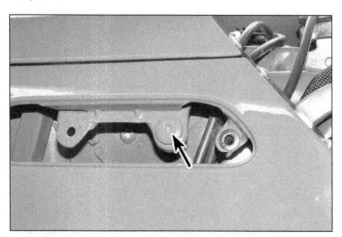

11.5 Remove the screw from each side, accessible through the side turn signal trim opening

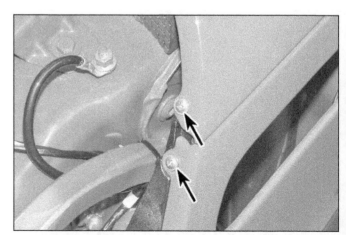

11.6 Remove the three screws (upper two shown, the third is directly below) at the front edge of the cowl panel

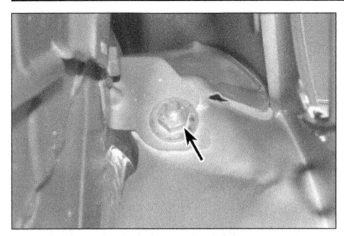

11.7 Remove the nut on each side behind the hood hinge mounting points

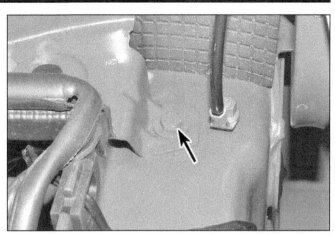

11.8 Remove the bolt on each side at the front outer edge of the cowl panel

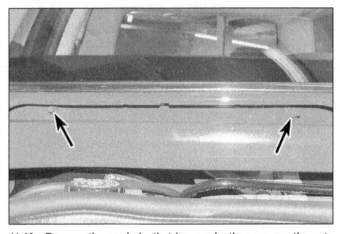

11.10a Remove the cowl plastic trim panels, then remove the nuts on the passenger's side . . .

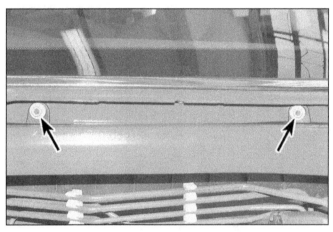

11.10b . . . and driver's side

8   Remove the bolt on each side at the front outer edge of the cowl panel **(see illustration)**.
9   Pry up the rear edges of the plastic cowl panel trims, and remove them.
10   Loosen the two nuts at the back of the driver's side cowl panel trim opening. Remove

the nuts at the rear of the passenger's side cowl panel trim opening, then loosen the nuts (on the same studs, behind the black plastic panel) securing the cowl panel to the firewall **(see illustrations)**.
11   Pull the rubber weatherstrip from the

edge of the door pillar, then remove the screws securing the trim panel to the pillar. With the screws removed, pull the trim panel from the pillar to release its retaining clips **(see illustrations)**.
12   Pull away the foam, and remove the Torx

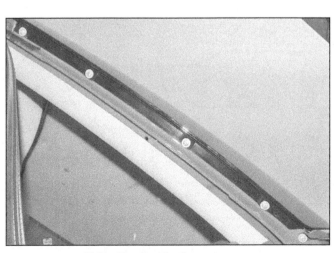

11.11a Remove the five screws . . .

11.11b . . . and pull the pillar trim outwards to release the retaining clips

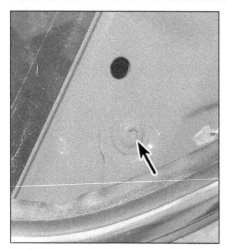

11.12 Remove the Torx screw at the base of the pillar

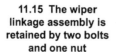

11.15 The wiper linkage assembly is retained by two bolts and one nut

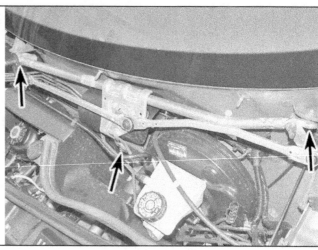

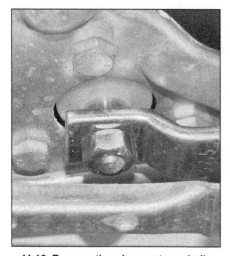

11.16 Remove the wiper motor spindle nut, then the three bolts securing the motor to the linkage

screw at the base of the windshield A-pillar **(see illustration)**.

13   Disconnect the washer tube at the junction adjacent to the brake master cylinder reservoir.

14   Pull the cowl panel forwards slightly, and lift it up and away from the vehicle. Note that the rear edge of the panel fits under the lower edge of the windshield. Take great care not to damage the paint or the windshield.

15   Remove the two bolts and one nut **(see illustration)**, then maneuver the wiper linkage assembly from position. Disconnect the electrical connector as the assembly is withdrawn.

16   If required, make alignment marks between the motor spindle and the arm, then remove the nut securing the arm to the motor spindle. Remove the three bolts and separate the motor from the linkage **(see illustration)**.

17   Installation is a reversal of the removal procedure, noting the following points:

a) Tighten the wiper motor mounting bolts securely.

b) Make sure that the wiper motor is in its parked position before installing the motor arm, and check that the wiper linkage is in line with the motor arm.

c) Use grease to lubricate the wiper spindle and linkages when reassembling.

Mk II models

Refer to illustrations 11.19, 11.20 and 11.21

18   Remove the wiper arms (see Section 10) and the cowl trim panels (see Chapter 11).

19   Remove the wiper linkage and motor mounting fasteners **(see illustration)**.

20   Disconnect the electrical connector **(see illustration)** and remove the motor and linkage assembly.

21   If required, pry the linkage arms from the crank arm, make alignment marks between the motor spindle and the crank arm, then remove the nut securing the arm to the motor spindle. Remove the three bolts and separate the motor from the linkage **(see illustration)**.

22   Installation is a reversal of the removal procedure, noting the following points:

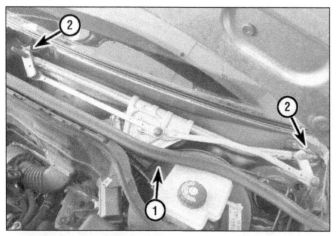

11.19 Wiper motor and linkage details (Mk II models)

1   Motor mounting nut (not visible)
2   Linkage mounting bolts

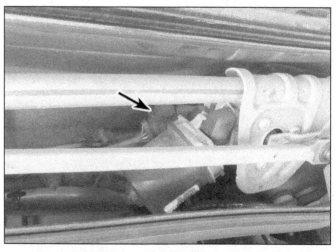

11.20 Wiper motor electrical connector

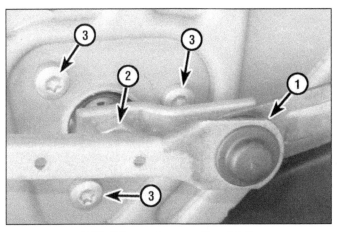

11.21 Wiper motor mounting details

1	Linkage arms
2	Crank arm/spindle nut

3   Motor mounting bolts

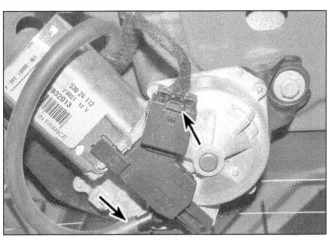

12.4 Disconnect the wiper motor washer hose and electrical connector

a) Tighten the wiper motor mounting bolts securely.
b) Make sure that the wiper motor is in its parked position before installing the motor arm, and check that the wiper linkage is in line with the motor arm.
c) Use grease to lubricate the wiper spindle and linkages when reassembling.

12   Tailgate wiper motor assembly - removal and installation

Refer to illustrations 12.4 and 12.5

Note: Hardtop model shown, Clubman rear door wiper motors similar.

1   Disconnect the battery negative cable (see Chapter 5).
2   Remove the tailgate wiper arm as described in Section 10.
3   Remove the tailgate inner trim panel (see Chapter 11).

4   Disconnect the washer hose and electrical connector from the wiper motor **(see illustration)**.
5   Unscrew the three mounting bolts and remove the wiper motor from inside the tailgate **(see illustration)**.
6   Installation is a reversal of the removal procedure. Make sure that the wiper motor is in its parked position before installing the wiper arm.

13   Washer fluid reservoir - removal and installation

1   Apply the parking brake and loosen the right front wheel bolts (Mk I models) or left front wheel bolts (Mk II models). Raise the front of the car and support it securely on jackstands. Remove the front wheel.
2   Remove the wheelwell liner for access to the reservoir (see Chapter 11).

Mk I models

Refer to illustrations 13.3 and 13.4

3   On Mk I models, reach behind and push the front edge of side turn signal surround trim from the right-hand fender **(see illustration 11.4)**, then remove the reservoir upper retaining bolt **(see illustration)**.
4   Unscrew the mounting bolts/nuts, and slide the reservoir forwards to release it from the inner fender panel **(see illustration)**.
5   Disconnect the washer hoses and wiring connectors from the washer pump and level sensor as the reservoir is being removed. Place a container beneath the reservoir to catch any spilled fluid.
6   Installation is the reverse of removal.

Mk II models

Refer to illustrations 13.8 and 13.9

7   Remove the left headlight (see Section 7).

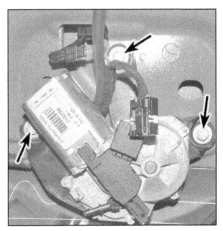

12.5 Remove the three bolts and detach the tailgate wiper motor

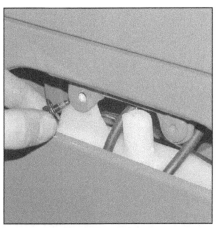

13.3 The upper reservoir bolt can be removed once the side turn signal trim is pushed from place

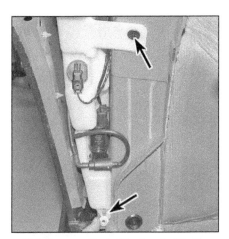

13.4 Remove the bolt/nut and slide the reservoir forwards (Mk I models)

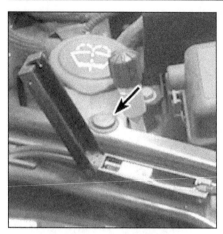

13.8 Remove this plastic fastener securing the top of the washer fluid reservoir

8   Remove the plastic fastener securing the top of the reservoir **(see illustration)**.
9   Disconnect the electrical connectors for the fog light, parking light, fluid level sensor and washer pump. Detach the hoses, then remove the mounting fasteners **(see illustration)**.
10   Carefully pry the bottom of the reservoir up to detach it from its lower mount, then guide the reservoir out.
11   Installation is the reverse of removal.

---

14 Audio unit - removal and installation

---

1   Disconnect the battery negative cable (see Chapter 5).

Mk I models

Refer to illustrations 14.3a and 14.3b

2   Remove the front center console as described in Chapter 11.
3   Remove the four screws securing the

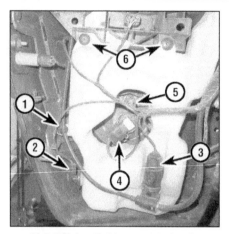

13.9 Washer fluid reservoir details

1   *Clip*
2   *Fluid level sensor electrical connector*
3   *Washer pump electrical connector*
4   *Fog light electrical connector*
5   *Parking light electrical connector*
6   *Reservoir mounting fasteners*

audio unit, then pull it from place. Noting their positions, disconnect the electrical connectors as the unit is withdrawn **(see illustrations)**.
4   Installation is a reversal of removal. With the leads reconnected to the rear of the unit, press it into position until the retaining clips are felt to engage.

Mk II models

5   Remove the main instrument (speedometer) panel (see Section 8).
6   Remove the screws and separate the radio from the speedometer/instrument panel.
7   Installation is the reverse of removal. If a new unit has been installed, it will have to be programmed by a dealer service department or other qualified repair shop equipped with the proper scan tool.

---

15 Speakers - removal and installation

Mk I models

Refer to illustrations 15.2a and 15.2b

1   Remove the front door inner trim panel (see Chapter 11) or luggage compartment side trim panel.
2   Unscrew the four cross-head screws and withdraw the speaker from the panel **(see illustrations)**. Disconnect the electrical connector on removal of the speaker.
3   Installation is a reversal of the removal procedure.

Mk II models

Door bass speaker

Refer to illustration 15.5

4   Remove the door trim panel (see Chapter 11).
5   Remove the screws, pull the speaker from the door, then disconnect the electrical connector **(see illustration)**.
6   Installation is the reverse of removal.

Door mid-range speaker

Refer to illustrations 15.7a, 15.7b, 15.8 and 15.9

7   If you're working on the driver's side, carefully pry the mirror switch from the speaker cover, then disconnect the electrical connector **(see illustrations)**.
8   Carefully pry off the speaker cover **(see illustration)**.
9   Remove the screws **(see illustration)**, separate the speaker from the door, then disconnect the electrical connector.

Tweeter speakers

10   If you're removing the front tweeter, remove the A-pillar trim panel (see Chapter 11). If you're removing the rear tweeter, remove the rear trim panel
11   Carefully pry the speaker from its retain-

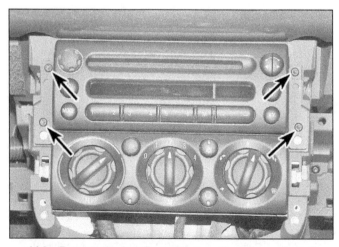

14.3a Remove the four Torx screws and pull the audio unit from place

14.3b Lift the locking catch and disconnect the electrical connector

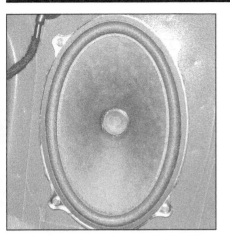

15.2a The rear speakers are behind the luggage compartment side trim panels . . .

15.2b . . . and the front ones are behind the door inner trim panels

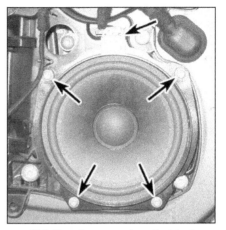

15.5 Door bass speaker electrical connector and mounting screws

ing tangs, then disconnect the electrical connector from the speaker.

12   Installation is the reverse of removal.

Rear woofer

13   On all except Clubman models, remove the rear trim panel. On Clubman models, remove the rear door panel.

14   Remove the speaker mounting screws, pull out the speaker, then disconnect the electrical connector.

15   Installation is the reverse of removal.

16   Airbag units - removal and installation

Warning: *Handle any airbag unit with extreme care, as a precaution against personal injury, and always hold it with the cover facing away from the body. If in doubt concerning any proposed work involving an airbag unit or its control circuitry, consult a MINI authorized dealer or other qualified specialist.*

Warning: *Stand any airbag in a safe place with the cover facing up, and do not expose*

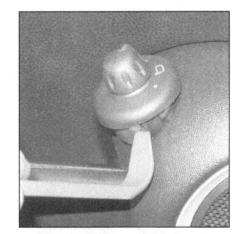

15.7a Pry the mirror switch from the speaker cover . . .

it to heat sources in excess of 212-degrees F (100-degrees C).

Warning: *Do not attempt to open or repair an airbag unit, or apply any electrical current to it. Do not use any airbag unit which is visibly damaged or which has been tampered with.*

15.7b . . . then disconnect the electrical connectors

Disarming

1   Disconnect the battery negative cable (see Chapter 5). Before proceeding, wait a minimum of 5 minutes as a precaution against accidental deployment of the airbag unit. This period ensures that any stored energy in the back-up capacitor is dissipated.

15.8 Pry off the speaker cover

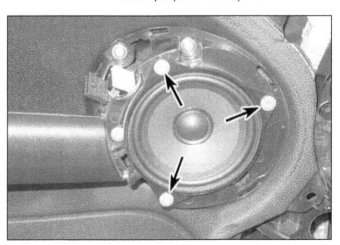

15.9 Door mid-range speaker mounting screws

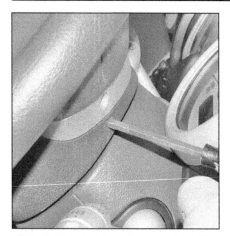

16.3 On the standard steering wheel, the airbag is retained by two Torx screws accessible through the back of the wheel

16.4 Pry up the locking catches and disconnect the electrical connectors

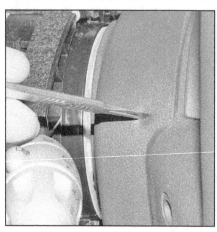

16.7 Make a small incision with a scalpel in the depression on each side of the steering wheel center boss

16.8a Insert a long, flat-bladed screwdriver into the hole . . .

16.8b . . . to push-in the retaining clip arm and release the airbag (shown with the airbag removed for clarity)

Driver's airbag

Mk I models, standard steering wheel

Refer to illustrations 16.3 and 16.4

2   Rotate the steering wheel so that one of the airbag mounting bolt holes (at the rear of the steering wheel boss) is visible above the steering column upper shroud.

3   Unscrew and remove the first mounting bolt **(see illustration)**, then turn the steering wheel 180-degrees and remove the remaining mounting bolt.

4   Carefully withdraw the airbag unit from the steering wheel far enough to disconnect the electrical connectors **(see illustration)**, then remove it from inside the vehicle. Place the airbag in a safe location with the cover facing up as soon as possible.

5   Installation is a reversal of the removal procedure, tightening the retaining bolts to 84 inch-pounds (10 Nm).

Mk I models, sport steering wheel

Refer to illustrations 16.7, 16.8a, 16.8b and 16.9

6   Position the steering wheel in the straight-ahead position, and engage the steering lock.

7   The airbag is secured to the steering wheel by two spring clips. To release these clips, identify the two depressions (one on each side) of the steering wheel center boss. If there are no holes there already, use a scalpel to make two 8 to 10 mm incisions in a cross pattern across each depression **(see illustration)**.

8   Insert a long screwdriver into each hole and release the airbag unit retaining clips **(see illustrations)**. Pull the airbag unit from the steering wheel.

9   To disconnect the airbag electrical connectors, pry up the center locking plate, then pry the plugs from their sockets **(see illustration)**. Place the airbag in a safe location with the cover facing up as soon as possible.

Installation is the reverse of removal, pushing the airbag unit into place until the retaining clips engage.

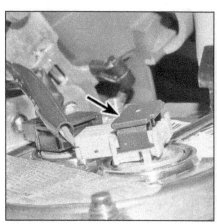

16.9 Pry up the locking plate, then disconnect the electrical connector

16.10 Insert a screwdriver into the hole in the bottom of the steering wheel, then push up on the retaining spring wire to release the airbag

16.11 Disconnect the airbag and horn electrical connectors

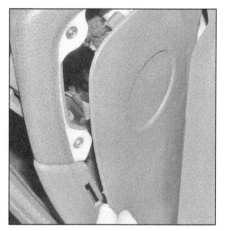

16.12a Pry the instrument panel end trims from place

Mk II models

Refer to illustrations 16.10 and 16.11

10   Insert a screwdriver into the hole in the underside of the steering wheel, feeling for the retaining spring wire. Push up on the spring wire with sufficient force to disengage the retaining spring from the airbag unit **(see illustration)**.

11   Disconnect the electrical connectors for the airbag and horn contact **(see illustration)**. Place the airbag in a safe location with the cover facing up as soon as possible. Installation is the reverse of removal, pushing the airbag unit into place until the retaining spring wires engage with the airbag.

Passenger's airbag

Note: *This procedure applies to Mk I models only, as it is necessary to remove the passenger's airbag for removal of the instrument panel, which is required for removal of the heat-*

ing and air conditioning blower motor. Removal for any other purpose is not recommended.

Models up to 07/2004

Refer to illustrations 16.12a, 16.12b, 16.14, 16.16, 16.17a, 16.17b, 16.18 and 16.19

12   Carefully pry the instrument panel end trims from place **(see illustration)**. Where necessary, disconnect the electrical connectors from the airbag de-activation switch. Remove the two screws on each end securing the instrument top panel trim **(see illustration)**.

13   On models with a steering column mounted instrument (tachometer), remove the two Torx screws and remove the instrument (see Section 8). Disconnect the electrical connector as the unit is withdrawn.

14   Lift the instrument top panel trim from place **(see illustration)**.

15   Remove the glove box as described in Chapter 11.

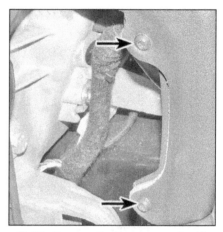

16.12b Remove the two screws at each end of the instrument panel

16   Remove the two screws and pull the rod out to release the retaining straps **(see illustration)**.

16.14 Lift the instrument top panel trim from place

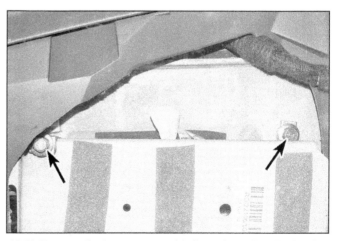

16.16 Remove the two screws and pull out the rod anchoring the retaining straps

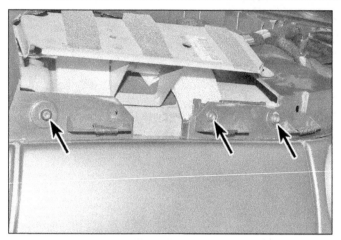

16.17a Remove the three screws at the top . . .

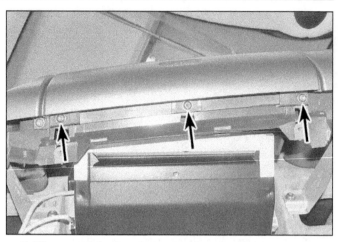

16.17b . . . and the three screws underneath, then remove the airbag cover

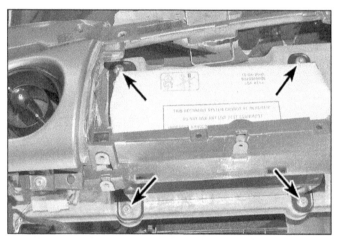

16.18 Remove the four airbag screws

16.19 Disconnect the airbag electrical connector

17   Remove the 6 screws and remove the airbag cover from the instrument panel (see illustrations).
18   Remove the 4 airbag unit mounting screws and lower the unit from the instrument panel (see illustration).
19   Disconnect the electrical connector (see illustration). Installation is the reverse of removal. Tighten the retaining bolts to 84 inch pounds (10 Nm).

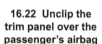

16.22 Unclip the trim panel over the passenger's airbag

Models from 07/2004

Refer to illustrations 16.22, 16.27a, 16.27b, 16.28, 16.29, 16.31, 16.32, 16.33a and 16.33b

20   Remove both A-pillar trim panels (see Chapter 11).
21   Remove the glove box (see Chapter 11).
22   Pull out the bottom edge of the trim panel over the passenger's airbag, and detach it from the retaining clips (see illustration).
23   Remove the steering column mounted instrument(s), if equipped (see Section 8).
24   Pull the driver's side instrument panel trim panel from its retaining clips (see illustration 8.3a).
25   Carefully pry the instrument panel end trims from place (see illustration 16.12a). Where necessary, disconnect the electrical connectors from the airbag de-activation switch.
26   Remove the two screws at each end securing the instrument panel cover (see illustration 16.12b).

16.27a Remove the four screws on the passenger's side . . .

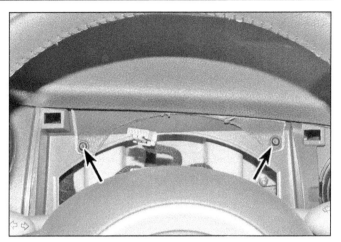

16.27b . . . and the two on the driver's side

27   Remove the screws securing the instrument panel cover: four on the passenger's side, and the two on the driver's side **(see illustrations)**.
28   Pry out the two caps in the center vent and remove the Torx screws beneath **(see illustration)**.
29   Remove the four long Torx screws in the windshield vent grille **(see illustration)**. **Note:** *When installing, the two outer screws must have the red protective plastic caps pushed onto their underside, to protect the wiring harness. Position these caps before replacing the instrument panel cover.*
30   On models with automatic air conditioning, carefully pry up the sunlight sensor in the center of the instrument panel, and disconnect the electrical connector.
31   Remove the two Torx bolts securing the airbag bracket **(see illustration)**. Note the routing of the restraint cables also secured by these bolts.
32   Lift off the instrument panel cover, feeding the airbag restraint cables out between

16.28 Pry out the two caps in the center vent

the airbag module and the panel lower section. Note the lug in the center of the cover which locates in the corresponding bracket on

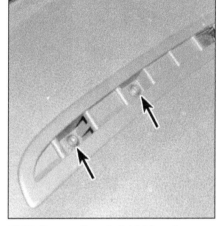

16.29 Remove the windshield vent screws (two left-hand ones shown)

the vehicle body **(see illustration)**.
33   Unlock and disconnect the airbag electrical connector, release the cable tie, then

16.31 Note the airbag cover restraint cables secured by the airbag bracket bolts

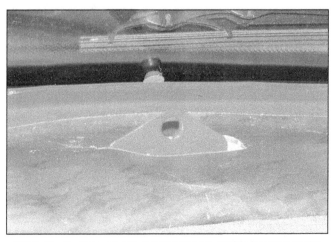

16.32 The instrument panel cover locates in this bracket on the vehicle body

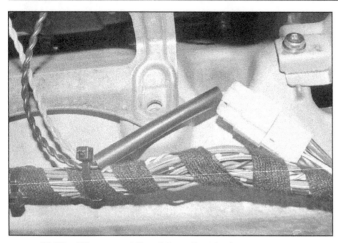

16.33a Disconnect the airbag electrical connector . . .

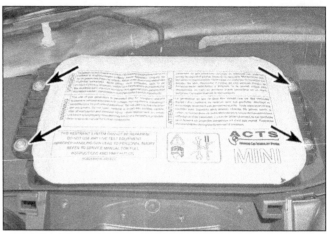

16.33b . . . then remove the four screws and remove the airbag

17.2 The clockspring is centralized when the blue indicator appears in the window

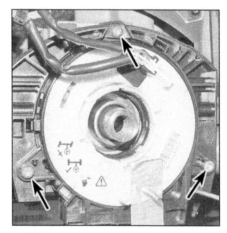

17.5 Clockspring retaining Torx screws

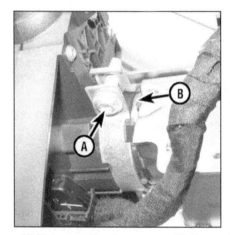

17.11 Loosen the clamp screw (A), then pry up the retaining tang (B) on each side and slide the unit from the steering column

remove the four screws and remove the airbag module **(see illustrations)**.

34 Installation is a reversal of removal. Tighten the retaining bolts to 84 inch pounds (10 Nm).

Side and side curtain airbags

35 These are not considered to be DIY operations, and should be referred to a MINI dealer service department.

---

17 Airbag clockspring - removal and installation

Mk I models

Removal

Refer to illustrations 17.2 and 17.5

1 Remove the steering wheel (see Chapter 10).

2 Use adhesive tape to secure the clockspring in the central position. The unit is centralized when the blue indicator appears in the window **(see illustration)**.

3 Remove the steering wheel upper and lower shrouds (see Chapter 11).

4 Disconnect the electrical connectors from the underside of the unit.

5 Remove the three screws, release the three retaining clips and detach the clockspring from the column switch housing **(see illustration)**.

Installation

6 Installation is a reversal of the removal procedure, noting the following points:

a) *Make sure that the front wheels are still pointing straight-ahead.*

b) *The clockspring must be installed in its central position, with the special alignment marks aligned (see Step 2).*

c) *If the central position has been lost, rotate the clockspring fully counter-clockwise, then rotate it fully clockwise and count the number of turns. Turn the clockspring half the number of revolutions counterclockwise. The unit is now centralized.*

Mk II models

Removal

Refer to illustration 17.11

7 Remove the steering wheel (see Chapter 10).

8 Use adhesive tape to secure the clockspring in the central position.

9 Remove the steering wheel upper and lower shrouds (see Chapter 11).

10 Disconnect the electrical connectors from the wiring harness.

11 Loosen the clamp screw, then pry the retaining tang from each side **(see illustration)** and slide the unit off the steering column.

Installation

12 Installation is a reversal of the removal procedure, noting the following points:

a) *Make sure that the front wheels are still pointing straight-ahead.*

b) *If the central position has been lost, rotate the clockspring fully counter-clockwise, then rotate it fully clockwise and count the number of turns. Turn the clockspring half the number of revolutions counterclockwise. The unit is now centralized.*

MINI wiring diagrams - 2006 and earlier models

Diagram 1

Key to symbols

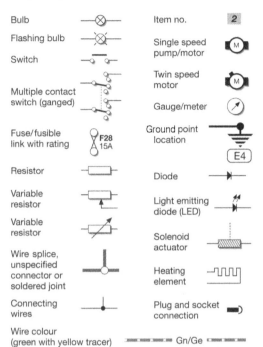

Bulb	—⊗—
Flashing bulb	—⊗—
Switch	—o o—
Multiple contact switch (ganged)	
Fuse/fusible link with rating	⦵F28 15A
Resistor	
Variable resistor	
Variable resistor	
Wire splice, unspecified connector or soldered joint	
Connecting wires	
Wire colour (green with yellow tracer)	▬▬▬ Gn/Ge ▬▬▬
Item no.	**2**
Single speed pump/motor	Ⓜ
Twin speed motor	Ⓜ
Gauge/meter	⊘
Ground point location	E4
Diode	—▶—
Light emitting diode (LED)	
Solenoid actuator	
Heating element	
Plug and socket connection	▬)

Dashed component outline denotes part of a larger item. Connectors may be show in two ways;

Multiple connectors — Sw(3)/1
Single connector — 2

Sw(3)/1 black 3 way connector, pin 1
2 single connector, pin 2

Typical engine compartment fusebox 6

Fuse	Rating	Circuit protected
FL1	50A	General control unit, heating & air conditioning blower, cigar lighter
FL2	50A	General control unit, steering angle sensor, clock, instrument cluster, navigation, on-board monitor, wash/wipe, horn, audio system
FL3	-	-
FL4	100A	Electric power steering pump
FL5	50A	Sunroof, antenna, rear wiper
FL6	40A	ABS/ASC/DSC
FL7	50A	Ignition switch
FL8	50A	General control unit, lights
FL9	50A	Engine cooling fan, heater blower
FL10	50A	Heated windshield
FL11	50A	Heating & air conditioning blower
FL12	50A	General control unit, lights
F1	5A	Engine control unit
F2	20A	Engine control unit, ignition coil, fuel injectors, crankshaft position sensor
F3	15A	Engine control
F4	15A	Engine control
F5	5A	Engine cooling fan
F6	30A	ABS/ASC/DSC
F7	30A	Air conditioning compressor
F8	30A	Heating & A/C blower, electric power steering pump, cooling fan
F9	20A	Front wiper
F10	15A	Fog lights

Typical passenger compartment fusebox 5

Fuse	Rating	Circuit protected
F1	30A	LH window motor
F2	5A	Steering angle sensor
F3	5A	Electric mirrors, clock
F4	5A	General control unit
F5	5A	Instrument cluster
F6	5A	Brake lights, clutch switch, EWS
F7	5A	Rain sensor, general control unit
F8	5A	Direction indicators, general control unit
F9	5A	Instrument cluster
F10	5A	Audio system
F11	5A	Front & rear washers
F12	20A	Heated seats
F13	5A	Reversing lights, automatic transmission
F14	10A	General control unit, interior lights
F15	20A	Sunroof
F16	30A	Heated rear window
F17	15A	Rear wiper
F18	5A	SRS system
F19	30A	RH window motor
F20	20A	Fuel pump
F21	10A	Instrument cluster
F22	15A	Navigation, on-board monitor
F23	20A	Front & rear washers
F24	5A	EWS, alarm
F25	30A	Headlight washer
F26	10A	Automatic transmission
F27	15A	Audio system
F28	15A	Horn
F29	5A	Air conditioning
F30	5A	Instrument cluster
F31	30A	Heating & A/C blower
F32	15A	Cigar lighter
F33	10A	ABS/ASC/DSC
F34	10A	Engine management main relay
F35	5A	Electric mirrors, heated washer jets
F36	5A	Heated windshield
F37	20A	Central locking
F38	-	-
F39	5A	Alternator, electric power steering pump
F40	5A	Instrument cluster, switch panel, steering angle sensor
F41	5A	Park distance control, electric power steering pump
F42	-	-

Engine fusebox

Passenger fusebox

Ground locations

E1	LH rear of engine compartment	E7	RH side of luggage compartment
E2	LH front of engine compartment	E8	RH door sill
E3	RH front of engine compartment	E9	LH door sill
E4	RH side of footwell	E10	Behind LH rear window trim panel
E5	LH door sill	E11	Behind RH rear window trim panel
E6	LH side of luggage compartment		

H33183

Wire colors

Bl	Blue	**Vi**	Violet
Br	Brown	**Ws**	White
Ge	Yellow	**Or**	Orange
Gr	Grey	**Rt**	Red
Gn	Green	**Sw**	Black
Rs	Pink	**Tr**	Transparent

Key to items

1 Battery
2 Ignition switch
3 Starter motor
4 Alternator
5 Passenger compartment fusebox
  a = horn relay
6 Engine fusebox
7 Luggage compartment fusebox
  (Cooper S only)

8 LH horn
9 RH horn
10 Steering wheel clock springs
11 Horn switch
12 LH button pad
13 RH button pad
14 General control unit

2006 and earlier models - Diagram 2

H33184

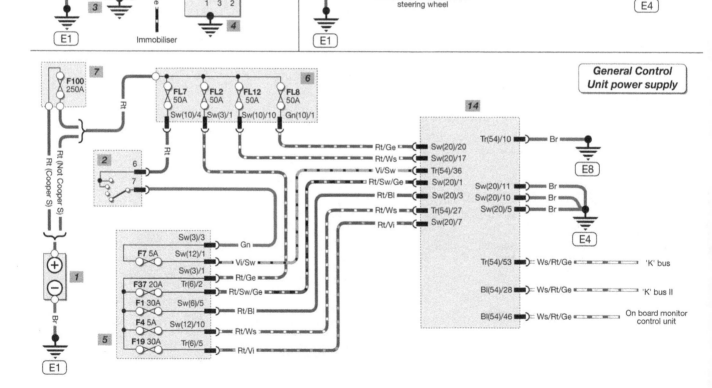

Wire colors

Bl	Blue	**Vi**	Violet
Br	Brown	**Ws**	White
Ge	Yellow	**Or**	Orange
Gr	Grey	**Rt**	Red
Gn	Green	**Sw**	Black
Rs	Pink	**Tr**	Transparent

Note:
See diagram 2 for General Control Unit power supply details.

Key to items

1 Battery
2 Ignition switch
5 Passenger compartment fusebox
6 Engine fusebox
7 Luggage compartment fusebox (Cooper S only)
14 General control unit
15 LH rear light unit
  a = stop light
  c = tail light
16 RH rear light unit
  a = stop light
  b = reversing light
  c = tail light

17 High level stop light
18 Reversing light switch (manual)
19 Transmission switch (auto)
20 Stop light switch
21 Multifunction switch
  a = side/headlight
  b = headlight flasher
22 LH front side light
23 RH front side light
24 Number plate light/tailgate switch
25 LH xenon headlight trigger
26 RH xenon headlight trigger

27 LH headlight unit
  a = main beam
  b = dip beam
28 RH headlight unit
  a = **high** beam
  b = **low** beam

2006 and earlier models - Diagram 3

H33185

Side, tail, number plate & headlights

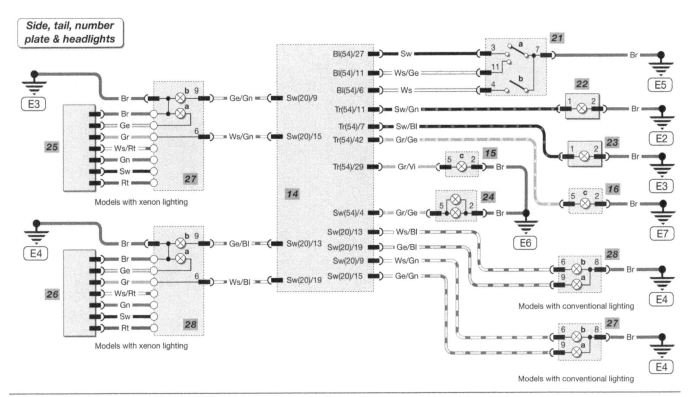

Stop & reversing lights

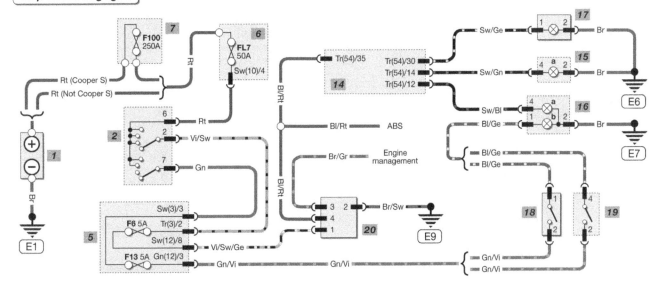

Wire colors

Bl	Blue	**Vi**	Violet
Br	Brown	**Ws**	White
Ge	Yellow	**Or**	Orange
Gr	Grey	**Rt**	Red
Gn	Green	**Sw**	Black
Rs	Pink	**Tr**	Transparent

Note:
See diagram 2 for General Control
Unit power supply details.

Key to items

1 Battery
2 Ignition switch
5 Passenger compartment fusebox
6 Engine fusebox
   a = foglight relay
   b = wiper relay (fast/slow)
   c = wiper relay (on/off)
7 Luggage compartment fusebox
   (Cooper S only)
14 General control unit
15 LH rear light unit
   d = direction indicator
   e = foglight

16 RH rear light unit
   d = direction indicator
   e = foglight
21 Multifunction switch
   c = direction indicator switch
30 LH front direction indicator
31 RH front direction indicator
32 LH indicator side repeater
33 RH indicator side repeater
34 Switch panel
   a = rear foglight switch & indicator
   b = front foglight switch & indicator

35 LH front foglight
36 RH front foglight
37 Wash/wipe switch
38 Front/rear washer pump
39 Front warher pump relay
40 Rear washer pump relay
41 Front wiper motor

H33186

Front & rear foglights

Direction indicators & hazard warning lights

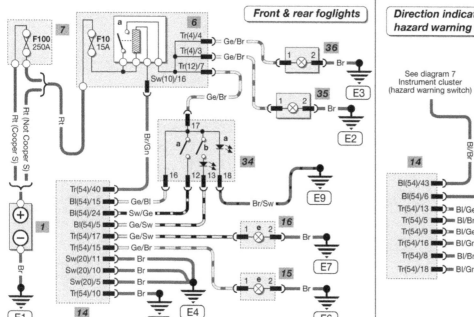
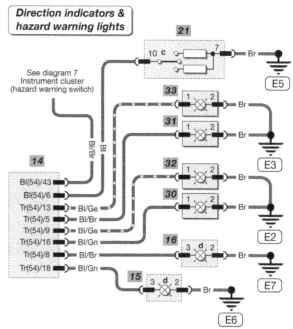

Front wash/wipe & rear washer

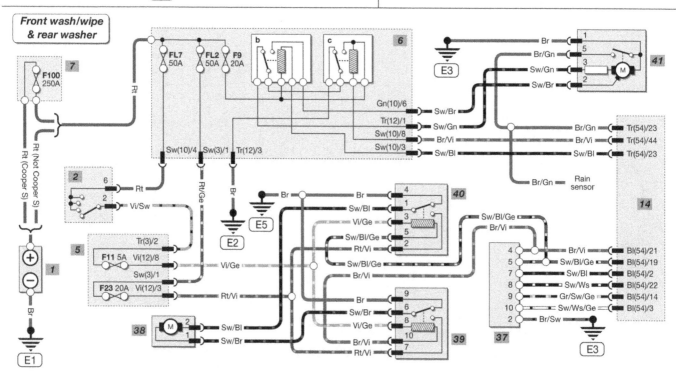

Wire colors

Bl	Blue	**Vi**	Violet
Br	Brown	**Ws**	White
Ge	Yellow	**Or**	Orange
Gr	Grey	**Rt**	Red
Gn	Green	**Sw**	Black
Rs	Pink	**Tr**	Transparent

Note:
See diagram 2 for General Control
Unit power supply details.

Key to items

1 Battery
2 Ignition switch
5 Passenger compartment fusebox
   b = rear wiper relay
   c = headlight washer relay
6 Engine fusebox
7 Luggage compartment fusebox
   (Cooper S only)
14 General control unit
37 Wash/wipe switch

45 Rear wiper motor
46 Headlight washer pump
47 LH mirror assembly
48 RH mirror assembly
49 Mirror control switch
50 Mirror 'fold in' control unit

2006 and earlier models - Diagram 5

H33187

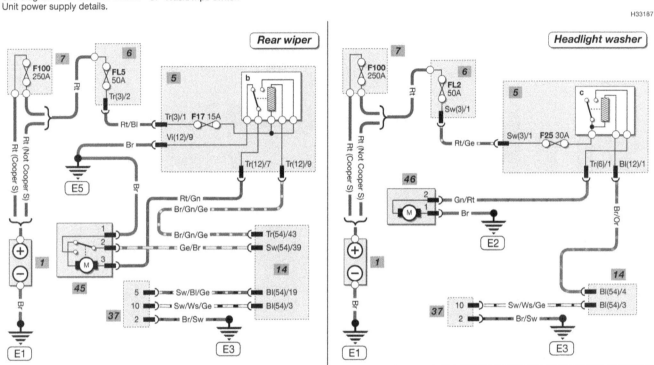

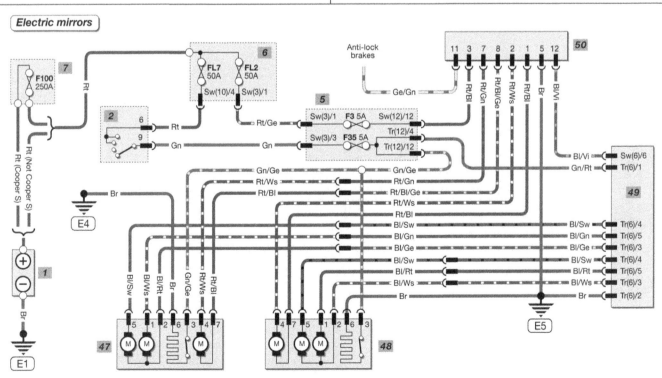

Wire colors

Bl	Blue	**Vi**	Violet
Br	Brown	**Ws**	White
Ge	Yellow	**Or**	Orange
Gr	Grey	**Rt**	Red
Gn	Green	**Sw**	Black
Rs	Pink	**Tr**	Transparent

Note:
See diagram 2 for General Control
Unit power supply details.

Key to items

1 Battery
2 Ignition switch
5 Passenger compartment fusebox
   d = heated rear window relay
   e = cigar lighter relay
6 Engine fusebox
   d = heater blower relay
7 Luggage compartment fusebox
   (Cooper S only)
14 General control unit

55 Heater blower switch
56 Heater blower resistors
57 Heater blower motor
58 Heated rear window
59 Heater control panel
60 RF filter
61 Heated windshield relay
62 LH heated windshield
63 RH heated windshield
64 Cigar lighter

2006 and earlier models - Diagram 6

H33188

Heater blower

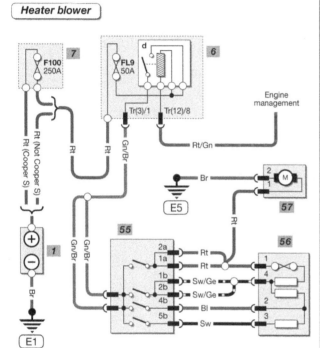

Heated rear screen

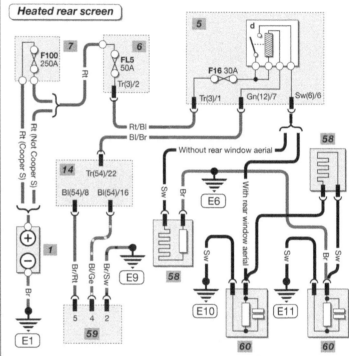

Heated front screen

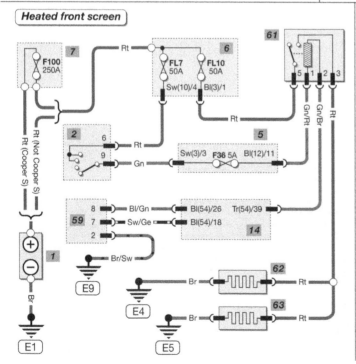

Cigar lighter

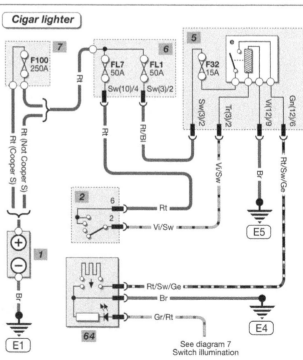

See diagram 7
Switch illumination

Wire colors

Bl	Blue	**Vi**	Violet
Br	Brown	**Ws**	White
Ge	Yellow	**Or**	Orange
Gr	Grey	**Rt**	Red
Gn	Green	**Sw**	Black
Rs	Pink	**Tr**	Transparent

Key to items

1 Battery
2 Ignition switch
5 Passenger compartment fusebox
6 Engine fusebox
7 Luggage compartment fusebox
  (Cooper S only)
21 Multifunction switch
  d = trip computer function switch
34 Switch panel
70 Instrument cluster control unit

71 Auxiliary instrument cluster
72 Outside air temperature sensor
73 Fuel gauge sender unit/fuel pump
  a = fuel level sensor 1
  b = fuel level sensor 2
74 Handbrake switch
75 Oil pressure switch
76 LH front pad wear sensor
77 RH front pad wear sensor

2006 and earlier models - Diagram 7

H33189

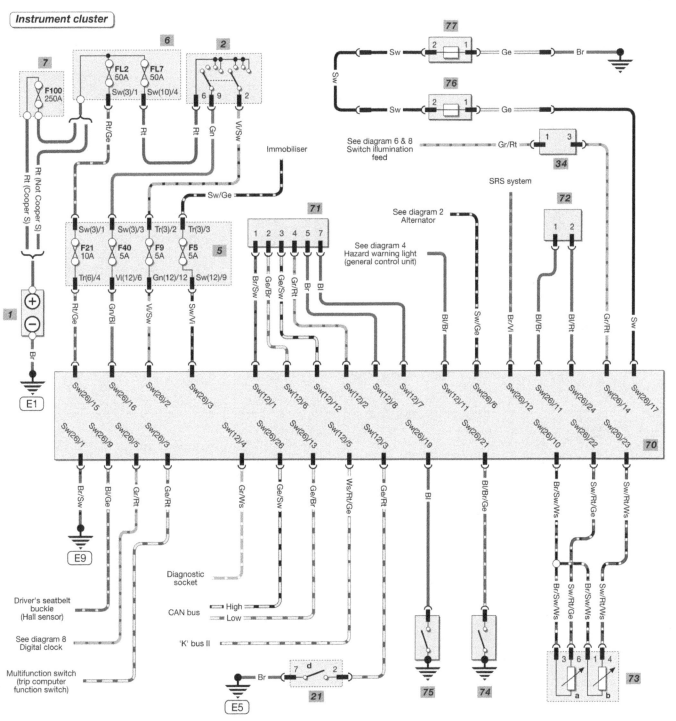

Wire colors

Bl	Blue	**Vi**	Violet
Br	Brown	**Ws**	White
Ge	Yellow	**Or**	Orange
Gr	Grey	**Rt**	Red
Gn	Green	**Sw**	Black
Rs	Pink	**Tr**	Transparent

Note:
See diagram 2 for General Control
Unit power supply details.

Key to items

1 Battery
2 Ignition switch
5 Passenger compartment fusebox
6 Engine fusebox
7 Luggage compartment fusebox
   (Cooper S only)
14 General control unit
80 LH vanity mirror light
81 RH vanity mirror light
82 Glovebox light
83 Luggage compartment light
84 Driver's footwell lighting

85 Passenger's footwell lighting
86 Front interior light
   a = LH map reading light
   b = RH map reading light
   c = illumination
87 Rear interior light
   a = interior light
   b = interior light switch
   c = LH map reading light
   d = RH map reading light
88 Digital clock
89 Audio unit

90 **Antenna** amplifier
91 LH front mid range speaker
92 LH front tweeter
93 RH front mid range speaker
94 RH front tweeter
95 LH rear speaker
96 RH rear speaker

2006 and earlier models - Diagram 8

H33190

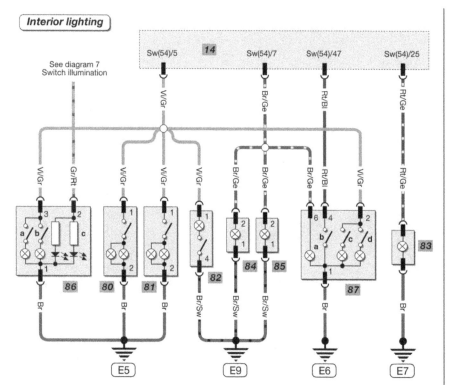

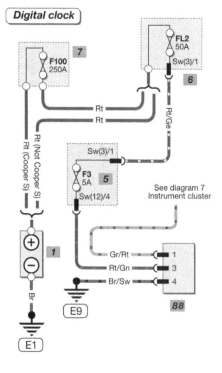

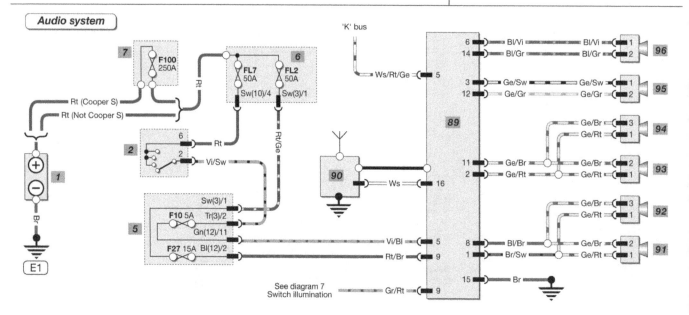

Wire colors

Bl	Blue	**Vi**	Violet
Br	Brown	**Ws**	White
Ge	Yellow	**Or**	Orange
Gr	Grey	**Rt**	Red
Gn	Green	**Sw**	Black
Rs	Pink	**Tr**	Transparent

Note:
See diagram 2 for General Control
Unit power supply details.

Key to items

1   Battery
5   Passenger compartment fusebox
6   Engine fusebox
7   Luggage compartment fusebox
    (Cooper S only)
14 General control unit
24 Number plate light/tailgate switch
34 Switch panel
    c = passenger's window switch
    d = driver's window switch
    e = centre lock switch

98   Passenger's window motor
99   Driver's window motor
100 Driver's door lock assembly
101 Passenger's door lock assembly
102 Filler flap lock motor
103 Tailgate lock motor
104 Remote control receiver

2006 and earlier models - Diagram 9

H33191

Electric windows

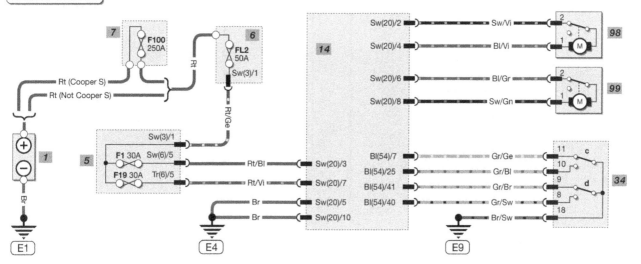

Central locking

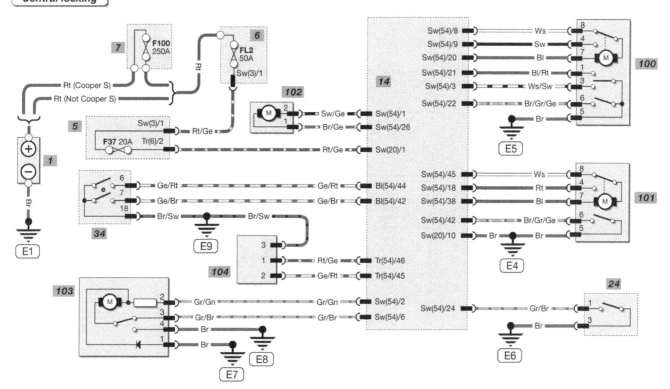

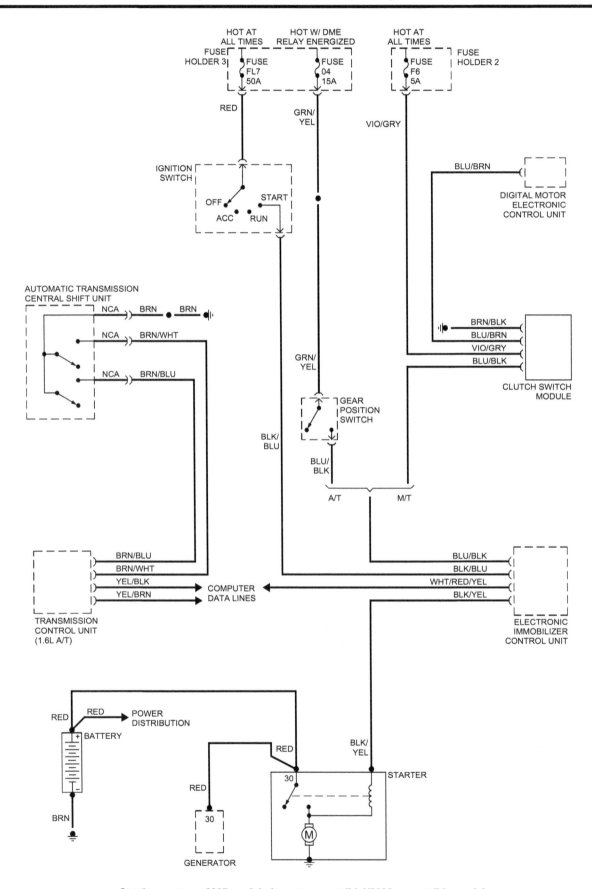

Starting system - 2007 models (except convertible)/2008 convertible models

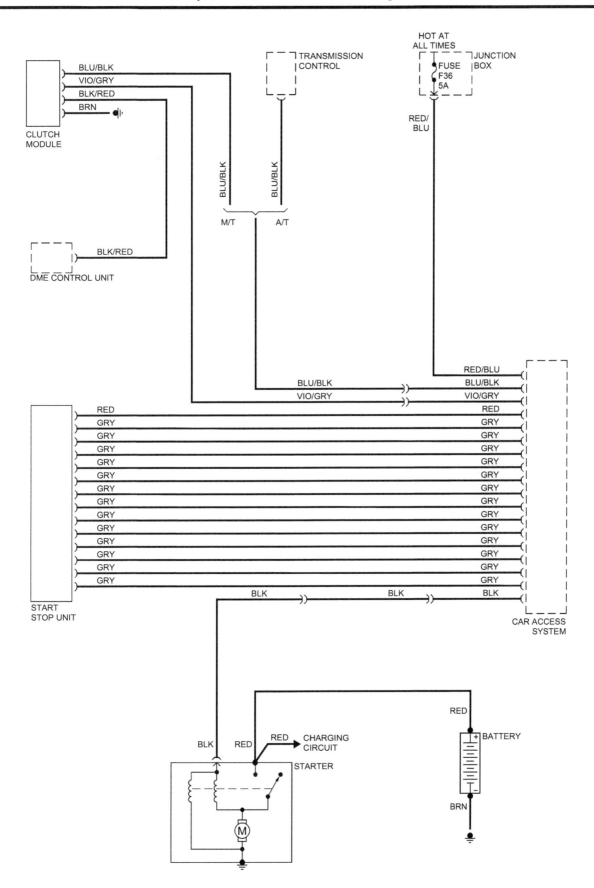

Starting system - 2008 models (except convertible)/all 2009 and later models

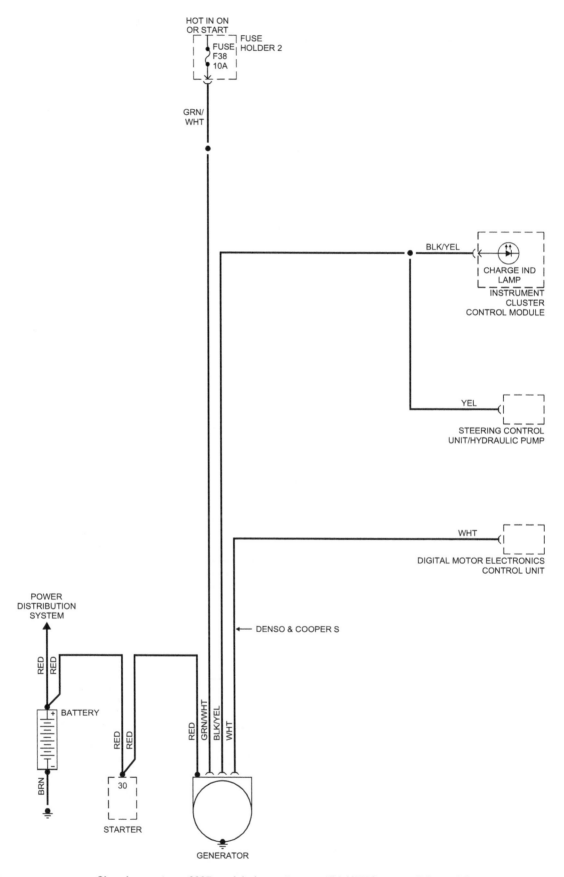

Charging system - 2007 models (except convertible)/2008 convertible models

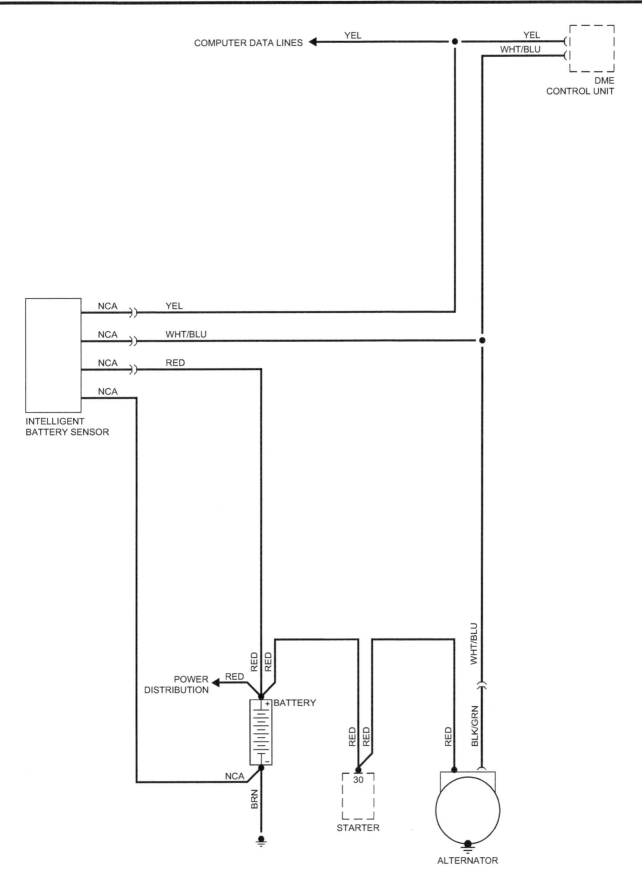

Charging system - 2008 models (except convertible)/all 2009 and later models

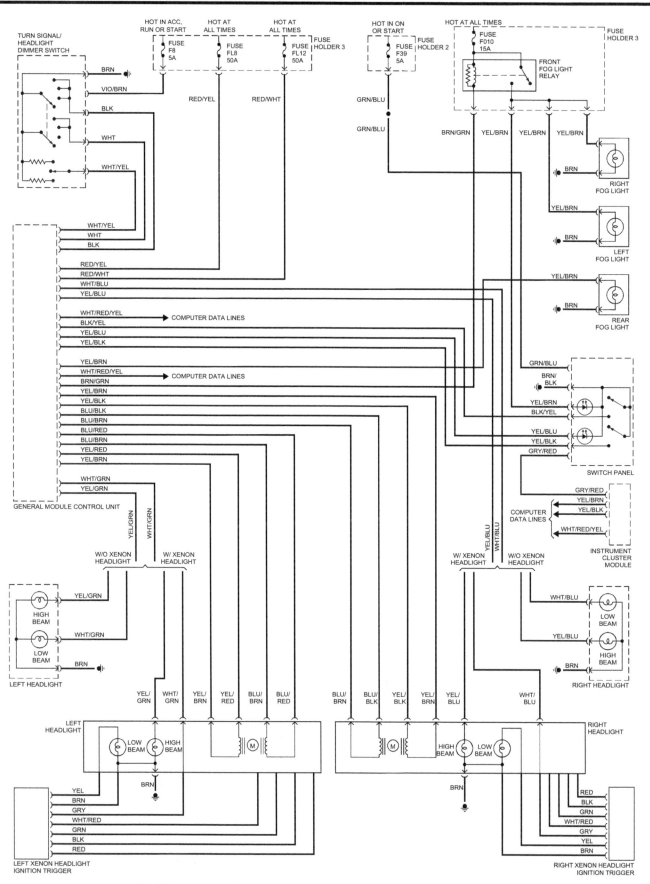

Headlight system - 2007 models (except convertible)/2008 convertible models

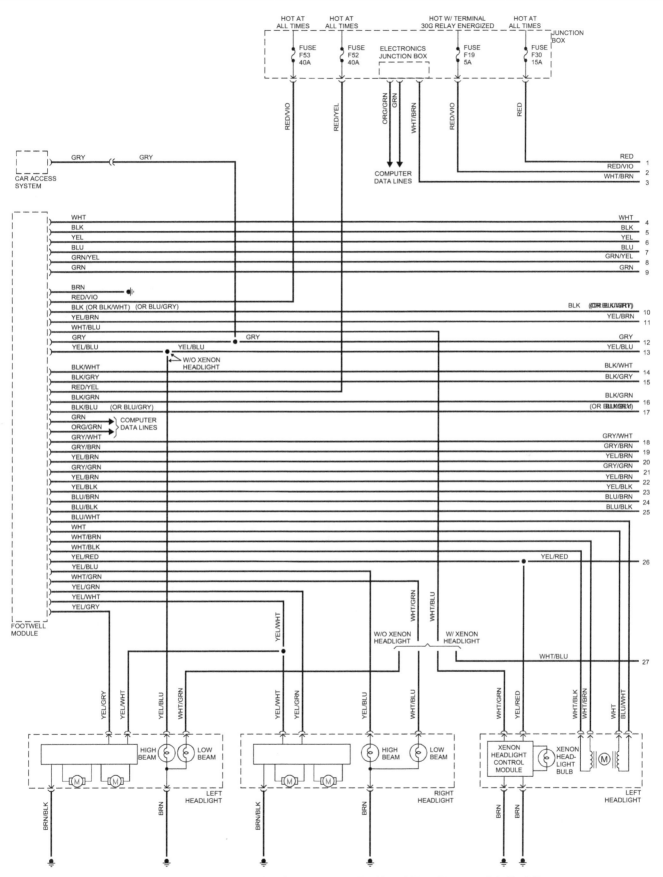

Headlight system - 2008 models (except convertible)/all 2009 and later models (1 of 2)

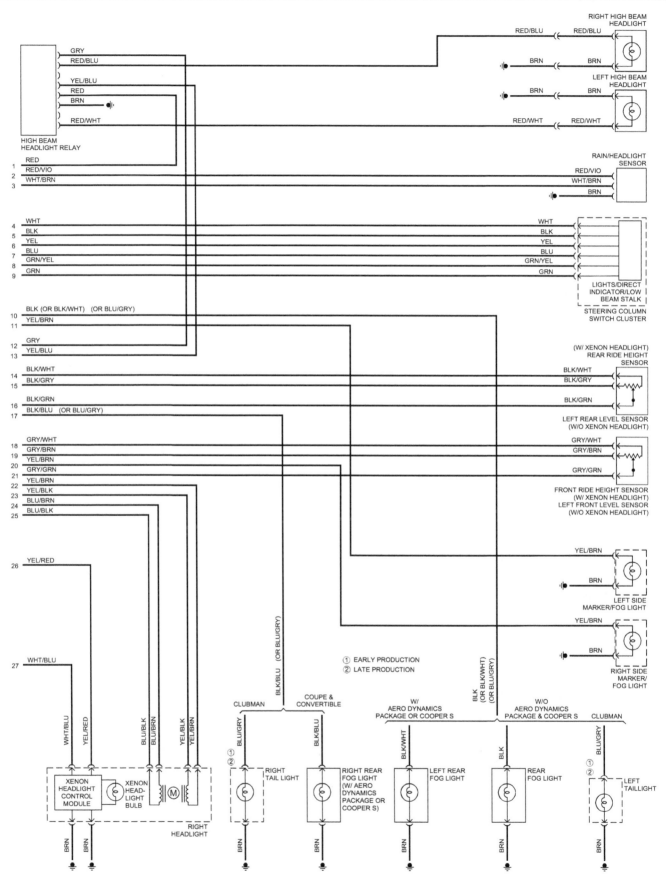

Headlight system - 2008 models (except convertible)/all 2009 and later models (2 of 2)

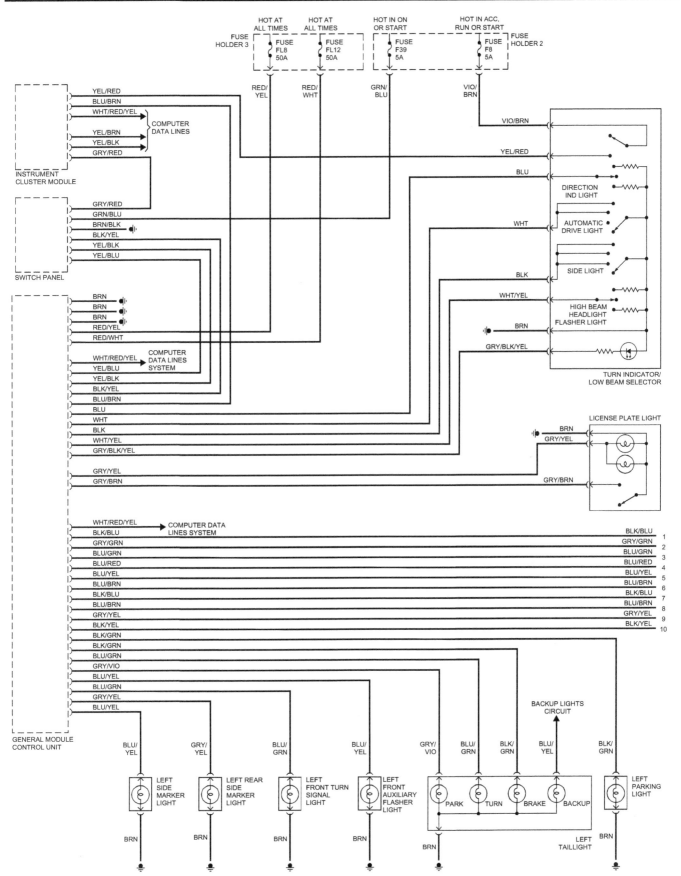

Exterior lighting system - 2007 models (except convertible)/2008 convertible models (1 of 2)

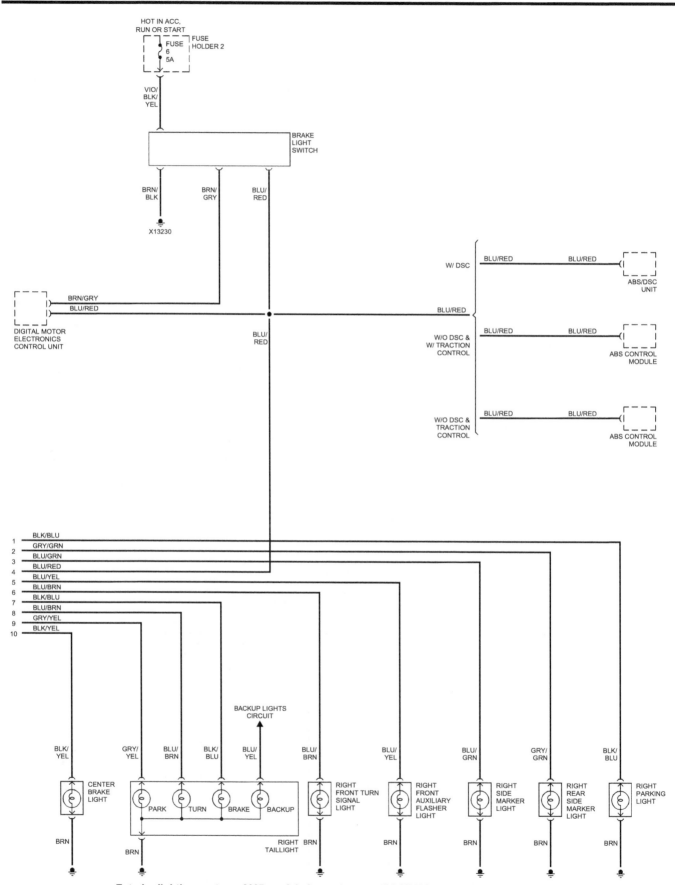

Exterior lighting system - 2007 models (except convertible)/2008 convertible models (2 of 2)

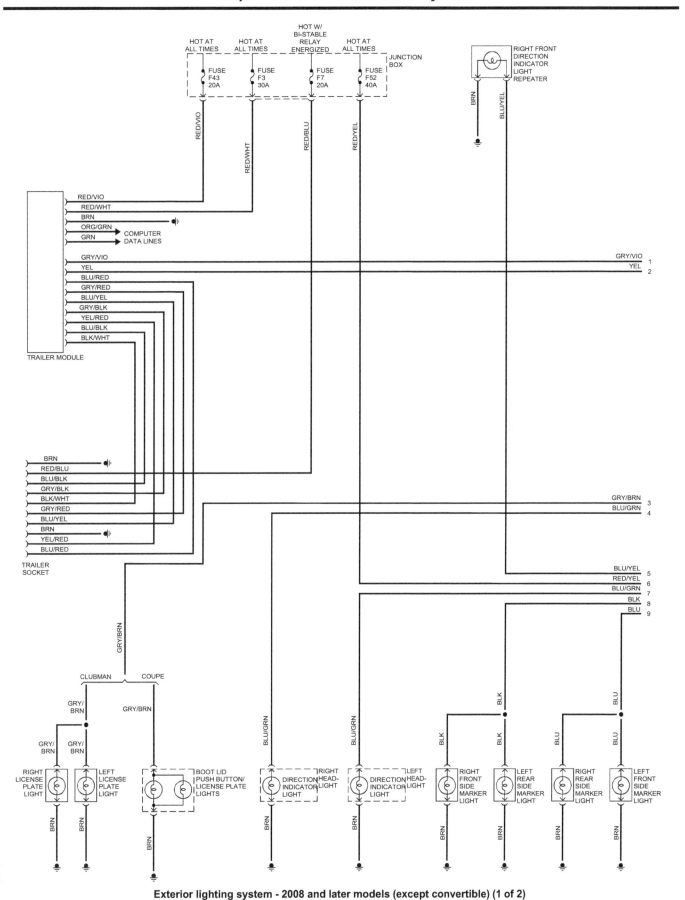

Exterior lighting system - 2008 and later models (except convertible) (1 of 2)

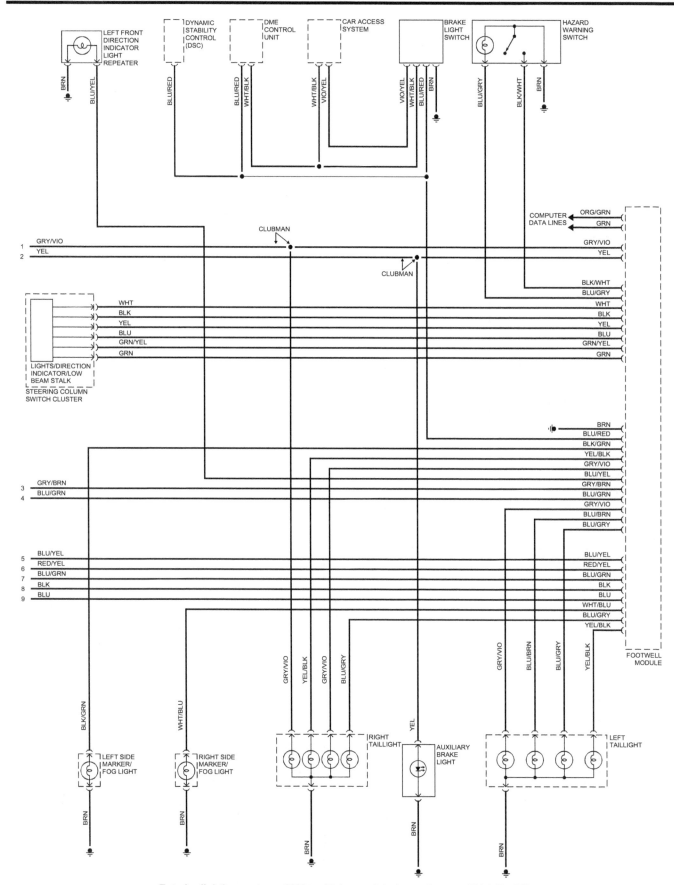

Exterior lighting system - 2008 and later models (except convertible) (2 of 2)

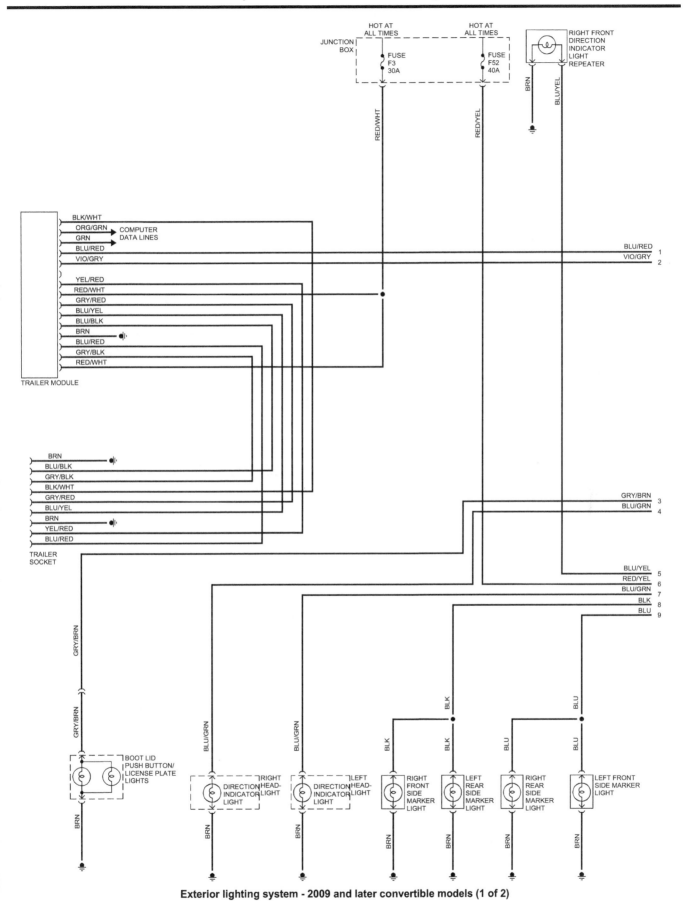

Exterior lighting system - 2009 and later convertible models (1 of 2)

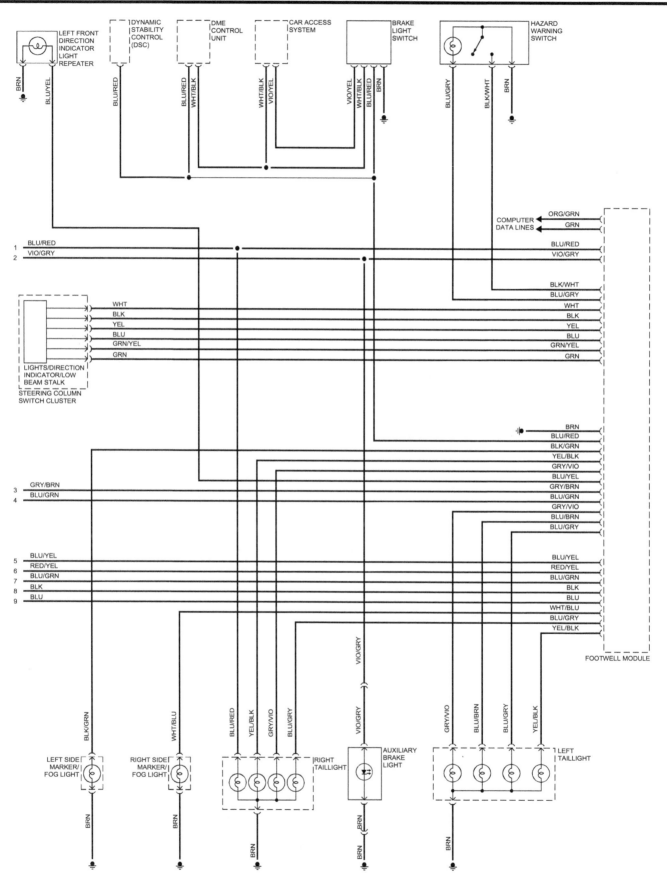

Exterior lighting system - 2009 and later convertible models (2 of 2)

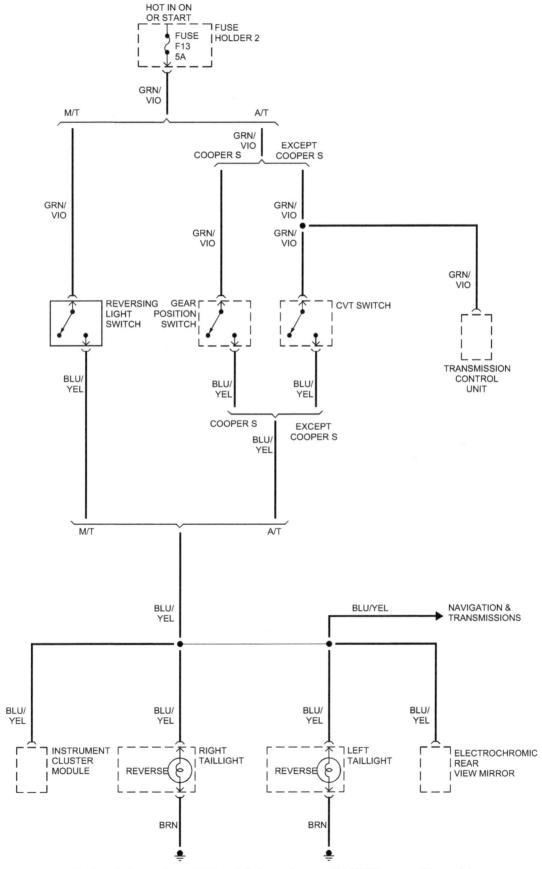

Back-up lights system - 2007 models (except convertible)/2008 convertible models

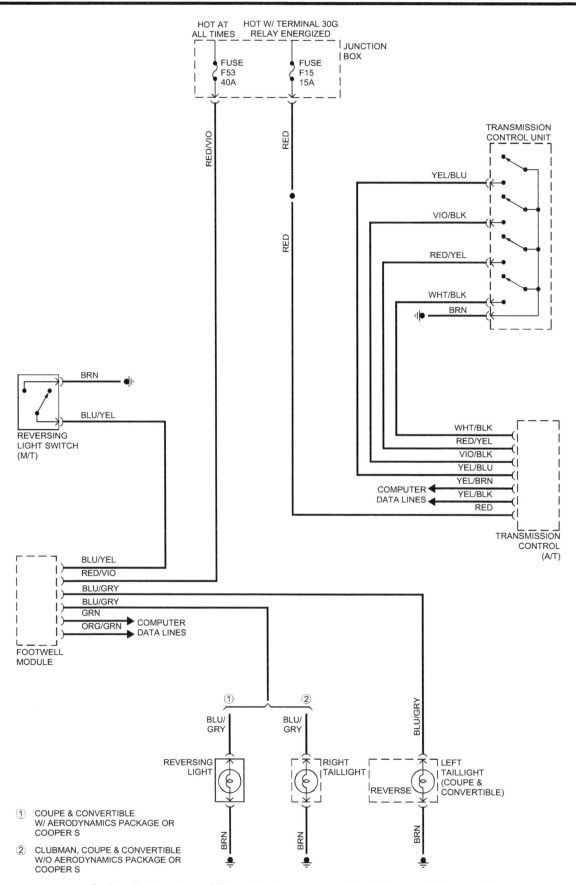

Back-up lights system - 2008 models (except convertible)/all 2009 and later models

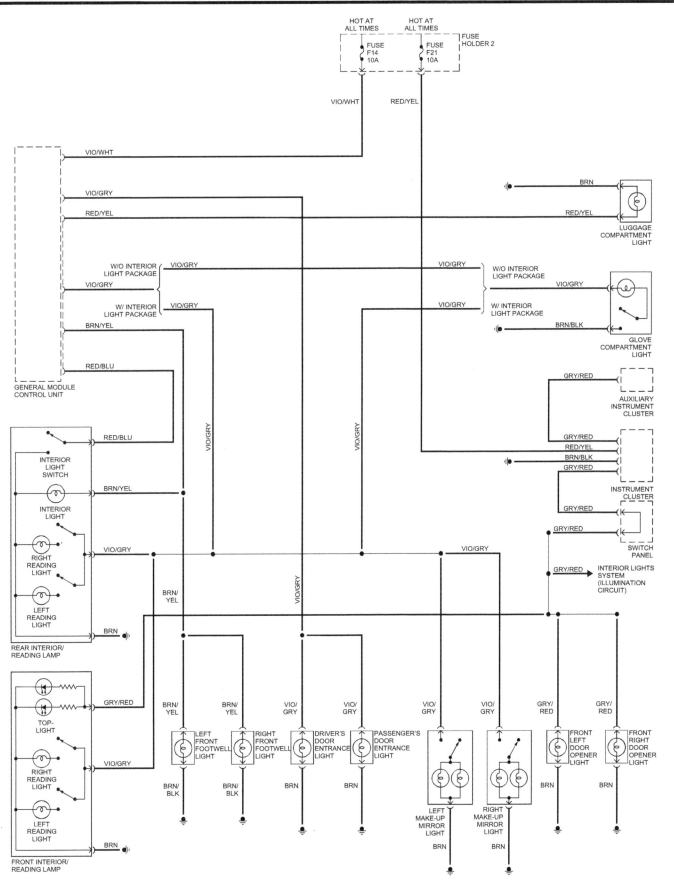

Courtesy lighting system - 2007 models (except convertible)

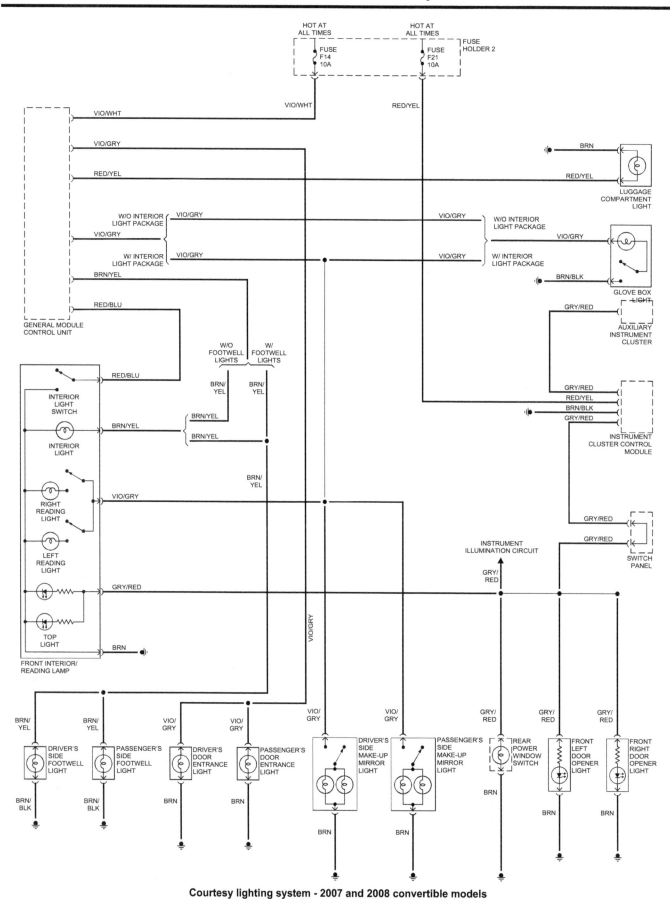

Courtesy lighting system - 2007 and 2008 convertible models

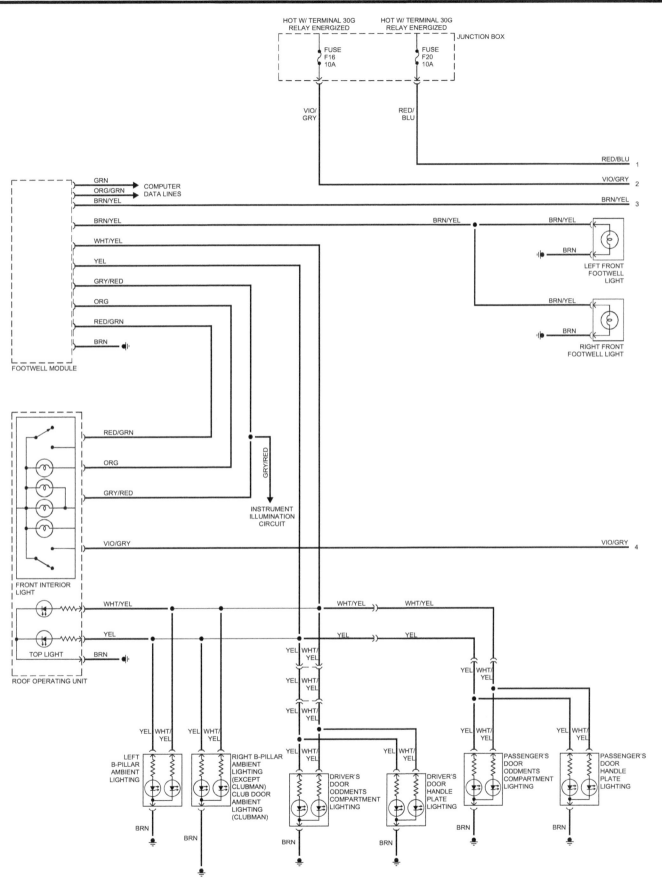

Courtesy lighting system (with interior lights package) - 2008 models (except convertible)/all 2009 and later models (1 of 2)

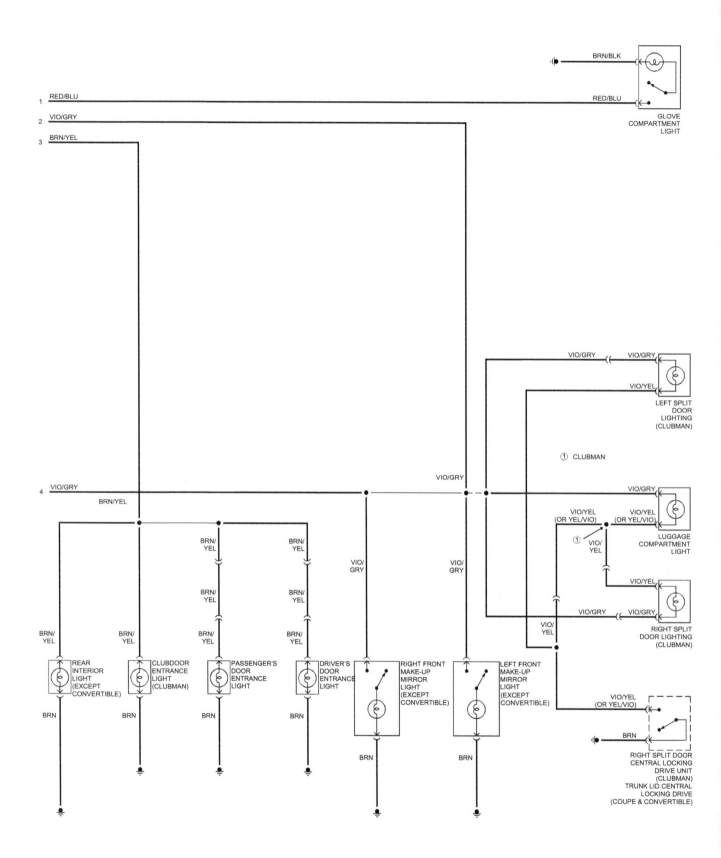

Courtesy lighting system (with interior lights package) - 2008 models (except convertible)/all 2009 and later models (2 of 2)

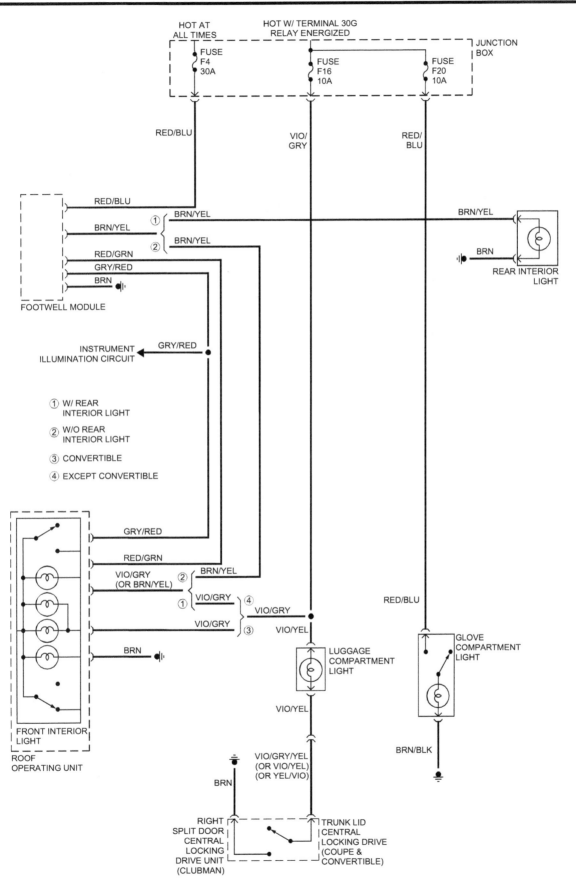

Courtesy lighting system (without interior lights package) - 2008 models (except convertible)/all 2009 and later models

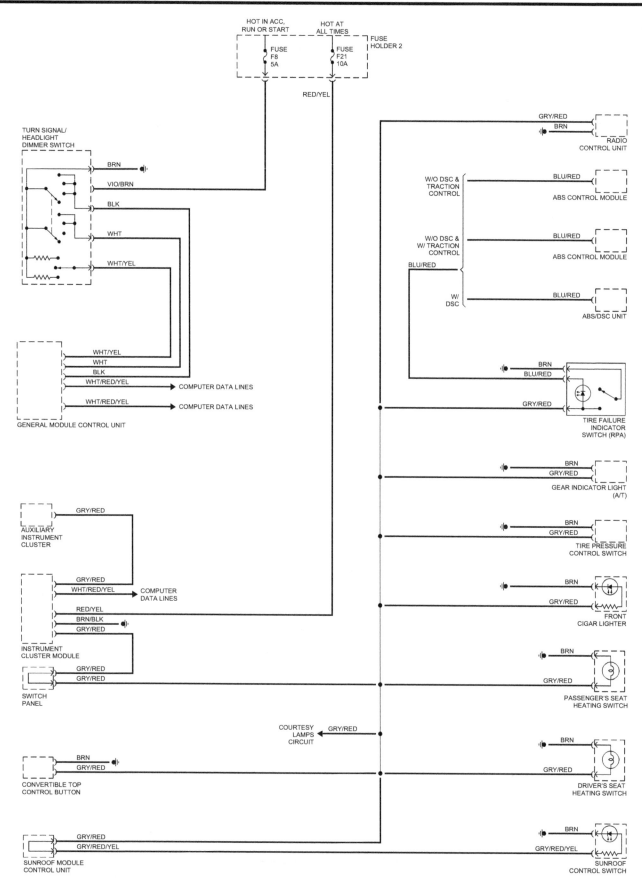

Instrument panel and switch illumination - 2007 models (except convertible)/2008 convertible models

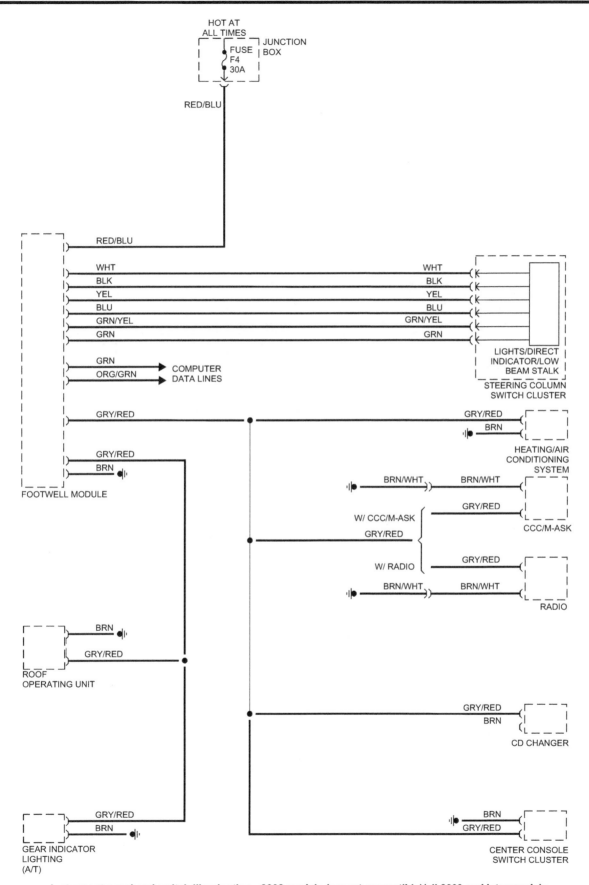

Instrument panel and switch illumination - 2008 models (except convertible)/all 2009 and later models

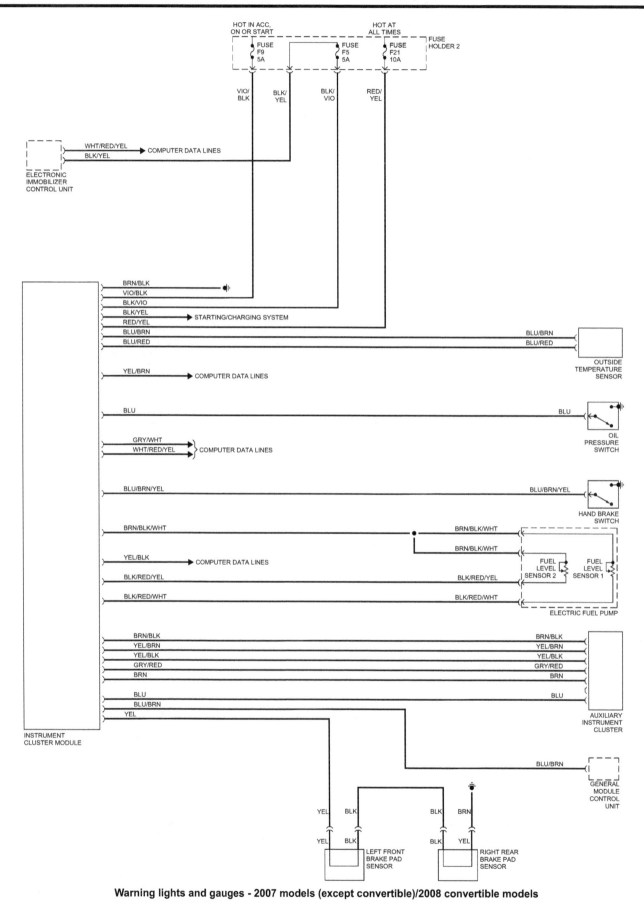

Warning lights and gauges - 2007 models (except convertible)/2008 convertible models

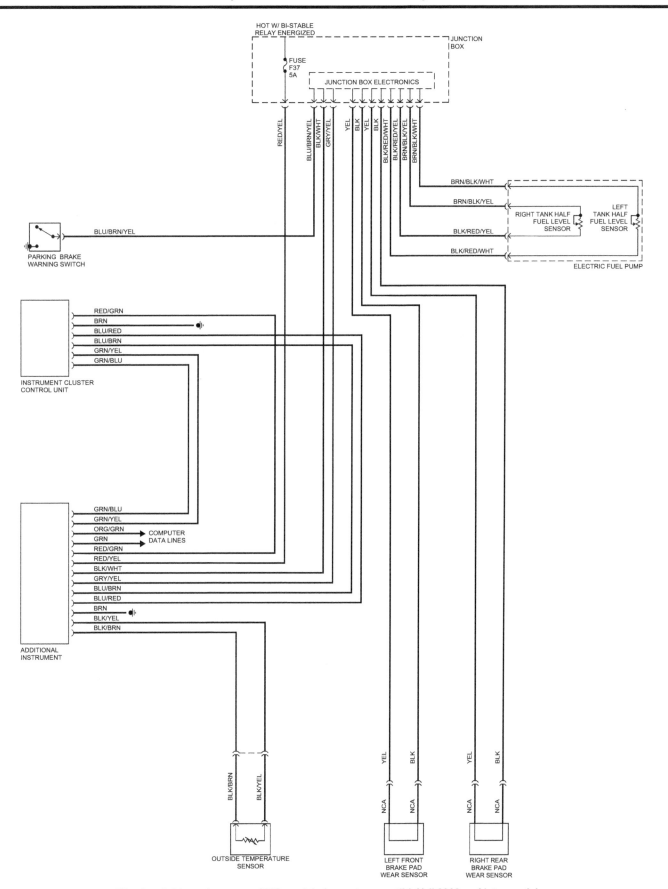

Warning lights and gauges - 2008 models (except convertible)/all 2009 and later models

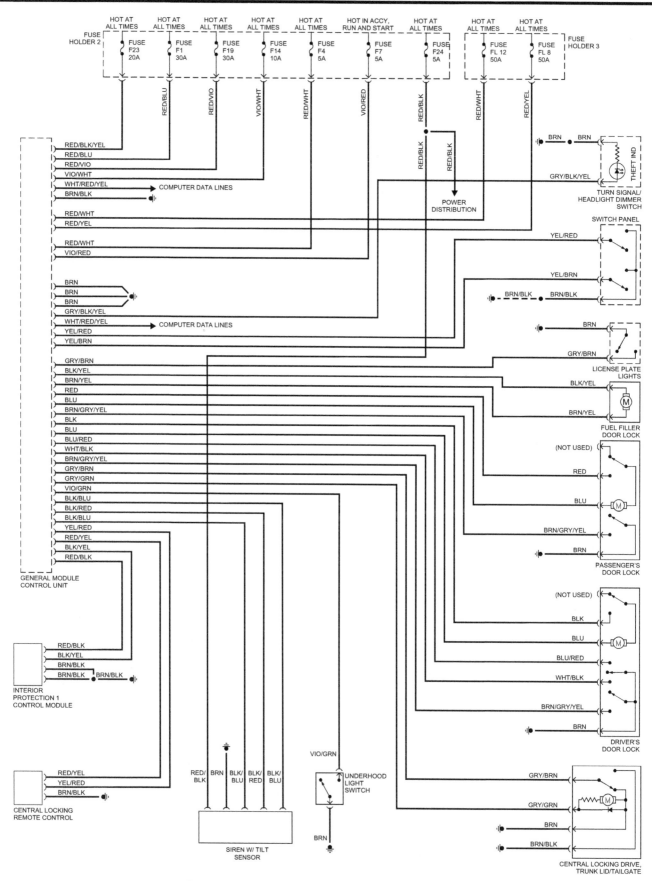

Power door lock system - 2007 models (except convertible)

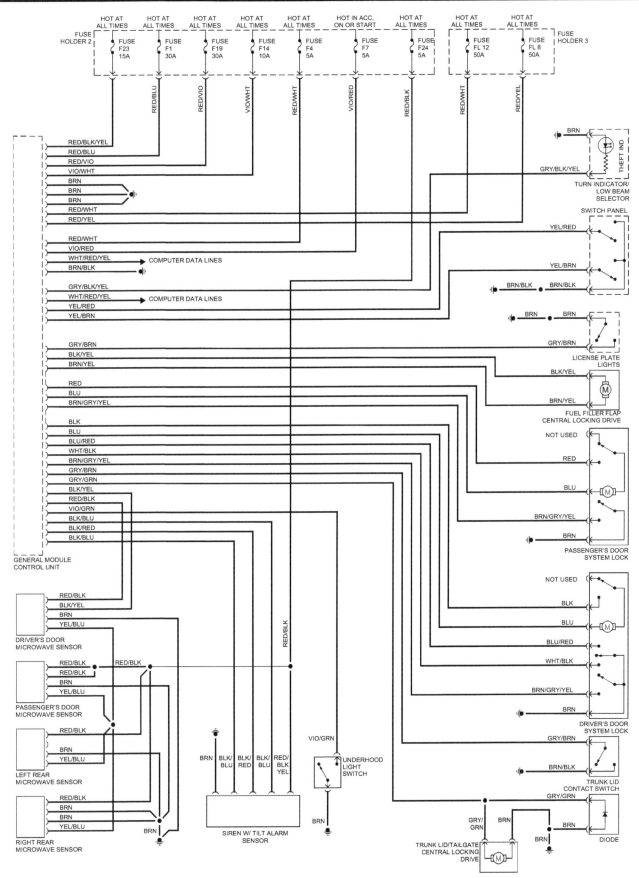

Power door lock system - 2007 and 2008 convertible models

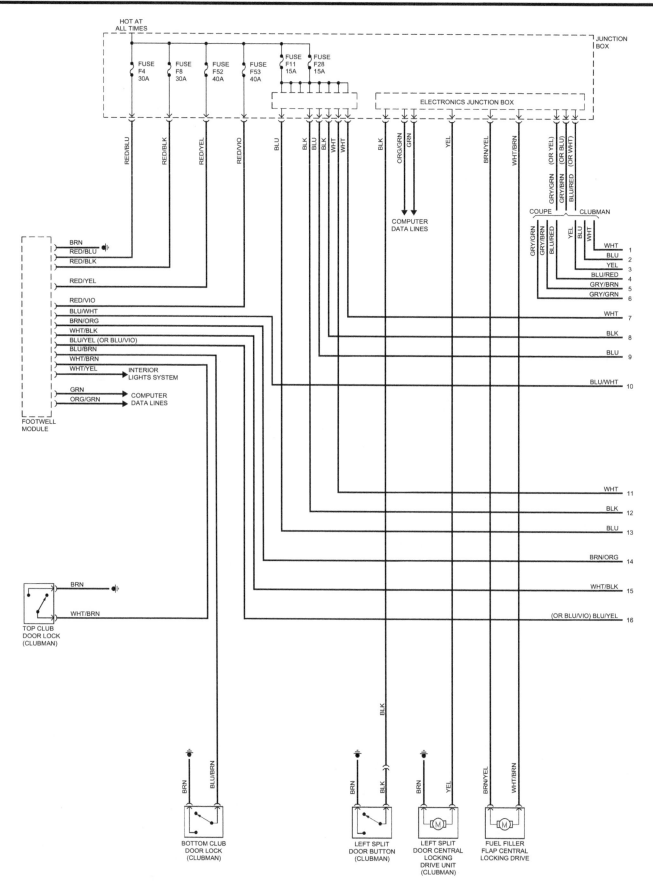

Power door lock system - 2008 and later models (except convertible) (1 of 2)

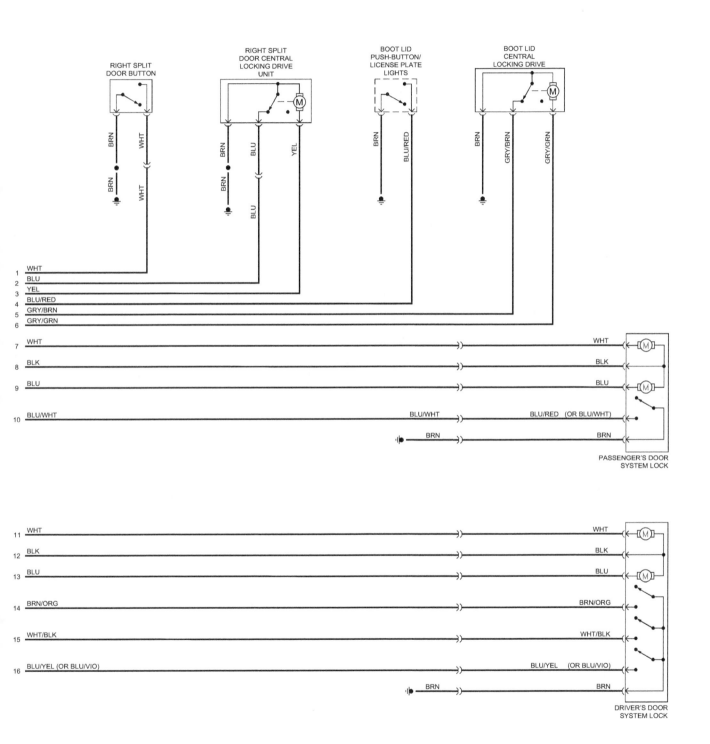

Power door lock system - 2008 and later models (except convertible) (2 of 2)

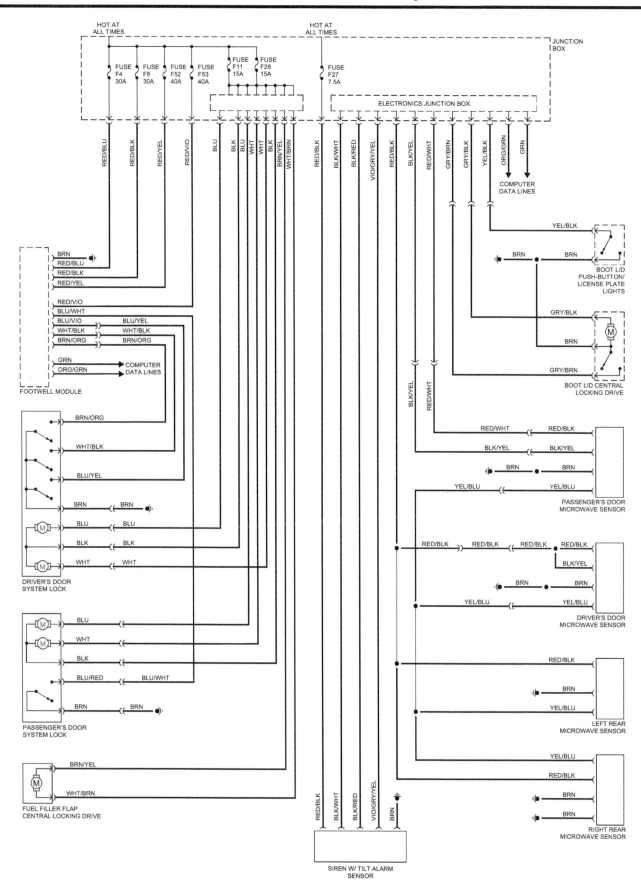

Power door lock system - 2009 and later convertible models

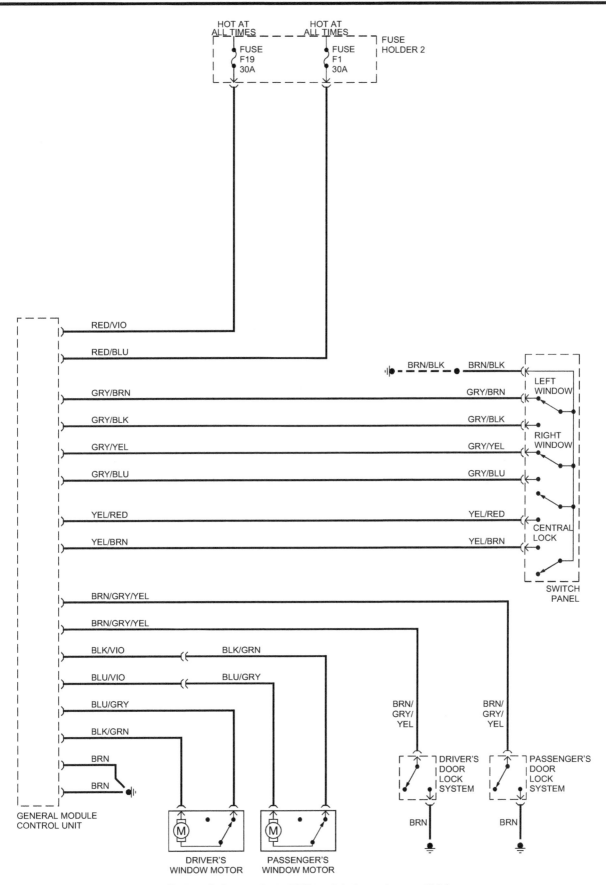

Power window system - 2007 models (except convertible)

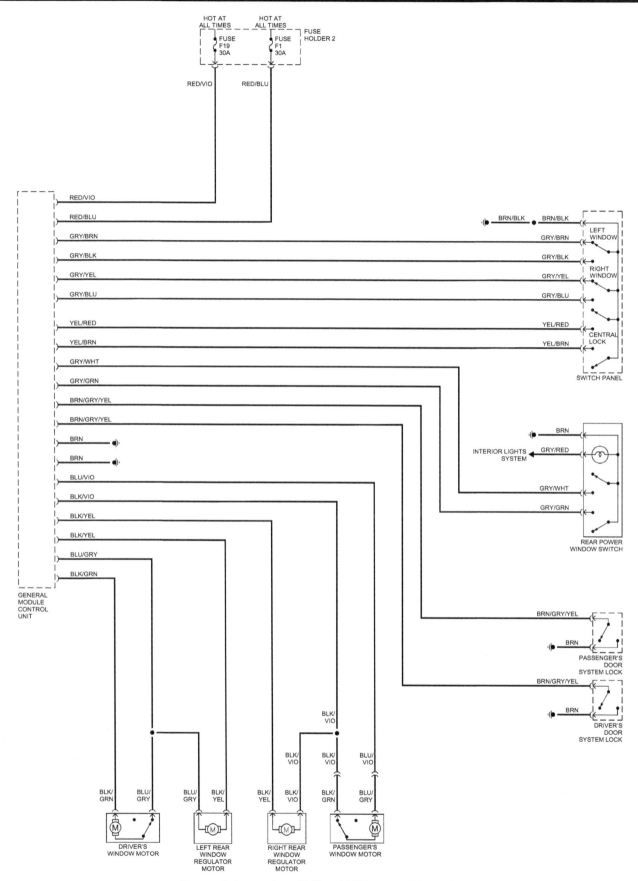

Power window system - 2007 and 2008 convertible models

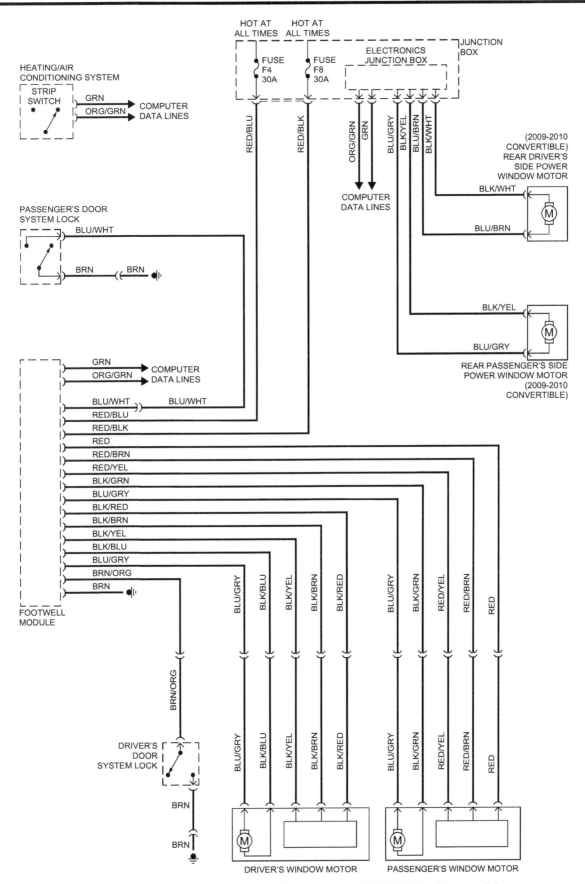

Power window system - 2008 models (except convertible)/all 2009 and later models

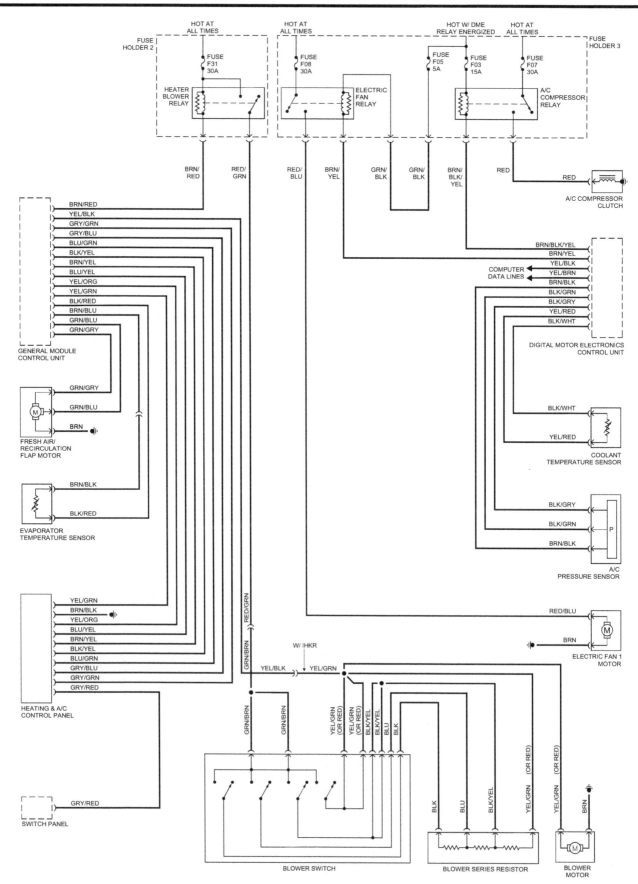

Air conditioning (manual) and engine cooling fan (single stage) circuit - 2007 models (except convertible)/2008 convertible models

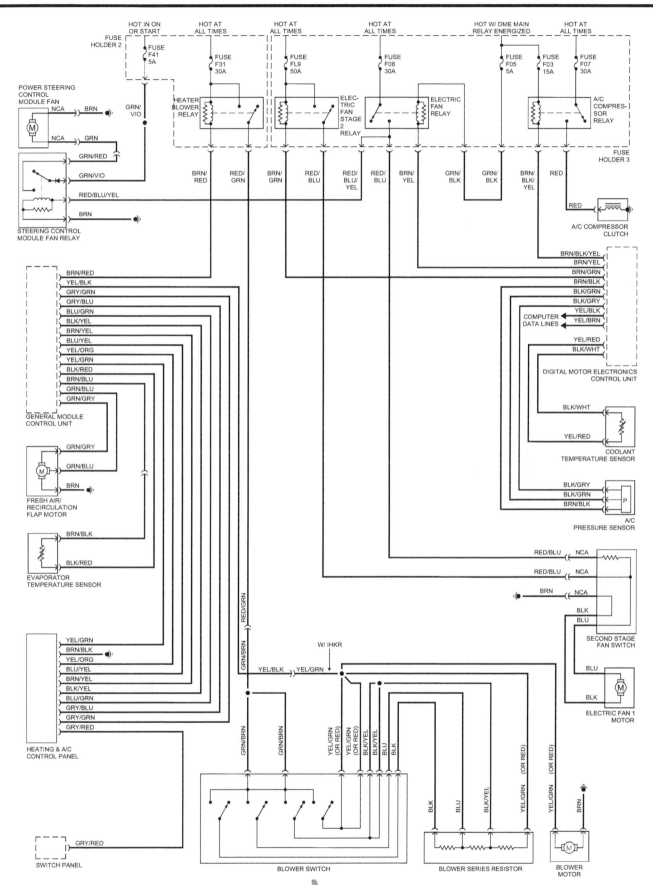

Air conditioning (manual) and engine cooling fan (dual stage) circuit - 2007 models (except convertible)/2008 convertible models

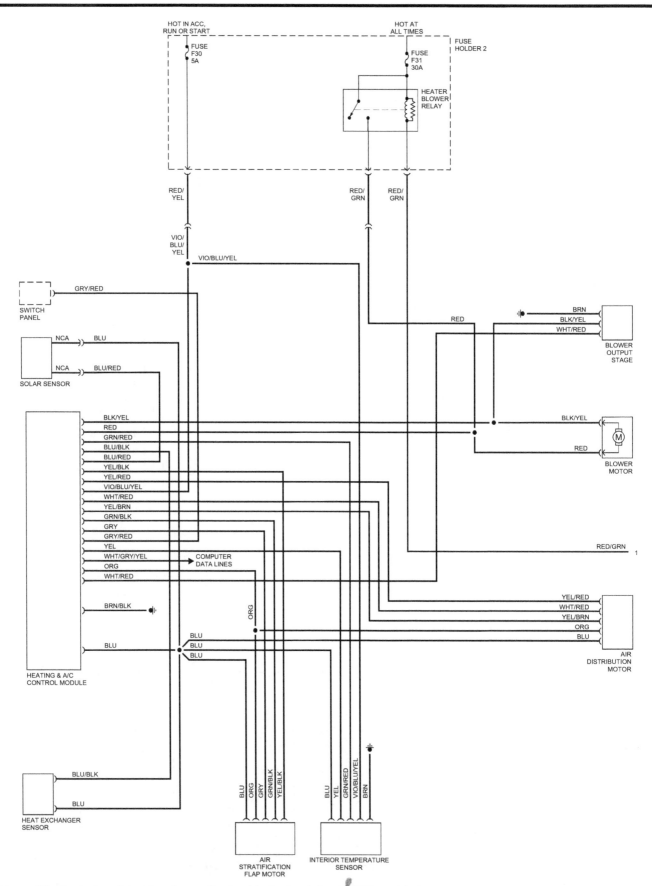

Air conditioning (automatic) and engine cooling fan (single stage) circuit - 2007 models (except convertible)/2008 convertible models (1 of 2)

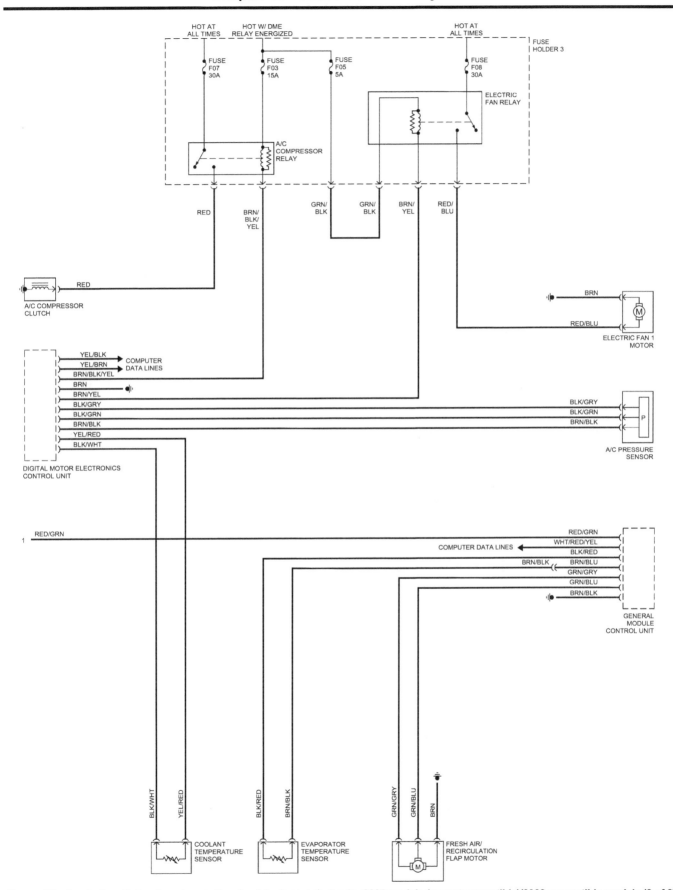

Air conditioning (automatic) and engine cooling fan (single stage) circuit - 2007 models (except convertible)/2008 convertible models (2 of 2)

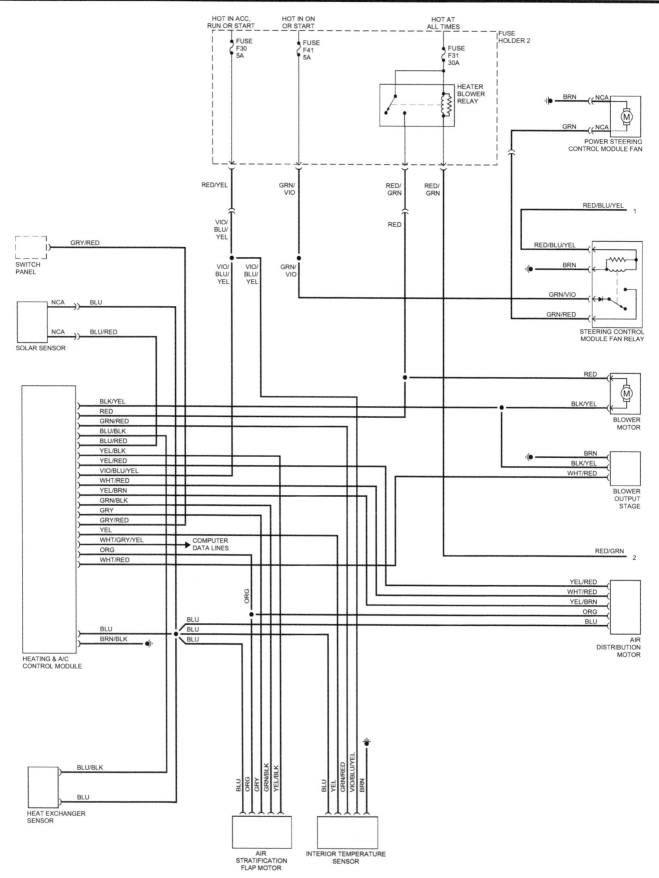

Air conditioning (automatic) and engine cooling fan (dual stage) circuit - 2007 models (except convertible)/2008 convertible models (1 of 2)

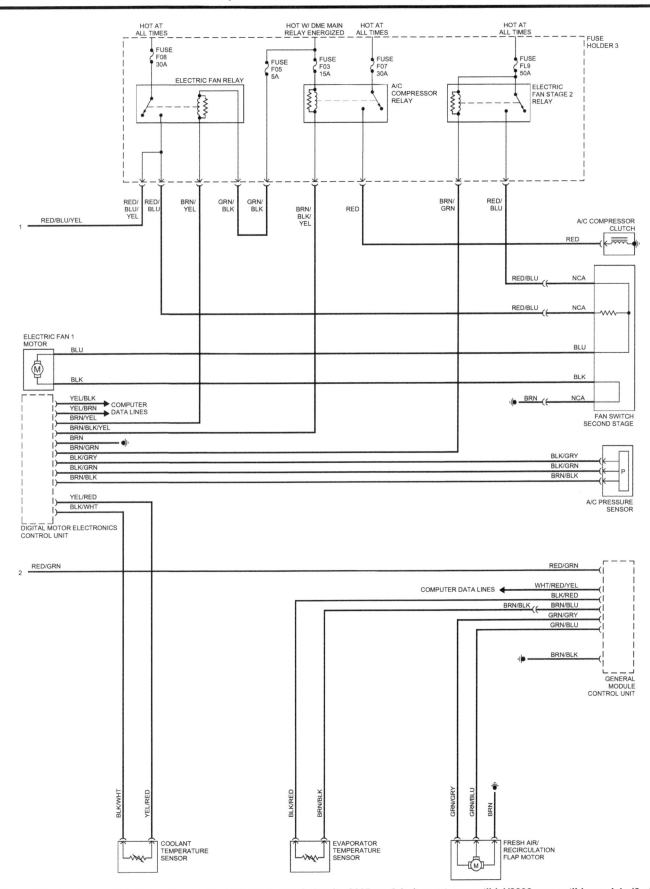

Air conditioning (automatic) and engine cooling fan (dual stage) circuit - 2007 models (except convertible)/2008 convertible models (2 of 2)

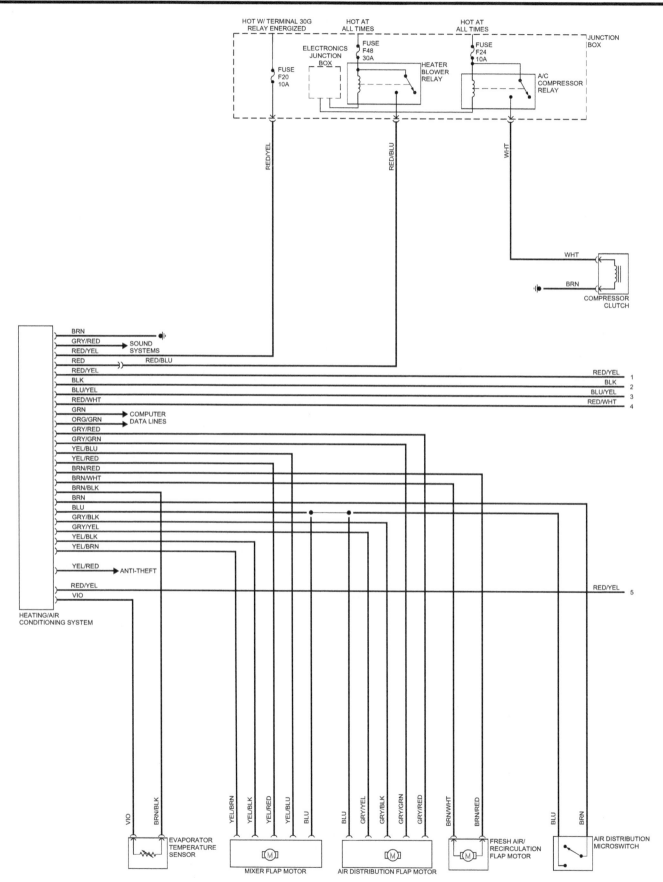

Air conditioning (automatic) and engine cooling fan - 2008 models (except convertible)/all 2009 and later models (1 of 2)

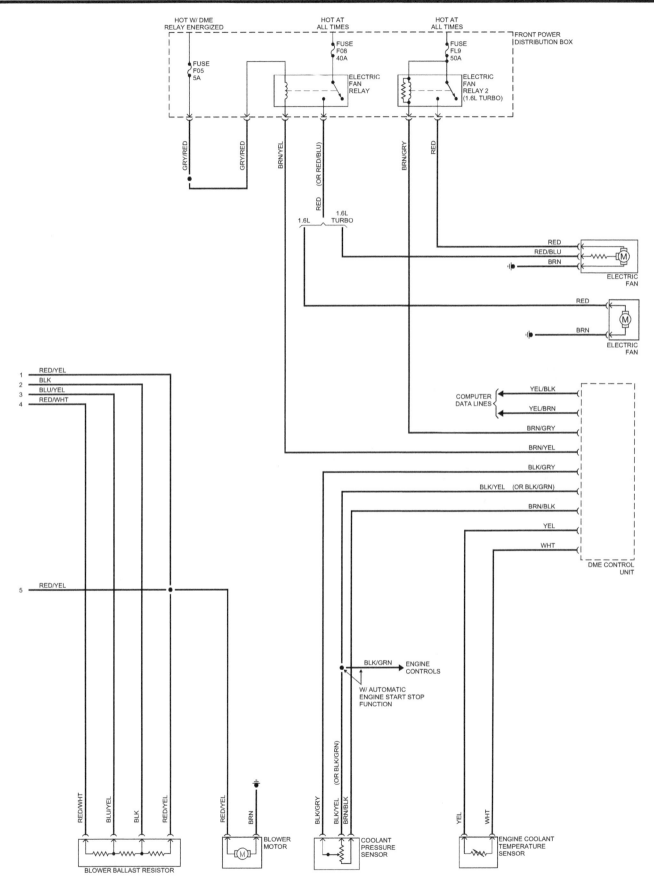

Air conditioning (automatic) and engine cooling fan - 2008 models (except convertible)/all 2009 and later models (2 of 2)

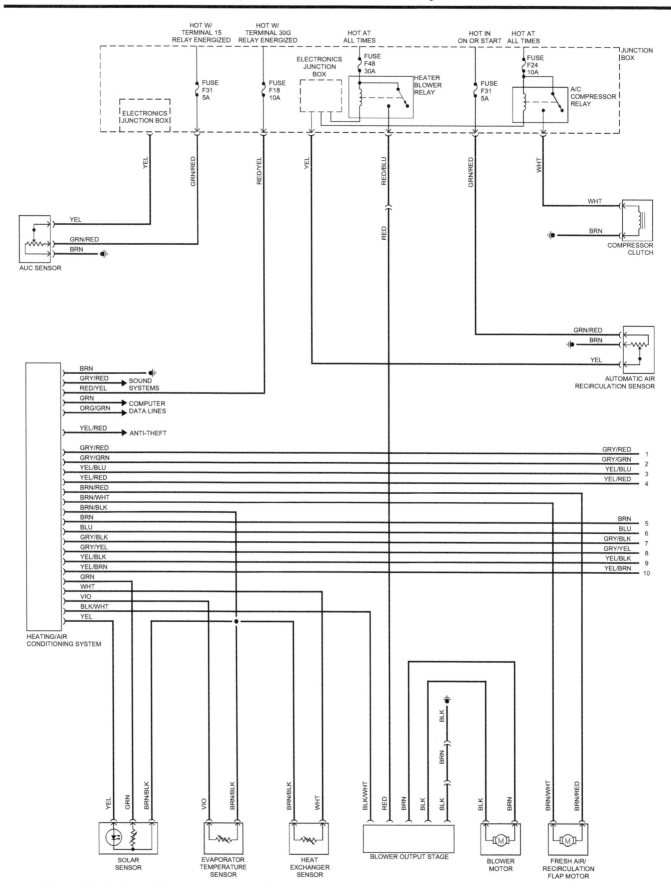

Air conditioning (manual) and engine cooling fan circuit - 2008 models (except convertible)/all 2009 and later models (1 of 2)

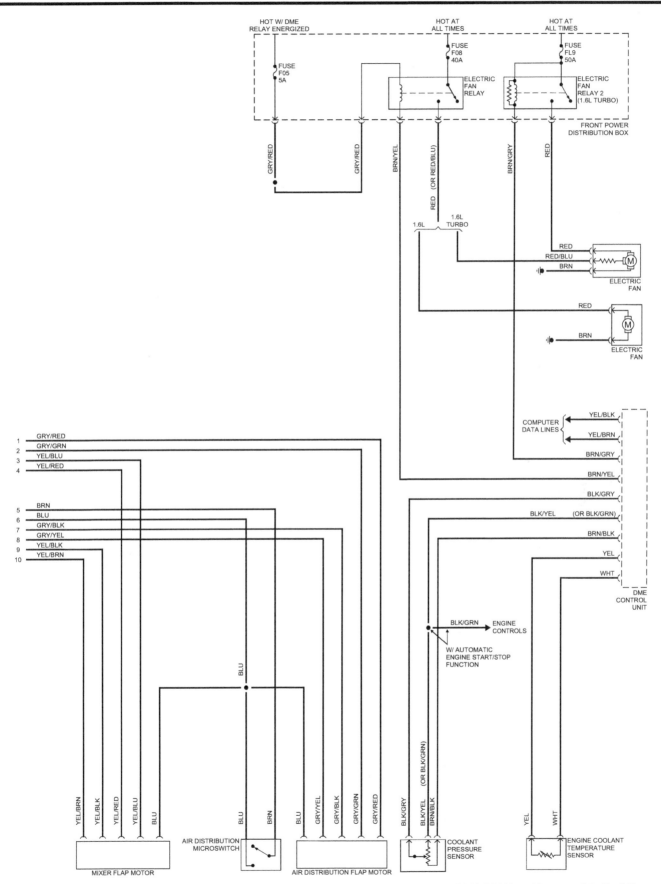

Air conditioning (manual) and engine cooling fan circuit - 2008 models (except convertible)/all 2009 and later models (2 of 2)

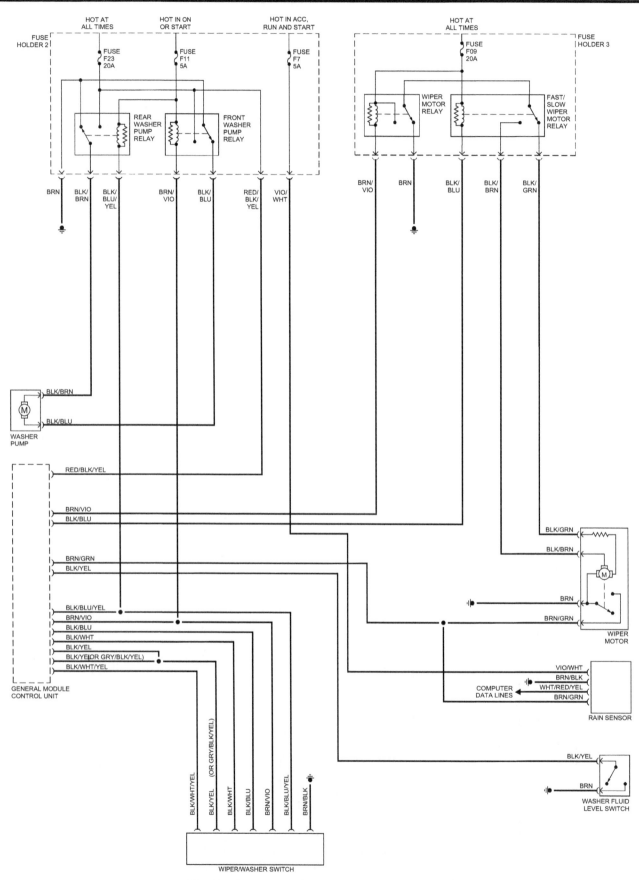

Wiper/washer system - 2007 models (except convertible)

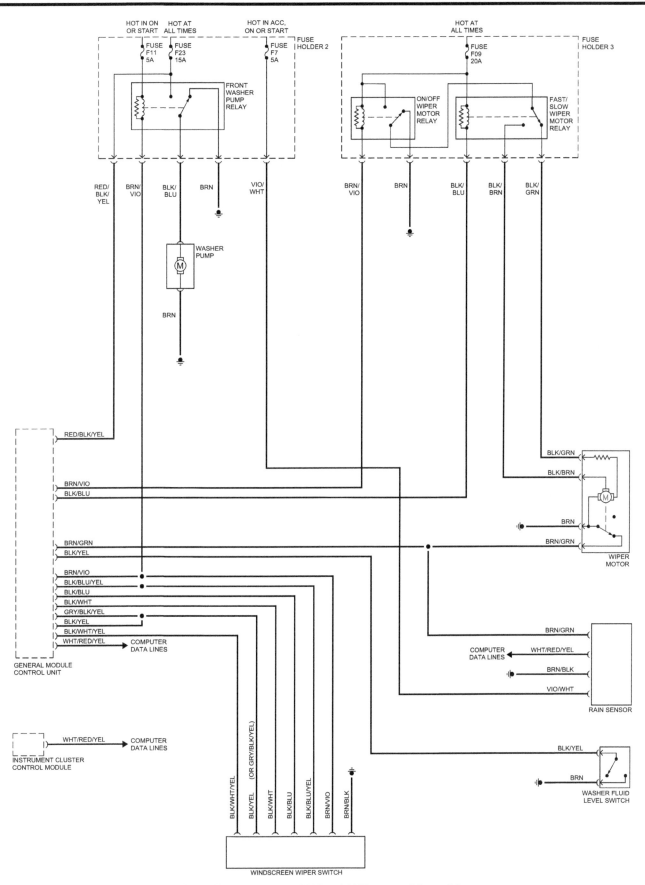

Wiper/washer system - 2007 and 2008 convertible models

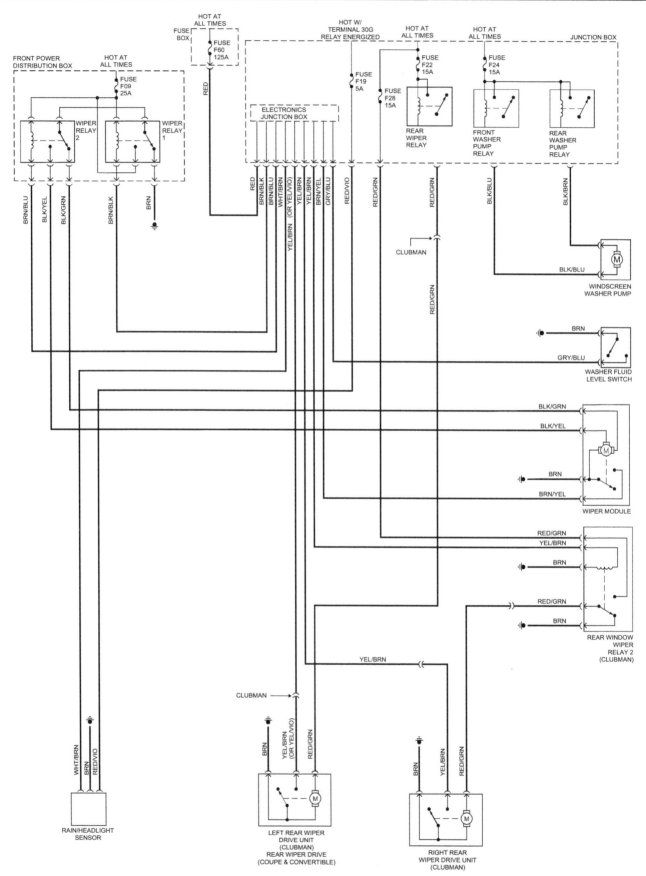

Wiper/washer system - 2008 models (except convertible)/all 2009 and later models

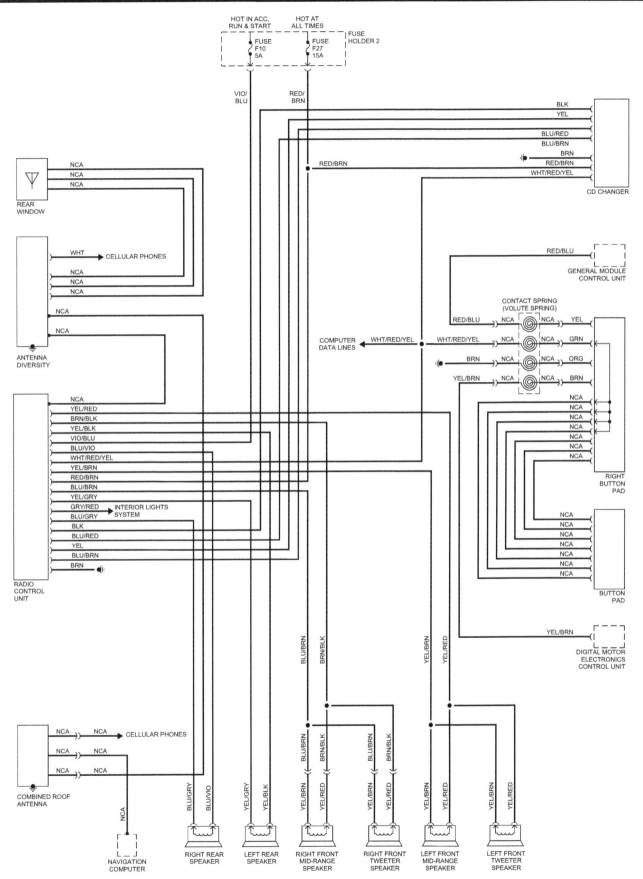

Audio system - 2007 models (except convertible)/2008 convertible models

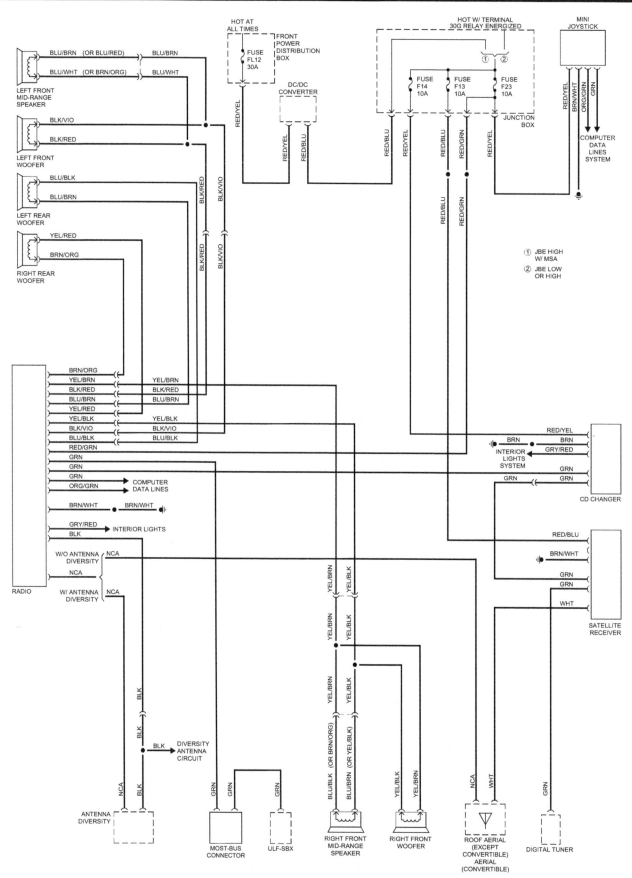

Audio system - 2008 and later models (except 2008 convertible)

Index

Notes

Haynes Automotive Manuals

NOTE: If you do not see a listing for your vehicle, please visit ***haynes.com*** *for the latest product information and check out our* ***Online Manuals!***

ACURA

12020 **Integra** '86 thru '89 **& Legend** '86 thru '90
12021 **Integra** '90 thru '93 **& Legend** '91 thru '95
Integra '94 thru '00 - *see HONDA Civic (42025)*
MDX '01 thru '07 - *see HONDA Pilot (42037)*
12050 **Acura TL** all models '99 thru '08

AMC

14020 **Mid-size models** '70 thru '83
14025 **(Renault) Alliance & Encore** '83 thru '87

AUDI

15020 **4000** all models '80 thru '87
15025 **5000** all models '77 thru '83
15026 **5000** all models '84 thru '88
Audi A4 '96 thru '01 - *see VW Passat (96023)*
15030 **Audi A4** '02 thru '08

AUSTIN-HEALEY

Sprite - *see MG Midget (66015)*

BMW

18020 **3/5 Series** '82 thru '92
18021 **3-Series** incl. Z3 models '92 thru '98
18022 **3-Series** incl. Z4 models '99 thru '05
18023 **3-Series** '06 thru '14
18025 **320i all 4-cylinder models** '75 thru '83
18050 **1500 thru 2002** except Turbo '59 thru '77

BUICK

19010 **Buick Century** '97 thru '05
Century (front-wheel drive) - *see GM (38005)*
19020 **Buick, Oldsmobile & Pontiac Full-size (Front-wheel drive)** '85 thru '05
Buick Electra, LeSabre and Park Avenue; **Oldsmobile** Delta 88 Royale, Ninety Eight and Regency; **Pontiac** Bonneville
19025 **Buick, Oldsmobile & Pontiac Full-size (Rear wheel drive)** '70 thru '90
Buick Estate, Electra, LeSabre, Limited, **Oldsmobile** Custom Cruiser, Delta 88, Ninety-eight, **Pontiac** Bonneville, Catalina, Grandville, Parisienne
19027 **Buick LaCrosse** '05 thru '13
Enclave - *see GENERAL MOTORS (38001)*
Rainier - *see CHEVROLET (24072)*
Regal - *see GENERAL MOTORS (38010)*
Riviera - *see GENERAL MOTORS (38030, 38031)*
Roadmaster - *see CHEVROLET (24046)*
Skyhawk - *see GENERAL MOTORS (38015)*
Skylark - *see GENERAL MOTORS (38020, 38025)*
Somerset - *see GENERAL MOTORS (38025)*

CADILLAC

21015 **CTS & CTS-V** '03 thru '14
21030 **Cadillac Rear Wheel Drive** '70 thru '93
Cimarron - *see GENERAL MOTORS (38015)*
DeVille - *see GENERAL MOTORS (38031 & 38032)*
Eldorado - *see GENERAL MOTORS (38030)*
Fleetwood - *see GENERAL MOTORS (38031)*
Seville - *see GM (38030, 38031 & 38032)*

CHEVROLET

10305 **Chevrolet Engine Overhaul Manual**
24010 **Astro & GMC Safari Mini-vans** '85 thru '05
24013 **Aveo** '04 thru '11
24015 **Camaro V8** all models '70 thru '81
24016 **Camaro** all models '82 thru '92
24017 **Camaro & Firebird** '93 thru '02
Cavalier - *see GENERAL MOTORS (38016)*
Celebrity - *see GENERAL MOTORS (38005)*
24018 **Camaro** '10 thru '15
24020 **Chevelle, Malibu & El Camino** '69 thru '87
Cobalt - *see GENERAL MOTORS (38017)*
24024 **Chevette & Pontiac T1000** '76 thru '87
Citation - *see GENERAL MOTORS (38020)*
24027 **Colorado & GMC Canyon** '04 thru '12
24032 **Corsica & Beretta** all models '87 thru '96
24040 **Corvette** all V8 models '68 thru '82
24041 **Corvette** all models '84 thru '96
24042 **Corvette** all models '97 thru '13
24044 **Cruze** '11 thru '19
24045 **Full-size Sedans** Caprice, Impala, Biscayne, Bel Air & Wagons '69 thru '90
24046 **Impala SS & Caprice and Buick Roadmaster** '91 thru '96
Impala '00 thru '05 - *see LUMINA (24048)*
24047 **Impala & Monte Carlo** all models '06 thru '11
Lumina '90 thru '94 - *see GM (38010)*
24048 **Lumina & Monte Carlo** '95 thru '05
Lumina APV - *see GM (38035)*
24050 **Luv Pick-up** all 2WD & 4WD '72 thru '82
24051 **Malibu** '13 thru '19
24055 **Monte Carlo** all models '70 thru '88
Monte Carlo '95 thru '01 - *see LUMINA (24048)*
24059 **Nova** all V8 models '69 thru '79
24060 **Nova and Geo Prizm** '85 thru '92
24064 **Pick-ups** '67 thru '87 - Chevrolet & GMC
24065 **Pick-ups** '88 thru '98 - Chevrolet & GMC
24066 **Pick-ups** '99 thru '06 - Chevrolet & GMC
24067 **Chevrolet Silverado & GMC Sierra** '07 thru '14
24068 **Chevrolet Silverado & GMC Sierra** '14 thru '19
24070 **S-10 & S-15 Pick-ups** '82 thru '93, **Blazer & Jimmy** '83 thru '94,
24071 **S-10 & Sonoma Pick-ups** '94 thru '04, including **Blazer, Jimmy & Hombre**
24072 **Chevrolet TrailBlazer, GMC Envoy & Oldsmobile Bravada** '02 thru '09
24075 **Sprint** '85 thru '88 **& Geo Metro** '89 thru '01
24080 **Vans - Chevrolet & GMC** '68 thru '96
24081 **Chevrolet Express & GMC Savana** Full-size Vans '96 thru '19

CHRYSLER

10310 **Chrysler Engine Overhaul Manual**
25015 **Chrysler Cirrus, Dodge Stratus, Plymouth Breeze** '95 thru '00
25020 **Full-size Front-Wheel Drive** '88 thru '93
K-Cars - *see DODGE Aries (30008)*
Laser - *see DODGE Daytona (30030)*
25025 **Chrysler LHS, Concorde, New Yorker, Dodge** Intrepid, **Eagle** Vision, '93 thru '97
25026 **Chrysler LHS, Concorde, 300M, Dodge** Intrepid, '98 thru '04
25027 **Chrysler 300** '05 thru '18, **Dodge Charger** '06 thru '18, **Magnum** '05 thru '08 **& Challenger** '08 thru '18
25030 **Chrysler & Plymouth Mid-size** front wheel drive '82 thru '95
Rear-wheel Drive - *see Dodge (30050)*
25035 **PT Cruiser** all models '01 thru '10
25040 **Chrysler Sebring** '95 thru '06, **Dodge Stratus** '01 thru '06 **& Dodge Avenger** '95 thru '00
25041 **Chrysler Sebring** '07 thru '10, **200** '11 thru '17 **Dodge Avenger** '08 thru '14

DATSUN

28005 **200SX** all models '80 thru '83
28012 **240Z, 260Z & 280Z** Coupe '70 thru '78
28014 **280ZX** Coupe & 2+2 '79 thru '83
300ZX - *see NISSAN (72010)*
28018 **510 & PL521 Pick-up** '68 thru '73
28020 **510** all models '78 thru '81
28022 **620 Series Pick-up** all models '73 thru '79
720 Series Pick-up - *see NISSAN (72030)*

DODGE

400 & 600 - *see CHRYSLER (25030)*
30008 **Aries & Plymouth Reliant** '81 thru '89
30010 **Caravan & Plymouth Voyager** '84 thru '95
30011 **Caravan & Plymouth Voyager** '96 thru '02
30012 **Challenger & Plymouth Sapporro** '78 thru '83
30013 **Caravan, Chrysler Voyager & Town & Country** '03 thru '07
30014 **Grand Caravan & Chrysler Town & Country** '08 thru '18
30016 **Colt & Plymouth Champ** '78 thru '87
30020 **Dakota Pick-ups** all models '87 thru '96
30021 **Durango** '98 & '99 **& Dakota** '97 thru '99
30022 **Durango** '00 thru '03 **& Dakota** '00 thru '04
30023 **Durango** '04 thru '09 **& Dakota** '05 thru '11
30025 **Dart, Demon, Plymouth Barracuda, Duster & Valiant** 6-cylinder models '67 thru '76
30030 **Daytona & Chrysler Laser** '84 thru '89
Intrepid - *see CHRYSLER (25025, 25026)*
30034 **Neon** all models '95 thru '99
30035 **Omni & Plymouth Horizon** '78 thru '90
30036 **Dodge & Plymouth Neon** '00 thru '05
30040 **Pick-ups** full-size models '74 thru '93
30042 **Pick-ups** full-size models '94 thru '08
30043 **Pick-ups** full-size models '09 thru '18
30045 **Ram 50/D50 Pick-ups & Raider and Plymouth Arrow Pick-ups** '79 thru '93
30050 **Dodge/Plymouth/Chrysler** RWD '71 thru '89
30055 **Shadow & Plymouth Sundance** '87 thru '94
30060 **Spirit & Plymouth Acclaim** '89 thru '95
30065 **Vans - Dodge & Plymouth** '71 thru '03

EAGLE

Talon - *see MITSUBISHI (68030, 68031)*
Vision - *see CHRYSLER (25025)*

FIAT

34010 **124 Sport Coupe & Spider** '68 thru '78
34025 **X1/9** all models '74 thru '80

FORD

10320 **Ford Engine Overhaul Manual**
10355 **Ford Automatic Transmission Overhaul**
11500 **Mustang '64-1/2 thru '70 Restoration Guide**
36004 **Aerostar Mini-vans** all models '86 thru '97
36006 **Contour & Mercury Mystique** '95 thru '00
36008 **Courier Pick-up** all models '72 thru '82
36012 **Crown Victoria & Mercury Grand Marquis** '88 thru '11
36014 **Edge** '07 thru '19 **& Lincoln MKX** '07 thru '18
36016 **Escort & Mercury Lynx** all models '81 thru '90
36020 **Escort & Mercury Tracer** '91 thru '02
36022 **Escape** '01 thru '17, **Mazda Tribute** '01 thru '11, **& Mercury Mariner** '05 thru '11
36024 **Explorer & Mazda Navajo** '91 thru '01
36025 **Explorer & Mercury Mountaineer** '02 thru '10
36026 **Explorer** '11 thru '17
36028 **Fairmont & Mercury Zephyr** '78 thru '83
36030 **Festiva & Aspire** '88 thru '97
36032 **Fiesta** all models '77 thru '80
36034 **Focus** all models '00 thru '11
36035 **Focus** '12 thru '14
36045 **Fusion** '06 thru '14 **& Mercury Milan** '06 thru '11
36048 **Mustang V8** all models '64-1/2 thru '73
36049 **Mustang II** 4-cylinder, V6 & V8 models '74 thru '78
36050 **Mustang & Mercury Capri** '79 thru '93
36051 **Mustang** all models '94 thru '04
36052 **Mustang** '05 thru '14
36054 **Pick-ups & Bronco** '73 thru '79
36058 **Pick-ups & Bronco** '80 thru '96
36059 **F-150** '97 thru '03, **Expedition** '97 thru '17, **F-250** '97 thru '99, **F-150 Heritage** '04 **& Lincoln Navigator** '98 thru '17
36060 **Super Duty Pick-ups & Excursion** '99 thru '10
36061 **F-150** full-size '04 thru '14
36062 **Pinto & Mercury Bobcat** '75 thru '80
36063 **F-150** full-size '15 thru '17
36064 **Super Duty Pick-ups** '11 thru '16
36066 **Probe** all models '89 thru '92
Probe '93 thru '97 - *see MAZDA 626 (61042)*
36070 **Ranger & Bronco II** gas models '83 thru '92
36071 **Ranger** '93 thru '11 **& Mazda Pick-ups** '94 thru '09
36074 **Taurus & Mercury Sable** '86 thru '95
36075 **Taurus & Mercury Sable** '96 thru '07
36076 **Taurus** '08 thru '14, **Five Hundred** '05 thru '07, **Mercury Montego** '05 thru '07 **& Sable** '08 thru '09
36078 **Tempo & Mercury Topaz** '84 thru '94
36082 **Thunderbird & Mercury Cougar** '83 thru '88
36086 **Thunderbird & Mercury Cougar** '89 thru '97
36090 **Vans** all V8 Econoline models '69 thru '91
36094 **Vans** full size '92 thru '14
36097 **Windstar** '95 thru '03, **Freestar & Mercury Monterey Mini-van** '04 thru '07

GENERAL MOTORS

10360 **GM Automatic Transmission Overhaul**
38001 **GMC Acadia** '07 thru '16, **Buick Enclave** '08 thru '17, **Saturn Outlook** '07 thru '10 **& Chevrolet Traverse** '09 thru '17
38005 **Buick Century, Chevrolet Celebrity, Oldsmobile Cutlass Ciera & Pontiac 6000** all models '82 thru '96
38010 **Buick Regal** '88 thru '04, **Chevrolet Lumina** '88 thru '04, **Oldsmobile Cutlass Supreme** '88 thru '97 **& Pontiac Grand Prix** '88 thru '07
38015 **Buick Skyhawk, Cadillac Cimarron, Chevrolet Cavalier, Oldsmobile Firenza, Pontiac J-2000 & Sunbird** '82 thru '94
38016 **Chevrolet Cavalier & Pontiac Sunfire** '95 thru '05
38017 **Chevrolet Cobalt** '05 thru '10, **HHR** '06 thru '11, **Pontiac G5** '07 thru '09, **Pursuit** '05 thru '06 **& Saturn ION** '03 thru '07
38020 **Buick Skylark, Chevrolet Citation, Oldsmobile Omega, Pontiac Phoenix** '80 thru '85
38025 **Buick Skylark** '86 thru '98, **Somerset** '85 thru '87, **Oldsmobile Achieva** '92 thru '98, **Calais** '85 thru '91, **& Pontiac Grand Am** all models '85 thru '98
38026 **Chevrolet Malibu** '97 thru '03, **Classic** '04 thru '05, **Oldsmobile Alero** '99 thru '03, **Cutlass** '97 thru '00, **& Pontiac Grand Am** '99 thru '03
38027 **Chevrolet Malibu** '04 thru '12, **Pontiac G6** '05 thru '10 **& Saturn Aura** '07 thru '10
38030 **Cadillac Eldorado, Seville, Oldsmobile Toronado & Buick Riviera** '71 thru '85
38031 **Cadillac Eldorado, Seville, DeVille, Fleetwood, Oldsmobile Toronado & Buick Riviera** '86 thru '93
38032 **Cadillac DeVille** '94 thru '05, **Seville** '92 thru '04 **& Cadillac DTS** '06 thru '10
38035 **Chevrolet Lumina APV, Oldsmobile Silhouette & Pontiac Trans Sport** all models '90 thru '96
38036 **Chevrolet Venture** '97 thru '05, **Oldsmobile Silhouette** '97 thru '04, **Pontiac Trans Sport** '97 thru '98 **& Montana** '99 thru '05
38040 **Chevrolet Equinox** '05 thru '17, **GMC Terrain** '10 thru '17 **& Pontiac Torrent** '06 thru '09

GEO

Metro - *see CHEVROLET Sprint (24075)*
Prizm - *'85 thru '92 see CHEVY (24060), '93 thru '02 see TOYOTA Corolla (92036)*
40030 **Storm** all models '90 thru '93
Tracker - *see SUZUKI Samurai (90010)*

(Continued on other side)

Haynes North America, Inc. • (805) 498-6703 • www.haynes.com

Haynes Automotive Manuals (continued)

NOTE: If you do not see a listing for your vehicle, please visit haynes.com for the latest product information and check out our Online Manuals!

GMC
- **Acadia** - *see GENERAL MOTORS (38001)*
- **Pick-ups** - *see CHEVROLET (24027, 24068)*
- **Vans** - *see CHEVROLET (24081)*

HONDA
- 42010 **Accord CVCC** all models '76 thru '83
- 42011 **Accord** all models '84 thru '89
- 42012 **Accord** all models '90 thru '93
- 42013 **Accord** all models '94 thru '97
- 42014 **Accord** all models '98 thru '02
- 42015 **Accord** '03 thru '12 & **Crosstour** '10 thru '14
- 42016 **Accord** '13 thru '17
- 42020 **Civic 1200** all models '73 thru '79
- 42021 **Civic 1300 & 1500 CVCC** '80 thru '83
- 42022 **Civic 1500 CVCC** all models '75 thru '79
- 42023 **Civic** all models '84 thru '91
- 42024 **Civic & del Sol** '92 thru '95
- 42025 **Civic** '96 thru '00, **CR-V** '97 thru '01 & **Acura Integra** '94 thru '00
- 42026 **Civic** '01 thru '11 & **CR-V** '02 thru '11
- 42027 **Civic** '12 thru '15 & **CR-V** '12 thru '16
- 42030 **Fit** '07 thru '13
- 42035 **Odyssey** all models '99 thru '10
- **Passport** - *see ISUZU Rodeo (47017)*
- 42037 **Honda Pilot** '03 thru '08, **Ridgeline** '06 thru '14 & **Acura MDX** '01 thru '07
- 42040 **Prelude CVCC** all models '79 thru '89

HYUNDAI
- 43010 **Elantra** all models '96 thru '19
- 43015 **Excel & Accent** all models '86 thru '13
- 43050 **Santa Fe** all models '01 thru '12
- 43055 **Sonata** all models '99 thru '14

INFINITI
- **G35** '03 thru '08 - *see NISSAN 350Z (72011)*

ISUZU
- **Hombre** - *see CHEVROLET S-10 (24071)*
- 47017 **Rodeo** '91 thru '02, **Amigo** '89 thru '94 & '98 thru '02 & **Honda Passport** '95 thru '02
- 47020 **Trooper** '84 thru '91 & **Pick-up** '81 thru '93

JAGUAR
- 49010 **XJ6** all 6-cylinder models '68 thru '86
- 49011 **XJ6** all models '88 thru '94
- 49015 **XJ12 & XJS** all 12-cylinder models '72 thru '85

JEEP
- 50010 **Cherokee, Comanche & Wagoneer Limited** all models '84 thru '01
- 50011 **Cherokee** '14 thru '19
- 50020 **CJ** all models '49 thru '86
- 50025 **Grand Cherokee** all models '93 thru '04
- 50026 **Grand Cherokee** '05 thru '19 & **Dodge Durango** '11 thru '19
- 50029 **Grand Wagoneer & Pick-up** '72 thru '91 Grand Wagoneer '84 thru '91, Cherokee & Wagoneer '72 thru '83, Pick-up '72 thru '88
- 50030 **Wrangler** all models '87 thru '17
- 50035 **Liberty** '02 thru '12 & **Dodge Nitro** '07 thru '11
- 50050 **Patriot & Compass** '07 thru '17

KIA
- 54050 **Optima** '01 thru '10
- 54060 **Sedona** '02 thru '14
- 54070 **Sephia** '94 thru '01, **Spectra** '00 thru '09, **Sportage** '05 thru '20
- 54077 **Sorento** '03 thru '13

LEXUS
- **ES 300/330** - *see TOYOTA Camry (92007, 92008)*
- **ES 350** - *see TOYOTA Camry (92009)*
- **RX 300/330/350** - *see TOYOTA Highlander (92095)*

LINCOLN
- **MKX** - *see FORD (36014)*
- **Navigator** - *see FORD Pick-up (36059)*
- 59010 **Rear-Wheel Drive Continental** '70 thru '87, **Mark Series** '70 thru '92 & **Town Car** '81 thru '10

MAZDA
- 61010 **GLC** (rear-wheel drive) '77 thru '83
- 61011 **GLC** (front-wheel drive) '81 thru '85
- 61012 **Mazda3** '04 thru '11
- 61015 **323 & Protegé** '90 thru '03
- 61016 **MX-5 Miata** '90 thru '14
- 61020 **MPV** all models '89 thru '98
- **Navajo** - *see Ford Explorer (36024)*
- 61030 **Pick-ups** '72 thru '93
- **Pick-ups** '94 thru '09 - *see Ford Ranger (36071)*
- 61035 **RX-7** all models '79 thru '85
- 61036 **RX-7** all models '86 thru '91
- 61040 **626** (rear-wheel drive) all models '79 thru '82
- 61041 **626 & MX-6** (front-wheel drive) '83 thru '92
- 61042 **626** '93 thru '01 & **MX-6/Ford Probe** '93 thru '02
- 61043 **Mazda6** '03 thru '13

MERCEDES-BENZ
- 63012 **123 Series Diesel** '76 thru '85
- 63015 **190 Series** 4-cylinder gas models '84 thru '88
- 63020 **230/250/280** 6-cylinder SOHC models '68 thru '72
- 63025 **280 123 Series** gas models '77 thru '81
- 63030 **350 & 450** all models '71 thru '80
- 63040 **C-Class:** C230/C240/C280/C320/C350 '01 thru '07

MERCURY
- 64200 **Villager & Nissan Quest** '93 thru '01
- *All other titles, see FORD Listing.*

MG
- 66010 **MGB** Roadster & GT Coupe '62 thru '80
- 66015 **MG Midget, Austin Healey Sprite** '58 thru '80

MINI
- 67020 **Mini** '02 thru '13

MITSUBISHI
- 68020 **Cordia, Tredia, Galant, Precis & Mirage** '83 thru '93
- 68030 **Eclipse, Eagle Talon & Plymouth Laser** '90 thru '94
- 68031 **Eclipse** '95 thru '05 & **Eagle Talon** '95 thru '98
- 68035 **Galant** '94 thru '12
- 68040 **Pick-up** '83 thru '96 & **Montero** '83 thru '93

NISSAN
- 72010 **300ZX** all models including Turbo '84 thru '89
- 72011 **350Z & Infiniti G35** all models '03 thru '08
- 72015 **Altima** all models '93 thru '06
- 72016 **Altima** '07 thru '12
- 72020 **Maxima** all models '85 thru '92
- 72021 **Maxima** all models '93 thru '08
- 72025 **Murano** '03 thru '14
- 72030 **Pick-ups** '80 thru '97 & **Pathfinder** '87 thru '95
- 72031 **Frontier** '98 thru '04, **Xterra** '00 thru '04, & **Pathfinder** '96 thru '04
- 72032 **Frontier & Xterra** '05 thru '14
- 72037 **Pathfinder** '05 thru '14
- 72040 **Pulsar** all models '83 thru '86
- 72042 **Roque** all models '08 thru '20
- 72050 **Sentra** all models '82 thru '94
- 72051 **Sentra & 200SX** all models '95 thru '06
- 72060 **Stanza** all models '82 thru '90
- 72070 **Titan pick-ups** '04 thru '10, **Armada** '05 thru '10 & **Pathfinder Armada** '04
- 72080 **Versa** all models '07 thru '19

OLDSMOBILE
- 73015 **Cutlass** V6 & V8 gas models '74 thru '88
- *For other OLDSMOBILE titles, see BUICK, CHEVROLET or GENERAL MOTORS listings.*

PLYMOUTH
- *For PLYMOUTH titles, see DODGE listing.*

PONTIAC
- 79008 **Fiero** all models '84 thru '88
- 79018 **Firebird** V8 models except Turbo '70 thru '81
- 79019 **Firebird** all models '82 thru '92
- 79025 **G6** all models '05 thru '09
- 79040 **Mid-size Rear-wheel Drive** '70 thru '87
- **Vibe** '03 thru '10 - *see TOYOTA Corolla (92037)*
- *For other PONTIAC titles, see BUICK, CHEVROLET or GENERAL MOTORS listings.*

PORSCHE
- 80020 **911** Coupe & Targa models '65 thru '89
- 80025 **914** all 4-cylinder models '69 thru '76
- 80030 **924** all models including Turbo '76 thru '82
- 80035 **944** all models including Turbo '83 thru '89

RENAULT
- **Alliance & Encore** - *see AMC (14025)*

SAAB
- 84010 **900** all models including Turbo '79 thru '88

SATURN
- 87010 **Saturn** all S-series models '91 thru '02
- **Saturn Ion** '03 thru '07- *see GM (38017)*
- **Saturn Outlook** - *see GM (38001)*
- 87020 **Saturn L-series** all models '00 thru '04
- 87040 **Saturn VUE** '02 thru '09

SUBARU
- 89002 **1100, 1300, 1400 & 1600** '71 thru '79
- 89003 **1600 & 1800** 2WD & 4WD '80 thru '94
- 89080 **Impreza** '02 thru '11, **WRX** '02 thru '14, & **WRX STI** '04 thru '14
- 89100 **Legacy** all models '90 thru '99
- 89101 **Legacy & Forester** '00 thru '09
- 89102 **Legacy** '10 thru '16 & **Forester** '12 thru '16

SUZUKI
- 90010 **Samurai/Sidekick & Geo Tracker** '86 thru '01

TOYOTA
- 92005 **Camry** all models '83 thru '91
- 92006 **Camry** '92 thru '96 & **Avalon** '95 thru '96
- 92007 **Camry, Avalon, Solara, Lexus ES 300** '97 thru '01

- 92008 **Camry, Avalon, Lexus ES 300/330** '02 thru '06 & **Solara** '02 thru '08
- 92009 **Camry, Avalon & Lexus ES 350** '07 thru '17
- 92020 **Celica Rear-wheel Drive** '71 thru '85
- 92020 **Celica Front-wheel Drive** '86 thru '99
- 92025 **Celica Supra** all models '79 thru '92
- 92030 **Corolla** all models '75 thru '79
- 92032 **Corolla** all rear-wheel drive models '80 thru '87
- 92035 **Corolla** all front-wheel drive models '84 thru '92
- 92036 **Corolla & Geo/Chevrolet Prizm** '93 thru '02
- 92037 **Corolla** '03 thru '19, **Matrix** '03 thru '14, & **Pontiac Vibe** '03 thru '10
- 92040 **Corolla Tercel** all models '80 thru '82
- 92045 **Corona** all models '74 thru '82
- 92050 **Cressida** all models '78 thru '82
- 92055 **Land Cruiser FJ40, 43, 45, 55** '68 thru '82
- 92056 **Land Cruiser FJ60, 62, 80, FZJ80** '80 thru '96
- 92060 **Matrix** '03 thru '11 & **Pontiac Vibe** '03 thru '10
- 92065 **MR2** all models '85 thru '87
- 92070 **Pick-up** all models '69 thru '78
- 92075 **Pick-up** all models '79 thru '95
- 92076 **Tacoma** '95 thru '04, **4Runner** '96 thru '02 & **T100** '93 thru '08
- 92077 **Tacoma** all models '05 thru '18
- 92078 **Tundra** '00 thru '06 & **Sequoia** '01 thru '07
- 92079 **4Runner** all models '03 thru '09
- 92080 **Previa** all models '91 thru '95
- 92081 **Prius** all models '01 thru '12
- 92082 **RAV4** all models '96 thru '12
- 92085 **Tercel** all models '87 thru '94
- 92090 **Sienna** all models '98 thru '10
- 92095 **Highlander** '01 thru '19 & **Lexus RX330/330/350** '99 thru '19
- 92179 **Tundra** '07 thru '19 & **Sequoia** '08 thru '19

TRIUMPH
- 94007 **Spitfire** all models '62 thru '81
- 94010 **TR7** all models '75 thru '81

VW
- 96008 **Beetle & Karmann Ghia** '54 thru '79
- 96009 **New Beetle** '98 thru '10
- 96016 **Rabbit, Jetta, Scirocco & Pick-up** gas models '75 thru '92 & Convertible '80 thru '92
- 96017 **Golf, GTI & Jetta** '93 thru '98, **Cabrio** '95 thru '02
- 96018 **Golf, GTI, Jetta** '99 thru '05
- 96019 **Jetta, Rabbit, GLI, GTI & Golf** '05 thru '11
- 96020 **Rabbit, Jetta & Pick-up** diesel '77 thru '84
- 96021 **Jetta** '11 thru '18 & **Golf** '15 thru '19
- 96023 **Passat** '98 thru '05 & **Audi A4** '96 thru '01
- 96030 **Transporter 1600** all models '68 thru '79
- 96035 **Transporter 1700, 1800 & 2000** '72 thru '79
- 96040 **Type 3 1500 & 1600** all models '63 thru '73
- 96045 **Vanagon Air-Cooled** all models '80 thru '83

VOLVO
- 97010 **120, 130 Series & 1800 Sports** '61 thru '73
- 97015 **140 Series** all models '66 thru '74
- 97020 **240 Series** all models '76 thru '93
- 97040 **740 & 760 Series** all models '82 thru '88
- 97050 **850 Series** all models '93 thru '97

TECHBOOK MANUALS
- 10205 **Automotive Computer Codes**
- 10206 **OBD-II & Electronic Engine Management**
- 10210 **Automotive Emissions Control Manual**
- 10215 **Fuel Injection Manual** '78 thru '85
- 10225 **Holley Carburetor Manual**
- 10230 **Rochester Carburetor Manual**
- 10305 **Chevrolet Engine Overhaul Manual**
- 10320 **Ford Engine Overhaul Manual**
- 10330 **GM and Ford Diesel Engine Repair Manual**
- 10331 **Duramax Diesel Engines** '01 thru '19
- 10332 **Cummins Diesel Engine Performance Manual**
- 10333 **GM, Ford & Chrysler Engine Performance Manual**
- 10334 **GM Engine Performance Manual**
- 10340 **Small Engine Repair Manual,** 5 HP & Less
- 10341 **Small Engine Repair Manual,** 5.5 thru 20 HP
- 10345 **Suspension, Steering & Driveline Manual**
- 10355 **Ford Automatic Transmission Overhaul**
- 10360 **GM Automatic Transmission Overhaul**
- 10405 **Automotive Body Repair & Painting**
- 10410 **Automotive Brake Manual**
- 10411 **Automotive Anti-lock Brake (ABS) Systems**
- 10420 **Automotive Electrical Manual**
- 10425 **Automotive Heating & Air Conditioning**
- 10435 **Automotive Tools Manual**
- 10445 **Welding Manual**
- 10450 **ATV Basics**

Over a 100 Haynes motorcycle manuals also available

10/22

Haynes North America, Inc. • (805) 498-6703 • www.haynes.com